Graphene Chemistry

Graphene Chemistry

Theoretical Perspectives

Edited by

DE-EN JIANG

Chemical Sciences Division, Oak Ridge National Laboratory, USA

and

ZHONGFANG CHEN

Department of Chemistry, University of Puerto Rico, USA

WILEY

This edition first published 2013
© 2013 John Wiley & Sons, Ltd

Registered office
John Wiley & Sons Ltd, The Atrium, Southern Gate, Chichester, West Sussex, PO19 8SQ, United Kingdom

For details of our global editorial offices, for customer services and for information about how to apply for permission to reuse the copyright material in this book please see our website at www.wiley.com.

The right of the author to be identified as the author of this work has been asserted in accordance with the Copyright, Designs and Patents Act 1988.

All rights reserved. No part of this publication may be reproduced, stored in a retrieval system, or transmitted, in any form or by any means, electronic, mechanical, photocopying, recording or otherwise, except as permitted by the UK Copyright, Designs and Patents Act 1988, without the prior permission of the publisher.

Wiley also publishes its books in a variety of electronic formats. Some content that appears in print may not be available in electronic books.

Designations used by companies to distinguish their products are often claimed as trademarks. All brand names and product names used in this book are trade names, service marks, trademarks or registered trademarks of their respective owners. The publisher is not associated with any product or vendor mentioned in this book. This publication is designed to provide accurate and authoritative information in regard to the subject matter covered. It is sold on the understanding that the publisher is not engaged in rendering professional services. If professional advice or other expert assistance is required, the services of a competent professional should be sought.

The publisher and the author make no representations or warranties with respect to the accuracy or completeness of the contents of this work and specifically disclaim all warranties, including without limitation any implied warranties of fitness for a particular purpose. This work is sold with the understanding that the publisher is not engaged in rendering professional services. The advice and strategies contained herein may not be suitable for every situation. In view of ongoing research, equipment modifications, changes in governmental regulations, and the constant flow of information relating to the use of experimental reagents, equipment, and devices, the reader is urged to review and evaluate the information provided in the package insert or instructions for each chemical, piece of equipment, reagent, or device for, among other things, any changes in the instructions or indication of usage and for added warnings and precautions. The fact that an organization or Website is referred to in this work as a citation and/or a potential source of further information does not mean that the author or the publisher endorses the information the organization or Website may provide or recommendations it may make. Further, readers should be aware that Internet Websites listed in this work may have changed or disappeared between when this work was written and when it is read. No warranty may be created or extended by any promotional statements for this work. Neither the publisher nor the author shall be liable for any damages arising herefrom.

Library of Congress Cataloging-in-Publication Data

Graphene chemistry : theoretical perspectives / edited by De-en Jiang, Zhongfang Chen.
    pages cm
  Includes bibliographical references and index.
  ISBN 978-1-119-94212-2 (hardback)
  1. Graphene.   I. Jiang, De-en, 1975–   II. Chen, Zhongfang, 1971–
  QD341.H9G693 2014
  546′.68142–dc23

                                                                                    2013015047

A catalogue record for this book is available from the British Library.

ISBN: 978-1-119-94212-2 (Hardback)

Set in 10/12pt Times by Aptara Inc., New Delhi, India.

Printed and bound in Singapore by Markono Print Media Pte Ltd

Contents

List of Contributors xv
Preface xix
Acknowledgements xxi

1 **Introduction** 1
  De-en Jiang and Zhongfang Chen

2 **Intrinsic Magnetism in Edge-Reconstructed Zigzag Graphene Nanoribbons** 9
  Zexing Qu and Chungen Liu

  2.1 Methodology 10
    2.1.1 Effective Valence Bond Model 10
    2.1.2 Density Matrix Renormalization Group Method 11
    2.1.3 Density Functional Theory Calculations 12
  2.2 Polyacene 12
  2.3 Polyazulene 14
  2.4 Edge-Reconstructed Graphene 17
    2.4.1 Energy Gap 17
    2.4.2 Frontier Molecular Orbitals 18
    2.4.3 Projected Density of States 19
    2.4.4 Spin Density in the Triplet State 20
  2.5 Conclusion 22
  Acknowledgments 23
  References 23

3 **Understanding Aromaticity of Graphene and Graphene Nanoribbons by the Clar Sextet Rule** 29
  Dihua Wu, Xingfa Gao, Zhen Zhou, and Zhongfang Chen

  3.1 Introduction 29
    3.1.1 Aromaticity and Clar Theory 30
    3.1.2 Previous Studies of Carbon Nanotubes 33
  3.2 Armchair Graphene Nanoribbons 34
    3.2.1 The Clar Structure of Armchair Graphene Nanoribbons 34
    3.2.2 Aromaticity of Armchair Graphene Nanoribbons and Band Gap Periodicity 37

	3.3	Zigzag Graphene Nanoribbons	40
		3.3.1 Clar Formulas of Zigzag Graphene Nanoribbons	40
		3.3.2 Reactivity of Zigzag Graphene Nanoribbons	40
	3.4	Aromaticity of Graphene	42
	3.5	Perspectives	44
	Acknowledgements	45
	References	45

4 Physical Properties of Graphene Nanoribbons: Insights from First-Principles Studies   51
Dana Krepel and Oded Hod

4.1	Introduction	51
4.2	Electronic Properties of Graphene Nanoribbons	53
	4.2.1 Zigzag Graphene Nanoribbons	53
	4.2.2 Armchair Graphene Nanoribbons	56
	4.2.3 Graphene Nanoribbons with Finite Length	58
	4.2.4 Surface Chemical Adsorption	60
4.3	Mechanical and Electromechanical Properties of GNRs	63
4.4	Summary	66
Acknowledgements	66
References	66

5 Cutting Graphitic Materials: A Promising Way to Prepare Graphene Nanoribbons   79
Wenhua Zhang and Zhenyu Li

5.1	Introduction	79
5.2	Oxidative Cutting of Graphene Sheets	80
	5.2.1 Cutting Mechanisms	80
	5.2.2 Controllable Cutting	83
5.3	Unzipping Carbon Nanotubes	85
	5.3.1 Unzipping Mechanisms Based on Atomic Oxygen	86
	5.3.2 Unzipping Mechanisms Based on Oxygen Pairs	88
5.4	Beyond Oxidative Cutting	91
	5.4.1 Metal Nanoparticle Catalyzed Cutting	92
	5.4.2 Cutting by Fluorination	95
5.5	Summary	96
References	96

6 Properties of Nanographenes   101
Michael R. Philpott

6.1	Introduction	101
6.2	Synthesis	103
6.3	Computation	103
6.4	Geometry of Zigzag-Edged Hexangulenes	104

6.5	Geometry of Armchair-Edged Hexangulenes		107
6.6	Geometry of Zigzag-Edged Triangulenes		110
6.7	Magnetism of Zigzag-Edged Hexangulenes		112
6.8	Magnetism of Zigzag-Edged Triangulenes		114
6.9	Chimeric Magnetism		115
6.10	Magnetism of Oligocenes, Bisanthene-Homologs, Squares and Rectangles		117
	6.10.1	Oligocene Series: $C_{4m+2}H_{2m+4}$ ($n_a = 1; m = 2, 3, 4 \ldots$)	117
	6.10.2	Bisanthene Series: $C_{8m+4}H_{2m+8}$ ($n_a = 3; m = 2, 3, 4\ldots$)	119
	6.10.3	Square and Rectangular Nano-Graphenes: $C_{8m+4}H_{2m+8}$ ($m = 2, 3, 4 \ldots$)	122
6.11	Concluding Remarks		122
Acknowledgment			123
References			124

7 Porous Graphene and Nanomeshes   129
Yan Jiao, Marlies Hankel, Aijun Du, and Sean C. Smith

7.1	Introduction		129
	7.1.1	Graphene-Based Nanomeshes	130
	7.1.2	Graphene-Like Polymers	130
	7.1.3	Other Relevant Subjects	131
		7.1.3.1 Isotope Separation	131
		7.1.3.2 Van der Waals Correction for Density Functional Theory	132
		7.1.3.3 Potential Energy Surfaces for Hindered Molecular Motions Within the Narrow Pores	133
7.2	Transition State Theory		134
	7.2.1	A Brief Introduction of the Idea	134
	7.2.2	Evaluating Partition Functions: The Well-Separated "Reactant" State	136
	7.2.3	Evaluating Partition Functions: The Fully Coupled 4D TS Calculation	137
	7.2.4	Evaluating Partition Functions: Harmonic Approximation for the TS Derived Directly from Density Functional Theory Calculations	138
7.3	Gas and Isotope Separation		139
	7.3.1	Gas Separation and Storage by Porous Graphene	139
		7.3.1.1 Porous Graphene for Hydrogen Purification and Storage	139
		7.3.1.2 Porous Graphene for Isotope Separation	140
	7.3.2	Nitrogen Functionalized Porous Graphene for Hydrogen Purification/Storage and Isotope Separation	140
		7.3.2.1 Introduction	140
		7.3.2.2 NPG and its Asymmetrically Doped Version for D_2/H_2 Separation – A Case Study	141
	7.3.3	Graphdiyne for Hydrogen Purification	144

7.4 Conclusion and Perspectives	147
Acknowledgement	147
References	147

8 Graphene-Based Architecture and Assemblies 153
Hongyan Guo, Rui Liu, Xiao Cheng Zeng, and Xiaojun Wu

8.1	Introduction	153
8.2	Fullerene Polymers	154
8.3	Carbon Nanotube Superarchitecture	156
8.4	Graphene Superarchitectures	160
8.5	C_{60}/Carbon Nanotube/Graphene Hybrid Superarchitectures	163
	8.5.1 Nanopeapods	163
	8.5.2 Carbon Nanobuds	165
	8.5.3 Graphene Nanobuds	168
	8.5.4 Nanosieves and Nanofunnels	169
8.6	Boron-Nitride Nanotubes and Monolayer Superarchitectures	171
8.7	Conclusion	173
Acknowledgments		173
References		174

9 Doped Graphene: Theory, Synthesis, Characterization, and Applications 183
Florentino López-Urías, Ruitao Lv, Humberto Terrones, and Mauricio Terrones

9.1	Introduction	183
9.2	Substitutional Doping of Graphene Sheets	184
9.3	Substitutional Doping of Graphene Nanoribbons	194
9.4	Synthesis and Characterization Techniques of Doped Graphene	196
9.5	Applications of Doped Graphene Sheets and Nanoribbons	200
9.6	Future Work	201
Acknowledgments		202
References		202

10 Adsorption of Molecules on Graphene 209
O. Leenaerts, B. Partoens, and F. M. Peeters

10.1	Introduction	209
10.2	Physisorption *versus* Chemisorption	210
10.3	General Aspects of Adsorption of Molecules on Graphene	212
10.4	Various Ways of Doping Graphene with Molecules	215
	10.4.1 Open-Shell Adsorbates	215
	10.4.2 Inert Adsorbates	217
	10.4.3 Electrochemical Surface Transfer Doping	220
10.5	Enhancing the Graphene-Molecule Interaction	221
	10.5.1 Substitutional Doping	221
	10.5.2 Adatoms and Adlayers	222

		10.5.3 Edges and Defects	224
		10.5.4 External Electric Fields	224
		10.5.5 Surface Bending	225
	10.6	Conclusion	226
	References		226

11 Surface Functionalization of Graphene   233
Maria Peressi

11.1	Introduction		233
11.2	Functionalized Graphene: Properties and Challenges		236
11.3	Theoretical Approach		237
11.4	Interaction of Graphene with Specific Atoms and Functional Groups		238
	11.4.1 Interaction with Hydrogen		238
	11.4.2 Interaction with Oxygen		240
	11.4.3 Interaction with Hydroxyl Groups		241
	11.4.4 Interaction with Other Atoms, Molecules, and Functional Groups		245
11.5	Surface Functionalization of Graphene Nanoribbons		247
11.6	Conclusions		248
References			249

12 Mechanisms of Graphene Chemical Vapor Deposition (CVD) Growth   255
Xiuyun Zhang, Qinghong Yuan, Haibo Shu, and Feng Ding

12.1	Background		255
	12.1.1 Graphene and Defects in Graphene		255
	12.1.2 Comparison of Methods of Graphene Synthesis		257
	12.1.3 Graphene Chemical Vapor Deposition (CVD) Growth		257
		12.1.3.1 The Status of Graphene CVD Growth	257
		12.1.3.2 Phenomenological Mechanism	260
		12.1.3.3 Challenges in Graphene CVD Growth	260
12.2	The Initial Nucleation Stage of Graphene CVD Growth		261
	12.2.1 C Precursors on Catalyst Surfaces		262
	12.2.2 The sp C Chain on Catalyst Surfaces		262
	12.2.3 The sp^2 Graphene Islands		263
	12.2.4 The Magic Sized sp^2 Carbon Clusters		264
	12.2.5 Nucleation of Graphene on Terrace versus Near Step		266
12.3	Continuous Growth of Graphene		271
	12.3.1 The Upright Standing Graphene Formation on Catalyst Surfaces		271
	12.3.2 Edge Reconstructions on Metal Surfaces		273
	12.3.3 Growth Rate of Graphene and Shape Determination		275
	12.3.4 Nonlinear Growth of Graphene on Ru and Ir Surfaces		276
12.4	Graphene Orientation Determination in CVD Growth		278
12.5	Summary and Perspectives		280
References			282

13 From Graphene to Graphene Oxide and Back 291
Xingfa Gao, Yuliang Zhao, and Zhongfang Chen

- 13.1 Introduction 291
- 13.2 From Graphene to Graphene Oxide 292
  - 13.2.1 Modeling Using Cluster Models 292
    - 13.2.1.1 Oxidative Etching of Armchair Edges 292
    - 13.2.1.2 Oxidative Etching of Zigzag Edges 293
    - 13.2.1.3 Linear Oxidative Unzipping 294
    - 13.2.1.4 Spins upon Linear Oxidative Unzipping 296
- 13.3 Modeling Using PBC Models 297
  - 13.3.1 Oxidative Creation of Vacancy Defects 297
  - 13.3.2 Oxidative Etching of Vacancy Defects 298
  - 13.3.3 Linear Oxidative Unzipping 299
  - 13.3.4 Linear Oxidative Cutting 300
- 13.4 From Graphene Oxide back to Graphene 302
  - 13.4.1 Modeling Using Cluster Models 302
    - 13.4.1.1 Cluster Models for Graphene Oxide 302
    - 13.4.1.2 Hydrazine De-Epoxidation 302
    - 13.4.1.3 Thermal De-Hydroxylation 307
    - 13.4.1.4 Thermal De-Carbonylation and De-Carboxylation 308
    - 13.4.1.5 Temperature Effect on De-Epoxidation and De-Hydroxylation 309
    - 13.4.1.6 Residual Groups of Graphene Oxide Reduced by Hydrazine and Heat 311
  - 13.4.2 Modeling Using Periodic Boundary Conditions 312
    - 13.4.2.1 Hydrazine De-Epoxidation 312
    - 13.4.2.2 Thermal De-Epoxidation 313
- 13.5 Concluding Remarks 314
- Acknowledgement 314
- References 314

14 Electronic Transport in Graphitic Carbon Nanoribbons 319
Eduardo Costa Girão, Liangbo Liang, Jonathan Owens, Eduardo Cruz-Silva, Bobby G. Sumpter, and Vincent Meunier

- 14.1 Introduction 319
- 14.2 Theoretical Background 320
  - 14.2.1 Electronic Structure 320
    - 14.2.1.1 Density Functional Theory 320
    - 14.2.1.2 Semi-Empirical Methods 320
  - 14.2.2 Electronic Transport at the Nanoscale 322
- 14.3 From Graphene to Ribbons 324
  - 14.3.1 Graphene 324
  - 14.3.2 Graphene Nanoribbons 325

14.4	Graphene Nanoribbon Synthesis and Processing		329
14.5	Tailoring GNR's Electronic Properties		330
	14.5.1 Defect-Based Modifications of the Electronic Properties		331
		14.5.1.1 Non-Hexagonal Rings	331
		14.5.1.2 Edge and Bulk Disorder	332
	14.5.2 Electronic Properties of Chemically Doped Graphene Nanoribbons		332
		14.5.2.1 Substitutional Doping of Graphene Nanoribbons	332
		14.5.2.2 Chemical Functionalization of Graphene Nanoribbons	333
	14.5.3 GNR Assemblies		334
		14.5.3.1 Nanowiggles	334
		14.5.3.2 Antidots and Junctions	335
		14.5.3.3 GNR Rings	335
		14.5.3.4 GNR Stacking	336
14.6	Thermoelectric Properties of Graphene-Based Materials		336
	14.6.1 Thermoelectricity		336
	14.6.2 Thermoelectricity in Carbon		336
14.7	Conclusions		338
Acknowledgements			339
References			339

15 Graphene-Based Materials as Nanocatalysts 347
Fengyu Li and Zhongfang Chen

15.1	Introduction		347
15.2	Electrocatalysts		347
	15.2.1 N-Graphene		348
	15.2.2 N-Graphene-NP Nanocomposites		350
	15.2.3 Non-Pt Metal on the Porphyrin-Like Subunits in Graphene		351
	15.2.4 Graphyne		352
15.3	Photocatalysts		353
	15.3.1 TiO_2-Graphene Nanocomposite		353
	15.3.2 Graphitic Carbon Nitrides (g-C_3N_4)		355
15.4	CO Oxidation		356
	15.4.1 Metal-Embedded Graphene		357
	15.4.2 Metal-Graphene Oxide		358
	15.4.3 Metal-Graphene under Mechanical Strain		359
	15.4.4 Metal-Embedded Graphene under an External Electric Field		360
	15.4.5 Porphyrin-Like Fe/N/C Nanomaterials		361
	15.4.6 Si-Embedded Graphene		361
	15.4.7 Experimental Aspects		361
15.5	Others		362
	15.5.1 Propene Epoxidation		362
	15.5.2 Nitromethane Combustion		362

15.6	Conclusion		363
Acknowledgements			364
References			364

16 Hydrogen Storage in Graphene 371
Yafei Li and Zhongfang Chen

16.1	Introduction			371
16.2	Hydrogen Storage in Molecule Form			373
	16.2.1	Hydrogen Storage in Graphene Sheets		373
	16.2.2	Hydrogen Storage in Metal Decorated Graphene		374
		16.2.2.1	Lithium Decorated Graphene	375
		16.2.2.2	Calcium Decorated Graphene	376
		16.2.2.3	Transition Metal Decorated Graphene	377
	16.2.3	Hydrogen Storage in Graphene Networks		377
		16.2.3.1	Covalently Bonded Graphene	378
	16.2.4	Notes to Computational Methods		381
16.3	Hydrogen Storage in Atomic Form			382
	16.3.1	Graphane		382
	16.3.2	Chemical Storage of Hydrogen by Spillover		383
16.4	Conclusion			386
Acknowledgements				386
References				386

17 Linking Theory to Reactivity and Properties of Nanographenes 393
Qun Ye, Zhe Sun, Chunyan Chi, and Jishan Wu

17.1	Introduction		393
17.2	Nanographenes with Only Armchair Edges		394
17.3	Nanographenes with Both Armchair and Zigzag Edges		397
	17.3.1	Structure of Rylenes	398
	17.3.2	Chemistry at the Armchair Edges of Rylenes	398
	17.3.3	Anthenes and Periacenes	402
17.4	Nanographene with Only Zigzag Edges		405
	17.4.1	Phenalenyl-Based Open-Shell Systems	406
17.5	Quinoidal Nanographenes		411
	17.5.1	Bis(Phenalenyls)	412
	17.5.2	Zethrenes	414
	17.5.3	Indenofluorenes	417
17.6	Conclusion		417
References			418

18 Graphene Moiré Supported Metal Clusters for Model Catalytic Studies 425
Bradley F. Habenicht, Ye Xu, and Li Liu

18.1	Introduction	425
18.2	Graphene Moiré on Ru(0001)	426

18.3	Metal Cluster Formation on g/Ru(0001)	430
18.4	Two-dimensional Au Islands on g/Ru(0001) and its Catalytic Activity	434
18.5	Summary	440
Acknowledgments		441
References		441

Index **447**

List of Contributors

Zhongfang Chen Department of Chemistry, Institute for Functional Nanomaterials, University of Puerto Rico, USA

Chunyan Chi Department of Chemistry, National University of Singapore, Singapore

Eduardo Costa Girão Departamento de Física, Universidade Federal do Piauí, Brazil

Eduardo Cruz-Silva Department of Physics, Applied Physics, and Astronomy, Rensselaer Polytechnic Institute, USA and Department of Polymer Science and Engineering, University of Massachusetts, USA

Feng Ding Institute of Textiles and Clothing, Hong Kong Polytechnic University, Hong Kong

Aijun Du Centre for Computational Molecular Science, Australian Institute for Bioengineering and Nanotechnology, The University of Queensland, Australia

Xingfa Gao Key Laboratory for Biomedical Effects of Nanomaterials and Nanosafety, Institute of High Energy Physics, Chinese Academy of Sciences, China

Hongyan Guo Department of Materials Science and Engineering, CAS Key Laboratory of Materials for Energy Conversion, and Hefei National Laboratory for Physical Science at the Microscale, University of Science and Technology of China, China

Bradley F. Habenicht Center for Nanophase Materials Sciences, Oak Ridge National Laboratory, USA

Marlies Hankel Centre for Computational Molecular Science, Australian Institute for Bioengineering and Nanotechnology, The University of Queensland, Australia

Oded Hod Department of Chemical Physics, School of Chemistry, The Raymond and Beverly Sackler Faculty of Exact Sciences, Tel-Aviv University, Israel

De-en Jiang Chemical Sciences Division, Oak Ridge National Laboratory, USA

Yan Jiao Centre for Computational Molecular Science, Australian Institute for Bioengineering and Nanotechnology, The University of Queensland, Australia and School of Chemical Engineering, The University of Queensland, Australia

Dana Krepel Department of Chemical Physics, School of Chemistry, The Raymond and Beverly Sackler Faculty of Exact Sciences, Tel-Aviv University, Israel

O. Leenaerts Department of Physics, University of Antwerp, Belgium

Fengyu Li Department of Physics and Department of Chemistry, Institute for Functional Nanomaterials, University of Puerto Rico, USA

Yafei Li Department of Chemistry, Institute for Functional Nanomaterials, University of Puerto Rico, USA

Zhenyu Li Hefei National Laboratory for Physical Sciences at the Microscale, University of Science and Technology of China, China

Liangbo Liang Department of Physics, Applied Physics, and Astronomy, Rensselaer Polytechnic Institute, USA

Chungen Liu Institute of Theoretical and Computational Chemistry, Key Laboratory of Mesoscopic Chemistry of the Ministry of Education (MOE), Nanjing University, China

Li Liu Department of Chemistry, Texas A&M University, USA

Rui Liu Department of Chemistry, University of Nebraska-Lincoln, USA

Florentino López-Urías Department of Physics, The Pennsylvania State University, USA and Advanced Materials Department, IPICYT, México

Ruitao Lv Department of Physics, The Pennsylvania State University, USA

Vincent Meunier Department of Physics, Applied Physics, and Astronomy, Rensselaer Polytechnic Institute, USA

Jonathan Owens Department of Physics, Applied Physics, and Astronomy, Rensselaer Polytechnic Institute, USA

B. Partoens Department of Physics, University of Antwerp, Belgium

F. M. Peeters Department of Physics, University of Antwerp, Belgium

Maria Peressi Department of Physics, University of Trieste, Trieste, Italy and CNR-IOM DEMOCRITOS National Simulation Center, Italy

Michael R. Philpott Visiting Scholar, Kenneth S. Pitzer Center for Theoretical Chemistry, University of California Berkeley, USA and Center for Computational Materials Science, Institute of Materials Research, Tohoku University, Japan

Zexing Qu Institute of Theoretical and Computational Chemistry, Key Laboratory of Mesoscopic Chemistry of the Ministry of Education (MOE), Nanjing University, China

Haibo Shu Institute of Textiles and Clothing, Hong Kong Polytechnic University, Hong Kong

Sean C. Smith Center for Nanophase Materials Sciences, Oak Ridge National Laboratory, USA

Bobby G. Sumpter Center of Nanophase Materials Sciences, Oak Ridge National Laboratory, USA

Zhe Sun Department of Chemistry, National University of Singapore, Singapore

Humberto Terrones Department of Physics, The Pennsylvania State University, USA and Departamento de Física, Universidade Federal do Ceará, Brazil

Mauricio Terrones Department of Physics, The Pennsylvania State University, USA and Department of Materials Science and Engineering and Materials Research Institute, The Pennsylvania State University, USA and Research Center for Exotic Nanocarbons (JST), Shinshu University, Japan

Dihua Wu Tianjin Key Laboratory of Metal and Molecule Based Material Chemistry, Key Laboratory of Advanced Energy Materials Chemistry (Ministry of Education), Computational Centre for Molecular Science, Institute of New Energy Material Chemistry, Nankai University, China

Jishan Wu Department of Chemistry, National University of Singapore, Singapore

Xiaojun Wu Department of Materials Science and Engineering, CAS Key Laboratory of Materials for Energy Conversion, and Hefei National Laboratory for Physical Science at the Microscale, University of Science and Technology of China, China

Ye Xu Center for Nanophase Materials Sciences, Oak Ridge National Laboratory, USA

Qun Ye Department of Chemistry, National University of Singapore, Singapore

Qinghong Yuan Institute of Textiles and Clothing, Hong Kong Polytechnic University, Hong Kong

Xiao Cheng Zeng Department of Chemistry, University of Nebraska-Lincoln, USA

Wenhua Zhang Hefei National Laboratory for Physical Sciences at the Microscale, University of Science and Technology of China, China

Xiuyun Zhang Institute of Textiles and Clothing, Hong Kong Polytechnic University, Hong Kong

Yuliang Zhao Key Laboratory for Biomedical Effects of Nanomaterials and Nanosafety, Institute of High Energy Physics, Chinese Academy of Sciences, China

Zhen Zhou Tianjin Key Laboratory of Metal and Molecule Based Material Chemistry, Key Laboratory of Advanced Energy Materials Chemistry (Ministry of Education), Computational Centre for Molecular Science, Institute of New Energy Material Chemistry, Nankai University, China

Preface

The book before you is about the chemistry of graphene from theoretical perspectives. The sp^2 carbon and its various manifestations are dear to all chemists' hearts. Graphene is no exception. No doubt, graphene is a new wonder material. Since its isolation in 2004, graphene has rapidly risen to be one of the hottest stars for various disciplines of basic and applied sciences due to its many exciting unusual characteristics. Over 8000 papers related to graphene were published in 2012, and graphene keeps occupying the headlines of scientific news media.

As stated by John Dewey, "Every great advance in science has issued from a new audacity of the imagination." The discovery of graphene was a strong proof. In 1947, Canadian theorist Philip Russell Wallace predicted the relativistic behavior of electrons of graphene using the tight-binding approximation, though no one believed then that such a monolayer could exist. The groundbreaking experiments on realizing graphene by mechanical cleavage in 2004 totally changed our minds, and led to the Nobel Prize for Physics in 2010 being given to Andre Geim and Konstantin Novoselov. This path immediately reminds us of the history of fullerenes, especially C_{60}, which was theoretically predicted by Eiji Osawa 1970, experimentally realized in 1985, and won Richard Smalley, Robert Curl and Harold Kroto the Nobel Prize for Chemistry in 1996.

Chemistry will play an increasingly important role in realizing graphene's potentials, as synthesizing targeted graphene nanosystems, scaling up the synthesis, controlling defects, introducing functional groups, making composites, and so on, all involve chemistry. This book intends to provide a comprehensive state-of-the-art understanding of graphene chemistry at the atomic level from theoretical perspectives, especially on the structure-property-function relationship. The diverse group of contributors are leading experts and their dedication has made this book possible. Computational and theoretical chemists who have not studied graphene before will find this book useful in giving them an overview of the field. Graduate students in computational nanosciences will find this book particularly helpful in learning how one addresses important problems in an interesting and hot topic. Experimentalists in graphene will benefit from the many concepts and interesting properties predicted here for many novel graphene systems; what's more, this book has two chapters with a significant experimental portion, which demonstrates how deeper insights can be obtained by joint experimental and theoretical efforts. Even experts in the theoretical studies of graphene will find this book interesting thanks to the diverse topics covered in this book.

Future discoveries, innovation, and development in graphene chemistry will continue to be accelerated by modern theoretical and computational studies. We hope that the readers in both theoretical and experimental communities will enjoy this book, and discover more of the magic power of the sp^2 carbon.

Acknowledgements

We thank generous financial support from the funding agencies: the Division of Chemical Sciences, Geosciences, and Biosciences, Office of Basic Energy Sciences, U.S. Department of Energy (D.J.); US Department of Defense and National Science Foundation (Z.C.). We are grateful to Sarah Hall of John Wiley & Sons, Ltd for initiating the idea of this book and working through the book proposal with us; and to Sarah Tilley of John Wiley & Sons, Ltd for helping us deliver this book. We thank all the authors who contributed to this book and the reviewers for their time; and Ms. Adi Shinar for creating the cover image.

1

Introduction

De-en Jiang[a] *and Zhongfang Chen*[b]
[a]*Chemical Sciences Division, Oak Ridge National Laboratory, USA*
[b]*Department of Chemistry, Institute for Functional Nanomaterials, University of Puerto Rico, USA*

From the breakthrough discovery in 2004 [1], to the award of the Nobel Prize for Physics to Geim and Novosolov in 2010, it took only six years for graphene to reach the pinnacle of scientific research. However, this is not the first time that sp^2 carbon has won a Nobel Prize. Remember that Smalley, Kroto, and Curl won the Nobel Prize for Chemistry in 1996 for their discovery of fullerenes in 1985 [2]; only that took 11 years. What's more interesting is what the late Smalley said in his Nobel lecture: "Carbon has this genius of making a chemically stable two-dimensional, one-atom-thick membrane in a three-dimensional world. And that, I believe, is going to be very important in the future of chemistry and technology in general." We have to admire his foresight of the explosion of interest in graphene 10 years later.

The fact that graphene won the Nobel Prize for Physics also reflects the community's focus on the physics aspect of graphene research. However, any large-scale application of graphene would undoubtedly rely on the chemistry of graphene. The expanding interest in graphene oxide through the Hummers method [3] or chemical-vapor deposition [4] approach to graphene synthesis are just two typical examples. The versatility of chemistry presents endless opportunities in graphene research.

Why do we focus on the theoretical perspectives of graphene chemistry? On one hand, there are books already discussing the experimental aspects of graphene chemistry; on the other hand, graphene provides an ideal proving ground for testing theoretical methods and computational imagination. In addition, sp^2 or graphitic carbons are the basis of many important materials, such as carbon fibers for building cars and planes; activated carbons

---

Graphene Chemistry: Theoretical Perspectives, First Edition. Edited by De-en Jiang and Zhongfang Chen.
© 2013 John Wiley & Sons, Ltd. Published 2013 by John Wiley & Sons, Ltd.

for supercapacitors; and graphite and hard carbons for lithium-ion batteries. Understanding graphene chemistry would also lay a foundation of understanding of complex carbonaceous materials. Therefore, the aim of this book is to deliver a comprehensive view of graphene chemistry from various theoretical and computational perspectives.

As a truly two-dimensional system, the honeycomb lattice of graphene has given rise to many interesting physical properties. However, no size is infinite in the real world and we eventually come to the edge of the graphene sheet just like the vast ocean greeting the shore line. Thus, we can expect that edge geometry will have a profound effect on the π-electronic structure.

A chemist relates to graphene by thinking about benzene: graphene is nothing but fused benzene rings. The geometric character is the same: each carbon atom has sp^2 hybridization and contributes one $2p_z$ electron for π bonding. However, when contrasting the electronic structure of benzene with that of graphene, we can immediately see the difference: the benzene molecule has a large HOMO-LUMO gap (Figure 1.1), while graphene is a zero-gap semiconductor where the conduction band and the valence band touch at one point at the Fermi level (Figure 1.2). How does the electronic structure evolve from a large-gap six-membered ring to a zero-gap lattice of infinite number of fused rings? This is a rather intriguing question. It turns out that the chemical details matter. Let's take acenes, or linearly fused benzene rings, as an example. Quite a few theoretical papers have been devoted to examining the evolution of the electronic structure with the number of benzene rings in these interesting systems, starting with the earlier studies by Whangbo et al.

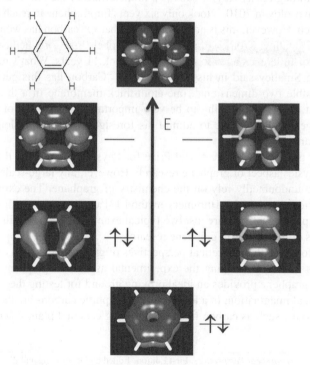

Figure 1.1 The π-orbitals of the benzene molecule

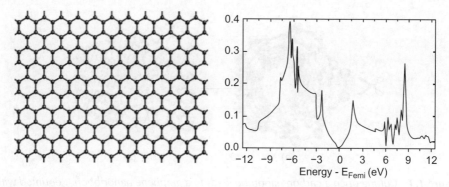

Figure 1.2 *The graphene lattice (left) and its electronic density of states (right) computed using density functional theory*

[5] and Kivelson et al. [6]. In the experimental community, researchers have been trying to synthesize ever longer acenes. The record so far seems to be a nonacene derivative which has nine fused rings as its core [7]. Chapter 2 examines acenes or polyacenes by applying the effective valence bond model and the density-matrix-renormalization-group method, compared to density functional theory (DFT). More often than not, the linearly fused six-membered rings may not be as happy as alternating five-membered and seven-membered rings. This leads to a class of molecules called fused-azulenes or polyazulenes which also forms on the graphene edge due to reconstruction. Chapter 2 addresses this system too.

Extending acenes infinitely results in a ribbon with two zigzag edges. The width of the ribbon can also be further increased while preserving the two zigzag edges. What is unique about the zigzag edge is the edge state, as predicted by Fujita et al. [8–9] from the tight-binding method. This pioneering study has inspired many follow-up theoretical investigations and experiments. DFT calculations in particular, revealed that the edge state also possesses a radical-like chemical reactivity [10]. Chapter 4 examines the electronic, mechanical, and electromechanical properties of both zigzag-edged and armchair-edged graphene ribbons.

The uniqueness of the zigzag edge not only manifests itself in ribbons but also in nanographenes. This is fundamentally related to the inherent instability of the zigzag edge. Furthermore, this may be due to the difficulty that π-electrons have in forming localized double bonds since the consecutive zigzag edges do not support bond-length variation well [11]. In contrast, this is not a problem for armchair edges. The geometry-related π-electron distribution is captured by the Clar's sextet rule which seems to be particularly useful in predicting stable nanographenes [12] and ribbons. Chapter 3 seeks to understand aromaticity of graphene and graphene nanoribbons from the perspective of Clar's sextet rule, while Chapter 4 also briefly discusses the electronic structure of nanographenes or finite graphene flakes as the authors call them. Chapter 6 focuses on nanographene exclusively where the authors interrogates C–C bond length variation, orbital energies, and magnetization with the nanographene's size.

A most natural way to relate carbon nanotubes and graphene ribbons is probably to think of unzipping a carbon nanotube to a graphene ribbon. Amazingly, this idea, put forward

4 Graphene Chemistry

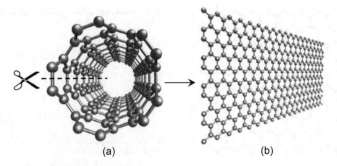

Figure 1.3 Cutting open a carbon nanotube leads to a graphene nanoribbon. Reprinted with permission from [10] © 2007 American Institute of Physics

for fun in 2007 [10] (Figure 1.3), was realized by two experiments two years later through plasma etching [13] and chemical oxidation [14]. Chapter 5 examines the mechanisms of various ways to cut open not only carbon nanotubes, but also graphene itself.

A straightforward way to utilize the one-atom thickness of graphene is to use it as a membrane, since a membrane's permeance is inversely proportional to its thickness. But a perfect graphene is impermeable to molecules as small as helium [15]. Hence, it is necessary to create holes to empower the graphene sheet for membrane separations; for example, the pore as shown in Figure 1.4 was computationally shown to be highly selective towards H_2 [16] and CO_2 [17]. More interestingly, this idea of using porous graphene to sieve gas molecules was recently experimentally confirmed through a clever setup [18]. Chapter 7 gives a detailed discussion of how to apply transition state theory to understand porous graphene and graphene nanomeshes for gas and isotope separations.

Using Lego bricks, a kid will let the imagination fly and build all sorts of things. Similarly, when given sp^2-carbon building blocks, namely, fullerene, carbon nanotubes, and

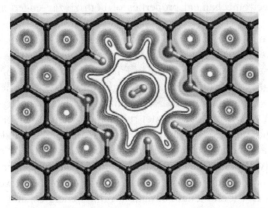

Figure 1.4 H_2 molecule inside a subnanometer pore in the graphene sheet. Electron density contours of both H_2 and the porous graphene are shown. Reprinted with permission from [16] © 2009 American Chemical Society

graphene, scientists can assemble them into sp^2-hybridized carbon-based superarchitectures, such as fullerene polymers, nanotube clathrates, nanopeapods, nanobuds, nanofunnels, and nanofoams. Chapter 8 reviews the recent efforts in designing the prototypes of such materials and discusses their potential applications in electronics and energy storage.

Doping is an important topic in terms of graphene chemistry, as it is the major approach to opening and tuning the electronic band gap of graphene for device applications, and for modifying the graphene surface to improve reactivity and its interface with other materials for desired applications. Chapter 9 reviews different ways to dope graphene and their resulting changes in graphene's properties, as well as the characterization of doped graphenes and their applications.

Interaction between graphene and a foreign atom, functional group, or molecule is also interesting. From a chemical perspective, this is directly related to functionalization of graphene, a hot topic in itself [19]. From a physical perspective, one immediately thinks about sensing. The preference of π-electron delocalization in graphene dictates that converting a sp^2 carbon to sp^3 won't be easy. However, this could happen in a pairwise fashion [20]. Chapter 10 examines in detail the physical adsorption of molecules on graphene, while Chapter 11 focuses on the chemical interaction of functional groups such as –OH with the graphene sheet.

Chemical vapor deposition is the method of choice for large-scale production of high-quality graphene. Various substrates especially copper have been used for CVD growth of graphene. Chapter 12 discusses the mechanistic insights into graphene CVD growth obtained by DFT computations for the nucleation stage and the kinetic Wulff plots for the growth stage.

Graphene oxide is the preferred medium by chemists and materials scientists for manipulating graphene due to its easy preparation and processability [21]. Due to the difficulties in experimentally characterizing the oxygen-containing groups, computational methods are of great help. Chapter 13 discusses the mechanistic insights into oxidation of graphene and reduction of graphene oxide from DFT computation.

Applications of graphene in electronic devices rely on our understanding how the electrons in graphene nanostructures respond to applied potentials. In other words, how do the factors governing the nanographene electronic structure (such as zigzag edges versus armchair edges, doping, boundary shape, etc.) manifest in electronic transport? Chapter 14 reviews recent progress in elucidating the electronic transport of graphene nanoribbons and their assemblies.

Graphene has many applications in energy-related fields, such as batteries [22–23], pseudo-capacitors [24–25] and electrochemical double-layer capacitors [26]. More excitingly, nitrogen doped graphene and carbon nanotubes have been found to be excellent fuel cell catalysts [27]. Chapter 15 covers a variety of topics related to catalysis by graphene, including oxygen reduction reaction, photocatalysis, and CO oxidation. Chapter 16 focuses on hydrogen storage by graphene-based materials.

Chapter 17 and Chapter 18 are unique in this book in that they have a significant experimental portion. This is to show the ever-increasing need to integrate computational insights into experimental discoveries. Chapter 17 is written by experts in synthesis of nanographenes who examine the challenge, strategy, and reward in making novel nanographene molecules of unique electronic and magnetic properties, often predicted by theory. Chapter 18 examines the growth of metal clusters on the moiré structure of graphene on the Ru(0001) surface: both surface science experiments and large-scale DFT calculations are used to understand catalysis in such a model system.

The chapters in this book aim to provide the reader a comprehensive snapshot of our present understanding of graphene chemistry from theoretical perspectives. We hope that both experimental and computational researchers will find reading through the chapters a rewarding experience.

References

[1] Novoselov, K. S.; Geim, A. K.; Morozov, S. V.; Jiang, D.; Zhang, Y.; Dubonos, S. V.; et al., "Electric field effect in atomically thin carbon films," *Science*, **2004**, 306, 666–669.
[2] Kroto, H. W.; Heath, J. R.; Obrien, S. C.; Curl, R. F.; and Smalley, R. E., "C-60 – Buckminsterfullerene," *Nature*, **1985**, 318, 162–163.
[3] Hummers, W. S. and Offeman, R. E., "Preparation of graphitic oxide," *J. Am. Chem. Soc.*, **1958**, 80, 1339–1339.
[4] Li, X. S.; Cai, W. W.; An, J. H.; Kim, S.; Nah, J.; Yang, et al., "Large-area synthesis of high-quality and uniform graphene films on copper foils," *Science*, **2009**, 324, 1312–1314.
[5] Whangbo, M. H.; Hoffmann, R. and Woodward, R. B., "Conjugated one and two-dimensional polymers," *Proc. R. Soc. London A*, **1979**, 366, 23–46.
[6] Kivelson, S. and Chapman, O. L., "Polyacene and a new class of quasi-one-dimensional conductors," *Phys. Rev. B*, **1983**, 28, 7236–7243.
[7] Purushothaman, B.; Bruzek, M.; Parkin, S. R.; Miller, A. F. and Anthony, J. E., "Synthesis and structural characterization of crystalline nonacenes," *Angew. Chem.-Int. Edit.*, **2011**, 50, 7013–7017.
[8] Fujita, M.; Wakabayashi, K.; Nakada, K. and Kusakabe, K., "Peculiar localized state at zigzag graphite edge," *J. Phys. Soc. Jpn.*, **1996**, 65, 1920–1923.
[9] Nakada, K.; Fujita, M.; Dresselhaus, G. and Dresselhaus, M. S., "Edge state in graphene ribbons: Nanometer size effect and edge shape dependence," *Phys. Rev. B*, **1996**, 54, 17954–17961.
[10] Jiang, D. E.; Sumpter, B. G. and Dai, S., "Unique chemical reactivity of a graphene nanoribbon's zigzag edge," *J. Chem. Phys.*, **2007**, 126, 134701.
[11] Jiang, D. E. and Dai, S., "Electronic ground state of higher acenes," *J. Phys. Chem. A*, **2008**, 112, 332–335.
[12] Jiang, D. E. and Dai, S., "Circumacenes versus periacenes: HOMO-LUMO gap and transition from nonmagnetic to magnetic ground state with size," *Chem. Phys. Lett.*, **2008**, 466, 72–75.
[13] Jiao, L. Y.; Zhang, L.; Wang, X. R.; Diankov, G. and Dai, H. J., "Narrow graphene nanoribbons from carbon nanotubes," *Nature*, **2009**, 458, 877–880.
[14] Kosynkin, D. V.; Higginbotham, A. L.; Sinitskii, A.; Lomeda, J. R.; Dimiev, A.; Price, B. K. and Tour, J. M., "Longitudinal unzipping of carbon nanotubes to form graphene nanoribbons," *Nature*, **2009**, 458, 872-U5.
[15] Bunch, J. S.; Verbridge, S. S.; Alden, J. S.; van der Zande, A. M.; Parpia, J. M.; Craighead, H. G. and McEuen, P. L., "Impermeable atomic membranes from graphene sheets," *Nano Lett.*, **2008**, 8, 2458–2462.

[16] Jiang, D. E.; Cooper, V. R. and Dai, S., "Porous graphene as the ultimate membrane for gas separation," *Nano Lett.*, **2009**, 9, 4019–4024.

[17] Liu, H.; Cooper, V. R.; Dai, S. and Jiang, D. E., "Windowed carbon nanotubes for efficient CO_2 removal from natural gas," *J. Phys. Chem. L.*, **2012**, 3, 3343–3347.

[18] Koenig, S. P.; Wang, L. D.; Pellegrino, J. and Bunch, J. S., "Selective molecular sieving through porous graphene," *Nat. Nanotechnol.*, **2012**, 7, 728–732.

[19] Sarkar, S.; Bekyarova, E. and Haddon, R. C., "Chemistry at the Dirac point: Diels-Alder reactivity of graphene," *Accounts Chem. Res.*, **2012**, 45, 673–682.

[20] Jiang, D. E.; Sumpter, B. G. and Dai, S., "How do aryl groups attach to a graphene sheet?," *J. Phys. Chem. B*, **2006**, 110, 23628–23632.

[21] Kim, J.; Cote, L. J. and Huang, J. X., "Two dimensional soft material: new faces of graphene oxide," *Accounts Chem. Res.*, **2012**, 45, 1356–1364.

[22] Xiao, J.; Mei, D. H.; Li, X. L.; Xu, W.; Wang, D. Y.; Graff, G. L.; et al., "Hierarchically porous graphene as a lithium-air battery electrode," *Nano Lett.*, **2011**, 11, 5071–5078.

[23] Jang, B. Z.; Liu, C. G.; Neff, D.; Yu, Z. N.; Wang, M. C.; Xiong, W. and Zhamu, A., "Graphene surface-enabled lithium ion-exchanging cells: next-generation high-power energy storage devices," *Nano Lett.*, **2011**, 11, 3785–3791.

[24] Wang, H. L.; Cui, L. F.; Yang, Y. A.; Casalongue, H. S.; Robinson, J. T.; Liang, Y. Y.; et al., "Mn_3O_4-graphene hybrid as a high-capacity anode material for lithium ion batteries," *J. Am. Chem. Soc.*, **2010**, 132, 13978–13980.

[25] Wang, H. L.; Casalongue, H. S.; Liang, Y. Y. and Dai, H. J., "$Ni(OH)_2$ nanoplates grown on graphene as advanced electrochemical pseudocapacitor materials," *J. Am. Chem. Soc.*, **2010**, 132, 7472–7477.

[26] El-Kady, M. F.; Strong, V.; Dubin, S. and Kaner, R. B., "Laser scribing of high-performance and flexible graphene-based electrochemical capacitors," *Science*, **2012**, 335, 1326–1330.

[27] Gong, K. P.; Du, F.; Xia, Z. H.; Durstock, M. and Dai, L. M., "Nitrogen-doped carbon nanotube arrays with high electrocatalytic activity for oxygen reduction," *Science*, **2009**, 323, 760–764.

Ferromagnetismus:
Eigenschaft entweder selbst ein statisches Magnetfeld zu verursachen oder von einem Pol eines äußeren Magnetfelds angezogen zu werden.

2

Intrinsic Magnetism in Edge-Reconstructed Zigzag Graphene Nanoribbons

Zexing Qu and Chungen Liu

Institute of Theoretical and Computational Chemistry, School of Chemistry and Chemical Engineering, Nanjing University, China

Graphene, a two-dimensional carbon-based monolayer honeycomb lattice, is considered to be a prospective material for future electronics [1–3]. One-dimensional nano-structure strips of graphene a few nanometers in width, known as graphene nanoribbons (GNR) have been chemically synthesized and studied in laboratory, [4, 5] which quickly attracted attention from theorists [4, 5].

Zigzag-terminated graphene nanoribbons (ZGNRs) are of particular interest, due to observed room-temperature ferromagnetism as well as indirect evidence of a magnetic edge state, suggesting potential applications for spintronics [2, 10–14]. However, numerous theoretical studies on pure and hydrogen-passivated ZGNRs using density functional theory (DFT) methods have predicted an anti-ferromagnetic (AFM) ground state with a minor band gap of approximately several meV between the lowest ferromagnetic (FM) and AFM states [15–18]. It seems that, instead of being an inherent feature of GNRs, room-temperature magnetism could come from easily generated edge or in-body structure defects.

Edge defects, which could shift the edge states from their Fermi energy level (E_F), play a critical role in tuning the electronic properties of graphene nanoribbons [17, 19–21]. Among them, a novel mechanism is the planar edge reconstructions on zigzag edges, resulted in the appearance of alternating pentagon-heptagon carbon rings on those edges. It is believed that the energy barrier involved in edge reconstruction is less than 1

Graphene Chemistry: Theoretical Perspectives, First Edition. Edited by De-en Jiang and Zhongfang Chen.
© 2013 John Wiley & Sons, Ltd. Published 2013 by John Wiley & Sons, Ltd.

eV [21]. Such reconstructed structures are usually denoted Rc-ZGNRs. As revealed by density functional theory (DFT) computations, a Rc-ZGNR edge is energetically more favorable by 0.35 eVÅ than a normal zigzag edge [19, 22]. The existence of different lengths of alternating pentagon-heptagon structure on ZGNR edges has already been confirmed by aberration-corrected transmission electron microscopy (TEM) observations [23, 24].

Furthermore, spin-polarized DFT calculations on one-edge reconstructed ZGNRs indicated an FM ground state, with a finite magnetic moment arising from the opposite (unreconstructed) zigzag edge [19], which supplies a yet to be confirmed theoretical foundation of the ferromagnetism of GNRs, since that the calculated energy gap between the lowest FM and AFM states is too small to maintain a stable magnetic ground state at room temperature [17]. At this stage, the origin of room-temperature magnetism of ZGNRs remains an open problem.

Although DFT is widely accepted as one of the reliable quantum chemical methods that is affordable for small to moderate-size molecules, it has been well documented that traditional DFT methods often have problems in dealing with extended π systems, due to the inherent defect on account of the long-range electron correlation, which usually leads to the underestimation of excitation energies [25–27]. This problem can at least be partly remedied by employing long-range corrected density functional theory (LC-DFT) methods [28–30].

In this work, comparative studies of the electronic structures of ZGNRs and Rc-ZGNRs are presented in valence bond formalism as well as density functional theory (DFT) methods. Our results will show that the edge reconstructions create considerable spin coupling between π electrons on edge and in-body carbon atoms, and open up a wide gap between the FM and AFM states, which is essential for understanding the up to room temperature ferromagnetic character of graphene nanoribbons. We will also demonstrate the importance of employing long-range corrected DFT (LC-DFT) methods in computing extended π systems like Rc-ZGNRs.

2.1 Methodology

2.1.1 Effective Valence Bond Model

The nonempirical Effective Valence Bond (EVB) model was suggested by Malrieu and coworkers [31, 32]. The Hamiltonian had been originally written as,

$$H = \sum_{i-j}[R_{ij} + g_{ij}(a_i^+ a_{\bar{j}}^+ a_j^- a_{\bar{i}}^- + a_{\bar{i}}^+ a_j^+ a_{\bar{j}}^- a_i^- \\ - a_i^+ a_{\bar{j}}^+ a_{\bar{j}}^- a_i^- - a_{\bar{i}}^+ a_j^+ a_j^- a_{\bar{i}}^-)] \qquad (2.1)$$

where R_{ij} is a parameter dependent on bonding atoms i and j, and g_{ij} accounts for the effective exchange coupling between bonded atoms i and j. The inter-atomic distance r_{ij} and the torsion angle θ_{ij} dependence of R_{ij} and g_{ij} can be found in the original paper by

Said et al. [31]. A simpler form of Hamiltonian that is easier to handle could be reformulated by use of spin operators [33],

$$H = \sum_{i-j} \left[R_{ij} + g_{ij} \left(2\mathbf{S}_i \cdot \mathbf{S}_j - \frac{1}{2} \right) \right] \qquad (2.2)$$

Geometry optimization can be carried out by minimizing the energy [31]. This model has proved to give reliable predictions on the low-lying states of conjugated hydrocarbons as well as conjugated diradical or polyradical systems at comparatively low expense for systems composed of less than 30 electrons [32, 34].

This semiempirical quantum chemical model has the advantages of explicit π-electron correlation as well as the consistent parameterization of the σ backbone and σ-π interaction energies, which are essential for computing the energy gradients. In return, it has proven to be a predictive tool which produces true spin eigenfunctions for the study of the low-lying states of conjugated hydrocarbons; in particular, for the study of spin coupling in radicular conjugated hydrocarbons [34, 35].

However, for even larger molecules, the dimension of configuration spaces becomes prohibitively large. Then it becomes computationally too costly to diagonalize the Hamiltonian matrix, even when large sparse matrix diagonalization techniques, such as Lanczos or Davidson methods, are employed. In such cases, DMRG method provides a valuable alternative that is both numerically accurate and computationally feasible (at least for one-dimensional systems). Exact diagonalization of the EVB Hamiltonian matrix is restricted to small and medium-size systems with no more than 30 conjugated atoms, while high precision numerical solution is applicable to larger systems with hundreds of atoms, in virtue of the density matrix renormalization group (DMRG) method [36–38].

2.1.2 Density Matrix Renormalization Group Method

The density matrix renormalization group method, which was first developed by White in 1992 [36, 37], is an extremely effective way of dealing with strongly correlated Hamiltonians in real active space. It has been used very successfully in the study of large conjugated hydrocarbons. A detailed and well-organized review of the DMRG algorithm could be found in a recent publication by Schollwöck [38]. Here we shall give a brief description of the technique.

In DMRG, the interaction between different fragments is taken into account with the use of a "superblock" **AB**, which is composed of one "system block" **A** and one "environment block" **B**. For a specific state of the superblock, the importance of each basis function in the system block (towards this superblock state) can be defined by the reduced density matrix ρ of the system block,

$$\rho_{ii'} = \sum_j \psi_{ij} \psi_{i'j}. \qquad (2.3)$$

where ψ_{ij} is the contribution of the direct product of basis $|i\rangle$ in the system block and $|j\rangle$ in the environment block to the specific superblock state of interest. In general, the target state is denoted as,

$$\Psi = \sum_{ij} \psi_{ij} |i\rangle |j\rangle. \qquad (2.4)$$

The diagonalization of ρ leads to a set of eigenvalues ω_α and eigenvectors u^α. According to the definition of the density matrix, the states corresponding to larger eigenvalues of density matrix ρ are the more probable configurations of the system block. Accordingly, the m largest eigenstates are retained, spanning a truncated space for the system block. All the operators (e.g., **H**) are transformed into this representation with lower dimension (m).

A typical real space DMRG computation is divided into two stages. In the first stage, an infinite system algorithm is adopted to generate a set of bases for various lengths of molecular fragment. One may start from a very small fragment of the whole molecule, taken as the smallest system block, and then enlarge it by a few atoms each time until the superblock reaches the size of the whole molecule. In the second stage, a finite system algorithm would be adopted iteratively, where the basis sets for various sizes system blocks is optimized for converged description of the superblock. We keep in mind that the DMRG method is not as efficient for quasi one-dimension systems like graphene nanoribbons studied in this work; enough states are retained in the DMRG calculations to ensure the convergence of the wavefunction and energy.

2.1.3 Density Functional Theory Calculations

The traditional spin-polarized DFT calculations are performed using the Vienna *ab initio* simulation package (VASP) [39–41], at the level of generalized-gradient approximation (GGA) for electron exchange and correlation. LC-DFT calculations are performed with the Gaussian 09 package [42], using CAM-B3LYP [28] and LC-ωPBE [30] functionals.

2.2 Polyacene

Polyacene can be regarded as the smallest member in ZGNR. Over the past two decades, small oligomers of polyacene attracted considerable interest from theoretical and experimental scientists due to their fascinating electronic properties originaing from their extended π conjugation [43–57]. However, larger oligomers are chemically unstable.[58, 59] The extended π electron conjugation leads to significant long-range electron correlations that must be dealt with very carefully. Within the theoretical chemistry community, there has been controversy over their geometries. Early theoretical studies at the level of Hückel molecular orbital (HMO) theory, as well as some recent theoretical investigations using the nonempirical valence bond (VB) method, or density functional theories (DFT) methods suggested that Peierls distortion might take place to remove the HOMO-LUMO degeneration at the polymer limit [33, 57, 60–72]. In contrast, other investigations showed that electron correlation effects could introduce an energy gap between the ground and the lowest excited states, that is large enough to make the Peierls distortion essentially unnecessary [53, 54, 73–76].

Theoretical investigation of the π-electron spin coupling in polyacene is also far beyond a trivial problem [77]. Small oligomers are indisputably believed to have a closed-shell ground state. Large oligomers, however, had been supposed at one time to have a triplet ground state [53], or have an even higher-spin ground state [72], with unpaired electrons occupying two or more nearly degenerate frontier π orbitals. However, more recent studies claim that large oligomers have an open-shell singlet (AFM) ground state with diradical or even polyradical characters [12, 54, 78, 79], and the transition from a closed-shell structure to an AFM structure takes place around octacene.

With the EVB calculations, the geometries of both the lowest singlet and triplet states are found to converge to very similar structures with increasing length of oligomers. For oligomers as large as [18]-acene, both states have essentially equal bond lengths along the zigzag edges, with the largest error being less than 0.001Å, for the transannular bond length, 0.004Å; similar to the conclusions formed by Bendikov et al. with DFT calculations [54]. Further increasing the length of the oligomers, the geometrical differences between two states becomes even smaller.

Geometry optimizations have revealed that larger oligoacenes should present a weakly coupled soliton-antisoliton characterized ground state that will probably show open-shell character. However, whether the spin-coupling between two solitons is ferromagnetic or anti-ferromagnetic is the next question to answer. Consistent with many DFT or even higher level studies, the EVB calculations suggest that the ground state is an open-shell singlet, which indicates an anti-ferromagnetic coupling between two solitons. The strength of the coupling should be reflected by the singlet-triplet energy gap. In Figure 2.1 the calculated S-T gap is calculated with various quantum chemical methods, for oligoacenes with up to $n = 27$. Published results of DMRG-CASSCF calculations are available for oligomers as large as dodecacene [78]. DMRG-PPP model calculations have been performed on oligomers as large as decacene with all carbon–carbon bonds being fixed at 1.397 Å. Spin-unrestricted B3LYP methods were employed first by Bendikov et al. to study oligomers as large as decacene [54], this was later extended to dodecacene by Hachmann et al. [78], and was further extended to tetradecacene by Qu et al. [79]. From VB, PPP, and CASSCF calculations, one can see a similar trend of the energy gap decreasing and approaching the polymer limit. With EVB, we can extrapolate the energy gap for long oligomers to around

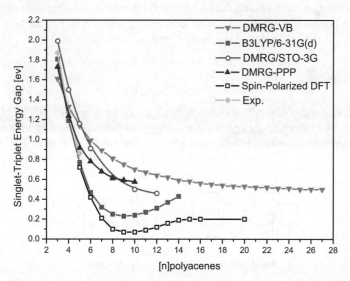

Figure 2.1 Singlet-triplet energy gap of acene oligomers calculated with different methods. Reprinted with permission from [79] © 2009 American Chemical Society

Figure 2.2 *Triplet state spin density distribution along one zigzag edge of [14]-acene*

0.47 eV for the polymer limit, which is comparable to an estimated value of 0.53 eV with the PPP method [80], *ab initio* CASSCF calculations predicted even smaller energy gaps as low as 0.14 eV [78]. However, extrapolation from data obtained with the spin-unrestricted DFT calculations is impossible since it results in a questionable evolution of the singlet–triplet energy gap beyond octacene, with the gap hitting a lowest value first and going up unexpectedly again.

Inspection of the spin distribution in the triplet state could also reveal the way in which the soliton–antisoliton pair is coupled. A spin distribution pattern for the triplet state T_1 of [14]-acene is illustrated in Figure 2.2, which shows how the spin densities vary along one zigzag edge. Very interestingly, the spin distribution pattern of each soliton is very similar to that of a neutral soliton in polyene [81]. However, the spin density, which spreads almost over the entire oligomer except for several terminal atoms, is distributed much more broadly than that of neutral solitons in polyene. In the picture of molecular orbital description, two SOMOs (which correspond to the two edge states in ZGNR) overlap only marginally at the transannular bonds [54], which implies a weak coupling between two solitons that would not only remove the degeneracy of the ground state (and thus make a Peierls distortion unnecessary), but also allow for an open-shell singlet ground state [82]. Besides, the spin polarizations are almost one order of magnitude smaller than those of the neutral solitons on polyene. This is likely to be related to the interchain coupling of the two solitons through the transannular bonds. Such interchain coupling also suppresses the negative spin polarization at the inter-edge connecting atoms, which leads to much greater averaged ρ_+/ρ_- ratio than that of the neutral solitons in polyene [81].

2.3 Polyazulene

Electronic spin-coupling in polyazulene, which is an isomerized structure of polyacene, is another story. The shortest member, azulene, is prone to thermally isomerize to naphthalene due to its much weaker aromaticity [83–86]. As illustrated in Figure 2.3, topological

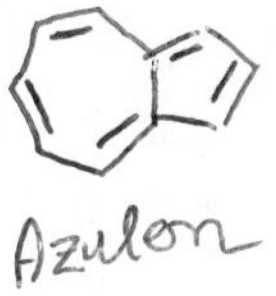

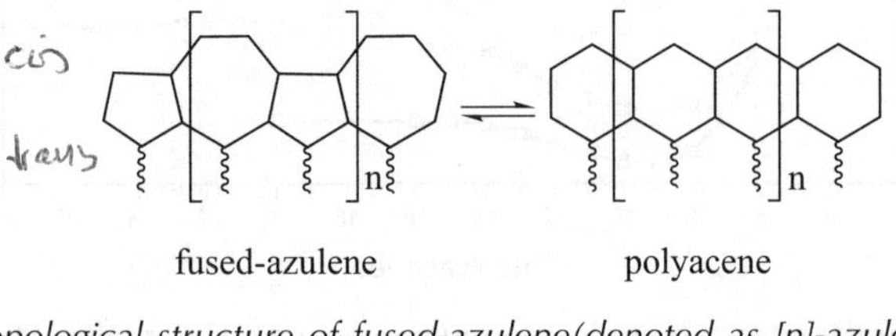

Figure 2.3 *Topological structure of fused-azulene(denoted as [n]-azulene) and polyacene. Reprinted with permission from [89] © 2011 American Institute of physics*

Table 2.1 Adiabatic singlet-triple energy gaps ΔE (kcal/mol) calculated with EVB and spin-unrestricted DFT methods

n	EVB model			DFT[f]	
	ΔE^a	ΔE^b	ΔE^c	$\langle S^2 \rangle^d$	ΔE^e
1	22.6	19.8	18.1		
2	17.1	14.5	6.1		
3	12.9	12.2	−0.8	0.60	−0.04
4	8.1	11.3	−1.9	1.01	−3.2
5	1.1	10.4	−1.5	1.00	−2.4
6	−1.0	−0.5	−0.2	0.98	−0.05

[a] EVB//B3LYP/6-31G(d).
[b] EVB//EVB.
[c] Triplet energy minus open shell broken symmetry singlet (BS) state energy.
[d] Spin contamination for the open-shell singlet ($\langle S^2 \rangle$).
[e] Energy gaps using the method suggested in References [90, 91].
[f] B3LYP/6-31G(d).

structure of a polyazulene chain is like a ladder composed of one *cis*-polyene chain and one *trans*-polyene chain, which are combined together by inter-chain C–C connections. Bending of the chain in the real structure is expected to take place due to the unbalanced strains between the top and bottom chains of the straight line structure. In contrast to polyacene, less attention has been paid to this structure moiety [87–89], which could be responsible for significant changes in the physical properties of graphene nanoribbons.

Although it has been well accepted that there existed a closed-shell to open-shell singlet ground state transformation in polyacene with increasing length of the chain, we still find to our surprise that all of the computationsal results listed in Table 2.1 indicate a similar transformation also takes place in polyazulene homologues, but in a different way. The geometries have been optimized for both values of S_z, so that the energy differences are adiabatic. According to EVB calculations, when $n \geq 6$ (smaller n observed in DFT calculations), the ground state of fused-azulene chains changes to triplet (ferromagnetic), though the singlet-triplet energy gaps are very small. Figure 2.2 also reports the mean values of the S^2 operator for the UDFT calculations, showing that the $S_z = 0$ solution becomes a half and half mixture of singlet and triplet states. The energies of the singlet state after spin decontamination are also given. For [6]-azulene, B3LYP predicted a negligible 0.2 kcal/mol energy gap between the triplet ground state and lowest singlet just above it. EVB calculations using structures either optimized by B3LYP or by EVB itself give slightly larger energy gaps. However, a significant difference is observed between results of EVB calculations on [6]-azulene when different geometries are employed, which should originate from the end effect of the oligomer. For large oligomers, when the end effects are comparatively much smaller, the results from both methods are coincident with each other.

The nature of ferromagnetic ground state in longer oligomers can be understood clearly by inspecting the frontier molecular orbitals (FMOs) which are nearly degenerate for large oligomers. The two near-degenerate FMOs of [6]-azulene shown in Figure 2.4(a), are calculated using the spin-unrestricted DFT method, indicating obviously

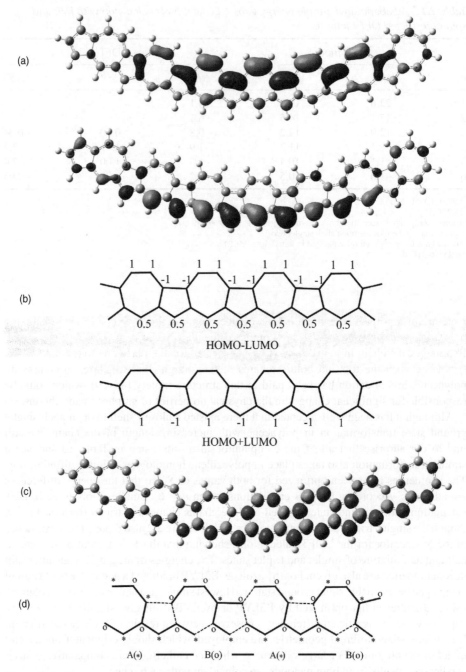

Figure 2.4 Two near-degenerate FMOs (a) calculated with UB3LYP/6-31G(d), and (b) determined by molecular graph theoretical methods. Ground state spin distribution of [6]-azulene (c) calculated with EVB. (d) Mapping of fused-azulene into a spin 1/2 chain. See colour version in the colour plate section. Reprinted with permission from [89] © 2011 American Institute of Physics

non-bonding characters. Looking at the central part of the chain, one of the FMOs is distributed predominantly on the bottom chain, with alternating signs on succeeding carbon atoms, which is similar to the case of polyacene [54]. The other FMO locates mainly on the top chain, but with a small portion of distributions on the carbon atoms indirectly connected with the top chain, while the coefficients of atomic orbitals are signed alternatively on pentagons and heptagons. These results can be rationalized by a simple molecular graphical analysis on the Hückel non-bonding MOs in periodical fused-azulene chain unit presented in Figure 2.4(b). In the case of polyacene, the open-shell character of the ground state could be explained by the electron correlation of the electrons singly occupying the non-bonding FMOs [54, 82, 92]. These MOs may be localized in the upper and lower side atoms respectively. Let us call SOMO a and b respectively these localized non-bonding MOs. Since they have zero amplitudes on the central atoms, their exchange integral, K_{ab}, which would favor a ferromagnetic coupling, is negligible. The spin polarization of the closed-shell MOs results in an anti-ferromagnetic ground state [54]. In contrast, in the case of fused-azulene, the direct exchange ferromagnetic contribution, K_{ab}, is nonzero, as easily seen when calculating it from a Hubbard Hamiltonian, since the upper side SOMO a has amplitudes on the atoms bearing the lower side SOMO b, making the parallel spin (ferromagnetic) state preferable. In small oligomers, the closed-shell ground state is due to a strong end effect leading to large energy separation of the two FMOs, which are either doubly occupied or unoccupied.

The pattern of spin polarization invoked by electron spin coupling is illustrated in Figure 2.4(c) for the triplet state of [6]-azulene. Obviously, the larger spin densities on central atoms indicates stronger radical character in central region of the fused-azulene chain. It is found that on both the top and bottom chains, the sign of spin changes alternatively on succeeding atoms, which results in alternative anti-ferromagnetic and ferromagnetic spin coupling on crossing bonds (indicated by either the same or opposite sign of spin on the two bonding atoms), which is very different from polyacene, in which all of the spin couplings at the cross chains are ferromagnetic in the triplet state [79].

2.4 Edge-Reconstructed Graphene

A series of one edge-reconstructed ZGNRs, approximately 3.5 nm in length and five layers of carbon chains in width, together with the corresponding undisturbed structure, are investigated to understand the spin coupling in these systems. The reconstruction takes place firstly in the middle of one zigzag edge, and then extends to the end of the edge as more hexagons are transformed to pentagons and heptagons.

2.4.1 Energy Gap

For convenience, we passivate all the edge atoms with hydrogen, which would not change the π energies near E_F [17, 19]. Applying EVB as well as spin-polarized DFT methods, we have carried out geometry optimizations for the ground state of ZGNRs as well as seven Rc-ZGNRs. In Table 2.2 we list the calculated adiabatic energy gaps at equilibrium ground state geometries, where n is the number of pentagon-heptagon units in Rc-ZGNRs, when $n = 7$, the whole zigzag edge is transformed to pentagon-heptagon moieties, while $n = 0$ is correspond to the unreconstructed ZGNR.

Table 2.2 Adiabatic energy gaps in unit of eV calculated by the EVB (ΔE_{T-S}) and DFT (ΔE_{FM-AFM}) methods on ground state geometries with different azulene unit

n	0	1	2	3	4	5	6	7
ΔE_{T-S}	0.13	0.12	−0.75	−0.74	−0.77	−0.75	−0.74	−0.74
ΔE_{FM-AFM}	0.11	−0.018	−0.012	−0.0067	−0.010	−0.034	−0.023	−0.020

In EVB calculations, the spin quantum number is strictly tracked to ensure the purity of the spin states, and the calculated two states can be assigned as singlet and triplet states, respectively. However, the spin-polarized DFT calculations have broken the spin symmetry, within the terminology in VASP package, the two states under investigation are named AFM and FM states, respectively. One might notice the differences in the physical meaning between the two sets of terminologies. However, in our case, they show qualitative equivalence between each other.

As shown in Table 2.2, for unreconstructed ZGNR, DFT and EVB calculations show comparable small energy gaps, which are 0.11 and 0.13 eV, respectively, with the singlet below the triplet state. Accordingly, it suggests that the ground state should be an open-shell singlet with a weak anti-ferromagnetic coupling, as a common view of previous studies [54, 79]. On the other hand, similar to the case of polyazulene [89], the transition from singlet to triplet ground state is observed both in the DFT and the EVB calculations, when enough number of consecutive hexagons rings at one zigzag edge are reconstructed to fused pentagon-heptagon moieties. Moreover, this transition takes place much earlier than in the case of polyazulene. Besides the similarities between the results of EVB and DFT methods, a surprising difference can also be seen in Table 2.2. Coinciding with the results of Dutta et al. [19], DFT calculations indicate a negligible energy gap between the FM and AFM states when one zigzag edge is partially to exhaustively reconstructed in this way. In contrast, the EVB calculations show a much greater energy gap around 0.75 eV, which implies that edge reconstruction should have aroused strong spin couplings among two or more low-lying electronic states in undisturbed ZGNRs.

It has been well documented that traditional DFT methods usually underestimate the energy gap for large π-conjugate systems due to the "delocalization error" [26, 93–95]. In comparison, demonstrative computations are performed on a Rc-ZGNR ($n = 2$) with long-range corrected density functionals, CAM-B3LYP and LC-ωPBE, as well as traditional density functionals, B3LYP and PBE1PBE, which are implemented in Gaussian 09 package. The molecular geometries have been taken directly from those optimized with traditional DFT methods using the VASP package. Interesting, the triplet-singlet energy gap is determined as 0.31 and 0.95 eV respectively, with CAM-B3LYP and LC-ωPBE functionals. In contrast, traditional DFT calculations with B3LYP (0.042 eV) and PBE1PBE (0.067 eV) functionals show much smaller energy gaps.

2.4.2 Frontier Molecular Orbitals

Two singly occupied frontier molecular orbitals (SOMOs) in an infinite length Rc-ZGNR are illustrated in Figure 2.5, which are nearly degenerate in the framework of Hückel Molecular Orbital (HMO) theory. SOMOa, which locates exclusively on half of the atoms in the bottom edge with alternating signs, is equivalent to the case of the unreconstructed

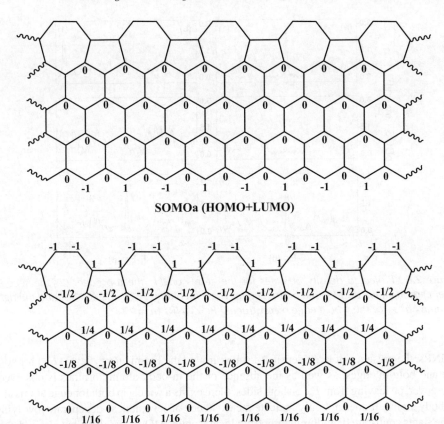

Figure 2.5 Comparative contribution of each atomic orbital in the two frontier SOMOs determined by molecular graph theoretical methods

ZGNRs. The SOMOb, whilst it is located exclusively on the top chain in the same way as the SOMOa located on the bottom chain in the case of ZGNRs, shows a quite different pattern of distribution in the case of the Rc-ZGNRs. Besides the top chain, considerable distributions on the in-body atoms are visible, with a gradual damping in crossing the ribbon. Obviously, if only the ribbon is not too wide, SOMOb should have non-negligible amplitudes on the bottom edge. Accordingly, due to the delocalization character of SOMOb, the direct exchange ferromagnetic contribution between the two SOMOs, K_{ab}, should have a nonzero value and the parallel spin (triplet) state should be preferred. Therefore, the prediction of a minor energy gap presented by DFT calculations seems to be questionable.

2.4.3 Projected Density of States

Further support for a substantial energy gap as well as the rationalization of above-room-temperature ferromagnetism are available through an analysis of the projected density of states (pDOS) of ZGNRs and Rc-ZGNRs, which are displayed in Figure 2.6. In the case of

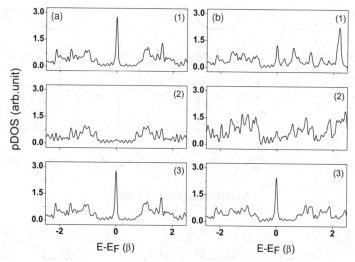

Figure 2.6 *Projected density of states for the atoms on (1) the top edge, (2) the second layer carbon chain, and (3) the bottom edge. The left column "**a**" is for ZGNRs, and the right column "**b**" is for the whole edge reconstructed Rc-ZGNRs (n = 7)*

ZGNRs, the presence of large density of states around the Fermi energy level (E_F) for both top and bottom edges, as well as the absence of the in-body atomic orbitals is observed. Moreover, on leaving from E_F on both sides, there exists a wide gap both for edge atoms and in-body atoms. It does not only indicate the existence of several near-degenerate low-lying edge states contributed predominantly from local frontier MOs near edges, but also indicate that these edge states, which seem like doping states, are responsible for the elimination of the energy gap in this system.

However, the pDOS of Rc-ZGNRs behaves in a different way except on the bottom edge. Obviously the the pDOS on the top edge decreases tremendously near the Fermi energy level when compared with that of ZGNRs, which is in accordance with the conclusion of stabilization of edge due to reconstruction, as was reported by Koskinen and Dutta [19, 22]. On the other hand, an impressive effect of edge reconstruction on the in-body state is also witnessed. The "band gap" reflected in the pDOS of in-body atoms is diminished, thus the low-lying electronic states are no longer charactered as edge states, more MOs are expected to have contributions, and the spin couplings among them are much more complicated instead of being zero in the case of ZGNRs. The traditional DFT methods' lack of long-range corrections often fails in handling such problems, and will misleadingly result in zero energy gaps, while the VB method which shows explicit multi-reference character is still efficient. Further discussion on the two theoretical methods will be given in the next section based on the analysis of the spin-distribution.

2.4.4 Spin Density in the Triplet State

The spin distribution could give us a direct picture of spin couplings in radicular systems. Those of the FM state of ZGNRs and Rc-ZGNRs, which are calculated by EVB and DFT

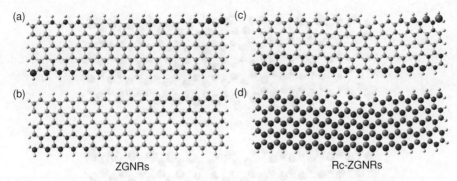

Figure 2.7 Spin distribution of ferromagnetic states for ZGNRs by (a) DFT and (b) EVB methods, and Rc-ZGNRs by (c) DFT and (d) EVB methods with a cutoff of 0.005

methods, are illustrated in Figure 2.7. For undisturbed ZGNRs, both methods show similar distribution patterns. Clearly, spin polarizations take place mainly on the two zigzag edges (the unsymmetrical distribution along one edge issue to the dangling bond at one end of both the top and bottom edges). In principle, from the viewpoint of single-reference picture, the spin-polarization is created by the averaged partial occupation of a few frontier MOs due to configuration interactions, while the doubly occupied MOs deep under the Fermi energy level do not make contributions. In the case of ZGNRs with the frontier MOs mainly localized on edge atoms, as has been discussed in the preceding section, the edge states are predominantly contributed to by electronic configurations involving hopping among edge states, which explains why the spin polarization is mainly localized on edge atoms [6, 79, 96, 97].

As has been reflected in Figure 2.7(c), which is based on the computation of Rc-ZGNRs with two pentagon-heptagon units on the top edge, the spin distribution given by DFT method is basically similar to the case of ZGNRs, which suggests a weak spin coupling among the original edge states even after edge reconstruction [19]. However, the EVB method presents an obviously different spin-polarization picture, which spans throughout the whole molecular backbone with alternating signs on succeeding atoms, except for some atoms in the azulene moieties (Figure 2.7d). This picture, which is maintained in all of the Rc-ZGNRs, suggests that the edge reconstruction has aroused strong spin coupling between the original locally distributed edge states and delocalized in-body states in Rc-ZGNRs. In return, spin-polarization will take place not only at the edges, but also on atoms deep in the body of the ribbon. Most probably, such kind of strong spin coupling will also imply the widening of the energy gap between the low-lying FM and AFM states.

It is worthy of notice that long-range corrected DFT computations also indicated strong spin-coupling in the high-spin ground state, as are shown in Figure 2.8. This explains the discord between the traditional and long-range corrected DFT methods in the prediction of energy gap, as is mentioned in the previous sections. However, the patterns of spin distribution computed using two LC-DFT methods still have non-negligible differences. Obviously, CAM-B3LYP indicates much weaker spin-coupling at the left half of the molecular plane, while LC-ωPBE presents a balanced spin-coupling at both sides. Since the spin

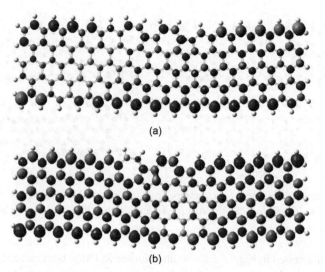

Figure 2.8 Spin distribution of triplet ground state for Rc-ZGNRs calculated with (a) CAM-B3LYP and (b) LC-ωPBE, with a cutoff of 0.005

distributions, as well as the energy gap obtained with LC-ωPBE, are much closer to those obtained with EVB method, we might tentatively suggest the LC-ωPBE functional for further study of graphene nanoribbons.

2.5 Conclusion

In our very recent studies, it was confirmed that the EVB method agrees well with high level *ab initio* methods in the description of long-range electron correlation in one-dimensional extended systems [79, 89]. By virtue of the DMRG technique, this nonempirical VB model is applicable to one-dimensional systems such as polyacene, polyazulene, or even narrow graphene nanoribbons, which allows for a systematic study of the electronic spin coupling in such systems. In this chapter, the EVB model is compared with the DFT method in the study of magnetic polarization in edge-reconstructed ZGNR. Both valence bond and DFT computations show that one-edge reconstruction from polyacene to polyazulene moieties could result in the anti-ferromagnetic to ferromagnetic state transformation for zigzag graphene nanoribbons. The essential disagreement between the valence bond and DFT calculations is the energy gap between the ferromagnetic and the anti-ferromagnetic states. While traditional DFT calculations present negligible energy preference of the ferromagnetic state over the anti-ferromagnetic state, the valence bond theory as well as long-range corrected DFT method, predicts the energy preference to be much larger, which implies the significant stabilization of the ferromagnetic electronic state due to edge reconstructions. This disparate discrepancy between traditional DFT and valence bond theories should mainly arise from the "delocalization error" inherent in traditional density functionals, which have been at least partly remedied with modern generation long-range corrected density functionals.

Acknowledgments

This work is supported by China NSF (Grant Nos. 20873058, 21173116), National Basic Research Program (Grant No. 2011CB808604), and the Scientific Research Foundation of Graduate at Nanjing University. Part of the computational work was finished on our Inspur TS10000 cluster, and IBM HS22 blade-centers at the High Performance Computing Center of Nanjing University.

References

[1] N. Tombros, C. Jozsa, M. Popinciuc, H. T. Jonkman, and B. J. van Wees, Electronic spin transport and spin precession in single graphene layers at room temperature, *Nature*, **448**(7153), 571–574 (2007).
[2] A. H. Castro Neto, F. Guinea, N. Peres, K. S. Novoselov, and A.K. Geim, The electronic properties of graphene, *Rev. Mod. Phys.*, **81**(1), 109–162 (2009).
[3] A.K. Geim, *Science*, Graphene: Status and Prospects, **324**(5934), 1530–1534 (2009).
[4] X. L. Li, X. R. Wang, L. Zhang, S. W. Lee, and H. J. Dai, Chemically derived, ultrasmooth graphene nanoribbon semiconductors, *Science*, **319**(5867), 1229–1232 (2008).
[5] X. R. Wang, Y. J. Ouyang, X. L. Li, H. L. Wang, J. Guo, and H. J. Dai, Room-temperature all-semiconducting sub-10-nm graphene nanoribbon field-effect transistors, *Phys. Rev. Lett.*, **100**(20), 206803 (2008).
[6] Y. W. Son, M. L. Cohen, and S. G. Louie, Half-metallic graphene nanoribbons, *Nature*, **444**(7117), 347–349 (2006).
[7] O. Hod, V. Barone, J. E. Peralta, and G. E. Scuseria, Enhanced half-metallicity in edge-oxidized zigzag graphene nanoribbons, *Nano Lett.* **7**(8), 2295–2299 (2007).
[8] E. J. Kan, Z. Y. Li, J. L. Yang, and J. G. Hou, Half-metallicity in edge-modified zigzag graphene nanoribbons, *J. Am. Chem. Soc.*, **130**(13), 4224–4225 (2008).
[9] S. Dutta, A. K. Manna, and S. K. Pati, Intrinsic half-metallicity in modified graphene nanoribbons, *Phys. Rev. Lett.*, **102**(9), 096601 (2009).
[10] J. Jung, T. Pereg-Barnea, and A. H. MacDonald, Theory of interedge superexchange in zigzag edge magnetism, *Phys. Rev. Lett.*, **102**(22), 227205 (2009).
[11] L. Pisani, J. A. Chan, B. Montanari, and N.M. Harrison, Electronic structure and magnetic properties of graphitic ribbons, *Phys. Rev. B*, **75**(6), 064418 (2007).
[12] D. E. Jiang, and S. Dai, Circumacenes versus periacenes: HOMO-LUMO gap and transition from nonmagnetic to magnetic ground state with size, *Chem. Phys. Lett.*, **466**(1–3), 72–75 (2008).
[13] V. L. J. Joly, M. Kiguchi, S. J. Hao, K. Takai, T. Enoki, R. Sumii, et al. Observation of magnetic edge state in graphene nanoribbons, *Phys. Rev. B*, **81**(24), 245428 (2010).
[14] Y. Wang, Y. Huang, Y. Song, X. Y. Zhang, Y. F. Ma, J. J. Liang, and Y. S. Chen, Room-temperature ferromagnetism of graphene, *Nano Lett.*, **9**(1), 220–224 (2009).
[15] Y. W. Son, M. L. Cohen, and S. G. Louie, Energy gaps in graphene nanoribbons, *Phys. Rev. Lett.*, **97**(21), 216803 (2006).
[16] G. Cantele, Y. S. Lee, D. Ninno, and N. Marzari, Spin channels in functionalized graphene nanoribbons, *Nano Lett.*, **9**(10), 3425–3429 (2009).

[17] J. Kunstmann, C. Özdoğan, A. Quandt, and H. Fehske, Stability of edge states and edge magnetism in graphene nanoribbons, *Phys. Rev. B*, **83**(4), 045414 (2011).

[18] D. E. Jiang, B. G. Sumpter, and S. Dai, First principles study of magnetism in nanographenes, *J. Chem. Phys.*, **127**(12), 124703 (2007).

[19] S. Dutta, and S. K. Pati, Edge reconstructions induce magnetic and metallic behavior in zigzag graphene nanoribbons, *Carbon*, **48**(15), 4409–4413 (2010).

[20] O. Voznyy, A. D. Güçlü, P. Potasz, and P. Hawrylak, Effect of edge reconstruction and passivation on zero-energy states and magnetism in triangular graphene quantum dots with zigzag edges, *Phys. Rev. B*, **83**(16), 165417 (2011).

[21] J. M. H. Kroes, M. A. Akhukov, J. H. Los, N. Pineau, and A. Fasolino, Mechanism and free-energy barrier of the type-57 reconstruction of the zigzag edge of graphene, *Phys. Rev. B*, **83**(16), 165411 (2011).

[22] P. Koskinen, S. Malola, and H. Häkkinen, Self-passivating edge reconstructions of graphene, *Phys. Rev. Lett.* **101**(11), 115502 (2008).

[23] P. Koskinen, S. Malola, and H. Häkkinen, Evidence for graphene edges beyond zigzag and armchair, *Phys. Rev. B*, **80**(7), 073401 (2009).

[24] C. O. Girit, J. C. Meyer, R. Erni, M. D. Rossell, C. Kisielowski, L. Yang, et al. Graphene at the edge: stability and dynamics, *Science*, **323**(5922), 1705–1708 (2009).

[25] A. Dreuw, and M. Head-Gordon, Failure of time-dependent density functional theory for long-range charge-transfer excited states: The zincbacteriochlorin-bacterlochlorin and bacteriochlorophyll-spheroidene complexes, *J. Am. Chem. Soc.*, **126**(12), 4007–4016 (2004).

[26] A. J. Cohen, P. Mori-Sanchez, and W. T. Yang, Insights into current limitations of density functional theory, *Science*, **321**(5890), 792–794 (2008).

[27] B. M. Wong, and T. H. Hsieh, Optoelectronic and excitonic properties of oligoacenes: substantial improvements from range-separated time-dependent density functional theory, *J. Chem. Theory Comput.*, **6**(12), 3704–3712 (2010)

[28] T. Yanai, D. Tew, and N. Handy, A new hybrid exchange-correlation functional using the coulomb-attenuating method (CAM-B3LYP), *Chem. Phys. Lett.*, **393**(1–3), 51–57 (2004).

[29] Y. Tawada, T. Tsuneda, S. Yanagisawa, T. Yanai, and K. Hirao, A long-range-corrected time-dependent density functional theory, *J. Chem. Phys.*, **120**(18), 8425–8433 (2004).

[30] J. Chai, and M. Head-Gordon, Systematic optimization of long-range corrected hybrid density functionals, *J. Chem. Phys.*, **128**(8), 084106 (2008).

[31] M. Said, D. Maynau, J. P. Malrieu, and M.-A. Garcia Bach, A nonempirical Heisenberg hamiltonian for the study of conjugated hydrocarbons. Ground-state conformational studies, *J. Am. Chem. Soc.*, **106**(3), 571–579 (1984).

[32] M. Said, D. Maynau, and J. P. Malrieu, Excited-state properties of linear polyenes studied through a nonempirical Heisenberg Hamiltonian, *J. Am. Chem. Soc.*, **106**(3), 580–587 (1984).

[33] M. A. Garcia-Bach, A. Penaranda, and D. J. Klein, Valence-bond treatment of distortions in polyacene polymers, *Phys. Rev. B*, **45**(19), 10891–10901 (1992).

[34] S. H. Li, J. Ma, and Y. S. Jiang, Is ferromagnetic spin coupling constant within homologous di- and triradicals?, *J. Phys. Chem.*, **100**(12), 4775–4780 (1996).

[35] M. J. Bearpark, F. Bernardi, S. Clifford, M. Olivucci, M. A. Robb, B. R. Smith, and T. Vreven, The azulene S_1 state decays via a conical intersection: A CASSCF study with MMVB dynamics, *J. Am. Chem. Soc.*, **118**(1), 169–175 (1996).
[36] S. R. White, Density matrix formulation for quantum renormalization groups, *Phys. Rev. Lett.*, **69**(19), 2863–1866 (1992).
[37] S. R. White, Density-matrix algorithms for quantum renormalization groups, *Phys. Rev. B*, **48**(14), 10345–13056 (1993).
[38] U. Schollwöck, The density-matrix renormalization group, *Rev. Mod. Phys.*, **77**(1), 259–315 (2005).
[39] G. Kresse, and J. Furthmüller, Efficient iterative schemes for ab initio total energy calculations using a plane-wave basis set, *Phys. Rev. B*, **54**(16), 11169–11186 (1996).
[40] G. Kresse, and J. Furthmüller, Efficiency of ab-initio total energy calculations for metals and semiconductors using a plane-wave basis set, *Comput. Mater. Sci.*, **6**(1), 15–50 (1996).
[41] G. Kresse, and J. Hafner, Ab initio molecular dynamics for liquid metals, *Phys. Rev. B*, **47**(1), 558–561 (1993).
[42] M. J. Frisch, G. W. Trucks, H. B. Schlegel, G. E. Scuseria, M. A. Robb, J. R. Cheeseman, et al., *G09*, Gaussian 09, Gaussian, Inc., Wallingford CT (2009).
[43] E. S. Kadantsev, M. J. Stott, and A. Rubio, Electronic structure and excitations in oligoacenes from ab initio calculations, *J. Chem. Phys.*, **124**(13), 134901 (2006).
[44] S. F. Nelson, Y.-Y. Lin, D. J. Gundlach, and T. N. Jackson, Temperature-independent transport in high-mobility pentacene transistors, *Appl. Phys. Lett.*, **72**(15), 1854–1856 (1998).
[45] V. Y. Butko, X. Chi, D. V. Lang, and A. P. Ramirez, Field-effect transistor on pentacene single crystal, *Appl. Phys. Lett.*, **83**(23), 4773–4775 (2003).
[46] D. Biermann, and W. Schmidt, Diels-Alder reactivity of polycyclic aromatic hydrocarbons. 1. Acenes and benzologs, *J. Am. Chem. Soc.*, **102**(9), 3163–3173 (1980).
[47] L. Sebastian, G. Weiser, and H. Bässler, Charge transfer transitions in solid tetracene and pentacene studied by electroabsorption, *Chem. Phys.*, **61**(1–2), 125–135 (1981).
[48] E. Heinecke, D. Hartmann, R. Müller, and A. Hese, Laser spectroscopy of free pentacene molecules (I): the rotational structure of the vibrationless $S_1 \leftarrow S_0$ transition, *J. Chem. Phys.*, **109**(3), 906–911 (1998).
[49] S. P. Park, S. S. Kim, J. H. Kim, C. N. Whang, and S. Im, Optical and luminescence characteristics of thermally evaporated pentacene films on Si, *Appl. Phys. Lett.*, **80**(16), 2872–2874 (2002).
[50] R. He, I. Dujovne, L. Chen, Q. Miao, C. F. Hirjibehedin, A. Pinczuk, C. Nuckolls, C. Kloc, and A. Ron, Resonant Raman scattering in nanoscale pentacene films, *Appl. Phys. Lett.*, **84**(6), 987–989 (2004).
[51] J. Lee, S. S. Kim, K. Kim, J. H. Kim, and S. Im, Correlation between photoelectric and optical absorption spectra of thermally evaporated pentacene films, *Appl. Phys. Lett.* **84**(10), 1701–1703 (2004).
[52] K. B. Wiberg, Properties of some condensed aromatic systems, *J. Org. Chem.*, **62**(17), 5720–5727 (1997).
[53] K. N. Houk, P. S. Lee, and M. Nendel, Polyacene and cyclacene geometries and electronic structures: bond equalization, vanishing band gaps, and triplet ground states contrast with polyacetylene, *J. Org. Chem.*, **66**(16), 5517–5521 (2001).

[54] M. Bendikov, F. Wudl, and D. F. Perepichka, Tetrathiafulvalenes, oligoacenenes, and their buckminsterfullerene derivatives: the brick and mortar of organic electronics, *Chem. Rev.*, **104**(11), 4891–4945 (2004).

[55] M. Bendikov, H. M. Duong, K. Starkey, K. N. Houk, E. A. Carter, and F. Wudl, Oligoacenes: Theoretical prediction of open-shell singlet diradical ground states, *J. Am. Chem. Soc.*, **126**(24), 7416–7417 (2004).

[56] J. P. Lowe, S. A. Kafafi, and J. P. LaFemina, Qualitative MO theory of some ring and ladder polymers, *J. Phys. Chem.*, **90**(25), 6602–6610 (1986).

[57] S. Kivelson, and O. L. Chapman, Polyacene and a new class of quasi-one-dimensional conductors, *Phys. Rev. B*, **28**(12), 7236–7243 (1983).

[58] M. M. Payne, S. R. Parkin, and J. E. Anthony, Functionalized higher acenes: hexacene and heptacene, *J. Am. Chem. Soc.*, **127**(22), 8028–8029 (2005).

[59] M. Kertesz, C. H. Choi, and S. Yang, Conjugated polymers and aromaticity, *Chem. Rev.*, **105**(10), 3448–3481 (2005).

[60] C. A. Coulson, Excited electronic levels in conjugated molecules. 1. Long wavelength ultra-violet absorption of naphthalene, anthracene and homologs, *Proc. Phys. Soc. London Sect. A*, **60**(339), 257–269 (1948).

[61] R. McWeeny, The diamagnetic anisotropy of large aromatic systems. 4. The polyacenes, *Proc. Phys. Soc. London Sect. A.*, **65**(394), 839–845 (1952).

[62] W. Moffitt, Configurational interaction in simple molecular orbital theory, *J. Chem. Phys.*, **22**(11), 1820–1829 (1954).

[63] L. Salem, and H. C. Longuet-Higgins, The alternation of bond lengths in long conjugated molecules. 2. The polyacenes, *Proc. R. Soc. London Ser. A*, **255**(1283), 435–443 (1960).

[64] M. Kimura, H. Kawaebe, K. Nishikawa, and S. Aono, Superconducting and other phases in organic high polymers of polyacenic carbon skeletons. 1. The method of sum of divergent perturbation series, *J. Chem. Phys.*, **85**(5), 3090–3096 (1986).

[65] M. R. Boon, Bond length alternation in infinitely long conjugated polyacenes, *Theor. Chim. Acta*, **23**(1), 109 (1971).

[66] M.-H. Whangbo, R. Hoffmann, and R. B. Woodward, Conjugated one and 2-dimensional polymers, *Proc. R. Soc. London Ser. A*, **366**(1724), 23–46 (1979).

[67] M. Kertesz, and R. Hoffmann, Higher-order peierls distortion of one-dimensional carbon skeletons, *Solid State Commun.*, **47**(2), 97–102 (1983).

[68] M. Kertesz, Y. S. Lee, and J. J. P. Stewart, Structure and electronic-structure of polyacene, *Int. J. Quantum Chem.*, **35**(2), 305–313 (1989).

[69] K. Tanaka, K. Ohzeki, S. Nankai, T. Yamabe, and H. Shirakawa, The electronic-structures of polyacene and polyphenanthrene, *J. Phys. Chem. Solids*, **44**(11), 1069–1075 (1983).

[70] A. L. S. da Rosa, and C. P. de Melo, *Phys. Rev. B*, Electronic-properties of polyacene, **38**(8), 5430–5437 (1988).

[71] I. Božović, Violation of the Peierls theorem in graphite chains, *Phys. Rev. B*, **32**(12), 8136–8143 (1985).

[72] M. C. dos Santos, Electronic properties of acenes: oligomer to polymer structure, *Phys. Rev. B***74**(4), 045426 (2006).

[73] P. Tavan, and K. Schulten, Correlation effects in the spectra of polyacenes, *J. Chem. Phys.*, **70**(12), 5414–5421 (1979).

[74] M. Baldo, A. Grassi, R. Pucci, and P. Tomasello, Electronic-structure of linear polyacenes in SCF RPA method *J. Chem. Phys.*, **77**(5), 2438–2444 (1982).

[75] M. Baldo, G. Piccitto, R. Pucci, and P. Tomasello, Semiconductor-like structure of infinite linear polyacene, *Phys. Lett. A*, **95**(3–4), 201–203 (1983).

[76] R. Pucci, and N. H. March, Mean π-electron energy and ionization-potential in linear polyacenes with many rings, and Hückel Parameters, *Phys. Lett. A*, **94**(1), 63–66 (1983).

[77] H. F. Bettinger, Electronic structure of higher acenes and polyacene: The perspective developed by theoretical analyses, *Pure Appl. Chem.*, **82**(4), 905–915 (2010).

[78] J. Hachmann, J. J. Dorando, M. Aviles, and G. Chan, The radical character of the acenes: A density matrix renormalization group study, *J. Chem. Phys.*, **127**(13), 134309 (2007).

[79] Z. X. Qu, D. W. Zhang, C. G. Liu, and Y. S. Jiang, Open-shell ground state of polyacenes: A valence bond study, *J. Phys. Chem. A*, **113**(27), 7909–7914 (2009).

[80] C. Raghu, Y. A. Pati, and S. Ramasesha, Structural and electronic instabilities in polyacenes: Density-matrix renormalization group study of a long-range interacting model, *Phys. Rev. B*, **65**(15), 155204 (2002).

[81] H. B. Ma, F. Cai, C. G. Liu, and Y. S. Jiang, Spin distribution in neutral polyene radicals: Pariser-Parr-Pople model studied with the density matrix renormalization group method, *J. Chem. Phys.*, **122**(10), 104909 (2005).

[82] W. T. Borden, and E. R. Davidson, Effects of electron repulsion in conjugated hydrocarbon diradicals, *J. Am. Chem. Soc.*, **99**(14), 4587–4594 (1977).

[83] L. T. Scott, and M. A. Kirms, Azulene thermal rearrangements carbon-13 labeling studies of automerization and isomerization to naphthalene, *J. Am. Chem. Soc.*, **103**(19), 5875–5879 (1981).

[84] L. T. Scott, Thermal rearrangements of aromatic compounds, *Acc. Chem. Res.*, **15**(2), 52–58 (1982).

[85] L. T. Scott, Azulene-to-naphthalene rearrangement. A comment on the kinetics, *J. Org. Chem.*, **49**(16), 3021–3022 (1984).

[86] A. Stirling, M. Lannuzzi, A. Laio, and M. Parrinello, Azulene-to-naphthalene rearrangement: The Car-Parrinello metadynamics method explores various mechanisms, *ChemPhysChem*, **5**(10), 1558–1568 (2004).

[87] B. A. Hess Jr., and L. J. Schaad, Hückel molecular orbital π resonance energies. Nonalternant hydrocarbons, *J. Org. Chem.*, **36**(22), 3418–3423 (1971).

[88] Z. X. Zhou, R. G. Parr, New measures of aromaticity: absolute hardness and relative hardness, *J. Am. Chem. Soc.*, **111**(19), 7371–7379 (1989).

[89] Z. X. Qu, S. S. Zhang, C. G. Liu, and J. P. Malrieu, A dramatic transition from nonferromagnet to ferromagnet in finite fused-azulene chain, *J. Chem. Phys.*, **134**(2), 021101 (2011).

[90] K. Yamaguchi, F. Jensen, A. Dorigo, K. N. Houk, A spin correction procedure for unrestricted Hartree-Fock and Møller-Plesset wavefunctions for singlet diradicals and polyradicals, *Chem. Phys. Lett.*, **149**(5–6), 537–542 (1988).

[91] K. Yamaguchi, Takahara, Y. Fueno, and T. K. N. Houk, Extended Hartree-Fock (EHF) theory of chemical reactions, *Theor. Chim. Acta.*, **73**(5–6), 337–364 (1988).

[92] A. Rajca, Organic diradicals and polyradicals: from spin coupling to magnetism? *Chem. Rev.*, **94**(4), 871–893 (1994).

[93] A. Ruzsinszky, J. P. Perdew, G. I. Csonka, O. A. Vydrov, and G. E. Scuseria, Density functionals that are one- and two- are not always many-electron self-interaction-free, as shown for H_2^+, He_2^+, LiH^+, and Ne_2^+, *J. Chem. Phys.*, **126**(10), 104102 (2007).

[94] Y. K. Zhang, and W. T. Yang, A challenge for density functionals: Self-interaction error increases for systems with a noninteger number of electrons, *J. Chem. Phys.*, **109**(7), 2604–2608 (1998).

[95] P. Mori-Sanchez, A. J. Cohen, and W. T. Yang, Many-electron self-interaction error in approximate density functionals, *J. Chem. Phys.*, **125**(20), 201102 (2006).

[96] K. Nakada, M. Fujita, G. Dresselhaus, and M. S. Dresselhaus, Edge state in graphene ribbons: Nanometer size effect and edge shape dependence, *Phys. Rev. B*, **54**(24), 17954–17961 (1996).

[97] M. Wimmer, I. Adagideli, S. Berber, D. Tománek, and K. Richter, Spin currents in rough graphene nanoribbons: Universal fluctuations and spin injection, *Phys. Rev. Lett.*, **100**(17), 177207 (2008).

3

Understanding Aromaticity of Graphene and Graphene Nanoribbons by the Clar Sextet Rule

Dihua Wu,[a] Xingfa Gao,[b] Zhen Zhou,[a] and Zhongfang Chen[c]

[a]Tianjin Key Laboratory of Metal and Molecule Based Material Chemistry, Key Laboratory of Advanced Energy Materials Chemistry (Ministry of Education), Computational Centre for Molecular Science, Institute of New Energy Material Chemistry, Nankai University, China
[b]Key Laboratory for Biomedical Effects of Nanomaterials and Nanosafety, Institute of High Energy Physics, Chinese Academy of Sciences, China
[c]Department of Chemistry, Institute for Functional Nanomaterials, University of Puerto Rico, USA

3.1 Introduction

The experimental realization of free-standing graphene in 2004 [1] has revolutionized modern science and technology. In particular, due to the high stability and ultrahigh intrinsic electron mobility, it is widely believed that graphene holds great promise as a replacement for silicon in the future, and to considerably improve the performance of many electronic devices, such as high-speed nano computer chips and biochemical sensors [2].

However, pristine graphene nanosheets have an "Achilles heel" for applications in electronics: they are semimetals and do not have the required band gap. One approach to introduce a band gap to graphene is to simply cut graphene sheets into graphene nanoribbons (GNRs) (Figure 3.1); this concept was proposed by the theoretical physicist, Fujita, and his coworkers in 1996 [3–6]. GNRs exhibit versatile electronic and magnetic properties dependent on the edge and shape: according to tight-binding computations, armchair GNRs

---

Graphene Chemistry: Theoretical Perspectives, First Edition. Edited by De-en Jiang and Zhongfang Chen.
© 2013 John Wiley & Sons, Ltd. Published 2013 by John Wiley & Sons, Ltd.

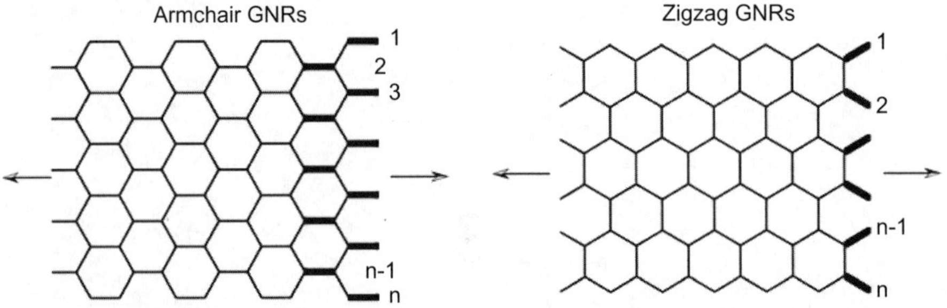

Figure 3.1 Structures of armchair graphene nanoribbons (left) and zigzag graphene nanoribbons (right). The thick line and the number on the right show the way to define the width of certain nanoribbons

with a ribbon width parameter of $w = 3n + 2$ (where n is an integer) are metallic, otherwise semiconducting, while zigzag GNRs are all metallic, independent on their width parameter w [3–6]. However, more accurate first-principles computations revealed that both armchair and zigzag nanoribbons have band gaps, and these band gaps are inversely dependent on the nanoribbon width [7–11]. These predictions were confirmed experimentally [12]. Where magnetic properties are concerned, armchair GNRs are nonmagnetic, in contrast, zigzag GNRs are characterized with special localized edge states and the corresponding flat energy bands at the Fermi level, which are ferromagnetically ordered at each edge but antiferromagnetically coupled to each other [4]. More interestingly, by means of density functional theory (DFT) computations, Son et al. pioneered the idea of driving zigzag GNRs into half-metals by applying a transverse electric field across the width [13], but this proposal was recently called into question [14].

Accompanying graphene's emergence is the revolution of device manufacturing technology: in particular the laser-carving method, which has been widely applied in silicon microelectronics, and may be replaced by this chemistry-driven self-assembly method of integrating desirable atomic/molecular components to modulate the properties of graphene and make devices. Recent experimental [15–19] and theoretical studies [20–25] have explored valuable chemical modifications of graphene.

However, the novel device design and manufacturing requires an easy model that incorporates the main electronic and chemical properties of graphene and its derivatives, such as graphene nanoribbons. Entirely composed of sp^2-carbon-atoms, the valence π-electrons of graphene dominate its electronic and chemical properties; therefore, an effective model of the π-bonding structure is essential for this purpose. Clar sextet theory provides such a simple and phenomenological model to represent the chemical bonding in polycyclic aromatic hydrocarbons (PAHs), fullerenes, and nanotubes, and is also applicable to graphene and its derivatives.

3.1.1 Aromaticity and Clar Theory

Though the notion "aromaticity" is frequently used in modern chemistry, until now, the term "aromaticity" does not have an unambiguous or quantitative definition. In fact, aromaticity is not an observable quantity; the definition by Schleyer and coworkers [26] given next conveys well the essence of the modern aromaticity concept.

> Aromaticity is a manifestation of electron delocalization in closed circuits, either in two or three dimensions. This results in energy lowering, often quite substantial, and a variety of unusual chemical and physical properties. These include a tendency towards bond length equalization, unusual reactivity, and characteristic spectroscopic features. Since aromaticity is related to induced ring currents, magnetic properties are particularly important for its detection and evaluation.

For a given aromatic system, the aromaticity can be roughly determined by the number of Kekulé structures (K); the more Kekulé structures, the higher the aromaticity. In theory, all the Kekulé structures of a molecule can be drawn and counted. Obviously though, this approach is not practical for large molecules and is prone to error; for example, the fullerene C_{60} has 12 500 Kekulé structures.

In 1964, Clar introduced a rule according to which the aromaticity of a PAH can be rationalized by counting the number of Clar sextets [27, 28]. This rule is now called the *Clar sextet rule*. According to this rule, the sp^2-carbon atoms of a closed-shell PAH can be formulated into two structural units that are linked by single bonds, namely the Clar sextet and the olefinic double bond, as shown in Figure 3.2.

In a Clar structure, the π-electron sextet is denoted by a circle, and the original definition is given as follows: (1) it is not allowed to draw circles in adjacent hexagons (two circles denote 12 π-electrons, two more than needed); (2) circles can be drawn in hexagons if the rest of the conjugated system has at least one Kekulé structure but without six-membered conjugated rings; (3) a Clar structure contains the maximum possible number of circles which can be drawn using (1) and (2).

Since each Clar sextet can be decomposed into two Kekulé valence structures of benzene rings, the number of Kekulé structures in a Clar structure with n π-sextets is 2^n. What naturally follows is that the stability of a PAH structure increases with the number of π-electron sextets, and the Clar representation accommodating the maximum possible number of Clar sextet best represents the chemical and physical properties (such as reactive positions in electrophilic aromatic substitution and bond lengths).

However, some molecules have more than a single Clar structure. For example, the Clar sextet can reside at either the left or the right ring of naphthalene, but these two rings have equal electron density due to resonance; thus the sextet is not on any one ring alone. To explain this, Clar proposed the "migrating Clar π-sextet"; there is only one sextet in naphthalene, but it migrates as denoted by an arrow. The overall local aromaticity of a ring is a superposition of all Clar structures acting for a given molecule, and the migration of

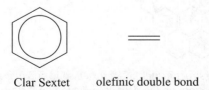

Clar Sextet   olefinic double bond

Figure 3.2 Clar structures consist of Clar sextets, olefinic double bonds and single bonds

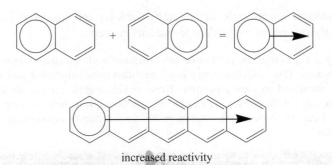

increased reactivity

Figure 3.3 *The migrating Clar π-sextet in naphthalene and higher acenes*

π-sextets decreases the stability (Figure 3.3). For example, linear acenes have only one Clar sextet ring, whose migration along all the rings leads to rather high reactivity for longer acenes: anthracene is reactive, tetracene is even more reactive, the violet pentacene is so reactive that it readily dimerizes and must be stored in nitrogen and in the dark, and heptacene and higher acenes have open-shell singlet ground state structures instead of closed-shell singlets and are elusive experimentally [29]. Similarly, [n]cyclacenes, where n = 6 or larger, as well as short zigzag (n,0) tubes, have open-shell singlet ground states [30].

The conceptual Clar sextets can be visualized experimentally and quantified theoretically. Experimentally, the measured STM images of large PAHs correlate with the Clar model [31]. Clar sextets can also be captured by different theoretical probes [26, 32]. For example, Schleyer and coworkers employed nucleus-independent chemical shifts (NICS) [33] to examine the electron delocalization of a series of D_{6h}-symmetric PAHs up to $C_{22}H_{42}$ [34], and found that the individual aromatic rings as in the Clar model can be easily distinguished by the computed NICS values. The sextet rings in the fully benzenoid PAHs (where all carbon atoms are members of a single sextet) show extreme NICS values, in comparison, those in the PAHs with migrating Clar sextets show intermediate NICS values for several rings (Figure 3.4).

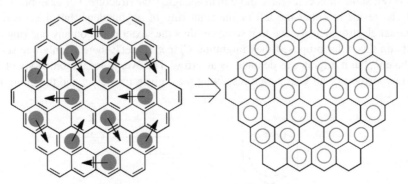

Figure 3.4 $C_{96}H_{24}$ *Kekulé structure (left): Clar sextet (solid dot) migration, which is responsible for the NICS aromaticity pattern on the right, is indicated by the arrows. See colour version in the colour plate section. Redrawn from Ref. [34]*

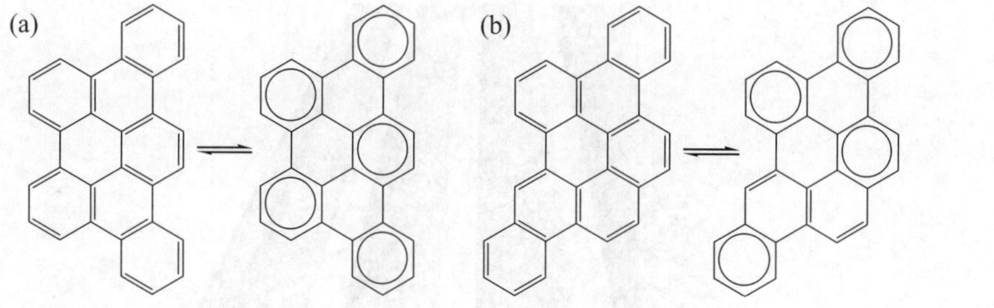

Figure 3.5 The structure of 1 : 12-2 : 3-10 : 11-tribenzoperylene (**1**,a) and 1 : 12-2 : 3-8 : 9-tribenzoperylene (**2**,b). Their Clar formulas are shown on the right

Consequently, the benzene rings in PAHs are either as the localized π sextet, as the migrating π sextet, as rings with localized double bond, or as the so-called "empty" rings, that is, without π-electrons. The algorithms to calculate all the different possibilities of Kekulé/Clar formulas are available from Ref. [35]. The Clar sextet rule predicts that the more sextets a PAH has, the higher its stability. The "fully-benzenoid" or "full Clar" PAHs whose structures can be represented with only sextets (i.e., no double bonds) possess particularly high stability and low reactivity [27, 28], which has stood the test of time [36, 37]. For example, 1 : 12-2 : 3-10 : 11-tribenzoperylene (**1**) and 1 : 12-2 : 3-8 : 9-tribenzoperylene (**2**) have very similar structures (Figure 3.5), but **1** can be represented by a fully benzenoid structure with five Clar sextets, while **2** can only accommodate four Clar sextets leaving three isolated olefinic double bonds. Such a small difference leads to significant consequence in their chemical properties: **2** readily reacts with maleic anhydride and chloranil for the Diels–Alder cycloaddition, while the all-benzenoid **1** does not react and is very stable [38].

3.1.2 Previous Studies of Carbon Nanotubes

Several research groups successfully applied the Clar sextet theory to explain the electronic properties and chemical reactivity of extended, defect free single-walled carbon nanotubes (CNTs).

Ormsby and King [39] developed rules for generating optimal Clar aromatic sextet valence bond (VB) models for CNTs with roll-up vectors (m,n) (Figure 3.6). They divided CNTs into two categories based on the value $R(n,m) = n - m$ modulo 3. When $R(n,m) = 0$, the CNTs are of a fully benzenoid Clar structure and the converse is also true. Considering that $R(n,m) = 0$ is necessary for metallic CNTs, we can get the conclusion that only fully benzenoid CNTs are metallic, and only potentially metallic CNTs are fully benzenoid. This behavior contrasts with that of planar PAHs, in which the fully benzenoid structures are known to have large HOMO-LUMO gaps (HOMO: large highest occupied molecular orbitial, LUMO: lowest unoccupied molecular orbital). When $R(n,m) = 1$ or 2, the CNTs are not fully benzenoid anymore but have a seam of double bond wraps about the otherwise benzenoid CNT, and these nanotubes are semiconducting. Notably these resonance hybrid of the Clar VB structures are not only supported by the computed nucleus independent chemical shifts (NICS) at the center of hexagons of a series of short,

Figure 3.6 *From left to right are Clar formulas of (12,9), (12,8), (12,7), and (19,0) CNTs, and their R(n,m) are 0, 1, 2, and 0, respectively. Reprinted with permission of [39] © 2004 American Chemical Society*

H-terminated chiral nanotubes, but also have been experimentally confirmed by the patterns occasionally observed by scanning tunneling microscopy (STM) images.

Matsuo *el al.* [40] demonstrated computationally that the different electronic properties and reactivity of finite-length armchair CNTs are due to the different aromaticities of the CNT fragment, classified as Clar, Kekulé, and incomplete-Clar types, in accordance with Clar theory. However, very recently, Martín-Martínez *et al.* [41] have pointed out that Matsuo *et al.*'s classification of armchair CNTs becomes unified with lengthening, and an alternated sequence of Clar and Kekulé domains occurs.

We should be very cautious when constructing models of CNT with finite lengths, the aromaticity of CNT may influence the electronic structure and makes the model incorrect. For example, short zigzag nanotube models were assumed to have a closed-shell singlet ground state, but actually they are of the open-shell singlet ground state [42]. Based on Clar theory, Mercuri *et al.* [43] proposed finite-length models of CNTs that can correctly represent their properties, and successfully applied them to study the reactivity of CNTs [44, 45].

Similar to CNTs, Clar sextet theory can also be applied to rationalize the stability, electronic, and magnetic properties of graphene and graphene nanoribbons [46–51], which we will introduce in following sections.

3.2 Armchair Graphene Nanoribbons

3.2.1 The Clar Structure of Armchair Graphene Nanoribbons

The Clar formulas of armchair graphene nanoribbons are dependent on the width of nanoribbons (width of GNRs is defined in Figure 3.1), which can be seen clearly in Figure 3.6. With increasing widths, the Clar formulas form a periodicity pattern and can be classified into three categories: fully benzenoid (Figure 3.7a and d), Kekulé (Figure 3.6b and e), and incomplete Clar structures (Figure 3.7c and f).

In fully benzenoid structures, no olefinic double bond exists, and the structures have bonding patterns very similar to two-dimensional graphene. For Kekulé structures, we can write out two types of VB structures that can maximize the number of Clar sextets, in either

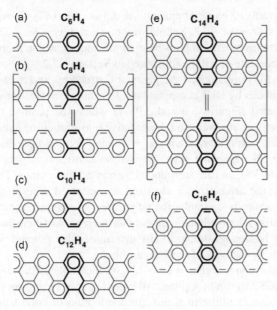

Figure 3.7 Clar formulas of armchair graphene nanoribbons with different widths. Reprinted with permission of [47] © 2008 Elsevier

of which are localized double bonds in one edge; therefore, the properties of their edge might be between benzenoid ring and localized double bond. For incomplete Clar structures, olefinic double bonds are in both edges, all the inner hexagons can form Clar sextets, and many equivalent Clar structures are available (an example is shown in Figure 3.8). In

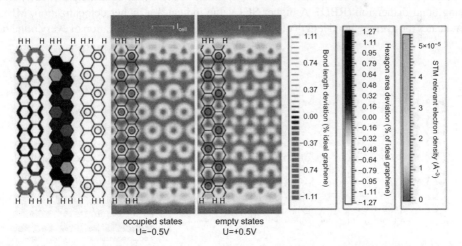

Figure 3.8 Bond length, hexagon area, Clar formulas, and simulated STM images of armchair graphene nanoribbons with w = 24 (fully benzenoid), the scales are shown in the right. Reprinted with permission of [48] © 2010 American Chemical Society

general, armchair graphene nanoribbons with width $w = 3n$ (n is a positive integer) have fully Clar structures, $w = 3n + 1$, Kekulé structures, and $w = 3n + 2$, incomplete Clar structures. Clar formulas of nanoribbons have significant impact on properties of armchair graphene nanoribbons, which will be illustrated in Section 3.2.2.

Wassmann et al. [48] investigated the geometric structures and π electron distribution of graphene nanoribbons by DFT computations. To examine π electron distribution, STM images were simulated (Figures 3.8 and 3.9), in which the positive bias measures the occupied states of a given structure and the negative bias probes the unoccupied state.

Armchair graphene nanoribbons with $w = 3n$ can form a fully benzenoid structure and are very stable. The aromatic sextet rings show shorter bond lengths and a smaller hexagon area (Figure 3.8), and are surrounded by non-aromatic rings. The simulated STM images indicate that the delocalized π electrons are occupied states in the system. As the fully benzenoid structure is so stable, the edge effect is minimized and other resonance structures with localized double bonds can only have very small contribution to the overall property. Such fully benzenoid structures also determine other properties such as magnetic response and chemical reactivity of these nanoribbons.

The graphene nanoribbons with $w = 3n + 2$ have incomplete-Clar structures, and many Clar formulas (Figure 3.9) can be written. All inner hexagons can form Clar sextets in one or more Clar formulas, resulting in a uniform distribution of bond lengths and hexagon areas. Interestingly, the STM images do not show a uniform π electron distribution. This phenomenon can be explained by the special pattern: in all Clar formulas, every third horizontal bond cannot be a localized double bond; if any of these bonds are assigned as a double bond, then the Clar structure cannot accommodate the maximum number of Clar sextets. Hence, in the STM image, the electron distribution has a pattern with two light regions followed by one dark region in the horizontal bond area.

Along the same lines, Martín-Martínez et al. [49] carefully examined the electronic structure and aromaticity of graphene nanoribbons using a series of delocalization and geometry analysis methods including six center (bond) index (SCI), mean bond length (MBL) and ring bond dispersion (RBD). A higher SCI value indicates a higher delocalization, MBL indicates the average bond length of a sextet, and RBD represents deviation of every bond in

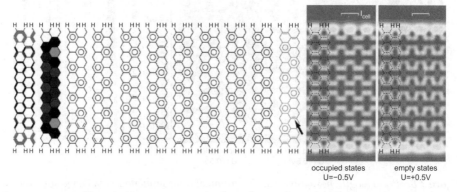

Figure 3.9 Bond length, hexagon area, Clar formulas, and simulated STM images of armchair graphene nanoribbons with w = 23 (incomplete Clar). Reprinted with permission of [48] © 2010 American Chemical Society

sextet to MBL. These analyses also identified three distinct classes of aromaticity patterns for armchair graphene nanoribbons as found by Wassmann et al., [48] which can be well explained within the framework of the Clar's sextet theory.

3.2.2 Aromaticity of Armchair Graphene Nanoribbons and Band Gap Periodicity

Clar formulas of certain armchair graphene nanoribbons determine the π electron delocalization character and thus greatly influence the electronic properties. In 1996, Fujita et al. [4] found that armchair GNRs are metallic only when ribbon width $w = 3n + 2$. In 2006, based on tight-binding computations, Ezawa [52] found that band gap of armchair GNRs oscillates with a periodicity of 3, which was confirmed by more accurate DFT computations [10,49].

The periodicity in the band gap can be easily seen by plotting the band gaps of armchair graphene nanoribbons as a function of the ribbon's width (Figure 3.10). Armchair graphene nanoribbons with $w = 3n + 1$ have Kekulé structures, and have the largest band gaps; nanoribbons with $w = 3n + 2$ have incomplete Clar formulas and have the smallest band gaps; nanoribbons with $w = 3n$ have complete Clar structures and their band gaps are in between.

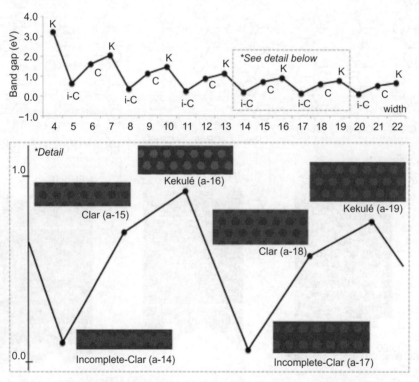

Figure 3.10 (Up) Band gaps of GNRs as a function of width. (Down) Relation between oscillation of the band gaps and the MBL patterns. Dark gray represents smaller MBL and light gray sextets represent larger MBL. Reprinted with permission of [49] © 2012 Wiley-VCH Verlag GmbH & Co. KGaA, Weinheim

Martin-Martinez *et al.* [49] used MBL to examine delocalization of π electrons in nanoribbons, and Figure 3.10 presents the relation between oscillation of the band gaps and the MBL patterns. Obviously, the incomplete-Clar structures have a uniform distributed MBL, which indicates the highest delocalization resulting in a nearly metallic electronic structure. For Kekulé structures, electrons are delocalized in Clar sextets; however, there is no obvious global delocalization. Between incomplete Clar structures and Kekulé structures are Clar structures in which the electrons are delocalized in Clar sextets and have some global delocalization along the ribbons.

We can get the same results by checking the Clar formulas [31]. Based on Clar sextet theory, when an aromatic system has more than one Clar formula, the overall properties are the combination of the contribution of all Clar formulas. As shown in Figure 3.11,

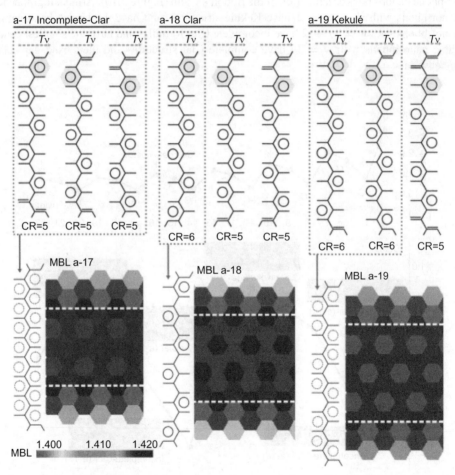

Figure 3.11 (Up) Clar structures of GNRs with ribbons of widths 17, 18, and 19. Clar structures in the dash lines are Clar formulas. (Down) The MBL images of corresponding GNRs. Reprinted with permission of [49] © 2012 Wiley-VCH Verlag GmbH & Co. KGaA, Weinheim

the incomplete Clar structures have many Clar formulas, leading to a uniform bond length distribution and a small band gap. For Clar structures, there is only one Clar formula. For Kekulé structures, there are two equivalent Clar formulas; the superposition of these two Clar formulas makes every sextet have an alternation between short and long bonds, resulting in wide band gap.

If we cancel the periodic boundary condition in GNRs and connect the two ends with chemical bonds, then relax the structure, a finite CNT is obtained. In another way, we can regard GNRs as finite CNTs with infinitely large diameters. In fact, finite CNTs are so similar to GNRs that finite armchair CNTs also have their "periodicity" not only in the Clar representations but also in the oscillation of HOMO-LUMO gaps [40], as shown in Figure 3.12. The difference from finite armchair GNRs is that finite armchair CNTs with complete Clar networks have the smallest HOMO-LUMO gaps, while CNTs with Kekulé networks have the largest HOMO-LUMO gaps, leaving the gaps for the incomplete Clar in between.

Since HOMO-LUMO gaps are an indicator of reactivity, we can expect that there is also a oscillation in reactivity, which was confirmed by Bettinger *et al.* [53] by examining the reaction energies between finite (5,5) CNTs and fluorine atoms or methylene functional groups. Due to the dependence of HOMO-LUMO gaps and reactivity on the tube length, Bettinger *et al.* [53] suggested that "the pronounced length dependence needs to be taken into account when probing the viability of sidewall functionalization computationally." The same consideration may also be taken into in the studies of GNRs when finite models are employed, as they have similar aromatic properties.

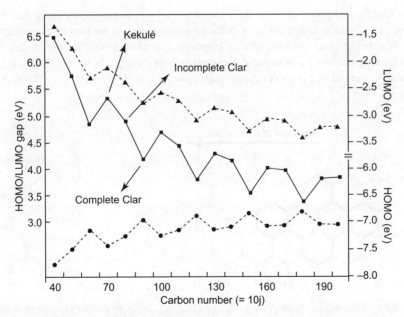

Figure 3.12 *HOMO/LUMO gap oscillation of the finite-length [5, 5]CNTs. Reprinted with permission of [40] © 2003 American Chemical Society*

3.3 Zigzag Graphene Nanoribbons

3.3.1 Clar Formulas of Zigzag Graphene Nanoribbons

The Clar formulas of a zigzag graphene nanoribbon are shown in Figure 3.13(a). Similar to the linear acenes, the migration of Clar sextets yields an infinite number of Clar formulas, which leads to low stability of such nanoribbons [48]. As the numbers of Clar sextets in zigzag graphene nanoribbons are minor, the structures of zigzag GNRs can be represented by the superposition of the two VB structures in Figure 3.13(b). The lack of Clar sextets makes zigzag GNRs much lower in stability as compared with armchair GNRs; therefore, it is very challenging to synthesize GNRs with zigzag edges.

It can be regarded that the structures of armchair GNRs and zigzag GNRs are both obtained by cutting graphene. The armchair edges can nearly retain the large π delocalization as in pristine graphene, while the zigzag edges cannot. Thus, for zigzag GNRs, the edges bring totally different properties to the system. Due to their unique edge structures, there are many interesting properties with zigzag GNRs. For example, the band gaps of armchair GNRs strongly depend on the ribbon width, in strong contrast, in zigzag GNRs, except for very narrow ones, band gaps are very small and are almost independent of ribbon width. In particular, there is a localized edge state at the zigzag edge, which results in a flat band feature in the band structure. Theoretical studies [3, 54, 55] showed that, within a certain ribbon width, though all carbons are sp^2 hybridized and three-coordinated, there is an antiferromagnetic interaction between the two zigzag edges.

3.3.2 Reactivity of Zigzag Graphene Nanoribbons

From a topological point of view, the aromaticity of GNRs originates from their edges; thus, we can predict whether certain GNRs are aromatic or not by simply examining their edges. Wassmann *et al.* [46] found that graphene nanoribbons that do not have partially filled edge states have a 1/3 benzenoid ring density near its edge (as shown in Figure 3.14a). When the benzenoid ring density near the edge is lower than 1/3 (as shown in Figure 3.14b),

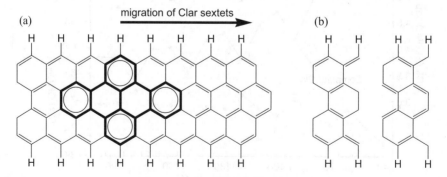

Figure 3.13 *Clar representation of a zigzag GNR. (a) Migration of Clar sextets yields an infinite number of Clar formulas. (b) Two Clar formulas representing a zigzag GNR. Redrawn following [48] with permission of American Chemical Society © 2010*

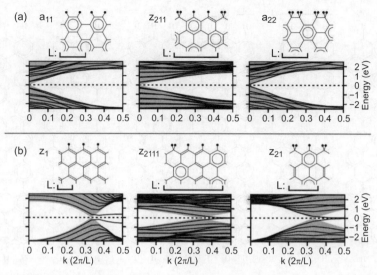

Figure 3.14 Different hydrogen-terminated edges of graphene nanoribbons and their band structures. (a) Structures which are aromatic and without edge states. (b) Structures which are nonaromatic and with edge states. The structures are periodic along the ribbon edge with periodicity L. The gray area corresponds to the electronic bands allowed in "bulk" graphene. Reprinted with permission of [46] © 2008 American Physical Society

there will be a tendency to reach such density; therefore, "edge states" will form. The "edge states" can be treated as free radicals and have great impact on the properties of GNRs.

By analyzing the computed edge formation energies, Wassmann et al. [46] found that, under certain hydrogen pressures, the GNRs having 1/3 benzenoid ring density near the edges are always more stable than those having lower benzenoid ring density near the edges. Note that edge states are unavoidable when the benzenoid ring density near the edges is lower than 1/3. Thus, we can conclude that the GNRs with edge states are less stable (or more reactive) than those without edge states, which is in agreement with our general chemical intuition.

In fact, the Clar structures of zigzag GNRs can also be represented by a fully benzenoid structure with radicals (Figure 3.15). Radicals make the system unstable; however, the conjugation with the fully benzenoid structures will stabilize these radicals. The free radicals in edges significantly influence the properties of zigzag GNRs. For example, as shown in Figure 3.15, there is one free radical in each edge of zigzag GNRs in a primitive cell, corresponding to a net spin moment of $\pm 1/3\ \mu_b$ per edge and per unit cell, and the spin polarized DFT computations reveal that the net spin moment on each side is indeed $\pm 0.3\ \mu_b$ [48].

Jiang et al. [9] investigated the reactivity of zigzag GNRs by examining the computed bond dissociation energy (BDE) for the bonds between H-terminated edge of zigzag GNRs with different widths and numbers of hydrogen atoms. Their DFT computations showed that the strength of a newly formed bond in a zigzag edge is about 60% of the bond between C_2H_5 radical and hydrogen. Therefore, Jiang et al. [9] proposed that there exists a "partial radical" character with a zigzag edge, which was calculated to be 0.14 e per carbon. To further understand this phenomenon, they computed the BDE values of different hydrogen

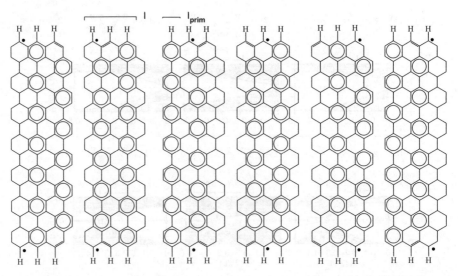

Figure 3.15 Clar formulas of zigzag graphene nanoribbons with free radical coverage (Figure 3.16). In the coverage of 1/6, though electrons are equally distributed in the edge before hydrogen addition, the magnetic moment decreases dramatically after hydrogenation. This indicates that the high reactivity of the partial radicals can be manifested collectively in reactions, despite their delocalized feature.

The "partial radical" suggested by Jiang *et al.* [9] is consistent with the Clar structures shown in Figure 3.15: on average, there is 1/6 electron for each carbon, which is in accordance with Jiang *et al.*'s results. Note that there are many Clar structures for zigzag GNRs, and the properties of these GNRs are represented by the superposition of all Clar structures; therefore, radicals are delocalized along the edge.

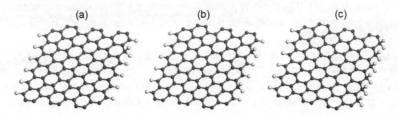

Figure 3.16 Structures of zigzag GNRs with hydrogen coverage of (a) 1/6, (b) 1/3 and (c) 1. The corresponding BDE values are 2.86, 2.73, and 1.93 eV, respectively

3.4 Aromaticity of Graphene

Graphene can be considered as an infinitely large PAH, and is sometimes called as "aromatic giant". Is graphene aromatic? How shall we understand the electron delocalization in graphene?

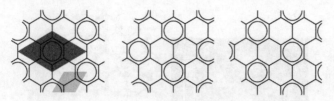

Figure 3.17 Three equal Clar formulas of graphene. Light gray area in the left Clar formula indicates the primitive cell, dark gray area represents a $(\sqrt{3} \times \sqrt{3})R30°$

Analyzing the Clar structures can give us some clue to this question. As illustrated in Figure 3.17, graphene has three equivalent Clar formulas, but the primitive cell (light gray area) cannot represents the Clar formulas; to correctly represent the Clar formulas, a $(\sqrt{3} \times \sqrt{3})R30°$ (rotate the primitive cell by 30° and multiple lattice constant by $\sqrt{3}$) supercell (dark gray area) is needed. Thus, in each Clar formula, one out of three hexagons is a Clar sextet. The superposition of these three Clar formulas results in a uniform distribution of π electrons; consequently, in graphene, every C–C bond has $2/3\,\pi$ electron and the bond order is 4/3. Every bond length and hexagon area are equal due to the uniform electron distribution.

These analyses are in good agreement with simulation results presented in Figure 3.18. As graphene has a symmetric band structure near Fermi level, the STM images in negative and positive bias are very similar.

Fujii et al. [56] experimentally investigated the π electron distribution in nanographene. The high-resolution STM image of a nanographene fragment in oxidized graphene (Figure 3.19a) clearly shows a $(\sqrt{3} \times \sqrt{3})R30°$ pattern (a schematic illustration is shown in Figure 3.19b). The results are in good agreement with investigation in PAHs and stacked graphene layers [57, 58].

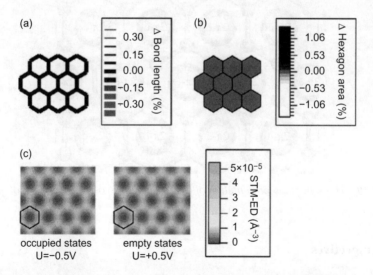

Figure 3.18 Bond length, hexagon area, and simulated STM images of graphene. Reprinted with permission of [48] © 2010 American Chemical Society

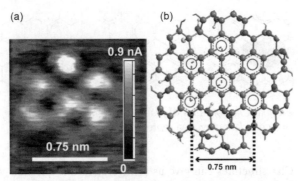

Figure 3.19 *The π state observed for nanographene fragment (A), and the schematic illustration of corresponding structure (B). Reprinted with permission of [56] © 2012 Wiley-VCH Verlag GmbH & Co. KGaA, Weinheim*

Is graphene aromatic? This question was recently addressed by Boldyrev and coworkers [50]. By carefully analyzing the chemical bonding with several theoretical methods (a fragmental approach, the adaptive natural density partitioning method, electron sharing indices, and nucleus-independent chemical shift indices), they concluded that graphene is truly aromatic, but its aromaticity differs from that of benzene, coronene, or circumcoronene: aromaticity in graphene is local with two π-electrons delocalized over every hexagon ring (Figure 3.20). This "six-center two- π" picture agrees with the earlier conclusion that "in graphene, every C–C bond has 2/3 π electrons", since there are six C–C bonds per ring but each C–C bond is shared by two hexagon rings.

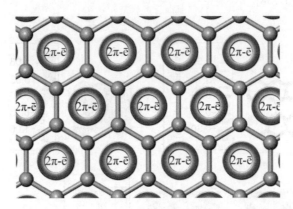

Figure 3.20 *The schematic presentation of the "six-center two-π" aromaticity of graphene*

3.5 Perspectives

As summarized earlier, the Clar sextet rule can greatly help us understand aromaticity and its related properties in graphene and nanoribbons. Recently, more and more researchers have managed to apply the Clar sextet rule in their studies related to graphene or graphene nanoribbons. Petersen analyzed the electronic structures of triangular, rectangular and honeycomb graphene antidot lattices [59]. Li et al. studied properties of C_4H nanosheets and their derivatives, in which the Clar sextet rule played a very important role [60]. Gao et al. used numbers of π-bonds surviving graphene addition reactions to estimate the thermodynamic stability of the addition products [61], and renovated the Clar model to make it applicable to heteroatom-doped sp^2 carbon systems with non-hexagonal rings [62]. Boldyrev et al. [63] rationalized the stability of the BC_3 honeycomb epitaxial sheet, which is the only experimentally known mixed B–C two-dimensional system out of all possibilities: it is the non-moving Clar sextets obtained by substituting C by B that make the BC_3 epitaxial sheet particularly stable.

However, the aromaticity concept and the Clar sextet rule have not received their deserved attention in graphene community. We believe that more efforts should be made to better understand aromaticity of graphene, nanoribbons, and their derivatives, and strongly encourage peers to apply the Clar sextet rule to rationalize and interpret graphene chemistry.

Acknowledgements

Support in China by NSFC (21273118), 973 Program (2012CB934001), and 111 Project (B12015), and in USA by Department of Defense (Grant W911NF-12-1-0083) and NSF (Grant EPS-1010094) is gratefully acknowledged.

References

[1] K. S. Novoselov, A. K. Geim, S. V. Morozov, D. Jiang, Y. Zhang, S. V. Dubonos, et al., Electric field effect in atomically thin carbon films, *Science*, **306**(5696), 666–669, (2004).

[2] A. K. Geim and K. S. Novoselov, The rise of graphene, *Nature Materials*, **6**(3), 183–191, (2007).

[3] M. Fujita, K. Wakabayashi, K. Nakada and K. Kusakabe, Peculiar localized state at zigzag graphite edge, *Journal of The Physical Society of Japan*, **65**, 1920, (1996).

[4] K. Nakada, M. Fujita, G. Dresselhaus and M. S. Dresselhaus, Edge state in graphene ribbons: Nanometer size effect and edge shape dependence, *Physical Review B*, **54**(24), 17954–17961, (1996).

[5] K. Wakabayashi, M. Sigrist and M. Fujita, Spin wave mode of edge-localized magnetic states in nanographite zigzag ribbons, *Journal of The Physical Society of Japan*, **67**, 2089–2093.

[6] K. Wakabayashi, M. Fujita, H. Ajiki and M. Sigrist, Electronic and magnetic properties of nanographite ribbons, *Physical Review B*, **59**(12), 8271–8282, (1999).

[7] Y.-W. Son, M. L. Cohen and S. G. Louie, Energy gaps in graphene nanoribbons, *Physical Review Letters*, **97**(21), 216803, (2006).

[8] D.-e. Jiang, B. G. Sumpter and S. Dai, First principles study of magnetism in nanographenes, *The Journal of Chemical Physics*, **127**(12), 124703, (2007).

[9] D.-e. Jiang, B. G. Sumpter and S. Dai, Unique chemical reactivity of a graphene nanoribbon's zigzag edge, *The Journal of Chemical Physics*, **126**(13), 134701, (2007).

[10] V. Barone, O. Hod and G. E. Scuseria, Electronic structure and stability of semiconducting graphene nanoribbons, *Nano Letters*, **6**(12), 2748–2754, (2006).

[11] L. Yang, C.-H. Park, Y.-W. Son, M. L. Cohen and S. G. Louie, Quasiparticle energies and band gaps in graphene nanoribbons, *Physical Review Letters*, **99**(18), 186801, (2007).

[12] M. Y. Han, B. Özyilmaz, Y. Zhang and P. Kim, Energy band-gap engineering of graphene nanoribbons, *Physical Review Letters*, **98**(20), 206805, (2007).

[13] Y. W. Son, M. L. Cohen and S. G. Louie, Half-metallic graphene nanoribbons, *Nature*, **444**, 347–349 (2006).

[14] M. Huzak. M. S. Deleuze and B. Hajgató, Half-metallicity and spin-contamination of the electronic ground state of graphene nanoribbons and related systems: An impossible compromise?, *Journal of Chemical Physics*, **135**, 104704 (2011).

[15] D. C. Elias, R. R. Nair, T. M. G. Mohiuddin, S. V. Morozov, P. Blake, M. P. Halsall, et al., Control of graphene's properties by reversible hydrogenation: evidence for graphane, *Science*, **323**(5914), 610–613, (2009).

[16] S. Ryu, M. Y. Han, J. Maultzsch, T. F. Heinz, P. Kim, M. L. Steigerwald and L. E. Brus, Reversible basal plane hydrogenation of graphene, *Nano Letters*, **8**(12), 4597–4602, (2008).

[17] J. Chattopadhyay, A. Mukherjee, C. E. Hamilton, J. Kang, S. Chakraborty, W. Guo, et al., Graphite epoxide, *Journal of the American Chemical Society*, **130**(16), 5414–5415, (2008).

[18] E. Bekyarova, M. E. Itkis, P. Ramesh, C. Berger, M. Sprinkle, W. A. d. Heer and R. C. Haddon, Chemical modification of epitaxial graphene: spontaneous grafting of aryl groups, *Journal of the American Chemical Society*, **131**(4), 1336–1337, (2009).

[19] S. Stankovich, D. A. Dikin, R. D. Piner, K. A. Kohlhaas, A. Kleinhammes, Y. Jia, et al., Synthesis of graphene-based nanosheets via chemical reduction of exfoliated graphite oxide, *Carbon*, **45**(7), 1558–1565, (2007).

[20] D. E. Jiang, B. G. Sumpter and S. Dai, How do aryl groups attach to a graphene sheet?, *Journal of Physical Chemistry B*, **110**(47), 23628–23632, (2006).

[21] J. O. Sofo, A. S. Chaudhari and G. D. Barber, Graphane: A two-dimensional hydrocarbon, *Physical Review B*, **75**(15), 153401, (2007).

[22] D. W. Boukhvalov, M. I. Katsnelson and A. I. Lichtenstein, Hydrogen on graphene: Electronic structure, total energy, structural distortions and magnetism from first-principle calculations, *Physical Review B*, **77**(3), 035427, (2008).

[23] E. R. Margine, M.-L. Bocquet and X. Blase, Thermal stability of graphene and nanotube covalent functionalization, *Nano Letters*, **8**(10), 3315–3319, (2008).

[24] F. OuYang, B. Huang, Z. Li, J. Xiao, H. Wang and H. Xu, Chemical functionalization of graphene nanoribbons by carboxyl groups on stone-wales defects, *The Journal of Physical Chemistry C*, **112**(31), 12003–12007, (2008).

[25] E.-j. Kan, Z. Li, J. Yang and J. G. Hou, Half-metallicity in edge-modified zigzag graphene nanoribbons, *Journal of the American Chemical Society*, **130**(13), 4224–4225, (2008).

[26] Z. Chen, C. S. Wannere, C. Corminboeuf, R. Puchta and P. v. R. Schleyer, Nucleus-independent chemical shifts (NICS) as an aromaticity criterion, *Chemical Reviews*, **105**(10), 3842–3888, (2005).
[27] E. Clar, *The aromatic Sextet*, John Wiley & Sons, Ltd, London and New York, (1972).
[28] E. Clar, *Polycyclic Hydrocarbons*, Academic Press, New York, (1964).
[29] M. Bendikov, H. M. Duong, K. Starkey, K. N. Houk, E. A. Carter and F. Wudl, Oligoacenes: theoretical prediction of open-shell singlet diradical ground states, *Journal of The American Chemical Society*, **126**(24), 7416–7417, (2004).
[30] Z. Chen, D. E. Jiang, X. Lu, H. F. Bettinger, S. Dai, P. v. R. Schleyer and K. N. Houk, Open-shell singlet character of cyclacenes and short zigzag nanotubes, *Organic Letters*, **9**(26), 5449–5452 (2007).
[31] I. Gutman, Z. Tomovic, K. Müllen and J. P. Rabe, On the distribution of pi-electrons in large polycyclic aromatic hydrocarbons, *Chemical Physics Letters*, **397**, 412–416 (2004).
[32] F. Feixas, E. Matito, J. Poater and M. Solà, On the performance of some aromaticity indices: A critical assessment using a test set, *Journal of Computational Chemistry*, **29**(10), 1543–1554 (2008).
[33] P. v. R. Schleyer, C. Maerker, A. Dransfeld, H. Jiao and N. J. R. v. E. Hommes, Nucleus-independent chemical shifts: a simple and efficient aromaticity probe, *Journal of the American Chemical Society*, **118**(26), 6317–6318 (1996).
[34] D. Moran, F. Stahl, H. F. Bettinger, H. F. Schaefer III and P. v. R. Schleyer, magnetic properties of large polybenzenoid hydrocarbons Towards graphite: magnetic properties of large polybenzenoid hydrocarbons, *Journal of the American Chemical Society*, **125**(22), 6746–6752 (2003).
[35] For example, C. P. Chou, H. A. Witek, *MATCH Commun. Math. Comput. Chem.* **68**, 3–30. (2012).
[36] M. D. Watson, A. Fechtenkötter and K. Müllen, Big is beautiful – "aromaticity" revisited from the viewpoint of macromolecular and supramolecular benzene chemistry, *Chemical Reviews*, **101**(5), 1267–1300, (2001).
[37] M. Randić, aromaticity of polycyclic conjugated hydrocarbons, *Chemical Reviews*, **103**(9), 3449–3606, (2003).
[38] E. Clar and M. Zander, 378. 1 : 12–2 : 3–10 : 11-Tribenzoperylene, *Journal of the Chemical Society (Resumed)*, 1861–1865, (1958).
[39] J. L. Ormsby and B. T. King, Clar valence bond representation of π-bonding in carbon nanotubes, *The Journal of Organic Chemistry*, **69**(13), 4287–4291, (2004).
[40] Y. Matsuo, K. Tahara and E. Nakamura, Theoretical studies on structures and aromaticity of finite-length armchair carbon nanotubes, *Organic Letters*, **5**(18), 3181–3184, (2003).
[41] F. J. Martín-Martínez, S. Melchor and J. A. Dobado, Clar–Kekulé structuring in armchair carbon nanotubes, *Organic Letters*, **10**(10), 1991–1994, (2008).
[42] Z. Chen, D. E. Jiang, X. Lu, H. F. Bettinger, S. Dai, P. v. R. Schleyer, K. N. Houk, Open-shell singlet character of cyclacenes and short zigzag nanotubes, *Organic Letters*, **10**(10), 5449–5452, (2007).
[43] M. Baldoni, A. Sgamellotti and F. Mercuri, Finite-length models of carbon nanotubes based on Clar sextet theory, *Organic Letters*, **9**(21), 4267–4270, (2007).

[44] M. Baldoni, D. Selli, A. Sgamellotti and F. Mercuri, Unraveling the reactivity of semiconducting chiral carbon nanotubes through finite-length models based on Clar sextet theory, *The Journal of Physical Chemistry C*, **113**(3), 862–866, (2009).

[45] F. Mercuri and A. Sgamellotti, First-principles investigations on the functionalization of chiral and non-chiral carbon nanotubes by Diels–Alder cycloaddition reactions, *Physical Chemistry Chemical Physics*, **11**(3), 563–567, (2009).

[46] T. Wassmann, A. P. Seitsonen, A. M. Saitta, M. Lazzeri and F. Mauri, Structure, stability, edge states, and aromaticity of graphene ribbons, *Physical Review Letters*, **101**(9), 96402, (2008).

[47] M. Baldoni, A. Sgamellotti and F. Mercuri, Electronic properties and stability of graphene nanoribbons: An interpretation based on Clar sextet theory, *Chemical Physics Letters*, **464**(4–6), 202, (2008).

[48] T. Wassmann, A. P. Seitsonen, A. M. Saitta, M. Lazzeri and F. Mauri, Clar's theory, π-electron distribution, and geometry of graphene nanoribbons, *Journal of The American Chemical Society*, **132**(10), 3440, (2010).

[49] F. J. Martín-Martínez, S. Fias, G. Van Lier, F. De Proft and P. Geerlings, Electronic structure and aromaticity of graphene nanoribbons, *Chemistry – A European Journal*, **18**(20), 6183–6194, (2012).

[50] F. J. Martín-Martínez, S. Melchor, A. Dobado, Edge effects, electronic arrangement, and aromaticity patterns on finite-length carbon nanotubes, *Physical Chemistry Chemical Physics* **13**(28), 12844–12857 (2011).

[51] I. Popov, K. Bozhenko and A. Boldyrev, Is graphene aromatic?, *Nano Research*, **5**(2), 117–123, (2012).

[52] M. Ezawa, Peculiar width dependence of the electronic properties of carbon nanoribbons, *Physical Review B*, **73**(4), 045432, (2006).

[53] H. F. Bettinger, Effects of finite carbon nanotube length on sidewall addition of fluorine atom and methylene, *Organic Letters*, **6**(5), 731–734, (2004).

[54] S. Okada and A. Oshiyama, Magnetic ordering in hexagonally bonded sheets with first-row elements, *Physical Review Letters*, **87**(14), 146803, (2001).

[55] H. Lee, Y.-W. Son, N. Park, S. Han and J. Yu, Magnetic ordering at the edges of graphitic fragments: Magnetic tail interactions between the edge-localized states, *Physical Review B*, **72**(17), 174431, (2005).

[56] S. Fujii and T. Enoki, Clar's Aromatic Sextet and π-Electron Distribution in Nanographene, *Angewandte Chemie International Edition*, **51**(29), 7236–7241, (2012).

[57] I. Gutman, Ž. Tomović, K. Müllen and J. P. Rabe, On the distribution of π-electrons in large polycyclic aromatic hydrocarbons, *Chemical Physics Letters*, **397**(4–6), 412–416, (2004).

[58] N. H. Cho, D. K. Veirs, J. W. A. III, M. D. Rubin, C. B. Hopper and D. B. Bogy, Effects of substrate temperature on chemical structure of amorphous carbon films, *Journal of Applied Physics*, **71**(5), 2243–2248, (1992).

[59] R. Petersen, T. G. Pedersen, A. P. Jauho, A. P. Clar sextet analysis of triangular, rectangular, and honeycomb graphene antidot lattices, *ACS Nano*, **5**(1), 523–529, (2011).

[60] Y. Li and Z. Chen, The patterned partially hydrogenated graphene (C_4H) and its one-dimensional analogues: a computational study, *The Journal of Physical Chemistry C*, **116**(7), 4526–4534, (2012).
[61] X. Gao, Y. Zhao, B. Liu, H. Xiang and S. B. Zhang, Π-Bond maximization of graphene in hydrogen addition reactions, *Nanoscale*, **4**, 1171–1176, (2012).
[62] X. Gao, S. B. Zhang, Y. Zhao and S. Nagase, A nanoscale jigsaw-puzzle approach to large π-conjugated systems, *Angewandte Chemie*, **122**(38), 6916–6919, (2010).
[63] I. A. Popov and A. I. Boldyrev, Deciphering chemical bonding in a BC3 honeycomb epitaxial sheet, *The Journal of Physical Chemistry C*, **116**(4), 3147–3152, (2012).

4

Physical Properties of Graphene Nanoribbons: Insights from First-Principles Studies

Dana Krepel and Oded Hod
Department of Chemical Physics, School of Chemistry, The Raymond and Beverly Sackler Faculty of Exact Sciences, Tel-Aviv University, Israel

4.1 Introduction

Graphene, a single layer of graphite, has been theoretically studied for over 60 years [1]. Nevertheless, the interest in graphene has gained a considerable boost since 2004 when it was first isolated as a stand-alone two-dimensional all-carbon network [2]. Follow-up experiments suggested that electrons in graphene behave like massless relativistic particles, which contribute to peculiar properties such as the anomalous quantum Hall effect [3, 4]. Graphene has further demonstrated a variety of intriguing properties including high electron mobility at room temperature [2, 5], exceptional thermal conductivity [6], and superior mechanical properties with a remarkable measured Young's modulus in the order of 1 TPa [7]. Moreover, the Dirac-like energy dispersion of its two-dimensional structure implies that graphene is a gapless semiconductor (or semi-metal), whose density of states vanishes linearly when approaching Fermi energy. As such, depending on the operating regime, graphene can be viewed as a unique material bridging the worlds of semiconductors (for example, it is possible to open a gap in a sample with the help of chemical modifications [8, 9], or lateral confinement [10–12]) and metals, with a finite density of electronic states at the Fermi energy (one can make graphene metallic, e.g., by chemical doping [13]). These

---

Graphene Chemistry: Theoretical Perspectives, First Edition. Edited by De-en Jiang and Zhongfang Chen.
© 2013 John Wiley & Sons, Ltd. Published 2013 by John Wiley & Sons, Ltd.

properties generated huge interest in the possible implementation of graphene as a central building block for a wide variety of miniaturized devices.

In recent years, major advances in the production of graphene have been achieved both in the fabrication of large graphene surfaces [14–18] and in the synthesis of low dimensional graphene derivatives [11, 19–28]. Numerous potential technological applications of large scale graphene structures have been suggested including: (1) touch screens; (2) smart windows; (3) flexible liquid crystal displays and organic light-emitting diodes; (4) energy storage devices such as supercapacitors and lithium ion batteries; [23, 29–38] as well as (5) transparent conducting films for solar cells. Similarly, nanoscale graphene derivatives show considerable technological promise with various potential applications ranging from chemical and biological detectors [39–41], to electronic and spintronic components [42–45], nanoscale interconnects [46, 47], and nanoelectromechanical systems (NEMS) [48–51].

Graphene nanoribbons (GNRs), which are narrow and elongated strips of graphene, are among the most studied graphene structures. There are several reasons for this: First, GNRs can be readily produced with a highly consistent and reproducible edge structure; [21, 22]. Second, they demonstrate unusual properties such as spin polarized edge states [52–58], an externally controllable half-metallic character [59], and a chemically reactive edge structure [60–62]; and third, they present an energy band-gap in their single-electron spectrum. This gap, which is a consequence of the electron confinement, presents three-fold oscillatory variations with the width of the nanoribbons which decay for large ribbon widths [63, 64]. This suggests that one may fabricate GNRs with desired band-gaps by controlling their transversal dimensions. Gaining such control is crucial for many of the aforementioned applications.

These intriguing graphene derivatives are usually classified by the structure of their edges; In Figure 4.1 we present a scheme of the two extreme crystallographic orientations of the GNRs axis, namely zigzag GNRs (ZZGNRs) and armchair GNRs (ACGNRs). Interestingly, despite the fact that they share the same sp^2 bonded hexagonal carbon structure, the mere difference in crystallographic orientation induces considerably different electronic and magnetic characteristics of the two systems.

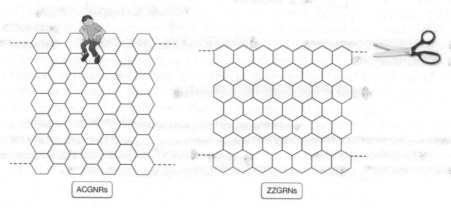

Figure 4.1 The two achiral GNR structures: armchair (left panel) and zigzag (right panel). Reproduced with permission from Kriat Beynaim and Israeli Center Of Chemistry Teachers: http://www.weizmann.ac.il/weizsites/blonder/files/2011/02/Graphen.pdf

The aim of this chapter is to briefly survey recent work discussing, on theoretical grounds, various properties of GNRs. We will discuss mainly the structure of these systems, their magnetic, and their electronic properties, while placing an emphasis on computational studies utilizing ground state density functional theory (DFT) [65, 66]. Several important topics such as the optical and thermal properties of graphene and its derivatives [18, 67–71] had to be excluded for the sake of conciseness. Furthermore, in light of the tremendous amount of information accumulated in this field during the past few years [72–79] and due to strict space limitations we stress that this chapter is mainly intended to serve as a starting point for the interested reader and does not convey a thorough account of the status of the field. Therefore, we apologize in advance for having to neglect many important contributions to the field.

4.2 Electronic Properties of Graphene Nanoribbons

The electronic properties of carbon based systems, such as carbon nanotubes, graphite, and graphene are dominated by their π-electronic structure which is determined by the sp^2 bonded hexagonal carbon network and the relevant boundary conditions. In the case of GNRs, the large aspect ratio induces considerable finite-size effects due to quantum confinement. Furthermore, near the edges the carbon network suffers from under-coordination, thus edge reconstruction [80–90] or unsaturated dangling bonds may appear that can lead to the occurrence of edge states [52–58, 64, 91–93].

Early studies based on tight-binding models already indicated that the electronic [52–55] and magnetic [54] properties of zigzag- and armchair-edged graphene nanoribbons may differ considerably. The band structure of wide ACGNRs was found to be similar to that of two-dimensional (2D) graphene where in both cases the conduction and valence bands approach the Dirac points. The main difference was identified as a small direct gap that develops in the case of the quasi-one-dimensional system whose size was found to reduce with increasing nanoribbon width. In the case of ZZGNRs, the edge structure was found to be responsible for the appearance of localized edge states close to the Fermi level which are manifested as a sharp peak in the density of states in the vicinity of the Fermi energy whose size decreases with increasing nanoribbon width [55]. Recently, this prediction of the existence of localized electronic states near the zigzag edges of graphene surfaces has received several experimental verifications [91, 94, 95]. In what follows, we present some of the consequences resulting from these pronounced differences in the electronic structure between graphitic systems bearing various edge structures.

4.2.1 Zigzag Graphene Nanoribbons

One way to view a ZZGNR is as an unrolled armchair carbon nanotube (ACCNT). While structurally correct, this analogy completely fails when trying to deduce the electronic properties of the unfolded structures from those of their tubular counterparts. Calculations based on first-principles methods revealed that ZZGNRs are all semiconducting with a width-dependent bandgap [56, 59, 96]. This is in complete contrast to ACCNTs which are known to be metallic. This major difference between the electronic characters of the two related systems is a direct consequence of the zigzag edge states mentioned earlier [97].

Figure 4.2 *Ground-state spin density of a hydrogen terminated ZZGNR. Shading stands for up and down spin density excess. Adapted with permission from [60] © 2007 American Chemical Society*

In order to understand the electronic ground state of ZZGNRs we recall that the minimal graphene unit-cell contains two carbon atoms which form the basis for the formation of two sub-lattices. In the hexagonal lattice of graphene these two carbon sub-lattices interpenetrate such that all the nearest-neighbors of an atom belonging to one sub-lattice are part of the complementary sub-lattice. In the ground state of ZZGNRs a unique magnetic ordering of spins occurs where the two sub-lattices bear opposite spin flavors arranged in an inter-site antiferromagnetically coupled network of spins (see Figure 4.2). Since the carbon atoms at the two zigzag edges belong to different sub-lattices the edge states carry opposite spins and the ground state is said to be of anti-parallel (or anti-ferromagnetic) nature [54]. This unique magnetic state, which has an overall zero magnetization but a microscopic spin polarization, is energetically favored over the nonmagnetic state or the ferromagnetic state with the parallel spin orientation at the two zigzag edges [98, 99]. It is this symmetry breaking between the two carbon sub-lattices that is responsible for the opening of an energy band-gap. The stability of the spin-polarized state was recently analyzed in chemical terms using Clar's sextet theory, which is a standard tool for analyzing aromaticity of organic compounds [85, 100–102].

The spatial separation between the two spin states localized around the opposite zigzag edges opens the way for unique control schemes over the electronic properties of the system. By applying an in-plane external electric field perpendicular to the ribbon's axis one can create an edge-localized gating effect where the states localized on one zigzag edge shift downwards in energy while those located at the opposite edge shift upward (see Figure 4.3) [59]. In this manner, the occupied and unoccupied bands for one spin flavor move closer in energy whereas those of the other spin flavor move apart. The resulting effect of the electric field on the spin-polarized band-gaps is presented in the right panel of Figure 4.3] In the absence of an electric field, the α- and β-spin states are energetically degenerate with identical band-gaps. The applied electric field increases the α-spin states band-gap while simultaneously decreasing the β-spin band-gap. For sufficiently large electric fields intensities the β-spin band-gap closes and a half-metallic behavior is achieved. Since the potential difference between two opposite zigzag edges is proportional to the distance between the zigzag edges, the critical electric field required to achieve the half-metallic state decreases with increasing nanoribbon width.

Following this proposal, several other investigations studied the appearance of half-metallicity in ZZGNRs. It is worth mentioning that the size of the gap estimated by various DFT methods was found to depend on the choice of the exchange-correlation functional approximation. For the B3LYP hybrid functional the absence of a half-metallic regime was suggested even at high electric fields [103]. However, several other theoretical studies reported that the non-local exchange term does not remove half-metallicity in ZZGNRs

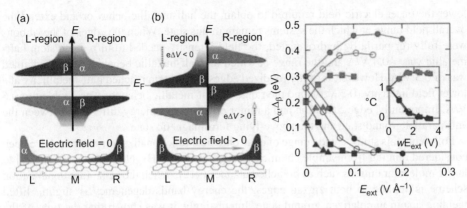

Figure 4.3 *Illustration of the spatial distribution of the ground state spin density and the corresponding density of states in the absence (left panel) and presence (middle panel) of an externally applied electric field. Right panel: α (light shading) and β (darker shading) spin-polarized band-gaps as a function of the field intensity. Reprinted with permission from [59] © 2006 Nature Publishing Group*

[104, 105]. Using the B3LYP functional, Kan et al. found a spin gap asymmetry caused by the applied electric field where the β-spin state is gapless [104]. The magnitude of the critical electric field required to transfer the system into a half-metallic state was, however, larger (about 0.7 V/Å) than that obtained in Ref. [59] using the local-density approximation (LDA), which is known to underestimate the size of the band-gap. For sufficiently large electric fields the half-metallicity was found to be destroyed resulting in a spin-unpolarized state. Similar results suggesting that at high electric field the system is spin-unpolarized were also reported in Ref. [60]. These results indicate that external electric fields can be used to control the magnetic properties of ZZGNRs with great potential for application in spin-related electronic devices.

Since, as discussed previously, many of the unique electronic and magnetic properties of ZZGNRs are associated with edge localized electronic states, edge defects [85, 106] and edge chemistry [62, 105, 107] can significantly alter the electronic properties of the ribbons. Intrigued by the interplay between the electric field and the spin degree of freedom, Hod et al. studied the influence of edge chemistry on the electronic and magnetic properties of ZZGNRs [60]. The authors postulated that during standard GNR fabrication processes the ribbon edges can become oxidized. Hence, various oxidation schemes were considered including hydroxyl, lactone, ketone, and ether groups. When examining the relative stability of the ground states of the different schemes, it was found that the oxidized ribbons are, in general, more stable than hydrogen terminated GNRs. Similar to the fully hydrogenated GNRs, all the studied oxidized structures exhibited a spin-polarized ground state with anti-parallel spin ordering localized at the opposite edges. Furthermore, the oxidized structures were found to have a band-gap comparable to that of the fully hydrogenated ribbon, indicating that their electronic character is generally preserved upon edge oxidation. Therefore, one would expect that these systems should behave as half-metals under the influence of an external electric field. Interestingly, edge oxidation has been found to lower the onset electric field required to obtain the half-metallic behavior and extend the overall field range at which the systems remain in this state. When the edges of the ribbons were fully (or partially) hydrogenated, the field intensity needed to turn the system half-metallic was ∼0.4 V/Å and the range at which the half-metallic behavior was maintained was of 0.3 V/Å. However, for fully oxidized edges, the system reached half-metallicity at a lower field intensity (0.2 V/Å) and the range of half-metallic behavior was doubled to 0.6 V/Å. Importantly, edge oxidation had a minor effect on the energy difference between the anti-parallel ground state and the above-lying ferromagnetic state.

Following this study, a wide range of edge chemical functionalization schemes have been considered ranging from simple chemical groups, such as NH_2, NO_2, and O [108–111] to larger molecular entities such as branched alkanes [112]. When the same functionalization scheme is applied to both zigzag edges, the energy bands degeneracy is slightly lifted yielding a spin-unpolarized ground state. Interestingly, it was shown that the type of the edge functionalization can be used as a control parameter for manipulating the size of the bandgap, the spin ordering along the zigzag edges, and the distribution of the molecular orbitals over the graphene lattice [111].

When functionalizing the two zigzag edges by different chemical groups a wealth of magnetic and electronic properties becomes accessible. It was shown that for graphene with one monohydrogenated zigzag edge and one dihydrogenated edge, spontaneous magnetization appears [113, 114]. Cervantes-Sodi et al. considered different functional groups terminating the dangling bonds at a single zigzag edge to induce spin-polarization in graphene [109, 110]. There, the edge functionalization was found to significantly modify the electronic structure of the graphene ribbon, particularly lifting the spin degeneracy. Moreover, in some cases, a significant spin gap asymmetry was achieved producing half-metallic behavior even in the absence of an external electric field. Similar results were shown in Ref. [115] where half-metallicity was achieved by terminating one zigzag edge by hydroxyl groups and the opposite one by chlorine atoms.

Another way to induce edge asymmetry in ZZGNRs is via the substitution of carbon atoms within the hexagonal lattice with atomic dopants such as boron and nitrogen atoms [105, 116, 117]. Upon substitution of edge carbon atoms spin ordering occurs due to the imbalance of majority and minority spins near the zigzag edges. The obtained band structure indicated a metallic behavior for wide enough nanoribbons [105, 116]. Additionally, it was found [118] that the substitutional doping of a single carbon atom at the zigzag edge by an O, B, or N atom can inject a hole or electron into the graphene π system depending on the impurity type. Therefore, such substitutions induce a localized impurity state close to the Fermi level while altering the transport properties of the system.

4.2.2 Armchair Graphene Nanoribbons

Similar to the case of ZZGNRs, an ACGNR can be viewed as an unrolled zigzag carbon nanotube (ZZCNT). Nevertheless, the replacement of the zigzag edges by their armchair counterparts eliminates the localized edge states and the spin ordering that they carry. Therefore, pristine ACGNRs are non-magnetic with a semiconducting character resulting from quantum confinement effects.

Here, the analogy between ACGNRs and ZZCNTs is carried much further where several studies have predicted pronounced band-gap oscillations as a function of the ACGNR width

[53,55,59,63,64,96,119–121]. The threefold period of these oscillations divides the family of ACGNRs into three subgroups according to their width with N = 3p, 3p + 1 and 3p + 2 (p is a positive integer), N being the number of zigzag chains across the width of the ribbon [58]. Within the framework of tight-binding calculations all ACGNRs with N = 3p + 2 present metallic character whereas the remaining ACGNRs show semiconducting behavior. Interestingly, when electronic correlation effects and long-range interactions are taken into account via DFT calculations, similar band-gap oscillations appear but with an important difference: all ACGNRs, including the N = 3p + 2 subgroup, are predicted to be semiconducting. It should be noted that the oscillatory bandgap behavior was predicted to have an envelope function decaying with the ACGNRs width, a behavior which has already been observed experimentally with measured band-gaps that were consistent with the results obtained by the DFT calculations [11, 12, 23].

In Figure 4.4(a), DFT results for the width dependent band-gap of hydrogen-passivated ACGNRs are presented [64]. For the range of widths covered by the DFT calculations (up to 3 nm) only partial quenching of the energy gap oscillations is observed. Since all GNRs are expected to reach the graphene limit of zero band-gap for sufficiently large widths, extrapolations can be used to study the wider ribbons regime. To this end, the points in Figure 4.4(a) belonging to each subgroup of ACGNRs are separately extrapolated to larger widths using an inverse power law with two fitting parameters [64]. This simple rule presents the correct asymptotic behavior and provides qualitative information of the electronic structure of ribbons beyond the range of widths studied by first-principles calculations.

From these results it is shown that, when considering ultra-narrow ACGNRs with widths up to 1nm, relatively large band-gaps of 1–3 eV are obtained. When ribbons of larger

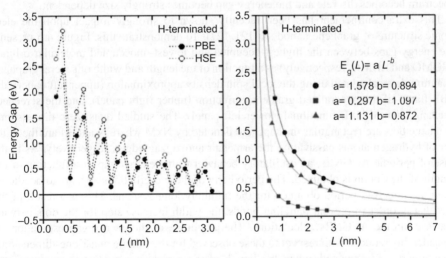

Figure 4.4 Left panel: Dependence of the band-gap on the width of hydrogen passivated ACGNRs obtained using the PBE (full circles) and HSE (open circles) functional approximations and the 6-31G** Gaussian basis set. Right panel: extrapolation of the HSE results towards larger ribbon width. Adapted with permission from [64] © 2006 American Chemical Society

widths are considered (L ∼ 10 nm), about 67% will present an energy gap of 0.2 eV while ∼33% of them will have a almost vanishing band-gap. When considering ribbons 10 times wider (L = 80 nm), the band gap is predicted to be smaller than 0.05 eV. This suggests that the lateral dimensions of GNRs can be used to tailor their electronic properties. As an example, in order to obtain an ACGNR with a band-gap comparable to that of Ge (0.67 eV) or InN (0.7 eV), nanoribbons as narrow as 2–3 nm should be fabricated. If a larger band-gap material such as Si (1.14 eV), InP (1.34 eV), or GaAs (1.43eV), is required the width of the GNR must be reduced to as low as 1–2 nm. It is interesting to note that when the first DFT calculations on these systems were carried out, it seemed impossible to produce nanoribbons as narrow as a few nanometers. These days, the fabrication of ultra-narrow GNRs with well-defined edge structures and the electrical measurement of nanoribbons with widths down to a few nanometers are performed routinely in many laboratories around the world [11, 12, 21–25, 122].

We note in passing that other routes to control the band-gap of graphene, including the possibility of breaking inversion symmetry by applying a noncentrosymmetric supperlattice of gate potentials [123], edge localized gate effects [124], symmetry broken substrates [125], and graphene antidot lattices [126] are further discussed in the literature.

4.2.3 Graphene Nanoribbons with Finite Length

As clearly demonstrated previously, finite size effects in general, and quantum confinement in particular, play a central role in dictating the electronic properties of nanoscale graphene structures. When the dimensionality of a graphene nanoribbon is reduced to form graphene dots of nanoscale dimensions the typical de-Broglie wavelength associated with Fermi electrons becomes comparable to the dimensions of the system. At this limit the energy spectrum becomes discrete and the energy gap becomes strongly size dependent.

This extra confinement has been recently shown to strongly impact upon the electronic structure of graphene nanodots [121, 127]. To demonstrate this, Figure 4.5 presents the energy gaps between the highest-occupied and lowest-unoccupied molecular orbitals (HOMO and LUMO, respectively) as a function of the length and width of several graphene quantum dots calculated using the local spin density approximation (upper left panel), the PBE flavor of the generalized gradient correction (upper right panel), and the screened-exchange hybrid HSE functional (lower left panel). The studied quasi-zero-dimensional graphene dots are rectangular in shape and denoted by NxM where N and M are the number of hydrogen atoms passivating the armchair and zigzag edges, respectively. As in the case of periodic ACGNRs, an oscillatory behavior of the energy gap as a function of the length of the ribbon is obtained. The periodicity of these oscillations appears to be altered from the threefold period obtained for the infinitely long systems. This is a result of the fact that in order to prevent dangling bonds, the width increase step for the dots is twice that used in the ACGNRs calculations. The amplitude of the energy gap oscillations is considerably reduced with respect to these observed for the infinite quasi-one-dimensional counterparts. As expected, when the length of the armchair edge (N) is increased, the oscillations amplitude increases as well. Furthermore, the HOMO-LUMO gap is generally inversely proportional to the width and the length of the finite GNR dots in accordance with the semi-metallic two-dimensional graphene limit. Therefore, if one wishes to tailor the energy gap of rectangular graphene nanodots, it is suggested to consider long armchair

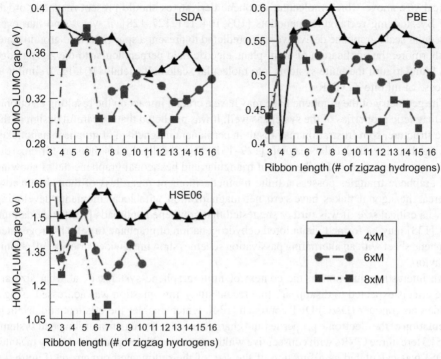

Figure 4.5 HOMO-LUMO gaps of rectangular graphene quantum dots calculated at the local spin density approximation (upper left panel), generalized gradient approximation (upper right panel) and screened hybrid exchange correlation functional approximation (lower left panel). Reprinted with permission from [127] © 2008 The American Physical Society

edges and short zigzag edges. This is expected to increase the amplitude of the energy gap oscillations while maintaining overall high gap values.

As mentioned previously, the effects of quantum confinement on the electronic properties of GNRs has been confirmed experimentally [20] where the band-gap was found to decrease with increasing nanoribbon width. Similar measurements of the band-gap of graphene dots were recently reported [19] using a scanning tunneling spectroscopy technique. Here, the graphene dots were labeled as armchair or zigzag systems according to the highest fraction of the edge type that they presented. Confinement effects were observed for all samples showing that the zigzag systems become metallic at a smaller lateral dimension than their armchair counterparts. Therefore, the experimentally observed behavior confirmed the theoretically predicted influence of the crystallographic orientation of the edges of nanoscale graphene systems on their electronic properties.

Apart from quantum confinement, edge effects, which play a central role in dictating the electronic structure of quasi-one-dimensional graphene nanoribbons, may become important in graphene quantum dots as well. With this respect, it is widely accepted that small molecular derivatives of graphene, such as different types of polyaromatic hydrocarbons, have a closed-shell nonmagnetic ground state. On the other hand, as discussed earlier, infinite ZZGNRs present a spin-polarized ground state. This suggests that there exists a critical size above which molecular graphene derivatives should present magnetic ordering. By studying rectangular nanodots [103, 104, 121, 127, 128], it was found that even molecular scale graphene derivatives are predicted to present a spin-polarized ground state. Furthermore, the application of an in-plane electric field perpendicular to the zigzag edge was found to turn the finite systems into molecular scale half-metals [104, 127], similar to the case of infinite ZZGNRs.

The geometry of the graphene flake may have a crucial impact on the resulting electronic and magnetic properties of the system as well. It was predicted that by careful design of the edge structure, spin frustration may result in permanent magnetism of graphene nanoflakes turning them into molecular magnets [129–131]. Specifically, Lieb's theorem for bipartite lattices was investigated for the case of triangular and hexagonal graphene flakes showing that graphene triangles possess a finite magnetic moment regardless of their dimensions whereas hexagonal flakes have zero net magnetization with local moments developing above a critical size. It was further suggested that graphene nanoroads [132] and quantum dots [133] may be formed via tailored dehydrogenation of graphane (a fully hydrogenated graphene sheet with an alternating passivation scheme) showing a wide variety of electronic behaviors.

An interesting question in the context of finite graphene systems is, at what size are edge effects expected to disappear? In a recent study, this question was addressed using a divide-and-conquer (D&C) DFT approach [134], which enables the efficient and accurate calculation of the electronic properties and charge transport through finite elongated systems [135]. Here, three GNRs with consecutive widths were considered and their density of states (DOS) was calculated as a function of the size of their elongated dimension. Figure 4.6 presents an example of the DOS of one such GNR at different ribbon lengths. It can be seen that for short rectangular graphene nanodots (up to 12 nm in length) the DOS is characterized by a discrete set of energy levels. When the ribbon is further elongated to 38 nm, a nearly constant DOS around the Fermi energy of the infinite system arises and typical van Hove singularities start to develop. For the system depicted in the figure, at a length of 72 nm most of the features that are related to the edge states disappeared and the DOS almost perfectly matches that of the respective infinite ACGNR.

4.2.4 Surface Chemical Adsorption

Another chemical alternative of modifying the electronic properties of graphene is via the adsorption of adatoms on its surface. In general, such adsorption may induce changes in the Fermi energy and dipole moment of the system accompanied with a work function shift [136], similar to the effects observed when placing graphene on metallic substrates [137, 138]. For example, a semiconducting to metal transition has been predicted for ACGNRs via the adsorption of manganese and boron atoms [139]. Here, the resulting system is expected to become ferromagnetic bearing a large magnetic moment. The dependence of the adsorption energy, net magnetic moment, and the electronic structure of the graphene–metal systems on the adsorption site has been also investigated for Ni atoms adsorbed on the surface of graphene nanoribbons [140]. For this system, it was also found that near zigzag edges the Ni atom forms more stable complexes than those formed at the middle of the ribbon.

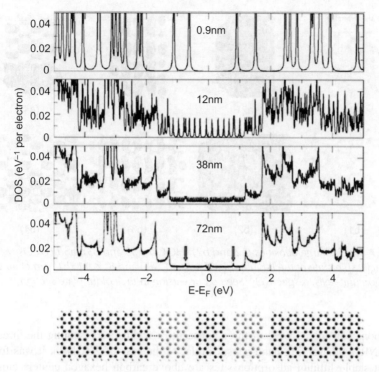

Figure 4.6 DOS of the finite GNR at different lengths. The gray curve in the lowermost DOS diagram corresponds to the DOS of the infinite system. Lower panel: a schematic representation of a finite elongated GNR. Reprinted with permission from [135] © 2007 The American Physical Society

The adsorption of alkali metal atoms on the surface of graphene has been further studied theoretically [141], showing that the favorable adsorption site is above the center of a graphene hexagon as opposed to directly above the carbon atoms. The adsorption energy was found to be affected by the atomic ionization energy, the radius of the metal ions, and the amount of charge transfer between the adsorbates and the graphene surface. For quasi-one-dimensional systems, the adsorption energy was also found to vary depending on the position of the adsorbates with respect to the nanoribbon edges where the strongest binding was obtained for edge adsorption sites. Among all alkali-metals, lithium has been identified as the most strongly binding atom [142]. When considering the interaction of lithium adatoms and graphene nanoribbons, the binding with ZZGNRs was found to be much stronger than that with ACGNRs occurring preferentially along the zigzag edges [143]. This enhancement was rationalized in terms of the larger number of electron acceptor states in ZZGNRs as compared to ACGNRs.

Despite this preference towards ZZGNRs, a recent study suggested that due to the controllable band-gap of ACGNRs lithium doping presents a possible route for bandgap engineering of graphitic systems [144]. Furthermore, it was suggested that the adsorbed lithium atoms serve as anchoring sites for the secondary adsorption of contaminant molecules. The

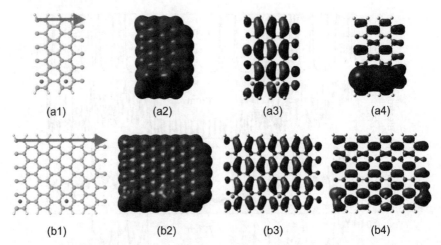

Figure 4.7 HSE charge densities (a2 and b2), HOCOs (a3 and b3), and LUCOs (a4 and b4) at the singlet spin state of the lithium doped 4 × 11 (a1) and 8 × 11 (b1) armchair graphene nanoribbon unit-cells, respectively. See colour version in the colour plate section

nanoribbons considered were of three consecutive widths, representing the three subsets of ACGNRs with varying band-gaps. Similar to the case of ZZGNRs, it was found that the most stable lithium adsorption sites are above carbon hexagon centers close to the edges of the ribbon. This was attributed to the reactive nature of the honeycomb lattice edges. When examining the changes in the electronic properties of the GNRs upon atomic lithium adsorption, the adsorption of the metal atom significantly decreased the band-gaps of all cell sizes turning them metallic for sufficiently large adsorbate densities. Figure 4.7 demonstrates the differences between high and low lithium density. At high densities, there is a considerable overlap between the charge densities of the lithium adatoms (panel a2). Furthermore, the lowest unoccupied crystalline orbital (LUCO; panel a4) indicates the formation of a lithium chain residing on-top of the graphene surface which may serve as a conducting channel turning the system metallic. As the adatom density decreases (panels b1–b4) the formation of such an atomic chain is prevented due to the increased distance between the lithium adatoms and the decreased charge density around each such atom.

One of the possible technological applications of molecular adsorption on GNRs can be found in the field of gas sensing and detection. Most molecular species interact rather weakly with the surface of pristine graphene mainly via physisorption dominated by dispersive interactions [145]. However, after exposure to some adsorbates, the intra-molecular charge transfer and re-hybridization of the molecular orbitals of graphene due to the interaction with adsorbates may lead to a significant change in the electronic properties of the graphene substrate. With this respect, the conductance of graphene was found to be sensitive to adsorption of several gas molecules such as NO_2, H_2O, NH_3, and CO [39]. Dopants within the graphene sheet can be used as docking sites for adsorbed molecules thus increasing the surface reactivity. Graphene doped with B, N, Al, or S atoms was found to present enhanced reactivity towards the binding of gas molecules that modify the conductivity of system [146]. Moreover, it has been reported that several organic molecules [147–152], and even biomolecules [153, 154] in which adatoms serve as anchoring sites, may induce significant change in the electronic structure of graphene.

4.3 Mechanical and Electromechanical Properties of GNRs

The mechanical properties of both monolayer and multi-layer graphene were studied extensively over the past few years. Experimentally, the mechanical properties of graphene were measured by using, for example, nanoindentation and mechanical vibration resonating techniques. In these experiments the reported Young's modulus of graphene ranged between 250 GPa and 1 TPa [7, 155–158]. Furthermore, making measurements by nanoindentation using an atomic force microscope (AFM), Lee et al. have shown that graphene has a breaking strength 200 times greater than steel [7].

The high mechanical stiffness of graphene, along with its high thermal conductivity, and high saturation current density makes graphene a promising candidate for serving as active components in future NEMS. Such devices are based on systems for which the structural and mechanical properties can be controlled via the application of external fields and vice versa. The first experimental realization of an electromechanical device based on graphene nanoribbons was presented in 2007 [159]. In this setup, single-atom thick layers of graphene have been suspended above predesigned trenches while bridging the gap between two gold electrodes. Optical and electrical actuation schemes were shown to induce GNR vibrational frequencies in the MHz regime. High charge sensitivities of the resonant frequencies suggested the possibility of using such devices as ultrasensitive mass and force detectors [159].

Following this pioneering experimental work achieving the fabrication and manipulation of GNRs as NEMS components, several other studies have explored the mechanical [155, 156, 160–163] and electromechanical properties of these systems [164–166]. Since the gapless linear dispersion of the band structure in the vicinity of the Dirac point results from the unique symmetry of the hexagonal carbon lattice, any local deformation of the lattice or imbalance of electrons of different spin flavors may lift the degeneracy of the two sub-lattices and induce an electronic band-gap. To this end, external strains may be used to break the hexagonal symmetry of the graphene lattice [167–169]. Naturally, a symmetric strain distribution preserves the hexagonal lattice symmetry and thus the band structure remains gapless [169]. Nevertheless, an asymmetrical strain breaks the lattice's translation symmetry while opening a gap around the Fermi energy [168, 169]. Such a strain, depending on its strength and direction, was found to shift the band crossing away from the K point and can thus be used to tune the electronic properties of the system [168].

The effects of applying a uniaxial strain on isolated GNRs were recently studied in detail using the generalized gradient approximation within the framework of DFT [170, 171]. It was found that the electronic properties of ZZGNRs are insensitive to uniaxial strain. Nevertheless, a recent study showed that relaxed ZZGNRs can exhibit unexpected symmetry dependent semiconductor-like behavior under a finite bias voltage [7]. In particular, asymmetric ZZGNRs were found to behave as conventional conductors with linear current-voltage characteristics while symmetric ZZGNRs carry relatively small currents at a given range of bias voltages. It was later shown that for symmetric ZZGNRs, shear strain breaks the mirror symmetry and thus greatly increases the conducting current. Conversely, uniaxial strain conserves the lattice symmetry thus the current remains small [172].

In contrast to ZZGNRs, the energy gap of ACGNRs displays an oscillatory pattern as a function of the applied strain [173]. In comparison to tight-binding calculations, it was deduced that the nearest-neighbor hopping terms between adjacent atomic sites within the carbon hexagonal lattice are mainly responsible for the observed electromechanical response. The ability to tune the band-gap of an ACGNR led to the idea that external strains may be used to tailor the properties of GNR-based tunneling field effect transistors (TFETs) [174]. It was found that strained wide ACGNRs present superior device performance as compared to their narrow counterparts. It was recently further suggested that the band-gap dependence of ACGNRs on an externally applied strain can be quite sensitive to the chemical decoration of the edges of the system [175]. Interestingly, the influence of uniaxial strain on the electronic properties of graphene has further gained recent experimental verification [176, 177].

As mentioned previously, many exotic quantum properties emerge when an electric field is applied to ZZGNRs. Focusing on the effect of an external electric field on the structural geometry of such nanoribbons, recent studies showed that a perpendicular electric field can induce bending of the skeleton of ZZGNRs [178, 179]. The extent of this bending at any particular field intensity was found to strongly depend on the spatial dimensions of the ribbon. The mechanism underlying this behavior was identified as field induced mixing of s and p orbitals through the second order Stark effect which modulates the electron–nuclear interaction in favor of a bent structure. Generally speaking, graphene nanoribbons are expected to be more mechanically sensitive to an external electric field than carbon nanotubes [180].

Another important structural deformation that graphene nanoribbons may undergo is spontaneous or induced torsion [181, 182]. Experimentally, scrolled edges of freestanding graphene sheets have been recently reported [183]. Spontaneous twisting was found to dominate sufficiently narrow bare GNRs ($w < 1.6$ nm) [181]. For wider ribbons or sheets, the twist abruptly vanishes and the instability was found to reduce into ripples localized near the edges [181, 184]. Chemistry at the edges of saturated graphene nanoribbons can induce out-of-plane deformations resulting in three dimensional helical structures. For example, DFT calculations, based on the local density approximation, predicted that fluorine terminated armchair ribbons are intrinsically twisted in contrast with the planar structure of their hydrogen terminated counterparts [185]. Twisting the ribbons of either termination was found to result in coupling of the conduction and valence bands leading to band-gap modulations. It was further shown that the band-gap of ACGNRs is inversely proportional to the strain energy and vanishes at certain twist angles [186].

The effects of induced bending and torsional deformations on the electronic properties of finite suspended graphene nanoribbons have also been addressed in a recent computational study [187]. Here, a large set of short armchair graphene nanoribbons, doubly clamped at their zigzag edges, was considered. The effects of a driven deformation due to an external tip were simulated by constraining a strip of hexagons at the central part of the ribbon so that it was either depressed or rotated with respect to the basal plane of the flat ribbon (see Figure 4.8).) To understand the mechanical response of GNRs towards externally applied stresses, the change in total energy (obtained via state-of-the-art DFT calculations) as a function of the depression depth and the torsional angle within the linear

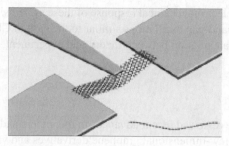

Figure 4.8 *Illustration of a GNR-based electromechanical device where a suspended GNR is subjected to a bending deformation induced by an external tip. Adapted with permission from [187] © 2009 American Chemical Society*

response regime was analyzed. In both types of deformations excellent parabolic fits were obtained for all systems considered indicating that they behave like classical springs at this regime. Furthermore, in the case of bending deformations, excellent correlation between the calculated values and predictions made by classical elasticity theory was found. The Young's modulus of the finite nanoscale GNR dots was extracted and the impressive large value of ∼7 TPa was obtained, exceeding the measured value for micrometer scale suspended graphene sheets [7] and the highest value calculated for CNTs with similar thickness parameters [188]. In order to evaluate their electromechanical response, large mechanical deformations were applied to the full set of finite GNRs studied. Plotting the band-gap as a function of the bending deformation (Figure 4.9 or the torsional angle

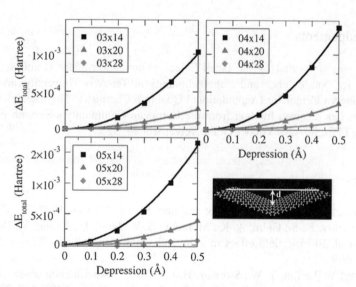

Figure 4.9 *Total energy changes of several suspended graphene nanoribbons due to an externally applied bending stress. Lines are parabolic fits indicating that all systems are within the linear response regime. Reproduced with permission from Nano Lett. **9**, 2619–2622 (2009) Copyright (2009) American Chemical Society*

demonstrated a strong electromechanical response of these systems. Moreover, evidence of dimension-dependent band-gap oscillations similar to the case of CNTs [189, 190], further demonstrated the intimate relation between theory and experiment.

4.4 Summary

In this chapter, we have reviewed recent advances in the understanding of the physical and chemical properties of low-dimensional graphene derivatives and various routes to control them. The latter span structural parameters, dimensional considerations, chemical modifications, and external perturbations such as electric fields and mechanical deformations. Apart from the basic scientific interest that they trigger, the unique physical properties of graphene nanoribbons and graphene nanodots along with the diverse possibilities to control them, opens the door for numerous potential applications. These include nanoscale transistors, spin filters, chemical, and biological detectors, and flexible electrodes for touch screens and solar cell applications.

We hope that whilst reading, the reader is already convinced that theoretical and computational methodologies have proven to be powerful predictive tools in this field. We re-emphasize, however, that this chapter by no means constitutes a thorough account of the vast amount of information accumulated in this field thus far. Instead, it should serve as a starting point for the intrigued reader. Finally, we note that the successful prediction of graphene's properties along with recent experimental advances in the fabrication of graphene-based devices mark the great potential in further exploring and exploiting many of this material's properties and indicate how much more there is to be learnt in this field.

Acknowledgements

This work was supported by the Israel Science Foundation under Grant No. 1313/08, the Center for Nanoscience and Nanotechnology at Tel-Aviv University, and the Lise Meitner-Minerva Center for Computational Quantum Chemistry. The research leading to these results has received funding from the European Community's Seventh Framework Programme FP7/2007-2013 under Grant Agreement 249225.

References

[1] Wallace, P. R. The band theory of graphite. *Physical Review* **1947**, *71*, 622–634.
[2] Novoselov, K. S.; Geim, A. K.; Morozov, S. V.; Jiang, D.; Zhang, Y.; Dubonos, S. V.; et al. Electric field effect in atomically thin carbon films. *Science* **2004**, *306*, 666–669.
[3] Zhang, Y. B.; Tan, Y. W.; Stormer, H. L.; Kim, P. Experimental observation of the quantum Hall effect and Berry's phase in graphene. *Nature* **2005**, *438*, 201–204.
[4] Novoselov, K. S.; Jiang, Z.; Zhang, Y.; Morozov, S. V.; Stormer, H. L.; Zeitler, U.; et al. Room-temperature quantum hall effect in graphene. *Science* **2007**, *315*, 1379–1379.

[5] Novoselov, K. S.; Geim, A. K.; Morozov, S. V.; Jiang, D.; Katsnelson, M. I.; Grigorieva, I. V.; et al. Two-dimensional gas of massless Dirac fermions in graphene. *Nature* **2005**, *438*, 197–200.

[6] Balandin, A. A.; Ghosh, S.; Bao, W. Z.; Calizo, I.; Teweldebrhan, D.; Miao, F.; Lau, C. N. Superior thermal conductivity of single-layer graphene. *Nano Letters* **2008**, *8*, 902–907.

[7] Lee, C.; Wei, X. D.; Kysar, J. W.; Hone, J. Measurement of the elastic properties and intrinsic strength of monolayer graphene. *Science* **2008**, *321*, 385–388.

[8] Sluiter, M. H. F.; Kawazoe, Y. Cluster expansion method for adsorption: Application to hydrogen chemisorption on graphene. *Physical Review B* **2003**, *68*, 085410.

[9] Sofo, J. O.; Chaudhari, A. S.; Barber, G. D. Graphane: A two-dimensional hydrocarbon. *Physical Review B* **2007**, *75*, 153401.

[10] Fujita, M.; Igami, M.; Nakada, K. Lattice distortion in nanographite ribbons. *Journal of the Physical Society of Japan* **1997**, *66*, 1864–1867.

[11] Han, M. Y.; Ozyilmaz, B.; Zhang, Y. B.; Kim, P. Energy band-gap engineering of graphene nanoribbons. *Physical Review Letters* **2007**, *98*, 206805.

[12] Chen, Z. H.; Lin, Y. M.; Rooks, M. J.; Avouris, P. Graphene nano-ribbon electronics. *Physica E-Low-Dimensional Systems & Nanostructures* **2007**, *40*, 228–232.

[13] McChesney, J. L.; Bostwick, A.; Ohta, T.; Seyller, T.; Horn, K.; Gonzalez, J.; Rotenberg, E. Extended van Hove singularity and superconducting instability in doped graphene. *Physical Review Letters* **2010**, *104*, 136803.

[14] Li, X. S.; Cai, W. W.; An, J. H.; Kim, S.; Nah, J.; Yang, D. X.; et al. Large-Area synthesis of high-quality and uniform graphene films on copper foils. *Science* **2009**, *324*, 1312–1314.

[15] Yan, Z.; Lin, J.; Peng, Z. W.; Sun, Z. Z.; Zhu, Y.; Li, L.; Xiang, C. S.; et al. Toward the synthesis of wafer-scale single-crystal graphene on copper foils. *ACS Nano* **2012**, *6*, 9110–9117.

[16] Wu, Y. M. A.; Fan, Y.; Speller, S.; Creeth, G. L.; Sadowski, J. T.; He, K.; et al. Large single crystals of graphene on melted copper using chemical vapor deposition. *ACS Nano* **2012**, *6*, 5010–5017.

[17] Terrones, M. Controlling the shapes and assemblages of graphene. *Proceedings of the National Academy of Sciences of the United States of America* **2012**, *109*, 7951–7952.

[18] Bae, S.; Kim, H.; Lee, Y.; Xu, X. F.; Park, J. S.; Zheng, Y.; et al. Roll-to-roll production of 30-inch graphene films for transparent electrodes. *Nature Nanotechnology* **2010**, *5*, 574–578.

[19] Ritter, K. A.; Lyding, J. W. The influence of edge structure on the electronic properties of graphene quantum dots and nanoribbons. *Nature Materials* **2009**, *8*, 235–242.

[20] Berger, C.; Song, Z. M.; Li, X. B.; Wu, X. S.; Brown, N.; Naud, C.; et al. Electronic confinement and coherence in patterned epitaxial graphene. *Science* **2006**, *312*, 1191–1196.

[21] Yang, X. Y.; Dou, X.; Rouhanipour, A.; Zhi, L. J.; Rader, H. J.; Mullen, K. Two-dimensional graphene nanoribbons. *Journal of the American Chemical Society* **2008**, *130*, 4216–4217.

[22] Cai, J. M.; Ruffieux, P.; Jaafar, R.; Bieri, M.; Braun, T.; Blankenburg, S.; et al. Atomically precise bottom-up fabrication of graphene nanoribbons. *Nature* **2010**, *466*, 470–473.

[23] Li, X. L.; Wang, X. R.; Zhang, L.; Lee, S. W.; Dai, H. J. Chemically derived, ultrasmooth graphene nanoribbon semiconductors. *Science* **2008**, *319*, 1229–1232.

[24] Kosynkin, D. V.; Higginbotham, A. L.; Sinitskii, A.; Lomeda, J. R.; Dimiev, A.; Price, B. K.; Tour, J. M. Longitudinal unzipping of carbon nanotubes to form graphene nanoribbons. *Nature* **2009**, *458*, 872–876.

[25] Jiao, L. Y.; Zhang, L.; Wang, X. R.; Diankov, G.; Dai, H. J. Narrow graphene nanoribbons from carbon nanotubes. *Nature* **2009**, *458*, 877–880.

[26] Wang, C.; Morton, K. J.; Fu, Z. L.; Li, W. D.; Chou, S. Y. Printing of sub-20 nm wide graphene ribbon arrays using nanoimprinted graphite stamps and electrostatic force assisted bonding. *Nanotechnology* **2011**, *22*, 445301.

[27] Murali, R.; Yang, Y. X.; Brenner, K.; Beck, T.; Meindl, J. D. Breakdown current density of graphene nanoribbons. *Applied Physics Letters* **2009**, *94*, 243114.

[28] Hicks, J.; Tejeda, A.; Taleb-Ibrahimi, A.; Nevius, M. S.; Wang, F.; Shepperd, K.; et al. A wide-bandgap metal-semiconductor-metal nanostructure made entirely from graphene. *Nature Physics* **2013**, *9*, 49–54.

[29] Bae, S.; Kim, S. J.; Shin, D.; Ahn, J. H.; Hong, B. H. Towards industrial applications of graphene electrodes. *Physica Scripta* **2012**, *T146*, 014024.

[30] Ghosh, S.; Sarker, B. K.; Chunder, A.; Zhai, L.; Khondaker, S. I. Position dependent photodetector from large area reduced graphene oxide thin films. *Applied Physics Letters* **2010**, *96*, 163109.

[31] Yin, Z. Y.; Sun, S. Y.; Salim, T.; Wu, S. X.; Huang, X. A.; He, Q. Y.; et al. Organic photovoltaic devices using highly flexible reduced graphene oxide films as transparent electrodes. *ACS Nano* **2010**, *4*, 5263–5268.

[32] Li, S. S.; Tu, K. H.; Lin, C. C.; Chen, C. W.; Chhowalla, M. Solution-Processable graphene oxide as an efficient hole transport layer in polymer solar cells. *ACS Nano* **2010**, *4*, 3169–3174.

[33] Liu, Q.; Liu, Z. F.; Zhong, X. Y.; Yang, L. Y.; Zhang, N.; Pan, G. L.; et al. Polymer photovoltaic cells based on solution-processable graphene and P3HT. *Advanced Functional Materials* **2009**, *19*, 894–904.

[34] Wang, H. L.; Cui, L. F.; Yang, Y. A.; Casalongue, H. S.; Robinson, J. T.; Liang, Y. Y.; et al. Mn3O4-graphene hybrid as a high-capacity anode material for lithium ion batteries. *Journal of the American Chemical Society* **2010**, *132*, 13978–13980.

[35] Paek, S. M.; Yoo, E.; Honma, I. Enhanced cyclic performance and lithium storage capacity of SnO$_2$/graphene nanoporous electrodes with three-dimensionally delaminated flexible structure. *Nano Letters* **2009**, *9*, 72–75.

[36] Wang, D. H.; Choi, D. W.; Li, J.; Yang, Z. G.; Nie, Z. M.; Kou, R.; et al. Self-assembled TiO$_2$-graphene hybrid nanostructures for enhanced li-ion insertion. *ACS Nano* **2009**, *3*, 907–914.

[37] Wang, Y.; Shi, Z. Q.; Huang, Y.; Ma, Y. F.; Wang, C. Y.; Chen, M. M.; Chen, Y. S. Supercapacitor devices based on graphene materials. *Journal of Physical Chemistry C* **2009**, *113*, 13103–13107.

[38] Chmiola, J.; Largeot, C.; Taberna, P. L.; Simon, P.; Gogotsi, Y. Monolithic carbide-derived carbon films for micro-supercapacitors. *Science* **2010**, *328*, 480–483.

[39] Schedin, F.; Geim, A. K.; Morozov, S. V.; Hill, E. W.; Blake, P.; Katsnelson, M. I.; Novoselov, K. S. Detection of individual gas molecules adsorbed on graphene. *Nature Materials* **2007**, *6*, 652–655.

[40] Kuila, T.; Bose, S.; Khanra, P.; Mishra, A. K.; Kim, N. H.; Lee, J. H. Recent advances in graphene-based biosensors. *Biosensors & Bioelectronics* **2011**, *26*, 4637–4648.

[41] Ohno, Y.; Maehashi, K.; Yamashiro, Y.; Matsumoto, K. Electrolyte-gated graphene field-effect transistors for detecting pH protein adsorption. *Nano Letters* **2009**, *9*, 3318–3322.

[42] Topsakal, M.; Sevincli, H.; Ciraci, S. Spin confinement in the superlattices of graphene ribbons. *Applied Physics Letters* **2008**, *92*, 173118.

[43] Li, Z. Y.; Qian, H. Y.; Wu, J.; Gu, B. L.; Duan, W. H. Role of symmetry in the transport properties of graphene nanoribbons under bias. *Physical Review Letters* **2008**, *100*, 206802.

[44] Candini, A.; Klyatskaya, S.; Ruben, M.; Wernsdorfer, W.; Affronte, M. Graphene spintronic devices with molecular nanomagnets. *Nano Letters* **2011**, *11*, 2634–2639.

[45] Soriano, D.; Munoz-Rojas, F.; Fernandez-Rossier, J.; Palacios, J. J. Hydrogenated graphene nanoribbons for spintronics. *Physical Review B* **2010**, *81*, 165409.

[46] Kim, R. H.; Bae, M. H.; Kim, D. G.; Cheng, H. Y.; Kim, B. H.; Kim, D. H.; et al. Stretchable, transparent graphene interconnects for arrays of microscale inorganic light emitting diodes on rubber substrates. *Nano Letters* **2011**, *11*, 3881–3886.

[47] Nasiri, S. H.; Moravvej-Farshi, M. K.; Faez, R. Stability analysis in graphene nanoribbon interconnects. *IEEE Electron Device Letters* **2010**, *31*, 1458–1460.

[48] Barton, R. A.; Ilic, B.; van der Zande, A. M.; Whitney, W. S.; McEuen, P. L.; Parpia, J. M.; Craighead, H. G. High, size-dependent quality factor in an array of graphene mechanical resonators. *Nano Letters* **2011**, *11*, 1232–1236.

[49] Barton, R. A.; Parpia, J.; Craighead, H. G. Fabrication and performance of graphene nanoelectromechanical systems. *Journal of Vacuum Science & Technology B* **2011**, *29*, 050801.

[50] van der Zande, A. M.; Barton, R. A.; Alden, J. S.; Ruiz-Vargas, C. S.; Whitney, W. S.; Pham, P. H. Q.; et al. Large-scale arrays of single-layer graphene resonators. *Nano Letters* **2010**, *10*, 4869–4873.

[51] Eichler, A.; Moser, J.; Chaste, J.; Zdrojek, M.; Wilson-Rae, I.; Bachtold, A. Nonlinear damping in mechanical resonators made from carbon nanotubes and graphene. *Nature Nanotechnology* **2011**, *6*, 339–342.

[52] Stein, S. E.; Brown, R. L. Pi-Electron properties of large condensed polyaromatic hydrocarbons. *Journal of the American Chemical Society* **1987**, *109*, 3721–3729.

[53] Tanaka, K.; Yamashita, S.; Yamabe, H.; Yamabe, T. Electronic-properties of one-dimensional graphite family. *Synthetic Metals* **1987**, *17*, 143–148.

[54] Fujita, M.; Wakabayashi, K.; Nakada, K.; Kusakabe, K. Peculiar localized state at zigzag graphite edge. *Journal of the Physical Society of Japan* **1996**, *65*, 1920–1923.

[55] Nakada, K.; Fujita, M.; Dresselhaus, G.; Dresselhaus, M. S. Edge state in graphene ribbons: Nanometer size effect and edge shape dependence. *Physical Review B* **1996**, *54*, 17954–17961.

[56] Pisani, L.; Chan, J. A.; Montanari, B.; Harrison, N. M. Electronic structure and magnetic properties of graphitic ribbons. *Physical Review B* **2007**, *75*, 064418.

[57] Lee, H.; Son, Y. W.; Park, N.; Han, S. W.; Yu, J. J. Magnetic ordering at the edges of graphitic fragments: Magnetic tail interactions between the edge-localized states. *Physical Review B* **2005**, *72*, 174431.

[58] Son, Y. W.; Cohen, M. L.; Louie, S. G. Energy gaps in graphene nanoribbons. *Physical Review Letters* **2006**, *97*, 216803.

[59] Son, Y. W.; Cohen, M. L.; Louie, S. G. Half-metallic graphene nanoribbons. *Nature* **2006**, *444*, 347–349.

[60] Hod, O.; Barone, V.; Peralta, J. E.; Scuseria, G. E. Enhanced half-metallicity in edge-oxidized zigzag graphene nanoribbons. *Nano Letters* **2007**, *7*, 2295-2299.

[61] Kan, E. J.; Li, Z. Y.; Yang, J. L.; Hou, J. G. Half-metallicity in edge-modified zigzag graphene nanoribbons. *Journal of the American Chemical Society* **2008**, *130*, 4224–4225.

[62] Radovic, L. R.; Bockrath, B. On the chemical nature of graphene edges: Origin of stability and potential for magnetism in carbon materials. *Journal of the American Chemical Society* **2005**, *127*, 5917–5927.

[63] Ezawa, M. Peculiar width dependence of the electronic properties of carbon nanoribbons. *Physical Review B* **2006**, *73*, 045432.

[64] Barone, V.; Hod, O.; Scuseria, G. E. Electronic structure and stability of semiconducting graphene nanoribbons. *Nano Letters* **2006**, *6*, 2748–2754.

[65] Kohn, W.; Sham, L. J. Self-consistent equations including exchange and correlation effects. *Physical Review* **1965**, *140*, A1133–A1138.

[66] Hohenberg, P.; Kohn, W. Inhomogeneous electron gas. *Physical Review* **1964**, *136*, B864–B871.

[67] Nair, R. R.; Blake, P.; Grigorenko, A. N.; Novoselov, K. S.; Booth, T. J.; Stauber, T.; et al. Fine structure constant defines visual transparency of graphene. *Science* **2008**, *320*, 1308–1308.

[68] Wang, F.; Zhang, Y. B.; Tian, C. S.; Girit, C.; Zettl, A.; Crommie, M.; Shen, Y. R. Gate-variable optical transitions in graphene. *Science* **2008**, *320*, 206–209.

[69] Kim, P.; Shi, L.; Majumdar, A.; McEuen, P. L. Thermal transport measurements of individual multiwalled nanotubes. *Physical Review Letters* **2001**, *87*, 215502.

[70] Ghosh, S.; Calizo, I.; Teweldebrhan, D.; Pokatilov, E. P.; Nika, D. L.; Balandin, A. A.; et al. Extremely high thermal conductivity of graphene: Prospects for thermal management applications in nanoelectronic circuits. *Applied Physics Letters* **2008**, *92*, 151911.

[71] Seol, J. H.; Jo, I.; Moore, A. L.; Lindsay, L.; Aitken, Z. H.; Pettes, M. T.; et al. Two-dimensional phonon transport in supported graphene. *Science* **2010**, *328*, 213–216.

[72] Geim, A. K.; Novoselov, K. S. The rise of graphene. *Nature Materials* **2007**, *6*, 183–191.

[73] Chen, F.; Tao, N. J. Electron transport in single molecules: from benzene to graphene. *Accounts of Chemical Research* **2009**, *42*, 429–438.

[74] Yazyev, O. V. Emergence of magnetism in graphene materials and nanostructures. *Reports on Progress in Physics* **2010**, *73*, 056501.

[75] Neto, A. H. C.; Novoselov, K. New directions in science and technology: two-dimensional crystals. *Reports on Progress in Physics* **2011**, *74*, 082501.

[76] Barone, V.; Hod, O.; Peralta, J. E.; Scuseria, G. E. Accurate prediction of the electronic properties of low-dimensional graphene derivatives using a screened hybrid density functional. *Accounts of Chemical Research* **2011**, *44*, 269–279.
[77] Terrones, H.; Lv, R. T.; Terrones, M.; Dresselhaus, M. S. The role of defects and doping in 2D graphene sheets and 1D nanoribbons. *Reports on Progress in Physics* **2012**, *75*, 062501.
[78] Andrei, E. Y.; Li, G. H.; Du, X. Electronic properties of graphene: a perspective from scanning tunneling microscopy and magnetotransport. *Reports on Progress in Physics* **2012**, *75*, 056501.
[79] Xin, S.; Guo, Y. G.; Wan, L. J. Nanocarbon networks for advanced rechargeable lithium batteries. *Accounts of Chemical Research* **2012**, *45*, 1759–1769.
[80] Rodrigues, J. N. B.; Gonçalves, P. A. D.; Rodrigues, N. F. G.; Ribeiro, R. M.; Lopes dos Santos, J. M. B.; Peres, N. M. R. Zigzag graphene nanoribbon edge reconstruction with Stone-Wales defects. *Physical Review B* **2011**, *84*, 155435.
[81] Huang, B.; Liu, M.; Su, N.; Wu, J.; Duan, W.; Gu, B.-l.; Liu, F. Quantum manifestations of graphene edge stress and edge instability: a first-principles study. *Physical Review Letters* **2009**, *102*, 166404.
[82] Koskinen, P.; Malola, S.; Häkkinen, H. Self-passivating edge reconstructions of graphene. *Physical Review Letters* **2008**, *101*, 115502.
[83] Bhowmick, S.; Waghmare, U. V. Anisotropy of the Stone-Wales defect and warping of graphene nanoribbons: A first-principles analysis. *Physical Review B* **2010**, *81*, 155416.
[84] Lee, G.-D.; Wang, C. Z.; Yoon, E.; Hwang, N.-M.; Ho, K. M. Reconstruction and evaporation at graphene nanoribbon edges. *Physical Review B* **2010**, *81*, 195419.
[85] Wassmann, T.; Seitsonen, A. P.; Saitta, A. M.; Lazzeri, M.; Mauri, F. Structure, stability, edge states, and aromaticity of graphene ribbons. *Physical Review Letters* **2008**, *101*, 096402.
[86] Koskinen, P.; Malola, S.; Häkkinen, H. Evidence for graphene edges beyond zigzag and armchair. *Physical Review B* **2009**, *80*, 073401.
[87] Girit, Ç. Ö.; Meyer, J. C.; Erni, R.; Rossell, M. D.; Kisielowski, C.; Yang, L.; et al. Graphene at the edge: stability and dynamics. *Science* **2009**, *323*, 1705–1708.
[88] Suenaga, K.; Koshino, M. Atom-by-atom spectroscopy at graphene edge. *Nature* **2010**, *468*, 1088–1090.
[89] Malola, S.; Häkkinen, H.; Koskinen, P. Comparison of Raman spectra and vibrational density of states between graphene nanoribbons with different edges. *The European Physical Journal D* **2009**, *52*, 71–74.
[90] Rakyta, P.; Kormányos, A.; Cserti, J.; Koskinen, P. Exploring the graphene edges with coherent electron focusing. *Physical Review B* **2010**, *81*, 115411.
[91] Kobayashi, Y.; Fukui, K.; Enoki, T.; Kusakabe, K. Edge state on hydrogen-terminated graphite edges investigated by scanning tunneling microscopy. *Physical Review B* **2006**, *73*, 125415.
[92] Okada, S. Energetics of nanoscale graphene ribbons: Edge geometries and electronic structures. *Physical Review B* **2008**, *77*, 041408R.
[93] Tian, J. F.; Cao, H. L.; Wu, W.; Yu, Q. K.; Chen, Y. P. Direct Imaging of Graphene Edges: Atomic Structure and Electronic Scattering. *Nano Letters* **2011**, *11*, 3663–3668.

[94] Kobayashi, Y.; Fukui, K.; Enoki, T.; Kusakabe, K.; Kaburagi, Y. Observation of zigzag and armchair edges of graphite using scanning tunneling microscopy and spectroscopy. *Physical Review B* **2005**, *71*, 193406.

[95] Niimi, Y.; Matsui, T.; Kambara, H.; Tagami, K.; Tsukada, M.; Fukuyama, H. Scanning tunneling microscopy and spectroscopy of the electronic local density of states of graphite surfaces near monoatomic step edges. *Physical Review B* **2006**, *73*, 085421.

[96] Yang, L.; Park, C. H.; Son, Y. W.; Cohen, M. L.; Louie, S. G. Quasiparticle energies and band gaps in graphene nanoribbons. *Physical Review Letters* **2007**, *99*, 186801.

[97] Brey, L.; Fertig, H. A. Electronic states of graphene nanoribbons studied with the Dirac equation. *Physical Review B* **2006**, *73*, 235411.

[98] Tapaszto, L.; Dobrik, G.; Lambin, P.; Biro, L. P. Tailoring the atomic structure of graphene nanoribbons by scanning tunnelling microscope lithography. *Nature Nanotechnology* **2008**, *3*, 397–401.

[99] Stampfer, C.; Guttinger, J.; Hellmueller, S.; Molitor, F.; Ensslin, K.; Ihn, T. Energy gaps in etched graphene nanoribbons. *Physical Review Letters* **2009**, *102*, 056403.

[100] Balaban, A. T.; Klein, D. J. Claromatic carbon nanostructures. *Journal of Physical Chemistry C* **2009**, *113*, 19123–19133.

[101] Baldoni, M.; Sgamellotti, A.; Mercuri, F. Finite-length models of carbon nanotubes based on clar sextet theory. *Organic Letters* **2007**, *9*, 4267–4270.

[102] Enoki, T. Role of edges in the electronic and magnetic structures of nanographene. *Physica Scripta* **2012**, *T146*, 014008.

[103] Rudberg, E.; Salek, P.; Luo, Y. Nonlocal exchange interaction removes half-metallicity in graphene nanoribbons. *Nano Letters* **2007**, *7*, 2211–2213.

[104] Kan, E. J.; Li, Z. Y.; Yang, J. L.; Hou, J. G. Will zigzag graphene nanoribbon turn to half metal under electric field? *Applied Physics Letters* **2007**, *91*, 243116.

[105] Dutta, S.; Pati, S. K. Half-metallicity in undoped and boron doped graphene nanoribbons in the presence of semilocal exchange-correlation interactions. *Journal of Physical Chemistry B* **2008**, *112*, 1333–1335.

[106] Xu, H. Y.; Heinzel, T.; Zozoulenko, I. V. Edge disorder and localization regimes in bilayer graphene nanoribbons. *Physical Review B* **2009**, *80*, 045308.

[107] Acik, M.; Chabal, Y. J. Nature of graphene edges: a review. *Japanese Journal of Applied Physics* **2011**, *50*, 070101.

[108] Gunlycke, D.; Li, J.; Mintmire, J. W.; White, C. T. Altering low-bias transport in zigzag-edge graphene nanostrips with edge chemistry. *Applied Physics Letters* **2007**, *91*, 112108.

[109] Cervantes-Sodi, F.; Csanyi, G.; Piscanec, S.; Ferrari, A. C. Edge-functionalized and substitutionally doped graphene nanoribbons: Electronic and spin properties. *Physical Review B* **2008**, *77*, 165427.

[110] Cervantes-Sodi, F.; Csanyi, G.; Piscanec, S.; Ferrari, A. C. Electronic properties of chemically modified graphene ribbons. *Physica Status Solidi B-Basic Solid State Physics* **2008**, *245*, 2068–2071.

[111] Zheng, H. X.; Duley, W. First-principles study of edge chemical modifications in graphene nanodots. *Physical Review B* **2008**, *78*, 045421.

[112] Konatham, D.; Striolo, A. Molecular design of stable graphene nanosheets dispersions. *Nano Letters* **2008**, *8*, 4630–4641.

[113] Kusakabe, K.; Maruyama, M. Magnetic nanographite. *Physical Review B* **2003**, *67*, 092406.
[114] Kudin, K. N. Zigzag graphene nanoribbons with saturated edges. *ACS Nano* **2008**, *2*, 516–522.
[115] Wu, M. H.; Wu, X. J.; Gao, Y.; Zeng, X. C. Materials design of half-metallic graphene and graphene nanoribbons. *Applied Physics Letters* **2009**, *94*, 223111.
[116] Nakamura, J.; Nitta, T.; Natori, A. Electronic and magnetic properties of BNC ribbons. *Physical Review B* **2005**, *72*, 205429.
[117] Li, Y.; Zhou, Z.; Shen, P.; Chen, Z. Spin gapless semiconductor–metal–half-metal properties in nitrogen-doped zigzag graphene nanoribbons. *ACS Nano* **2009**, *3*, 1952–1958.
[118] Martins, T. B.; da Silva, A. J. R.; Miwa, R. H.; Fazzio, A. sigma- and pi-defects at graphene nanoribbon edges: Building spin filters. *Nano Letters* **2008**, *8*, 2293–2298.
[119] Wakabayashi, K.; Fujita, M.; Ajiki, H.; Sigrist, M. Electronic and magnetic properties of nanographite ribbons. *Physical Review B* **1999**, *59*, 8271–8282.
[120] Miyamoto, Y.; Nakada, K.; Fujita, M. First-principles study of edge states of H-terminated graphitic ribbons. *Physical Review B* **1999**, *59*, 9858–9861.
[121] Shemella, P.; Zhang, Y.; Mailman, M.; Ajayan, P. M.; Nayak, S. K. Energy gaps in zero-dimensional graphene nanoribbons. *Applied Physics Letters* **2007**, *91*, 042101.
[122] Wang, X. R.; Ouyang, Y. J.; Li, X. L.; Wang, H. L.; Guo, J.; Dai, H. J. Room-temperature all-semiconducting sub-10-nm graphene nanoribbon field-effect transistors. *Physical Review Letters* **2008**, *100*, 206803.
[123] Tiwari, R. P.; Stroud, D. Tunable band gap in graphene with a noncentrosymmetric superlattice potential. *Physical Review B* **2009**, *79*, 205435.
[124] Yao, W.; Yang, S. A.; Niu, Q. Edge States in Graphene: From gapped flat-band to gapless chiral modes. *Physical Review Letters* **2009**, *102*, 096801.
[125] Giovannetti, G.; Khomyakov, P. A.; Brocks, G.; Kelly, P. J.; van den Brink, J. Substrate-induced band gap in graphene on hexagonal boron nitride: Ab initio density functional calculations. *Physical Review B* **2007**, *76*, 073103.
[126] Pedersen, T. G.; Flindt, C.; Pedersen, J.; Mortensen, N. A.; Jauho, A. P.; Pedersen, K. Graphene antidot lattices: Designed defects and spin qubits. *Physical Review Letters* **2008**, *100*, 136804.
[127] Hod, O.; Barone, V.; Scuseria, G. E. Half-metallic graphene nanodots: A comprehensive first-principles theoretical study. *Physical Review B* **2008**, *77*, 035411.
[128] Jiang, D. E.; Sumpter, B. G.; Dai, S. First principles study of magnetism in nanographenes. *Journal of Chemical Physics* **2007**, *127*, 124703.
[129] Fernandez-Rossier, J.; Palacios, J. J. Magnetism in graphene nanoislands. *Physical Review Letters* **2007**, *99*, 177204.
[130] Ezawa, M. Metallic graphene nanodisks: Electronic and magnetic properties. *Physical Review B* **2007**, *76*, 245415.
[131] Ezawa, M. Graphene nanoribbon and graphene nanodisk. *Physica E-Low-Dimensional Systems & Nanostructures* **2008**, *40*, 1421–1423.
[132] Singh, A. K.; Yakobson, B. I. Electronics and Magnetism of Patterned Graphene Nanoroads. *Nano Letters* **2009**, *9*, 1540–1543.
[133] Singh, A. K.; Penev, E. S.; Yakobson, B. I. Vacancy Clusters in Graphane as Quantum Dots. *ACS Nano* **2010**, *4*, 3510–3514.

[134] Hod, O.; Peralta, J. E.; Scuseria, G. E. First-principles electronic transport calculations in finite elongated systems: A divide and conquer approach. *Journal of Chemical Physics* **2006**, *125*, 114704.

[135] Hod, O.; Peralta, J. E.; Scuseria, G. E. Edge effects in finite elongated graphene nanoribbons. *Physical Review B* **2007**, *76*, 233401.

[136] Chan, K. T.; Neaton, J. B.; Cohen, M. L. First-principles study of metal adatom adsorption on graphene. *Physical Review B* **2008**, *77*, 235430.

[137] Giovannetti, G.; Khomyakov, P. A.; Brocks, G.; Karpan, V. M.; van den Brink, J.; Kelly, P. J. Doping graphene with metal contacts. *Physical Review Letters* **2008**, *101*, 026803.

[138] Khomyakov, P. A.; Giovannetti, G.; Rusu, P. C.; Brocks, G.; van den Brink, J.; Kelly, P. J. First-principles study of the interaction and charge transfer between graphene and metals. *Physical Review B* **2009**, *79*, 195425.

[139] Gorjizadeh, N.; Farajian, A. A.; Esfarjani, K.; Kawazoe, Y. Spin and band-gap engineering in doped graphene nanoribbons. *Physical Review B* **2008**, *78*, 155427.

[140] Rigo, V. A.; Martins, T. B.; da Silva, A. J. R.; Fazzio, A.; Miwa, R. H. Electronic, structural, and transport properties of Ni-doped graphene nanoribbons. *Physical Review B* **2009**, *79*, 075435.

[141] Choi, S. M.; Jhi, S. H. Self-assembled metal atom chains on graphene nanoribbons. *Physical Review Letters* **2008**, *101*, 266105.

[142] Rytkonen, K.; Akola, J.; Manninen, M. Sodium atoms and clusters on graphite by density functional theory. *Physical Review B* **2004**, *69*, 205404.

[143] Uthaisar, C.; Barone, V.; Peralta, J. E. Lithium adsorption on zigzag graphene nanoribbons. *Journal of Applied Physics* **2009**, *106*, 113715.

[144] Krepel, D.; Hod, O. Lithium adsorption on armchair graphene nanoribbons. *Surface Science* **2011**, *605*, 1633–1642.

[145] Ortmann, F.; Schmidt, W. G.; Bechstedt, F. Attracted by long-range electron correlation: Adenine on graphite. *Physical Review Letters* **2005**, *95*, 186101.

[146] Dai, J. Y.; Yuan, J. M.; Giannozzi, P. Gas adsorption on graphene doped with B, N, Al, and S: A theoretical study. *Applied Physics Letters* **2009**, *95*, 232105.

[147] Chen, W.; Chen, S.; Qi, D. C.; Gao, X. Y.; Wee, A. T. S. Surface transfer p-type doping of epitaxial graphene. *Journal of the American Chemical Society* **2007**, *129*, 10418–10422.

[148] Lu, Y. H.; Chen, W.; Feng, Y. P.; He, P. M. Tuning the electronic structure of graphene by an organic molecule. *Journal of Physical Chemistry B* **2009**, *113*, 2–5.

[149] Manna, A. K.; Pati, S. K. Tuning the electronic structure of graphene by molecular charge transfer: a computational study. *Chemistry-An Asian Journal* **2009**, *4*, 855–860.

[150] Kang, H. S. Theoretical study of binding of metal-doped graphene sheet and carbon nanotubes with dioxin. *Journal of the American Chemical Society* **2005**, *127*, 9839–9843.

[151] Zhang, Y. H.; Zhou, K. G.; Xie, K. F.; Zeng, J.; Zhang, H. L.; Peng, Y. Tuning the electronic structure and transport properties of graphene by noncovalent functionalization: effects of organic donor, acceptor and metal atoms. *Nanotechnology* **2010**, *21*, 065201.

[152] Zhang, Y.-H.; Zhou, K.-G.; Xie, K.-F.; Zhang, H.-L.; Peng, Y.; Wang, C.-W. Tuning the magnetic and transport properties of metal adsorbed graphene by co-adsorption with 1,2-dichlorobenzene. *Physical Chemistry Chemical Physics* **2012**, *14*, 11626–11632.

[153] Gowtham, S.; Scheicher, R. H.; Ahuja, R.; Pandey, R.; Karna, S. P. Physisorption of nucleobases on graphene: Density-functional calculations. *Physical Review B* **2007**, *76*, 033401.

[154] Zhang, Z. X.; Jia, H. S.; Ma, F.; Han, P. D.; Liu, X. G.; Xu, B. S. First principle study of cysteine molecule on intrinsic and Au-doped graphene surface as a chemosensor device. *Journal of Molecular Modeling* **2011**, *17*, 649–655.

[155] Poot, M.; van der Zant, H. S. J. Nanomechanical properties of few-layer graphene membranes. *Applied Physics Letters* **2008**, *92*, 063111.

[156] Frank, I. W.; Tanenbaum, D. M.; Van der Zande, A. M.; McEuen, P. L. Mechanical properties of suspended graphene sheets. *Journal of Vacuum Science & Technology B* **2007**, *25*, 2558–2561.

[157] Gomez-Navarro, C.; Burghard, M.; Kern, K. Elastic properties of chemically derived single graphene sheets. *Nano Letters* **2008**, *8*, 2045–2049.

[158] Garcia-Sanchez, D.; van der Zande, A. M.; Paulo, A. S.; Lassagne, B.; McEuen, P. L.; Bachtold, A. Imaging mechanical vibrations in suspended graphene sheets. *Nano Letters* **2008**, *8*, 1399–1403.

[159] Bunch, J. S.; van der Zande, A. M.; Verbridge, S. S.; Frank, I. W.; Tanenbaum, D. M.; Parpia, J. M.; et al. Electromechanical resonators from graphene sheets. *Science* **2007**, *315*, 490–493.

[160] Wei, D. C.; Liu, Y. Q.; Zhang, H. L.; Huang, L. P.; Wu, B.; Chen, J. Y.; Yu, G. Scalable synthesis of few-layer graphene ribbons with controlled morphologies by a template method and their applications in nanoelectromechanical switches. *Journal of the American Chemical Society* **2009**, *131*, 11147–11154.

[161] Poot, M.; van der Zant, H. S. J. Mechanical systems in the quantum regime. *Physics Reports-Review Section of Physics Letters* **2012**, *511*, 273–335.

[162] Scharfenberg, S.; Rocklin, D. Z.; Chialvo, C.; Weaver, R. L.; Goldbart, P. M.; Mason, N. Probing the mechanical properties of graphene using a corrugated elastic substrate. *Applied Physics Letters* **2011**, *98*, 091908.

[163] Liu, X.; Metcalf, T. H.; Robinson, J. T.; Houston, B. H.; Scarpa, F. Shear modulus of monolayer graphene prepared by chemical vapor deposition. *Nano Letters* **2012**, *12*, 1013–1017.

[164] Milaninia, K. M.; Baldo, M. A.; Reina, A.; Kong, J. All graphene electromechanical switch fabricated by chemical vapor deposition. *Applied Physics Letters* **2009**, *95*, 183105.

[165] Palacios, T.; Hsu, A.; Wang, H. Applications of graphene devices in rf communications. *IEEE Communications Magazine* **2010**, *48*, 122–128.

[166] Shi, Z. W.; Lu, H. L.; Zhang, L. C.; Yang, R.; Wang, Y.; Liu, D. H.; et al. Studies of graphene-based nanoelectromechanical switches. *Nano Research* **2012**, *5*, 82–87.

[167] Sevincli, H.; Topsakal, M.; Ciraci, S. Superlattice structures of graphene-based armchair nanoribbons. *Physical Review B* **2008**, *78*, 245402.

[168] Mohr, M.; Papagelis, K.; Maultzsch, J.; Thomsen, C. Two-dimensional electronic and vibrational band structure of uniaxially strained graphene from ab initio calculations. *Physical Review B* **2009**, *80*, 205410.

[169] Gui, G.; Li, J.; Zhong, J. X. Band structure engineering of graphene by strain: First-principles calculations. *Physical Review B* **2008**, *78*, 075435.

[170] Faccio, R.; Denis, P. A.; Pardo, H.; Goyenola, C.; Mombru, A. W. Mechanical properties of graphene nanoribbons. *Journal of Physics-Condensed Matter* **2009**, *21*, 285304.

[171] Sun, L.; Li, Q. X.; Ren, H.; Su, H. B.; Shi, Q. W.; Yang, J. L. Strain effect on electronic structures of graphene nanoribbons: A first-principles study. *Journal of Chemical Physics* **2008**, *129*, 074704.

[172] Wang, J. Y.; Liu, Z. F.; Liu, Z. R. First-principles study of the transport behavior of zigzag graphene nanoribbons tailored by strain. *AIP Advances* **2012**, *2*, 012103.

[173] Lu, Y.; Guo, J. Band Gap of Strained Graphene Nanoribbons. *Nano Research* **2010**, *3*, 189–199.

[174] Kang, J. H.; He, Y.; Zhang, J. Y.; Yu, X. X.; Guan, X. M.; Yu, Z. P. Modeling and simulation of uniaxial strain effects in armchair graphene nanoribbon tunneling field effect transistors. *Applied Physics Letters* **2010**, *96*, 252105.

[175] Peng, X. H.; Velasquez, S. Strain modulated band gap of edge passivated armchair graphene nanoribbons. *Applied Physics Letters* **2011**, *98*, 023112.

[176] Ni, Z. H.; Yu, T.; Lu, Y. H.; Wang, Y. Y.; Feng, Y. P.; Shen, Z. X. Uniaxial strain on graphene: raman spectroscopy study and band-gap opening. *ACS Nano* **2008**, *2*, 2301–2305.

[177] Huang, M. Y.; Yan, H. G.; Chen, C. Y.; Song, D. H.; Heinz, T. F.; Hone, J. Phonon softening and crystallographic orientation of strained graphene studied by Raman spectroscopy. *Proceedings of the National Academy of Sciences of the United States of America* **2009**, *106*, 7304–7308.

[178] Wang, Z. Alignment of graphene nanoribbons by an electric field. *Carbon* **2009**, *47*, 3050–3053.

[179] Chattopadhyaya, M.; Alam, M. M.; Chakrabarti, S. On the microscopic origin of bending of graphene nanoribbons in the presence of a perpendicular electric field. *Physical Chemistry Chemical Physics* **2012**, *14*, 9439–9443.

[180] Wang, Z.; Devel, M. Electrostatic deflections of cantilevered metallic carbon nanotubes via charge-dipole model. *Physical Review B* **2007**, *76*, 195434.

[181] Bets, K. V.; Yakobson, B. I. Spontaneous twist and intrinsic instabilities of pristine graphene nanoribbons. *Nano Research* **2009**, *2*, 161–166.

[182] Zhang, D. B.; Dumitrica, T. Effective-tensional-strain-driven bandgap modulations in helical graphene nanoribbons. *Small* **2011**, *7*, 1023–1027.

[183] Gass, M. H.; Bangert, U.; Bleloch, A. L.; Wang, P.; Nair, R. R.; Geim, A. K. Free-standing graphene at atomic resolution. *Nature Nanotechnology* **2008**, *3*, 676–681.

[184] Shenoy, V. B.; Reddy, C. D.; Ramasubramaniam, A.; Zhang, Y. W. Edge-stress-induced warping of graphene sheets and nanoribbons. *Physical Review Letters* **2008**, *101*, 245501.

[185] Gunlycke, D.; Li, J. W.; Mintmire, J. W.; White, C. T. Edges bring new dimension to graphene nanoribbons. *Nano Letters* **2010**, *10*, 3638–3642.

[186] Sadrzadeh, A.; Hua, M.; Yakobson, B. I. Electronic properties of twisted armchair graphene nanoribbons. *Applied Physics Letters* **2011**, *99*, 013102.

[187] Hod, O.; Scuseria, G. E. Electromechanical properties of suspended graphene nanoribbons. *Nano Letters* **2009**, *9*, 2619–2622.

[188] Huang, Y.; Wu, J.; Hwang, K. C. Thickness of graphene and single-wall carbon nanotubes. *Physical Review B* **2006**, *74*, 245413.

[189] Cohen-Karni, T.; Segev, L.; Srur-Lavi, O.; Cohen, S. R.; Joselevich, E. Torsional electromechanical quantum oscillations in carbon nanotubes. *Nature Nanotechnology* **2006**, *1*, 36–41.

[190] Nagapriya, K. S.; Berber, S.; Cohen-Karni, T.; Segev, L.; Srur-Lavi, O.; Tomanek, D.; Joselevich, E. Origin of torsion-induced conductance oscillations in carbon nanotubes. *Physical Review B* **2008**, *78*, 165417.

5

Cutting Graphitic Materials: A Promising Way to Prepare Graphene Nanoribbons

Wenhua Zhang and Zhenyu Li
Hefei National Laboratory for Physical Sciences at the Microscale,
University of Science and Technology of China, China

5.1 Introduction

As already shown in Chapter 4, graphene nanoribbons (GNRs) can be expected to play an important role in electronics and spintronics applications of graphene [1–5]. As a result, a method for mass production of GNRs is very desirable. Since electronic structures of GNRs depend strongly on their width and edge structure, a satisfactory synthesis path of GNRs should provide the flexibility of controlling sample shape and size. Generally, there are two kinds of method to get graphene nanostructures; that is, bottom-up approaches and top-down approaches. The former has the advantage of precisely controlling of the sample quality. For example, starting from small molecules, GNRs with fixed chemical formula can be synthesized by specially designed chemical reactions [6]. The cost of precise control is that the obtained structure is entirely determined by precursor molecules. New reactions should be designed for GNRs with different shapes. On the other hand, when high quality mother materials are available, top-down approaches are expected to provide sample parameter control with a greater flexibility.

Cutting graphitic materials is a typical top-down approach to preparing graphene nanostructures. Besides physical "scissors" like electronic unwrapping [7] and mechanical cutting [8], one commonly used cutting agent is oxygen. As observed in experiment, when

---

Graphene Chemistry: Theoretical Perspectives, First Edition. Edited by De-en Jiang and Zhongfang Chen.
© 2013 John Wiley & Sons, Ltd. Published 2013 by John Wiley & Sons, Ltd.

graphitic layers are oxidized in strong acid and oxidant, they are cut into smaller pieces [9, 10]. Catalyzed by transition metal nanoparticles, hydrogen can also be used to cut graphitic materials [11]. Besides these two mainstream means, lithium intercalation [12] and catalytic cutting under microwave radiation [13] are also reported in the literature for material cutting purposes. Different starting materials can be used to get graphene nanostructures. Among them, carbon nanotubes are expected to be a good choice to be cut into GNRs with controllable widths [14].

With a huge amount of research effort on graphitic material cutting, a precise control of all properties of the product is still very challenging [15]. To reach this goal, it is essential to understand the underlying cutting mechanisms. Recently, theoretical works based on density functional theory (DFT) calculations have made important progress in this direction. In this chapter, we will briefly review these progresses. We first focus on the oxidative cutting of two dimensional graphene sheets in Section 5.2. Then, in Section 5.3, mechanisms for carbon nanotube cutting are discussed. Besides oxidative cutting, we also introduce other methods, namely metal nanoparticle catalyzed cutting in Section 5.4. Finally, a perspective summary is given in the last section.

5.2 Oxidative Cutting of Graphene Sheets

Intuitively, the most straightforward way of top-down synthesizing GNRs is cutting two-dimensional (2D) graphene sheets into stripes of different widths and lengths. An atomic scissor is needed for this purpose. The clue to such a scissor has been identified in graphite oxidation, an important aspect of graphite chemistry. During oxidation of graphite, a significant size decrease has been observed [16, 17], and cracks were found in dark field optical microscope images of graphene oxide (GO) [10]. Reducing the starting flake size leads to faster oxidation reactions, while the final GO flake size does not depend on the initial size [16]. Oxidation also leads to wrinkle structures, which can then be cut with atomic force microscopic (AFM) tip [18]. To develop a controllable way of cutting graphene into GNRs via oxidation, it is essential to understand the atomic mechanism of oxidative cutting.

5.2.1 Cutting Mechanisms

One possible cutting mode is "tear-from-the-edge", that is, the oxidation process starts from the edge of a graphene sheet and the crack goes inward from the edge. The possibility of this process is evaluated from a thermodynamic point of view by Li et al. [19]. At a hydroxyl saturated armchair edge, a carbonyl pair can form after cutting a furthest outside carbon bond, such as bond between carbons A6 and A7 in Figure 5.1(a). We name this carbonyl pair as cp(A6,A7), and its optimized structure is shown in Figure 5.1(b). Cutting an edge carbon bond in column B and forming a corresponding carbonyl pair is energetically less favorable. For example, cp(B7,B8) is about 1.08 eV higher in energy than cp(A6,A7). To tear the graphene sheet starting from cp(A6,A7), we need to cut the next inside carbon bond and form a carbonyl pair cp(C6,C7). However, this process is about 0.63 eV energetically less favorable compared to cutting edge carbon bonds. Therefore, the crack is expected to go inward only after most edge carbon bonds are cut. NMR measurements suggest that

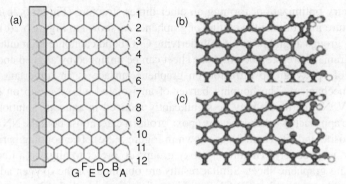

Figure 5.1 (a) Geometric model to describe a graphene edge. Atoms in the shadow area were fixed during geometry optimization; (b) A carbonyl pair at the edge, and (c) another carbonyl pair formed one step inward. Reprinted with permission from [19] © 2009 American Chemical Society

only part of the edge sites are oxidized to carboxyl [17], which indicates it is hard to tear the graphene sheet from the edge under the experimental oxidative environment.

If graphene cutting is not starting from the edge, it must be initiated at a specific point of the interior part of graphene plane, which can be either a defect or a special pattern of oxidative groups on graphene. Li et al. [10] suggested that a line alignment of epoxy groups can act as a cutting initiator. With cluster modes ($C_{24}H_{12}$ and $C_{54}H_{18}$), they investigated the possibility of formation of an epoxy chain on graphene. It was found that when an isolated atomic oxygen sitting on a C–C bond, its bond length was stretched from 1.42 Å to 1.58 Å (Figure 5.2a). When two epoxy groups locate at a same carbon hexagon, they prefer to occupy opposite ends of the hexagonal ring instead of occupying nearest-neighboring (NN) sites. Interestingly, the C–C bonds underlying the opposite-end epoxy dimer are broken

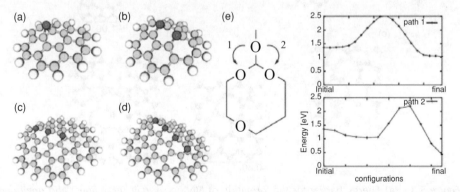

Figure 5.2 (a), (b), (d), (d): Attachment of one, two, three, and four epoxy groups on a graphene cluster; (e) Energy barrier of atomic oxygen hopping on an oxidized graphene cluster for two different paths. Reprinted with permission from [10] © 2006 The American Physical Society

upon geometry optimization, forming an ether dimer structure (Figure 5.2b). Therefore, such a structure may initiate the cutting of graphene. As shown in Figure 5.2(c) and 5.2(d), when epoxy groups align linearly, the underlying C–C bond cannot remain intact under the concerted strain. Therefore, the graphene sheet can be unzipped by aligned epoxy groups.

Kinetics of epoxy group diffusion on graphene surface is an important part of the unzipping mechanism. The hopping barrier of an isolated epoxy group on graphene is about 0.9 eV. Notice that it may be significantly reduced in aqueous solution. When an epoxy group approaches another nearby epoxy group, it can choose both the NN site (path 1) and the opposite-end site (path 2), as shown in Figure 5.2(e). The hopping barriers are 1.1 and 0.83 eV, respectively. Therefore, epoxy groups do prefer to align in a line and unzip the underlying graphene sheet. Similar results are obtained for the oxygen adsorption on single wall carbon nanotube (SWCNT), which is in contrast with the fluorinated nanotube [20, 21] where C–C bonds stay intact. This is consistent with experimental observation of the oxidation induced break of CNTs [9]. Also, notice that the formation of unzipped ether chains has a strong implication for the magnetic properties of graphene [22], which may lead to novel spintronic applications.

Sun *et al.* recently revisited this problem using a periodic model [23]. Surprisingly, they found that the energy of unzipped ether is 0.20 eV less stable than an NN epoxy dimer (Figure 5.3a), which is in contrast to the result based on cluster models. The diffusion epoxy barrier to form an NN epoxy dimer is 0.73 eV, which is comparable to that of an

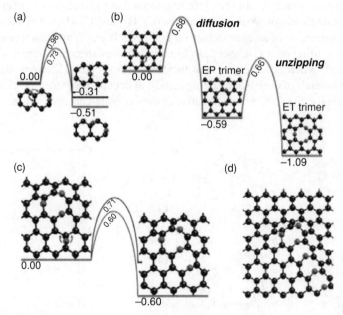

Figure 5.3 (a) Energy barrier for the formation of NN epoxy pair (gray lines) and unzipped ether (dark lines); (b) Energy profile of the diffusion barrier of forming NN epoxy trimer and the unzipping barrier of formation of NN ether trimer; (c) Energy barrier for the diffusion of the fourth oxygen atom; (d) Schematic structure of an ether chain. Reprinted with permission from [23] © 2012 American Chemical Society

isolated epoxy group and 0.23 eV lower than that of unzipped ether. Therefore, on the basis of periodic models, the NN epoxy dimer is preferred to unzipped ether. This difference compared to cluster-model results comes from the lattice constraint of periodic structure. Sun *et al.* then considered the third atomic oxygen on a graphene sheet with NN epoxy dimer that preexisted. It turns out that an NN epoxy trimer can be formed by conquering an energy barrier of 0.68 eV (Figure 5.3b). With strain created by three neighboring epoxy groups, the underlying C–C bond finally breaks with an energy gain of 0.50 eV. The NN ether trimer formed is about 0.6 eV lower than a linear ether chain. Interestingly, as shown in Figure 5.3(c), the next atomic oxygen prefers to form a linear ether chain rather than an NN epoxy structure from both energetic and kinetic points of view. An epoxy chain on the graphene sheet can then form starting from the trimer center (Figure 5.3d). Therefore, the NN ether trimer acts as a nucleation center for oxidative cutting of graphene. After the nucleation, ether chains are easy to form along three directions with angles of 120°.

A reader may ask why forming the ether chain is called unzipping, which does not really cause a breakup of the sheet. Actually, the formation of an epoxy chain on a graphene sheet only weakens its fracture stress by about 16% [24]. Therefore, linear alignment of epoxy groups is not the complete mechanism of graphene cutting. Based on DFT calculations, Li *et al.* [19] proposed an unzipping mechanism with an intermediate species named the *epoxy pair* (EP), where two atomic oxygens bind with two neighboring carbon atoms from opposite sides of the graphene. It is suggested that the formation of an EP on an ether chain (Figure 5.4a) is more energetically favorable by 2.71 eV for the first EP and 0.78 eV for the second EP than the formation of an isolated epoxy group. For a short ether chain, the energy gain is comparable for extending the chain and forming EPs on the chain.

EP can then be converted to a more stable configuration called the *carbonyl pair* (CP) with an energy gain of 0.48 eV. The energy barrier for the transition from an isolated EP on an ether chain to a CP is about 0.76 eV. Importantly, this barrier is lowered to 0.45 eV with the presence of a neighboring EP and the sequence dissociation of the neighboring EP only needs to conquer a small barrier of 0.26 eV. Thus, once the EP to CP transition is triggered, the graphene sheet can be easily cut along the epoxy chain.

The EP based cutting mechanism requires that both sides of the graphene sheet are available for oxidation. However, in some cases, for example, on graphite or a multiwall carbon nanotube wall, it is not easy to form such a configuration. Sun *et al.* [23] suggested an alternative reaction path for those cases. When an epoxy group approaches an epoxy chain by diffusion, an ether-epoxy pair can be formed (Figure 5.4b), with an energy barrier of 0.51 eV. This ether-epoxy pair can then be converted to a CP with a very small barrier (0.08 eV). When CP is formed at all sites along the epoxy chain, the graphene sheet is cut.

5.2.2 Controllable Cutting

Although studied with the motivation of preparing GNRs, the oxidative unzipping of graphene sheet does not provide us the controllability of cutting graphene into GNRs. Due to the symmetry of the honeycomb structure of graphene, it is hard to control the direction of epoxy chains and thus we expected graphene to be cut into pieces of small quantum dots [16–19] rather than nanoribbons. A clever protocol has been proposed by Ma *et al.* [25] to realize GNR synthesis based on oxidative cutting of graphene. By simply applying

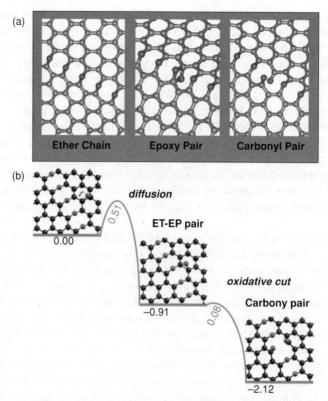

Figure 5.4 (a) Geometric structure of an ether chain, an isolated epoxy pair (EP), and an carbonyl pair (CP); (b) Energy barrier for the formation of ether-epoxy (ET-EP) pair and carbonyl pair along the ether chain. Reprinted with permission from [19,23] © 2009 and 2012 American Chemical Society

external strain (0 ~ 6%) along a specific direction, orientation-selective cutting of graphene is predicted due to the symmetry breaking.

With external strain along the armchair direction, a C–C bond in this direction is elongated and thus more active. Therefore, an epoxy chain is expected to align with a 90° angle relative to the strain direction (Figure 5.5a). Taking the lattice symmetry into account, another possible direction of an epoxy chain has 30° angle to the strain direction (Figure 5.5b). When external strain is along the zigzag direction, the epoxy chain can align with a 60° (Figure 5.5c) or 0° (Figure 5.5d) angle to the strain direction. As expected, the order of binding energy of an epoxy chain follows 90°>60°>30°>0°, and the difference increases with the increasing of the strain (Figure 5.5e). The binding energy difference is about 0.1–0.2 eV during the variation of strain from 2–5%, which is large enough to make an epoxy chain to prefer one direction. Thus, parallel epoxy chains can be formed on graphene under a reasonable external strain.

Based on the EP intermediated cutting mechanism [19], the unzipping goes via two steps: ep + ep → ep + cp → cp + cp. As shown in Figure 5.5(f), for 90° parallel epoxy chains

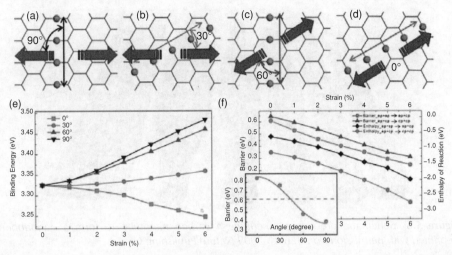

Figure 5.5 (a), (b) External strain (dark arrows) along the armchair direction with 90° and 30° aligned epoxy chains; (c), (d) External strain along zigzag direction with 60° and 0° aligned epoxy chains; (e) Binding energy change corresponding to the applied strain from 0~6%; (f) Reaction barriers and enthalpies changes of the first and second graphene cutting steps corresponding to the applied strain. Inset: Reaction barrier of the first cutting step for $\theta = 0, 30, 60,$ and $90°$ at a strain of $x = 3\%$. Reprinted with permission from [25] © 2012 Wiley-VCH Verlag GmbH & Co. KGaA, Weinheim

both reaction barrier and enthalpy decreases with the increase of external strain. The main reason is that unzipping can release the strain energy. Notice that even with low reaction barriers, the cutting reaction can still be controlled by means of the reactant concentration or reaction temperature. For a same magnitude of strain, the alignment of the epoxy chain has a big effect on the reaction barrier. The change of the reaction barrier of the first-step cutting along a different angle can be well fitted by a cosine function, as shown in the inset of Figure 5.5(f). Therefore, it is expected the graphene can be cut along a special direction depending on the direction of external strain, which can be applied by putting graphene on a flexible polymer film and then bending or stretching the film.

5.3 Unzipping Carbon Nanotubes

For the purpose of getting GNRs from cutting graphitic materials, starting from carbon nanotubes (CNTs) is geometrically a more natural choice compared to 2D graphene sheets. If a CNT can be unzipped along a specific direction, typically the axis direction, a GNR is obtained (Figure 5.6). The advantage of such an approach is that if the radius of the CNT can be controlled, for example, by the radius of the catalytic metal nanoparticle used in CNT growth, the width of the obtained GNR is also controllable. Tour *et al.* have realized this idea by chemical attacking of CNTs by oxidation using concentrated sulfuric acid with $KMnO_4$ [26, 27]. Understanding of the atomic mechanisms of CNT cutting is important to prepare high quality GNRs, for example, by precisely controlling the cutting direction.

Figure 5.6 Schematic procedure of cutting a carbon nanotube to form a nanoribbon. Reprinted with permission from [26] © 2009 Nature Publishing Group

5.3.1 Unzipping Mechanisms Based on Atomic Oxygen

The mechanisms obtained for 2D graphene sheets can be easily extended to CNTs. Using short open-ended nanotube models, Li *et al.* [28] have studied the unzipping of CNTs into GNRs upon oxidation. For an armchair CNT, the first oxygen atom prefers to adsorb at site A as shown in Figure 5.7(a), since the corresponding carbon atoms have a relatively large density of states at the Fermi level. After the adsorption of the first oxygen atom, the second oxygen prefers to adsorb at the parallel site (D), which finally leads to the formation of epoxy chain along the tube (Figure 5.7b). Then, similar to the mechanism proposed in Ref. [19], the formation of EP is preferred at site A. The energy barrier for converting this

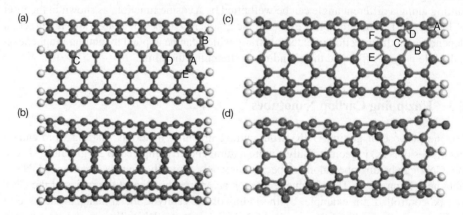

Figure 5.7 Structure of (a) a armchair carbon nanotube saturated by hydrogen atoms and (b) an epoxy chain along longitude direction on it; Structure of (c) a zigzag carbon nanotube saturated by hydrogen atoms and (d) an epoxy chain on it. Adsorption sites of atomic oxygen are marked in (a) and (c). Reprinted with permission from [28] © 2012 The Royal Society of Chemistry

EP to a CP is 0.59 eV. It is lowered to 0.12 eV for the formation of the second CP. Therefore, an armchair CNT is expected to be easily unzipped along the tube direction upon oxidation.

A zigzag CNT model is then considered. Atomic oxygen prefers to adsorb at the edge site A, as shown in Figure 5.7(c). In contrast with armchair CNTs, the second oxygen atom on the zigzag CNT will adsorb to the same site A, forming an EP and then spontaneously transferring to a CP to break the C–C bond. The energy difference between site C and site D for the adsorption of the third oxygen is only 0.12 eV, and it is even reduced to 0.06 eV for the fourth oxygen (between site E and site F). Therefore, the preference to form an epoxy chain is not very strong on a zigzag CNT wall. If oxidative cutting does happen for zigzag CNTs, it will lead to nanographene formation with irregular edges. Thus, avoiding use of zigzag CNTs is recommended for producing high-quality GNRs.

As already shown in the last section, oxygen adsorption energetics may be different for cluster models and periodic models. Therefore, it is also desirable to study CNT unzipping with an infinite long tube model. At the same time, if the tube is not open-ended, its interior wall will not available for oxidation, where a different mechanism is required. Guo et al. [29] have reported such a study with the periodic boundary condition.

Generally speaking, when an oxygen atom adsorbs on CNT, it will occupy a C–C bond possibly in three directions labeled as α, $\alpha + 60°$, $\alpha - 60°$, where α is the minimum angle a C–C align direction can make with the nanotube longitudinal axis (Figure 5.8a), which

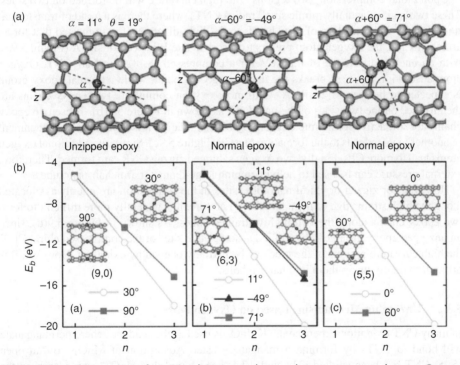

Figure 5.8 (a) Formation of unzipped epoxy and normal epoxy groups when an O atom adsorbed on different sites of a chiral (6, 3) SWNT; (b) Binding energy E_b between O and SWNT versus the O atom number n for different oxygen alignment directions. Reprinted with permission from [29] © 2010 The American Physical Society

is equal to 30° minus the nanotube chiral angle θ. Typically, an unzipped epoxy group will be formed in the α direction, while a normal epoxy group forms in the other two directions. For one oxygen atom, the adsorption energy of the α direction is slightly lower. With the alignment of sequential oxygens along the same direction, the energy difference increases significantly (Figure 5.8b). Therefore, an unzipped epoxy chain is energetically preferable along the direction close to the tube direction. When more and more epoxy groups adsorb in the same direction and form a chain, normal epoxy groups will become unzipped groups. In this case, a simple elastic theory model still favors forming epoxy chains in the α direction.

When an unzipped epoxy chain is formed on the CNT wall, the attachment of an extra atomic oxygen leads to the formation of a CP. As a consequence, the CNT will be opened along the energy–optimum α direction in an oxidative environment, and transform into a narrow zigzag GNR. Notice that when normal epoxy groups instead of unzipped epoxy groups existing on the CNT wall, the extra atomic oxygen will take away the epoxy oxygen atom to form an oxygen molecule rather than break the C–C bond.

Obviously, the energy difference between difference directions will decrease with the increase of the diameter of CNTs and eventually disappear when CNT becomes a 2D graphene sheet. Therefore, it becomes difficult to cut large CNTs into zigzag GNRs along a specific direction. Guo *et al.* suggested that, when a radial compression is applied to the CNT, its high curvature parts will be easier to be oxidized [30].

Upon radial compression, two regions with high curvature will be formed on two sides. These two regions actually mimic small radius CNTs, where there is a trend of transformation from sp^2 hybridization of C–C bonds to sp^3 hybridization. It is suggested that for an armchair CNT, the oxygen atom prefers to adsorb at site B instead of site A (Figure 5.9a), with an energy difference of 0.86 eV for a compressed (8, 8) armchair CNT. Oxygen atom adsorption at site B breaks the underlying C–C bond, while a normal epoxy group is expected to be formed at site A. With more oxygen coming in, two epoxy chains are then formed on the two sides of the nanotube, as shown in Figure 5.9(b). These two epoxy chains are still active for atomic oxygen. By MD simulation it is found that the attachment of another oxygen leads to the formation of a CP (Figure 5.9c). Similarly, additional oxygen atoms lead to more CPs. Finally, two oxygen saturated zigzag GNRs are formed. Therefore, external pressure can be used to help unzipping large-diameter armchair nanotubes.

However, for zigzag or chiral nanotube, sites in the α direction are not always located along the high strain edge of the nanotube. Therefore, it is generally more difficult to form two epoxy chains along the two highly strained edges. As shown in Figure 5.9(d), since an unzipped epoxy group can be formed at some edge sites of the compressed chiral CNT, the high curvature regions of the tube can be partially opened by oxidation. However, it is difficult to totally unzip them into nanoribbons.

5.3.2 Unzipping Mechanisms Based on Oxygen Pairs

In many CNT oxidation experiments, $KMnO_4$ is used as the oxidant, where permanganate will bond to CNTs by forming a manganese ester. Adsorption of MnO_4^- on an open (5, 5) CNT has been studied [31], and it has been found that MnO_4^- can adsorb on the open tube and the adsorption energy decreases with the increasing of the distance to the open end (Figure 5.10a). The C–C bond underlying B1–B4 (as labeled in Figure 5.10b)

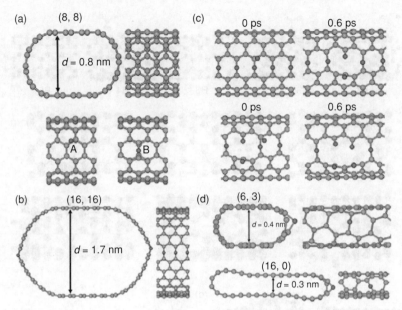

Figure 5.9 (a) Compressed armchair (8, 8) SWNT with two oxygen adsorption sites labeled as A and B; (b) Compressed (16, 16) armchair SWNT with unzipped epoxy chains formed on the high curvature regions; (c) Snapshots of MD simulations of one and another atomic oxygen interacting with the unzipped epoxy sites of the compressed (8, 8) SWNT at a temperature of 1000K; (d) Chiral (6, 3) and zigzag (16, 0) SWNT with oxygen adsorbed at the high curvature sides of the nanotubes. Reprinted with permission from [30] © 2010 American Chemical Society

site is elongated by 0.2–0.3 Å, which suggests a partial unzipping of the nanotube. After adsorbing on CNT surface, the two Mn–O bonds in MnO_4^- close to the nanotube become weakened, which are stretched by about 11% compared to the other two Mn–O bonds [32]. Therefore, in an oxidative environment, MnO_4^- is expected to be further decomposed through the cleavage of these two active Mn–O bonds, possibly accompanied by breaking of the underlying C–C bond. It is thus desirable to also study the cutting mechanism based on oxygen pairs instead of atomic oxygens. It turns out that the interaction of MnO_4^- can be approximated as a singlet O_2, which leads to the quite similar density of states for an underlying C–C near the Fermi level with that of MnO_4^- adsorption [32].

Rangel et al. [31] studied the sequential addition of oxygen pairs to the CNT wall. Without losing generality, the first oxygen pair is put on the B3 site. The first oxygen pair is expected to weaken adjacent bonds. Therefore, the next oxygen pair should attach to site B2 or B4. With more oxygen pairs, C–C bonds in the middle of the CNT attached by oxygen pairs are broken. Edge C–C bonds stretch but do not break. It is also found that if the CNT is hydrogen-saturated, the B1 bond can be opened while the external B5 end bond is not cleaved due to its large bonding strength. A zigzag nanotube is simulated by a limited long (6, 0) tube, the complete unzipping of CNT along a spiral direction is shown in Figure 5.10(d).

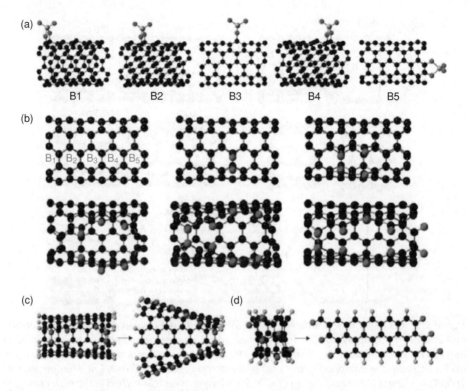

Figure 5.10 *(a) Adsorption sites of MnO_4^- anions on an open (5, 5) carbon nanotube; (b) Adsorption of oxygen pairs on the nanotube; (c) Schematic diagram of the unzipping of the nanotube; (d) Schematic diagram of the unzipping of a hydrogen passivated (6, 0) carbon nanotube. The adsorption sites are labeled in (b). See colour version in the colour plate section. Reprinted with permission from [31] © 2009 American Institute of Physics*

Zhang et al. [32] systematically studied orientation selectivity in CNT unzipping. For armchair nanotubes (Figure 5.11a), the adsorption of a singlet O_2 on the C–C bond perpendicular to the axis (P1) and on the C–C bond tilted to the axis (T1) have similar energies. However, the energy of the produced dione of the former (P2) is far more stable than that of the latter (T2). The dissociation barrier of P1 is as low as 0.12 eV, 0.40 eV lower than that of T1 (Figure 5.11c). The cluster model is used to investigate the adsorption and dissociation of O_2 at edge sites. The calculated energy barrier from the oxygen dimer (F1) to a dione (F2) is 0.59 eV, which is 0.47 eV higher than the dissociation in the middle. Thus, the unzipping process starts from the middle of the armchair nanotube. As expected, such a process becomes more and more difficult with the increasing of diameter of armchair nanotube.

A (8, 0) tube is used as an example of zigzag CNTs (Figure 5.11b). There are also two directions for O_2 adsorption: parallel to the axis (P1) and tilted to the axis (T1). The tilted O_2 adsorption is about 0.3 eV less favorable in energy than the parallel one. However, T1 dissociates to a more stable configuration T2 with an energy barrier less than that of P1 dissociation (Figure 5.11c). The dissociation barrier of P1 on a zigzag tube is 0.84 eV. At H-saturated edge site, the dissociation barrier from F1 to F2 is remarkably reduced to 0.16 eV, which indicates that for zigzag tube the unzipping starts from the edge. The

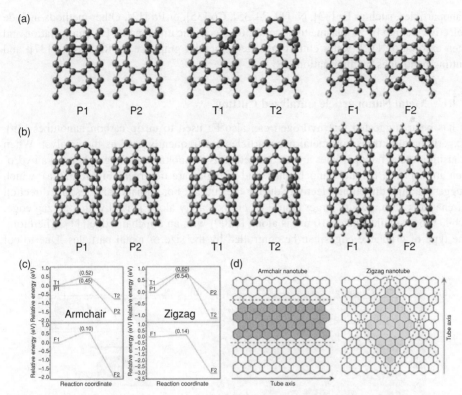

Figure 5.11 (a) Adsorption configuration of O_2 perpendicular (P1) or tilted (T1) to the axis in the middle-section or edge sites (F1) of an armchair nanotube and corresponding dissociation products (P2), (T2) and (F2); (b) Adsorption of O_2 parallel (P1) or tilted (T1) to the axis in the middle-section or edge site (F1) of a zigzag nanotube and corresponding dissociation products (P2), (T2) and (F2); (c) Dissociation paths of adsorption of O_2 with different adsorption configurations for the armchair (left) and zigzag (right) nanotubes; (d) Schematic diagram of the unzipping of armchair (left) and zigzag (right) nanotubes. Reprinted with permission from [32] © 2010 Royal Society of Chemistry

unzipping energy barrier for zigzag CNTs also does not change a lot with the increasing of tube diameter.

The difference between armchair and zigzag CNTs has a strong implication for the unzipping products. With a random attacking of MnO_4^-, an armchair nanotube will be cut into several stripes parallel to the axis. However, for zigzag tubes two helical directions are possible, which makes for those being cut into several uncontrollable shapes such as nanoflakes or short GNRs as shown in Figure 5.11(d).

5.4 Beyond Oxidative Cutting

Although oxidative cutting of graphitic materials is very promising and has been studied by many groups, there are several other means to cut graphene and CNTs. For example, graphitic materials can also be cut by hydrogen gas with the help of transition-metal nanoparticles, such as Fe [33], Ni [11, 34, 35], Co [35], or Pd [13]. Other methods include selective etching by Ar plasma treatment [36], electronic unwrapping [7], intercalation and then exfoliation [12], water cutting of heteroepitaxial graphene on Ru surface [37], and cutting nanotubes by fluorination [38].

5.4.1 Metal Nanoparticle Catalyzed Cutting

It has been suggested that hydrogen can also be used to unzip carbon nanotubes [39]. Experimentally, transition metal nanoparticles are frequently used as the catalyst. When transition metal nanoparticles are used together as a nanoknife to cut graphene in a hydrogen atmosphere, parallel nanochannels and triangles are observed (Figure 5.12a), which suggests a high direction selectivity during the cutting. For Ni nanoparticles, the direction of channels with a width larger than ~10 nm is mainly along [11$\bar{2}$0] with a zigzag edge, while those smaller than ~10 nm is along [1$\bar{1}$00] with an armchair edge [11]. Therefore, the type of graphene edges can be controlled by the size of metal particles. Due to cut

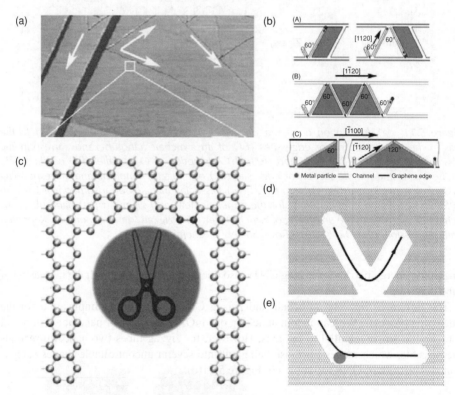

Figure 5.12 (a) STM images of nanochannels in various directions; (b) Different cutting models derived from experimental observations; (c) Schematic diagram of the formation of a zigzag edged channel simulated; (d) and (e) Simulated turning process of catalyst particle approaching a free edge and a void in order to minimize the adhesion energy. Reprinted with permission from [11] © 2008 Springer

direction selectivity, several cutting modes are observed (Figure 5.12b): parallelogram shaped, equilateral triangle, right-angled triangle, and isosceles triangle. Since nanoribbons with different widths and edges [3–5] and triangular nanographene islands [40, 41] exhibit novel electronic and magnetic properties, a controlled nanocutting of graphene has great potential for various applications.

To theoretically understand the metal nanoparticle catalyzed cutting process, an empirical kinetic Monte Carlo simulation has been performed. Different dissolution probabilities have been set to armchair and zigzag edge atoms, and the former ($P_{AM} = 1$) is much larger than the latter ($P_{ZZ} = 0.1$). With such a simple model, zigzag channels were formed in most cases after catalytic etching (Figure 5.12c). For smaller particles, considering the non-perfect contact between nanoparticle surface and graphene edge, a smaller dissolution possibility ($P_{AM} < 1$) is used. In this way, an armchair channel can be formed, agreeing with experiment. When moving inside a graphene lattice, the metal nanoparticle prefers to keep a straight channel. However, when the nanoparticle meets a free edge (or a wide channel), it will make a 60° turn (Figure 5.12d) due to the adhesive interaction between the nanoparticle and graphene edge. When the channel ahead is very narrow, or there is a small vacancy defect (Figure 5.12e), the particle will make a 120° turn.

At the first-principles level, since nanoparticles are difficult to model, simplified models are used to study transition metal aided cutting. For example, using one Cu atom to describe the catalyst, the possible cutting of a short (5, 5) CNT by molecular hydrogen was studied by Wang *et al.* [42]. Without a catalyst, although the cutting process is energy favorable (−0.73 eV), a high reaction barrier (3.11 eV) prohibits the proceeding of this reaction (Figure 5.13a). When a Cu atom is presented on CNT, the cutting barrier can be remarkably reduced. The Cu atom prefers to adsorb at the bridge site of a C–C bond at the end of the tube, which is 0.11 eV more stable than in the middle part of the tube.

With the help of this Cu atom, cutting of the CNT can proceed in the following way: the Cu atom inserts into a C–C bond to form a C–Cu–C structure with an energy barrier

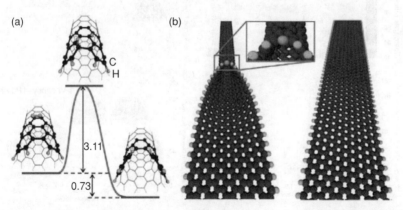

Figure 5.13 (a) Energy profile of the unzipping the first C–C bond by molecular hydrogen; (b) Schematic diagram of partial and full unzipping of the CNT by single a Cu atom and hydrogen molecules. Reprinted with permission from [42] © 2011 Wiley-VCH Verlag GmbH & Co. KGaA, Weinheim

of 1.16 eV; A H_2 molecule adsorbs on the Cu atom, with a significantly weakened H–H bond; One of the H atoms diffuses to a bridging C atom, along with the breaking of the corresponding C–Cu bond; The remain H on the Cu atom diffuses to the other bridging C atom, which releases the catalytic Cu atom; the Cu atom moves inwards to the second C–C bond. By repeating these steps, the second C–C bond can be broken similarly. Then, the whole SWNT can eventually be cut into a smooth zigzag nanoribbon (Figure 5.13b). The unzipping process of each step is exothermic and the barriers keep decreasing with the forward progression of cutting. The driving force of the unzipping process comes from the curvature energy of CNT. The further the Cu atom goes along the cutting direction during one step, the more curvature energy is released. Therefore, it is preferable to cut along the zigzag direction, for which Cu moves 0.246 nm and is longer than 0.213 nm in the armchair direction. It has been found that the energy barrier for C–C bond breaking increases with the increase of the CNT diameter, and armchair tubes are relatively more easily unzipped than chiral ones. Other transition metal atoms, such as Mn, Fe, Co, Ni, Pd, and Pt, can also be used as the catalysts for CNT unzipping.

Using the ReaxFF reactive Force Field [43], the unzipping of a (7, 7) nanotube is investigated with the effects of vacancies, oxygen, and metal catalysts considered [13]. An artificial cutting process is simulated by successively applying deformation stress. For perfect nanotube, a high energy is required to open the nanotube and all the C–C bonds break simultaneously (Figure 5.14b). If a vacancy presents on the tube wall, defect propagation is observed (Figure 5.14c) and the energy barrier is greatly reduced. When a single Pd atom occupies the vacancy site, the energy barrier is similar with that of vacancy

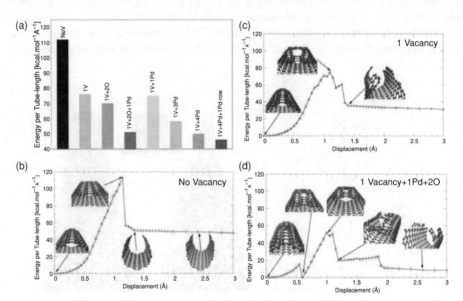

Figure 5.14 (a) Energy barrier necessary to unzip a single-walled carbon nanotube; (b) Energy diagram for a SWNT with no vacancy, (c) with one vacancy, and (d) with one vacancy decorated with two oxygen atoms and one Pd atom. Reprinted with permission from [13] © 2009 Elsevier

only (Figure 5.14a). The energy barrier is still relatively high in these two processes but it is suggested that if more Pd atoms such as 3Pd, 4Pd, 4Pd + Pd row, or mixed Pd + 2O (Figure 5.14d) present on the tube wall, the energy barrier becomes very low. Thus, the transition metal catalysts and oxygen next to vacancies are believed to be important for the catalytic unzipping of carbon nanotubes.

5.4.2 Cutting by Fluorination

Except for oxidative cutting and metal nanoparticle cutting of graphene, fluorination could be another method for preparing GNRs [38]. With a $C_{54}H_{18}$ cluster model, the attachment of F is investigated [44]. When the cluster is strongly fluorinated, severe distortion is observed. For example, if the F atoms are added along one zigzag row, the cluster bent from planarity to almost 88°. After bending, the C–C bond length is about 1.60 Å, which is similar to the normal F-substituted C–C single bond 1.58 Å. No cracking of the graphene sheet is observed in this case with a first-principles molecular dynamics simulation at 1500 K. The main reason is that the limited length scale of the cluster model graphene has a large flexibility in releasing the strain of fluorination. To restrict such a flexibility, an infinite graphene model should be adopted.

For an armchair GNR, two types of F_2 addition (2R and S2R) can be considered, as shown in Figure 5.15. Upon fluorination, the longest C–C bond length increases from 1.54 Å to 1.70–1.72 Å with the increase of attached F atoms. In an MD simulation at 1500 K, it further increases to 1.92 Å. It is suggested that the unzipping of graphene by fluorination starts from the most strained middle sites (marked in red in Figure 5.15) and then transfer to the edge and the whole process is exothermic. The driving force is the repulsive between F atoms and the strain.

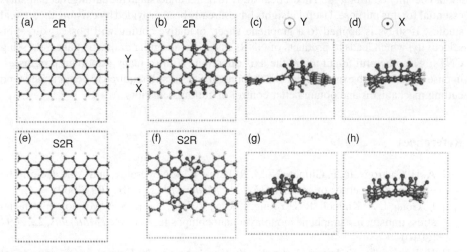

Figure 5.15 (a) Configuration 2R with F_2 addition to an armchair graphene nanoribbon (AGR), and (b–d) top and two side views of the fully fluorinated structure of 2R; (e) Configuration S2R with F_2 addition to the armchair nanoribbon, and (f–h) top and two side views of the fully fluorinated structure of S2R. Reprinted with permission from [44] © 2010 American Chemical Society

5.5 Summary

In this chapter, theoretical investigations of the cutting of graphitic materials have been briefly reviewed. In oxidative cutting, alignment of epoxy groups plays an important role, which unzips the underlying C–C bonds. Further oxidation then generates carbonyl pairs and breaks up the graphitic layer, either through an epoxy pair intermediate species when both surfaces of a single graphene sheet are available, or directly when only one surface is available. An oxygen dimer has also been used as an elementary unit in oxidation to mimic the attacking of permanganate ions. In this way, cutting proceeds directly without forming epoxy groups. However, we notice that even if using $KMnO_4$ as oxidant, the produced graphene oxide has mainly epoxy groups on its surface. Therefore, the oxidation process is expected to be more complicated than simply providing an oxygen dimer. Defects are expected to play an important role in oxidative cutting. However, theoretical study in this direction is still very rare.

Transition metal nanoparticle catalyzed etching and cutting is another important way to get a graphene nanostructure. Theoretically, the metal nanoparticle is either simplified to one or several metal atoms, or abstracted to a structureless sphere. Using metal atoms, the catalysis effect of transition metal on graphene/CNT cutting is clearly demonstrated. With an empirical sphere, direction selectivity observed in experiment has also been reproduced by kinetic Monte Carlo simulations. The open question is how to improve the controllability of cutting processes? For example, by using specially designed substrate [45]. Notice that such a metal catalyzed cutting process is an inversion process of graphene/CNT growth by chemical vapor deposition, which will be discussed in Chapter 13.

The main challenge in this field is to realize controllable cutting to produce high quality GRNs with smooth edges. Ideally, the width and edge structure of the produced GNRs should be fully controllable. Theoretical study to better understand the cutting mechanism is essential for this purpose. Useful insights have already been provided by existing theoretical studies. If stress is applied to a graphene sheet, oxidative cutting will possess direction selectivity, which leads to products of GNRs instead of irregular graphene flakes. Pressing CNTs is also helpful to cut along the axis direction. With so many important progresses, intense effort on both experimental and theoretical sides is still required to better understand cutting mechanisms and obtain better control of cutting behaviors.

References

[1] A. H. Castro Neto, F. Guinea, N. M. R. Peres, K. S. Novoselov, and A. K. Geim, The electronic properties of graphene, *Review Modern Physics*, **81**, 109 (2009).

[2] L. Gong, I. A. Kinloch, R. J. Young, I. Riaz, R. Jalil, and K. S. Novoselov, Interfacial stress transfer in a graphene monolayer nanocomposite, *Advanced Materials*, **22**, 2694 (2010).

[3] Y. Kobayashi, K, Fukui, T. Enoki, K. Kusakabe and Y. Kaburagi, Observation of zigzag and armchair edges of graphite using scanning tunneling microscopy and spectroscopy, *Physical Review B*, **71**, 193406 (2005).

[4] Y. Son, M. L. Cohen and S. G. Louie, Half-Metallic graphene nanoribbons, *Nature*, **444**, 347–349 (2006).

[5] Y. Son, M. L. Cohen and S. G. Louie, Energy gap in graphene nanoribbons, *Physical Review Letters*, **97**, 216803 (2006).
[6] J. Cai, P. Ruffieux, R. Jaafar, M. Bieri, T. Braun, and S. Blankenburg, *et al.*, Atomically precise bottom-up fabrication of graphene nanoribbons, *Nature*, **466**, 470 (2010).
[7] K. Kim, A. Sussman, and A. Zettl, Graphene nanoribbons obtained by electrically unwrapping carbon nanotubes, *ACS Nano*, **4**, 1362 (2010).
[8] M. C. Paiva, W. Xu, M. F. Proenc, R. M. Novais, E. Lægsgaard and F. Besenbacher, Unzipping of functionalized multiwall carbon nanotubes induced by STM, *Nano Letters*, **10**, 1764 (2010).
[9] J. Liu, A. G. Rinzler, H. Dai, J. H. Hafner, R. K. Bradley, P. J. Boul, *et al.*, Fullerene pipes, *Science*, **280**, 1253 (1998).
[10] J. L. Li; K. N. Kudin; M. J. McAllister; R. K. Prud'homme; I. A. Aksay and R. Car, Oxygen-driven unzipping of graphitic material, *Physical Review Letters*, **96**(17), 176101–176104 (2006).
[11] L. Ci, Z. Xu, L. Wang, W. Gao, F. Ding, K. F. Kelly, *et al.*, Controlled nanocutting of graphene, *Nano Research*, **1**, 116–122 (2008).
[12] A. G. Cano-Márquez, F. J. Rodríguez-Macas, J. Camopos-Delgado, C. G. Espinosa-González, F. Tristán-López, D. Ramírez-González, *et al.*, Ex-MWNTs: Graphene sheets and ribbons produced by lithium intercalation and exfoliation of carbon nanotubes, *Nano Letters*, **9**, 1527–1533 (2009).
[13] I. Janowska, O. Frsen, T. Jacob, P. Vennegues, D. Begi, D. M.-J. Ledoux and C. Pham-Huu, Catalytic unzipping of carbon nanotubes to few-layer graphene sheets under microwaves irradiation. *Applied Catalysis A: General*, **371**, 22–30 (2009).
[14] D. V. Kosynkin, A. L. Higginbotham, A. Sinitskii, J. R. Lomeda, A. Dimiev, B. K. Price and J. M. Tour, Longitudinal unzipping of carbon nanotubes to form graphene nanoribbons, *Nature*, **458**, 872 (2009).
[15] M. Terrones, Sharpening the chemical scissors to unzip carbon nanotubes: crystalline graphene nanoribbons, *ACS Nano*, **4**, 1775–1781 (2010).
[16] M. J. McAllister; J.-L. Li; D. H. Adamson, H. C. Schniepp, A. A. Abdala, J. Liu, *et al.*, Single sheet functionalized graphene by oxidation and thermal expansion of graphite, *Chemistry of Materials*, **19**, 4386 (2007).
[17] H.-K. Heong, Y. P. Lee, R. J. W. E. Lahaye, M.-H. Park, K. H. An, I. J. Kim, *et al.*, Evidence of graphitic stacking of graphite oxides, *Journal of American Chemical Society*, **130**, 1362 (2008).
[18] S. Fujii and T. Enoki, Cutting of oxidized graphene into nanosized pieces, *Journal of American Chemical Society*, **132**, 10034 (2010).
[19] Z. Y. Li, W. H. Zhang, Y. Luo, J. L. Yang and J. G. Hou, How graphene is cut upon oxidation. *Journal of the American Chemical Society*, **131**(18), 6320–6321(2009).
[20] K. N. Kudin, H. F. Bettinger and G. E. Scuseria, Fluorinated single-wall carbon nanotubes, *Physical Review B*, **63**(4), 045413 (2001).
[21] K. N. Kudin, G. E. Scuseria, and B. I. Yakobson, C(2)F, BN and C nanoshell elasticity from ab inito computations, *Physical Review B*, **64**(23), 235406 (2001).
[22] X. F. Gao, L. Wang, Y. Ohtsuka, D.-E. Jiang, Y. L. Zhao, S. Nagase and Z. F. Chen, Oxidation unzipping of stable nanographene into joint spin-rich fragments, *Journal of American Chemical Society*, **131**, 9663 (2009).

[23] T. Sun and S. Fabris, Mechanisms for oxidative unzipping and cutting of graphene, *Nano Letters* **12**, 17–21 (2012).

[24] J. T. Paci, T. Belytschko, and G. C. Schatz, Computational studies of the structure, behavior upon heating and mechanical properties of graphite oxide, *The Journal of Physical Chemistry C*, **111**(49), 18099–18111 (2007).

[25] L. Ma, J. L. Wang and F. Ding, Strain-induced orientation-selective cutting of graphene into graphene nanoribbons on oxidation, *Angewandte Chemie International Edition*, **124**, 1187–1190 (2012).

[26] D. V. Kosynkin, A. L. Higginbotham, A. Sinitskii, J. R. Lomeda, A. Dimiev, B. K. Price, and J. M. Tour, Longitudinal unzipping of carbon nanotubes to form graphene nanoribbons, *Nature*, **458**, 872–876 (2009).

[27] A. Sinitskii, A. Dimiev, D. V. Kosynkin and J. M. Tour, Graphene nanoribbon devices produced by oxidative unzipping of carbon nanotubes, *ACS Nano*, **9**, 5405–5413 (2010).

[28] F. Li, E. Kan, R. Lu, C. Xiao, K. Deng and H. Su, Unzipping carbon nanotubes into nanoribbons upon oxidation: A first-principles study, *Nanoscale*, **4**, 1254–1257 (2012).

[29] Y. Guo, L. Jiang and W. Guo, Opening carbon nanotubes into zigzag graphene nanoribbons by energy-optimum oxidation, *Physical Review B*, **82**, 115440 (2010).

[30] Y. Guo, Z. Zhang and W. Guo, Selective oxidation of carbon nanotubes into zigzag graphene nanoribbons, *The Journal of Physical Chemistry C*, **114**, 14729–14733 (2010).

[31] N. L. Rangel, J. C. Sotelo and J. M. Seminario, Mechanisms of carbon nanotubes unzipping into graphene ribbons, *The Journal of Chemical Physics*, **131**, 031105 (2009).

[32] H. Zhang, M. Zhao, T. He, X. Zhang, Z. Wang, Z. Xi, et al., Orientation-selective unzipping of carbon nanotubes, *Physical Chemistry Chemical Physics*, **12**, 13674–13680 (2010).

[33] S. S. Datta, D. R. Strachan, S. M. Khamis and A. T. C. Johnson, Crystallographic etching of few-layer graphene, *Nano Letters*, **8**(7), 1912–1915 (2008).

[34] L. Ci, L. Song, D. Jariwala, A. L. Elias, W. Gao, M. Terrones, and P. M. Ajayan, Graphene shape control by multistage cutting and transfer, *Advanced Materials*, **21**, 4487–4491 (2009).

[35] A. L. Elías, A. R. Botello-Méndez, D. Meneses-Rodríguez, V. J. González, D. Ramírez-González, L. Ci, E. Muñoz-Sandoval, et al., Longitudinal cutting of pure and doped carbon nanotubes to form graphitic nanoribbons using metal clusters as nanoscalpels, *Nano Letters*, **10**, 366–372(2010).

[36] L. Y. Jiao, L. Zhang, X. R. Wang, G. Diankov, and J. H. Dai, Narrow graphene nanoribbons from carbon nanotubes, *Nature*, **458**, 877–880 (2009).

[37] X. Feng, S. Maier and M. Salmeron, Water splits epitaxial graphene and intercalates, *Journal of the American Chemical Society*, **134**, 5562 (2012).

[38] Z. Gu, H. Peng, R. H. Hauge, R. E. Smalley and J. L. Margrave, Cutting single-wall carbon nanotubes through fluorination, *Nano Letters*, **2**, 1009–1013 (2002).

[39] L. Tsetseris and S. T. Pantelides, Graphene nano-ribbon formation through hydrogen-induced unzipping of carbon nanotubes, *Applied Physics Letters*, **99**, 143119 (2011).

[40] W. L. Wang, S. Meng and E. Kaxiras, Graphene nanoflakes with large spin, *Nano Letters*, **8**, 241–245 (2008).
[41] J. Fernández-Rossier and J. J. Palacios, Magnetism in graphene nanoislands, *Physical Review Letters*, **99**, 177204 (2007).
[42] J. L. Wang, L. Ma, Q. Yuan, L. Zhu and F. Ding, Transition-metal-catalyzed unzipping of single-walled carbon nanotubes into narrow graphene nanoribbons at low temperature, *Angewandte Chemie International Edition*, **50**, 8041–8045 (2011).
[43] A. C. T. van Duin, S. Dasgupta, F. Lorant and W. A. Goddard III, ReaxFF: A reactive force field for hydrocarbons, *The Journal of Physical Chemistry A*, **105**(41), 9396–9409 (2001).
[44] M. Wu, J. S. Tse and J. Z. Jiang, Unzipping of graphene by fluorination, *The Journal of Physical Chemistry Letters*, **1**, 1394–1397 (2010).
[45] T. Tsukamoto and T. Ogino, Control of graphene etching by atomic structures of the supporting substrate surfaces, *The Journal of Physical Chemistry C*, **115**, 8580 (2011).

6

Properties of Nanographenes

Michael R. Philpott
Kenneth S. Pitzer Center for Theoretical Chemistry, Department of Chemistry,
University of California Berkeley, USA
and
Center for Computational Materials Science, Institute of Materials Research,
Tohoku University, Japan

6.1 Introduction

Eldredge and Gould [1] pointed out that modern science progresses like the "punctuated equilibrium" hypothesis of the evolution of life on planet Earth. Major discoveries were followed by an equilibration process as hundreds joined the race, pushed the envelope and added important details. A punctuation in 2004 saw Geim and Novoselov [2] demonstrate that micron-sized basal sheets of graphite one atom thick (graphene) could be isolated by "scotch tape" peeling of single crystals. Subsequent measurements showed graphene to have completely unexpected properties including high conductivity and "massless" Dirac electrons near the K-points of the Brillouin zone [3]. For once, deserved recognition came fast [4].

A notable focus on theoretical studies of graphene has been on the electronic properties of graphene nano-ribbons (GNRs). This has been driven by device applications and aided by band structure modeling using density functional theory (DFT). In this article a different focus is offered; we describe DFT predicted properties of small nano-sized fragments of graphene and provide a chemical perspective, considering them as gigantic polyaromatic hydrocarbon (PAH) molecules. On the edges of nano-graphenes, also called graphene nano-dots (GNDs), a dichotomy occurs (weak reconstruction) because the interior carbon atoms have the trigonal C–C bonding of the infinite graphene sheet whereas on the edge, half the

---

Graphene Chemistry: Theoretical Perspectives, First Edition. Edited by De-en Jiang and Zhongfang Chen.
© 2013 John Wiley & Sons, Ltd. Published 2013 by John Wiley & Sons, Ltd.

Table 6.1 Working pi-electrons

C_6-ring	pi-electrons /ring
graphene	2
zigzag edge	2.167
zigzag apex	2.333
armchair edge	2.333
oligocene edge	2.333
triangulene apex	2.5
armchair apex	2.5
oligocene apex	2.667
benzene	6

carbons have terminal C–H bonds and a pi-electron shared with two neighbor carbons and not three as in the interior. This changes the aromaticity of edge rings. We have performed DFT calculations on many series of GNDs to discern trends that provide predictions for the largest members and these results are summarized in this report. We assume H termination at edges and no interior vacancies.

Simple chemical concepts provide perspectives useful for understanding some aspects of bonding and magnetism in graphene. All the carbon atoms of a graphene fragment belong to either one, two or three hexagonal C_6-rings. Table 6.1 lists the number of pi-electrons at work in rings at various locations (divide each pi-electron by the number of bonds it occupies and sum across the ring atoms).

Benzene and graphene are special. In benzene the sigma bonds are compressed by the high degree of pi-bonding. In graphene each pi-electron is shared among three bonds suggesting an ability to transmit or respond collectively to disturbances. In graphene and PAHs the network of carbons is a bi-lattice, known as an *alternant structure* in molecular orbital (MO) theory. Every atom has neighbors of the other lattice. If the two lattices have a different population (nonequivalent) the network is intrinsically magnetic, meaning there are unpaired spins [5]. In valence bond (VB) theory this means there is a non-Kekulé bond diagram [6]. If the lattices are equivalent, then as we will discuss magnetism can still occur on long zigzag edges. In an unsupported GND the network of pi- and sigma-bonds constitutes a Mandelbrot 2^+-dimensional system and though the system may conduct charge (conducting polymer) it is not metallic and there is no screening of charge as in a three-dimensional plasma. The edge reconstruction due to bond dichotomy polarizes interior bonds impressing on them a bond length pattern resembling the edge pattern. This phenomenon, evident in high symmetry, decreases quickly with distance. A useful result from DFT calculations of PAH molecules is the near linear dependence of C–C bond length on the highest valence electron charge density in a C–C bond [7].

The carbon networks of GNDs differ in shape, edge type and apex. In this report we describe calculations done using the same methodology, xc-functional and parameters for GNDs with six (hexangulenes), three (triangulenes), five edges and four edges (two armchair and two zigzag edges). As we proceed we refer to relevant prior work that has appeared in the literature. For the most part we consider only zigzag and armchair (crenellated) edges, and apexes containing four, two or one carbon atoms with attached hydrogen. A comparison of some zigzag edge geometries is given in the summary.

6.2 Synthesis

Synthesis and experiments on nanographenes are appearing. Clar [8] and Fetzer [9] have described the occurrence, synthesis and spectra of many large PAHs. The synthesis and reactivity of small triangulenes ($m < 4$) has been well studied [8]. Similarly for the oligocenes, systems with zigzag index m less than eight (8) have been synthesized [8, 10]. The larger and longer systems will need to be kept isolated or protected. One can speculate that in time a UHV isolation technique will be developed or some magical method like the water drop for protecting electrochemical surfaces [11]. Large size graphenes are believed to be involved in the combustion of carbon and the formation of soot particles [12, 13]. Progress with crenellated edges has been spectacular. Muellen and coworkers have described; the synthesis of very large PAHs [14], possible single molecule electronics systems [15], the *de novo* synthesis by surface dehydrogenation of large armchair hexangulenes (e.g., $C_{222}H_{42}$) [16], armchair ribbons from bi-anthracene precursors on stepped metal surfaces [17], and others by surface chemistry assembly [18]. Further in the future, there is the possibility of routinely performing nano-tailoring using lithography or local probes to cut, stitch stamp, and mold. We return to this topic again briefly in the concluding remarks (see Section 6.11).

6.3 Computation

The main computational tool used in the calculations has been the Vienna *ab initio* simulation package (VASP), a robust method designed for periodic solid and surface calculations that can also be used for clusters in large simulation cells. The documentation for this code and the forum are available on the web [19, 20]. We will mention only some important details pertinent for the results discussed here. The VASP code preforms *ab initio* plane wave based DFT calculations. It uses PAW frozen core pseudo-potentials and a spin polarized generalized gradient approximation for the exchange-correlation energy functional [21]. For the most part we use the simple non hybrid xc-functional PW91 parameterized by Perdew and coworkers [22]. The basis set is huge and can be systematically improved by increasing size of the cut-off energy for the plane waves. There are no basis-set superposition errors. Atoms in different periodic cells were 1.5 nm or more apart. The Brillouin zone integration was done at the gamma point. The geometry optimizations were continued until no Kohn-Sham levels contained fractional charge and the forces acting on each atom were close to or less than 0.05 meV/pm. We regularly calculated geometry, total energy and isometric surfaces of total charge and spin density for spin states $S = 0, 1, 2, \ldots$ as needed. The charge and spin densities were analyzed using the Vaspview (free software) program and geometry using Gaussview or VisIt [23]. States with integral spin and no fractional orbital charge could generally be found by decreasing the level width input parameter and narrowing energy and convergence factors. Nevertheless there are organics with a high spin ground state, for example, Closs radicals [24], so it is important for GNDs with large

6.4 Geometry of Zigzag-Edged Hexangulenes

Pioneering extended Hückel calculations of PAHs with thousands of atoms were reported long ago by Brown and Stern [26]. In Figure 6.1, the diagram on the right-hand side shows the series of zigzag edged hexangulenes (notation: z-GND or hx-m-zg) with formulae $C_{6m**2}H_{6m}$ ($m = 2, 3 \ldots 10$). We draw on our own geometry results calculated for molecules with zigzag index $m = 9, 10$ ($C_{486}H_{54}$ and $C_{600}H_{60}$) [27–29]. The basic geometry consists of nested hexagonal rows of carbon atoms joined every two carbons to rows on either side. There two types of C–C bond: (1) parallel bonds running zigzag fashion around the hexagonal rows and (2) perpendicular bonds that radiate like spokes in a hexagonal wheel joining the nested hexagonal rows alternately to the zigzag rows on either side. Figure 6.2 displays a quadrant of the hexagon showing how the C–C bonds are labeled. Nested zigzag rows for small and large row indices are shown by the dotted lines. Examples of parallel and perpendicular C–C bonds are identified by the arrows. The symbol **Q** in the apex ring signifies that in isometric surface plots of electron density this ring has a quinoidal structure. A quinoidal bond structure is depicted separately in the cartoon in the small inset panel (also labeled **Q**) at the end of the C_2'' axis in the upper right-hand corner of the figure.

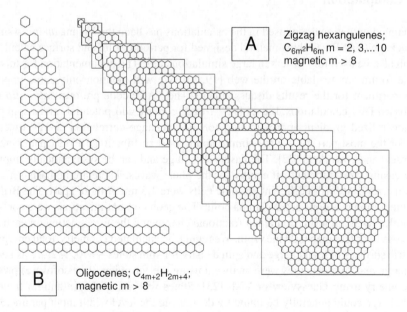

Figure 6.1 Diagrams: panel A, zigzag edged hexangulenes $C_{6m**2}H_{6m}$ ($m = 2, 3 \ldots 10$); panel B, oligocenes $C_{4m+2}H_{2m+4}$ ($m = 2, 3, 4 \ldots$). Adapted with permission from [29] © 2009 American Institute of Physics

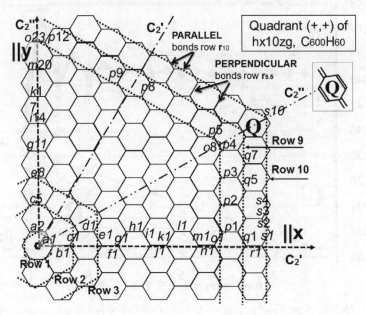

Figure 6.2 Quadrant of hexangulene $C_{600}H_{60}$ showing CC bond labels, zigzag row labels, quinoidal apex Q, nested hexagonal rows of carbon (dotted line) near center and perimeter, arrows indicating parallel and perpendicular bonds. Adapted with permission from [29] © 2009 American Institute of Physics

The geometry of $C_{486}H_{54}$ and $C_{600}H_{60}$ has remarkable features. Figure 6.3 and Figure 6.4 depict the main results. Figure 6.3 shows a plot of the lengths of C–C bonds versus their row indices, r_n (n = 1,2,... 10) for transverse bonds and $r_{n+1/2}$ (n = 1, 2 ... 9) for perpendicular bonds. Curves are drawn through the data points to guide the eye. The curves show an apparent oscillation between values for perpendicular bonds ($r_{n+1/2}$) that rise monotonically to C–C = 144.4pm and transverse (zigzag rows r_n) bonds that show increasing dispersion in their value range starting at **r7** four rows from the perimeter **r10**. Both sets have the same interior value (CC = 142.3pm). Perpendicular bonds are almost everywhere longer than the zigzag row bonds they connect. The left-hand side inset panel in Figure 6.3 shows the sector of CC bonds used in the figure and the vertically "squashed" triangular segment shows the stick-ball skeleton with total valence charge super-imposed on the stick bonds. This part of the figure shows high charge along the perimeter (highest on apex bond labeled **s10** in Figure 6.2) declining to resolved charge centroids on the interior bonds. Figure 6.4 shows how the CC bonds vary in length along their rows. The largest variations are in the parallel (zigzag) bond set in the perimeter row r_{10}. Rather remarkably the interior zigzag rows show oscillations that follow the pattern set by the perimeter diminishing with distance. We note that as shown in the small inset panel in Figure 6.2, quinoidal structures can be joined together by short C=C "double" bonds. Looking at the valence charge density partly depicted in Figure 6.3 we see that the difference in parallel (zigzag bonds) and perpendicular bonds is the result of a quinoidal polarization of valence charge radiating out from each apex. In a quinoidal structure the C_6-ring is distorted so that

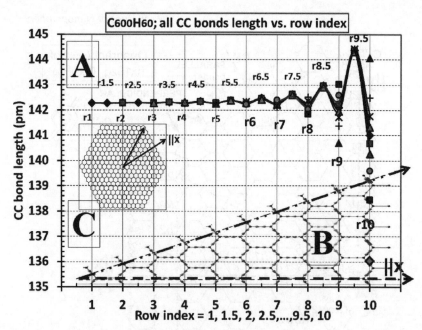

Figure 6.3 Hexangulene $C_{600}H_{60}$. Panel A, plot of all C–C bond lengths versus row indices r_n ($n = 1, 2, \ldots 10$) of transverse bonds and $r_{n+1/2}$ ($n = 1, 2 \ldots 9$) of perpendicular bonds. Curves are drawn through data to guide the eye. Panel B, "compressed" triangular half sector showing with total valence charge superimposed on stick bonds. Panel C, diagram showing location of the half sector of panel B. Adapted with permission from [29] © 2009 American Institute of Physics

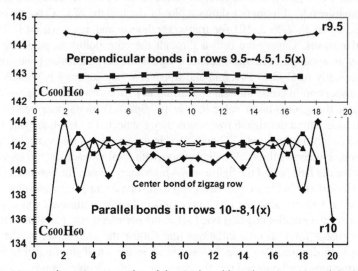

Figure 6.4 Hexangulene $C_{600}H_{60}$. Plot of the CC bond length vs. position of the bond on the row. Top panel, perpendicular bonds $r_{9.5} - r_{4.5}$, $r_{1.5}$ (point ×) are constant within their row. Bottom panel, parallel bonds $r_{10} - r_8$, $r1$ (points ×-×) show large variation at the apex (r_{10}) or at the row end. Adapted with permission from [29] © 2009 American Institute of Physics

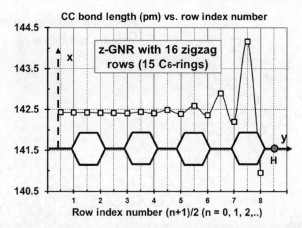

Figure 6.5 Graphene nanoribbon. Variation of CC bond lengths plotted versus the bond row index starting from the center of a 16 zigzag row GNR. Perpendicular bonds (0.5, 1.5, 2.5 ...) show a monotonic increase and parallel bonds (1, 2, 3, ... 10) a monotonic decrease towards the edge. Adapted with permission from [27] © 2009 American Institute of Physics

two parallel bonds are short and remaining four are long. The succession of quinoidally coupled structures gives rise to the strong bond alternation radiating along the perimeter edge from each apex. Shorter C–C bonds have higher valence electron density. Starting from a short C=C double bond (136 pm, see -♦- end point data in Figure 6.4) at each apex the C–C bonds alternate with diminishing length differences to the middle of the edge whereupon the differences increase again until the next apex is reached.

We close with a comment on the geometry of zigzag edged GNRs [27]. Figure 6.5 shows the variation of CC bond lengths plotted against the row index starting from the center of a 16 zigzag row z-GNR (15 rings wide along y-axis, one ring along x-axis). The cell constant in the parallel x-direction was fixed at the graphene value. All bonds can relax subject to this constraint. Figure 6.5 has the main characteristic of the hexangulene molecule in Figure 6.3. The effect is weaker in the z-GNR because of the periodic boundary condition with one C_6-ring along the x-axis. Nevertheless the oscillation in bond lengths occurs with perpendicular bonds showing a monotonic increase towards the edge and parallel bonds showing a monotonic decrease towards the edge.

In this section the examples described permit two important conclusions: (1) The interior of a zigzag edged hexangulene ends where the oscillations begin, about four rows from the perimeter in Figure 6.3, and (2) for zigzag edges the variation of bond lengths due to quinoidal valence charge variations along the perimeter results in a decaying pattern of quinoidal bond length variations imprinted on the interior bonds.

6.5 Geometry of Armchair-Edged Hexangulenes

We use the notation a-GND to identify armchair edged molecules. We describe DFT calculations of valence charge and geometric properties of armchair (crenellated) edged

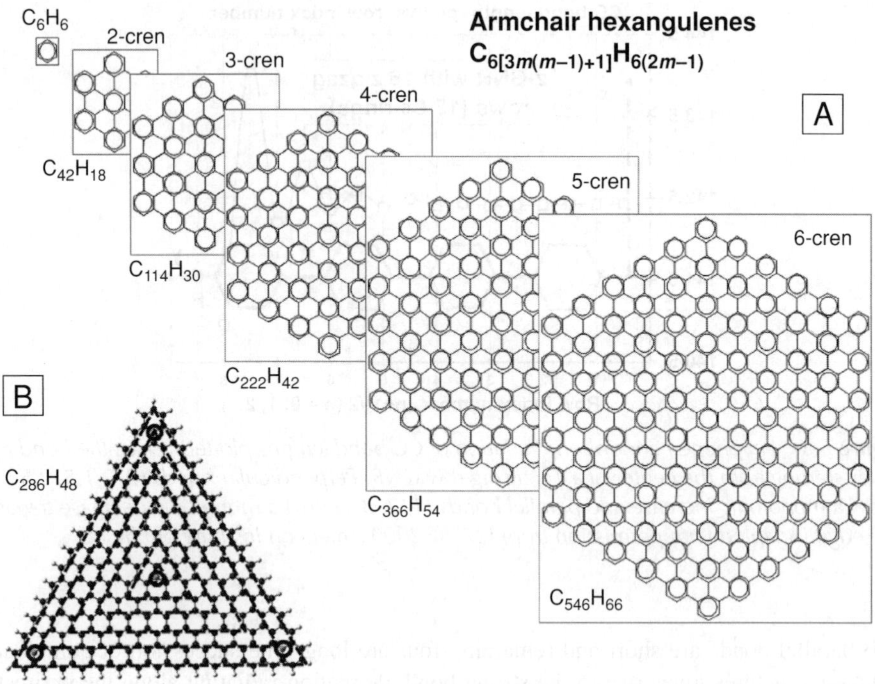

Figure 6.6 Panel A: *Diagram of the armchair hexangulene series $C_{6[3m(m-1)+1]}H_{6(2m-1)}$ ($m = 2, 3\ldots 6$) in the Clar notation for fully benzenoid hydrocarbons. Panel B: Diagram of tri-15-zg, $C_{286}H_{48}$, the $m = 15$ member of the triangulene series: $C_{m**2+4m+1}\ H_{3m}$. Overlays, nested triangular rows and atoms (circles ○) described in text. Adapted with permission from [29] © 2009 Elsevier*

nanographenes with formulae $C_{6[3m(m-1)+1]}H_{6(2m-1)}$ ($m = 2, 3\ldots 6$), the largest being $C_{546}H_{66}$ [30]. Panel A in Figure 6.6 depicts the series as a cartoon using the Clar–Robinson notation [8, 31] for fully benzenoid hydrocarbons. The fully benzenoid hydrocarbon concept was introduced by Clar [8]. It assumes that great stabilization occurs when a PAH has a Kekulé bond structure consisting of *independent* embedded C_6 aromatic rings. Experimentally small fully benzenoid PAH have large optical band gaps and so are colorless. In simple valence bond (VB) theory each benzenoid ring has value factor two (from two Kekulé structures). Accordingly the molecule 6-cren in Figure 6.6 has 2^{91} such structures. This seemingly powerful idea ignores another "overpowering" feature in valence bond theory, namely the even more "inflationary" increase in ionic VB structures with geometric size. The importance of ionic structures in calculation of benzene properties was known before the advent of MO theory [31]. Crystallographic studies have showed that the C_6-rings are not all the same in fully benzenoid compounds. On the other hand there have been STM studies of graphene edges that imaged extended fully benzenoid patterns [33–35]. Our DFT calculations [30] detected a vestige of an aromatic sextet-like structure in the apex region. In total valence charge density plots there was one ring within the apex zone with almost equal C–C bonds surrounded by longer bonds also of equal length. Since this is a

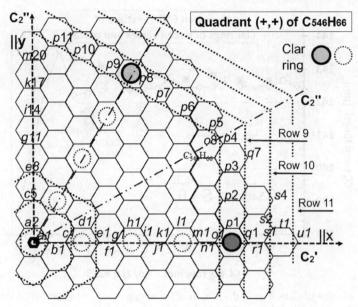

Figure 6.7 The $(+,+)$ quadrant of armchair hexangulene $C_{546}H_{66}$. Clar rings are shown by the full circles between zigzag rows **r8** and **r9**. Vestigial Clar rings by broken circles on the C_2' axes. Adapted with permission from [29] © 2009 Elsevier

hallmark of the Clar benzenoid model we call these structures Clar rings. Figure 6.7 shows the $(+,+)$ quadrant of $C_{546}H_{66}$. The location of the Clar ring near the apex is shown as a shaded heavy circle between zigzag rows **r8** and **r9**.

For armchair hexangulenes plots of all CC bonds against the row index $(n+1)/2 = 1$, 1.5, 2, 2.5, ... produced a ragged diagram when compared to zigzag edged hexangulene previous shown in Figure 6.3. This occurs because of the great variation in bond lengths within the apex of this a-GND. A plot of zigzag bonds along arm chair rows showed large variations with an apparent weak correlation between successive rows. Plots of bonds *connecting* successive armchair rows (called perpendicular bonds in z-hexangulenes) had some similarity to the z-GND plots. Upon re-examination of the data [36] it was found that when C–C bonds along and on either side of the C_2' axes (for example see the axis parallel ∥x) in Figure 6.7 are plotted as in Figure 6.8 we could detect the signature of the apex Clar ring and in addition a sequence of weaker rings diminishing in resolution towards the center. Bonds connected along the axis are shown as diamonds (–◊– C_6-ring bonds) and squares (–□– or ■ exterior C–C bonds connecting adjacent C_6-rings along the x-axis). Unconnected squares represent the C–C bonds radiating out from the Clar rings. The complete list of bonds in Figure 6.7 starting at the *apex* and working towards the *center* are: u1, t1, s1; (r1); q1(q2),p1,(o2)o1; (n1); m1(m2),l1,(k2)k1; (j1); i1(i2),h1(g2)g1; (f1); e1(e2),d1,(c2)c1; (b1); a1(b2)a2. The first three bonds u1, t1, s1 belong to the apex ring and, as shown in Figure 6.8, this ring does not have the Clar characteristics. Bonds in parenthesis are long C–C bonds attached to the exterior of the Clar ring; bonds between semi-colons (;) and not in parenthesis belong to a Clar ring. This plot is a counterpart of Figure 6.3

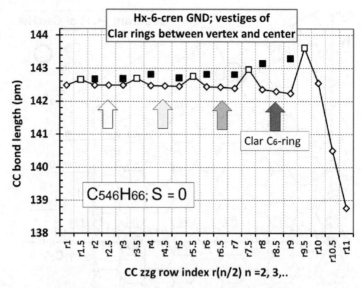

Figure 6.8 Vestige of Clar-rings in the armchair hexangulene $C_{546}H_{66}$. Bonds of the Clar-rings are denoted by an unfilled diamond symbol (–◊–). Exterior long bonds attached to the Clar-ring are denoted by square symbol (–□– for perpendicular bonds on the axis and ■ for parallel bonds off the axis). See the text for a complete list of bonds

and shows that the polarization of bonds due to edge bonding diminishes smoothly and quickly into the interior. All C–C ring and connector bonds converge to the graphene value irrespective of orientation.

In summary, we note that DFT calculations predict decaying vestigial Clar-like structures in the armchair hexangulenes. However, why they occur so strongly in the STM images is a mystery. It might be a consequence of graphene responding locally or collectively to the probe tip or the probe tip having acquired a special or extra atomic structure.

6.6 Geometry of Zigzag-Edged Triangulenes

The smallest triangulenes have been studied experimentally since the 1940s [8] and theoretically because of their spin in some of the first papers describing MO theory. Panel B in Figure 6.6 shows a cartoon of the tri-15zg, formula $C_{286}H_{48}$, the largest m = 15 member of the series: $C_{m**2+4m+1} H_{3m}$ (m = 2 ... 15) described in published work [7, 27, 36]. Highlighted are the triangular sets of zigzag rows and four atoms (inside circles ○) that are at special points, three within each apex join three separate zigzag lattices and the central atom in an inverted "Y" bond configuration. The apex carbon atoms have an electronic structure (revealed by isometric surface plots of the total charge density) that is distinctly different from the oligocenes and hexacenes. This is a result of overall D_{3h} symmetry and local site symmetry C_{2v} with an axis through the middle apical C atom. The topology of isometric surfaces of the total charge density of the three corner carbon atoms was found to resemble that around the apical C atom in pyrene [27].

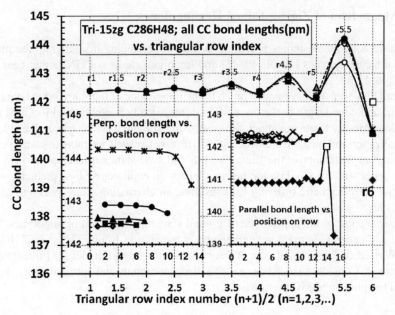

Figure 6.9 Triangulene, tri-15-zg, $C_{286}H_{48}$. Main panel, all CC bond lengths plotted versus their row index, parallel bonds on rows $r_1, r_2, r_3, \ldots r_6$ and perpendicular bonds on rows $r_{1.5}, r_{2.5}, \ldots, r_{5.5}$. Sub-panel left-hand side, variation of perpendicular bonds on rows $r_{n+\frac{1}{2}}$ ($n = 1, 2, \ldots 5$) versus position on row. Sub-panel right-hand side, variation of parallel bonds on rows r_n ($n = 1, 2, \ldots 6$). Adapted with permission from [35] © 2010 American Institute of Physics

In the triangulenes the majority α-spin lattice has attached H atoms on the edge and minority β-spin at the single apex carbon. DFT bond length calculations, summarized for tri-15zg in Figure 6.9, show that the C–C bond lengths fall into three distinct groups: core, apex and edge, irrespective of whether the molecular center is a single atom ("Y", m = 2 + 3n; inverted "Y", m = 3 + 3n where n = 0, 1, 2, ...) or a C_6-ring (m = 4 + 3n, n = 0, 1, 2 ...). The uniformity of zigzag-edge C–C bonds reduces the near perimeter reconstruction, so that the graphene core extends an extra row, almost to the penultimate triangular row of carbon atoms. Impressed on the interior core bonds starting at the center is a small increasing length oscillation. The sub-panels in Figure 6.9 show, with the exception of bonds at the ends of a row, that the variations along the zigzag parallel bond rows r_n (n = 1, 2, ... 6) and along the perpendicular bond rows $r_{n+\frac{1}{2}}$ (n = 1, 2, ... 5) are weaker than those for z- or a-hexangulenes. The perimeter C–C bonds that join at the apex are the shortest in the molecule and the edge carbons are separated from interior atoms by the longest bonds (on row $r_{5+\frac{1}{2}}$)... The C–C bonds in the high spin section of the edges r_6 are uniform in length and longer than perimeter C–C bonds in the zigzag edged linear acenes, hexangulenes, annulenes and benzene [28, 29]. This is attributed to the large number of edge localized non-bonding molecular orbitals (NBMOs) that sequestered a quarter of the valence charge removing it from effective bonding. Around the perimeter all the C–C bonds $\approx$141 pm, while perpendicular bonds pointing inwards from the perimeter have lengths $\approx$144 pm.

6.7 Magnetism of Zigzag-Edged Hexangulenes

There have been many studies providing new and overlapping insights into the magnetization of nanographenes [28, 30, 40–45] and important studies of GNRs that bear on the former [46, 47].

Overall these reports have added much to the importance of topology, edge type (zigzag vs. chair) and edge length to the discussion of intrinsically magnetic multi-radical ground state. Concurrently there was much progress using complete active space, density matrix renormalization and spin-flip methodologies to resolve a related and more tractable problem, namely the magnetization of the oligocenes [48–50]. One of the earliest reports on magnetic hexangulenes was for $C_{384}H_{48}$ the hexangulene with eight edge rings, predicted to have sub-lattice ferromagnetic ordering [38] resulting in alternating edge spins and a singlet ground state.

A basic feature of the appearance of ground state magnetization in oligocenes, hexangulenes and general alternant nanographenes is the change from a spin paired HOMO to spin polarized disjoint orbital pair. A very simple two-orbital model [48] illustrates how this happens. Consider two orthogonal one-electron orbitals and spin functions: $\{\varphi 1, \varphi 2, \alpha, \beta\}$. The spin paired MO picture is:

$\varphi_{\pm} = (1/\sqrt{2})(\varphi_1 \pm \varphi_2)$ .... (+ homo, - lumo).
$\Psi = \frac{1}{2}(\varphi_1 + \varphi_2)(\varphi_1 + \varphi_2)(\alpha\beta - \beta\alpha)$ .... (ground state, spin paired singlet)
$\Psi = \frac{1}{2}\mathrm{Det}|\varphi_+, \varphi_-| \times \{\alpha\alpha, \beta\beta, (\alpha\beta + \beta\alpha)\}$ ..... (triplet state sub-levels, S = 1 diradical)

The disjointed orbital picture starts from the original orbital spin functions:

$\Psi = \frac{1}{2}\mathrm{Det}|\varphi_1\alpha, \varphi_2\beta| = \frac{1}{2}(\varphi_1\varphi_2 + \varphi_2\varphi_1)(\alpha\beta - \beta\alpha)$ . . (ground S = 0 diradical singlet)
$\Psi = \frac{1}{2}\mathrm{Det}|\varphi_1, \varphi_2| \times \{\alpha\alpha, \beta\beta, (\alpha\beta + \beta\alpha)\}$ .... (triplet sub-levels, S = 1 diradical).

In a diradical ground state the "original" HOMO and LUMO are combined. This elementary bit of formalism provides a small clue to the transition from spin paired to disjoint orbitals and the appearance of zigzag edge magnetism. Hexagonally shaped zigzag edged GNDs have equal sub-lattices and magnetism is not expected. The scenario may play out as follows. In small zigzag index m systems the strong coulombic interactions and lower frontier level density cause opposite spin electrons to pair off. However, on long zigzag edges (zigzag index $\geq \approx 9$) magnetization arises from exchange interactions acting in a regime of reduced coulombic interactions. The highest HOMOs mix with lowest LUMOs and disjoint orbitals result with alpha- and beta-spin electrons occupying the different sub-lattices. Stronger electron correlations in the "double" bonds at the apexes and in the partial double bonds near the apexes reduce spin density such that in DFT calculations the magnetization changes smoothly from near zero to an extreme at the center of the edge where the C–C bond distances differences are narrowly bounded. Since the DFT results for oligocenes appear to track the more advanced methods [48–50], this provides a modicum of confidence in DFT results for the GNDs.

A large number of band structure calculations of graphene nano-ribbons (GNR) have appeared and the existence of edge magnetism appears substantiated. What is not clear

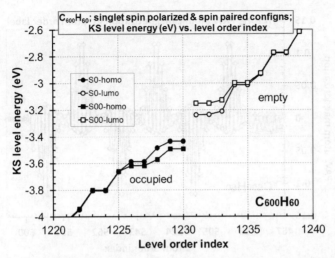

Figure 6.10 Kohn–Sham energy levels of hexangulene hx10zg, $C_{600}H_{60}$. Spin paired $S = 0$ configuration data: HOMOs –•– and LUMOs –○–. Singlet spin polarized ground configuration $S = 00$ data: HOMOs -■-, LUMOs -□-

is exactly how the edge magnetization of elongated GNDs tends to an extensive (length dependent) property.

The smallest hexagonal GNDs with spin polarized ground states appear as the single state (S = 0) of a six-radical with the magnetization appearing as if there were alternating majority spin per edge. Results for hydrogen addition in the interior or multiply added on different edges of hx-GNDs are consistent with the six-radical hypothesis. Similar results are calculated when small intrinsically magnetic protrusions are attached to GND edges [51–53].

Figure 6.10 shows the Kohn–Sham energy levels for hx10zg ($C_{600}H_{60}$) in the spin-paired state (denoted S = 0 in the legend) lying approximately 0.12 eV higher in total energy above the spin-polarized singlet ground state (denoted S = 00). The spin-paired levels lie within the spin-polarized S = 00 ground state gap and both manifolds converge beyond five levels (10 electrons). Hexagonal symmetry causes frequent appearance of the benzene pi-level pattern: a triple of levels (degenerate pair plus one non-degenerate level) below and above (inverted order) the gap. Changing from xc-function PW91 to the PBEknw1 (improves magnetics) increases the total energy difference to 0.20 eV and increases the S = 00 band gap from 0.34 to 0.43 eV.

Figure 6.11 plots the PAW magnetization on atoms versus their order index around the entire perimeter (row r_{10}). There are 19 carbons per edge and 114 (6 × 19) on the perimeter starting at C(487) and ending at C(600). The carbons on row r_9 end at C(486), hydrogen atoms start at H(601). An order index label error has occurred near C(595). The maximum spin is 0.115 units. The spin on an edge of $C_{600}H_{60}$ compares well with an island of spin surrounding a central carbon with an attached H attached to a centrally located C atom [49]. The entire spin on an edge can be quenched to a small value by attaching H atom to the mid-point atom [49]. Likewise when a small intrinsically magnetic C_3H group is appended at the mid edge the affected edge magnetization is quenched [50]. Multiple mid-edge H

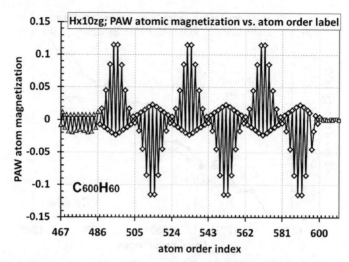

Figure 6.11 Magnetization of hexangulene hx10zg, $C_{600}H_{60}$. Plot of PAW atomic magnetization versus the atom order index around the perimeter on row r_{10}. There are 19 carbon atoms per edge starting at C(487) and ending at C(600). Hydrogen atoms start at H(601) and carbons on row r_9 end at C(486). Adapted with permission from [51] © 2011 American Institute of Physics

additions quench locally leaving a net spin on the GND. In these cases the magnetism on untreated edge remains intact with little change. In total these results support the hex-radical hypothesis [51]. Namely that the system reacts magnetically as if there were effectively one alternating majority spin per sector.

6.8 Magnetism of Zigzag-Edged Triangulenes

The synthesis of small triangulenes has been well documented [8]. They have an extreme proclivity to polymerize. Their structure, especially their "zero energy" levels, have been the subject of theoretical studies since the invention of MO theory in the 1950s. The zigzag edged triangulenes are alternant PAHs with D_{3h} symmetry and series formulae $C_{m**2+4m+1}$ H_{3m+3} ($m = 2, 3 \ldots$). They are magnetic because the three-fold symmetry axis insures that the sub-lattices are not equivalent. The α-spin sub-lattice has carbon atoms with attached H atoms at the edges and the β-spin sub-lattice has hydrogen atoms only on the apex carbon atoms. The net spin $S = \frac{1}{2}(m-1)$ comes from $(m - 1)$ unpaired electrons occupying a set of NBMOs in a gap above the lower energy spin-paired orbitals. There is a corresponding set of unoccupied NBMOs lying higher. Both NBMO sets fall in the middle of a large gap. Figure 6.12 compares the KS levels of the smallest triangulene tri-2-zg ($m = 2$, one spin) and a large triangulene tri-15-zg ($m = 15$, 14 spins). The KS level order indices of the tri-2zg molecule were matched to that of tri-15zg. The KS energy levels were not altered. The energy difference between the index matched levels is approximately 0.1eV. Ignoring the NBMO and NBMO* levels, the HOMO–LUMO gaps are 4.84 eV and 1.83 eV for tri-2zg and tri-15zg respectively. The gap containing the NBMOs shrinks greatly with size, due to the box confinement size. The gap between the two NBMO manifolds decreases

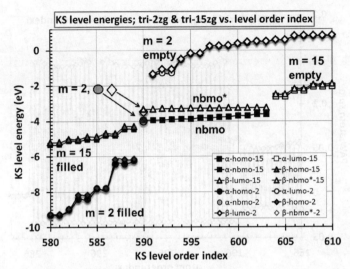

Figure 6.12 Comparison of KS levels of a small triangulene tri-02-zg $C_{13}H_9$ ($m = 2$, one spin) and a large triangulene, tri-15-zg $C_{286}H_{48}$ ($m = 15$, 14 spins). The KS level order indices of the tri-2zg molecule are matched to that of tri-15zg. The KS level energies were not altered. Note the flat band of 14 occupied nbmo and empty nbmo* levels for the large triangulene

with edge length. If these molecules could be configured in a device in a way that does not greatly disturb the isolated molecule level order then, as noted by many, these molecules could function as spin valves for transient electrons [35,42,52]. Electrical leads are a major problem for implementation of this idea.

Figure 6.13 shows a spin picket fence representation of perimeter carbon edge spins for tri-15zg. The coordinate unit (30 atoms) equals the number on each edge. The three carbons C194, C195 and C196 with α-spin 0.06 are the special atoms inside the apexes joining three zigzag rows. Atoms C166 to C193 are part of row $\mathbf{r}_5$. On the edges the PAW α-atomic spin is close to 0.15 and the β-atomic spin a uniform 0.03. The low α-spins (0.12) occur at the edge ends next to the apex. The PAW spin is effectively equal to 0.15 on all edge carbons with a hydrogen atom. Since in these DFT calculation the PAW sphere captured about one half of the total spin one can scale this by factor two to give approximately 0.3. In the large zigzag index limit this number is consistent with total spin divided by the edge atoms carrying the primary spin (molecular spin $= m - 1$, number of primary spin atoms 3m, approximate ratio 0.3). Assuming all the spin density is localized on edge (not apex) carbons with attached hydrogen (α-sub-lattice) then 0.3 is likely the highest spin attainable in any nanographene.

6.9 Chimeric Magnetism

In this section we briefly describe magnetism on the same contiguous hexagonal network that arises by two distinct processes. We call this chimeric magnetism. An alternant carbon network of zigzag edged GNDs with a triangular protrusion can be partitioned into a region where sub-lattices appear as if they are equivalent (parent moiety) and a region where

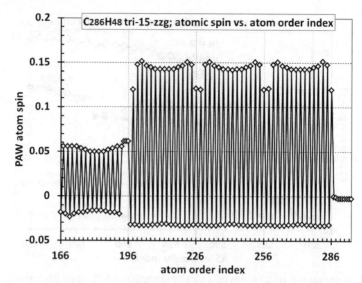

Figure 6.13 *Triangulene tri-15-zg, $C_{286}H_{48}$. Spin picket fence representation of perimeter carbon C197-C286 spins. The ordinate unit (30 atoms) equals the number on each edge. The three carbons C194, C195 and C196 with α-spin 0.06 are the atoms inside the apex joining edge and interior zigzag rows*

they are not (protrusion moiety). The magnetization of the protrusion edges is intrinsic due to non-equivalent sub-lattices and the magnetization of the parent is not. Nevertheless magnetization of parent edges does occur in a regime of reduced coulomb interactions brought on by long edges with zigzag index $m \geq 9$.

Accordingly both types of magnetization may co-exist on the same contiguous network. We show here one example [53] based on the hexangulene $C_{486}H_{54}$ (hx09zg, zigzag index $m = 9$). Figure 6.14 shows the geometry and spin of a parent with a large triangular protrusion that has a base of six (6) rings. As an isolated triangulene the protrusion would support five (5) unpaired spins. The inset panel plots KS level energy (eV) against the level order index on either side of the HOMO–LUMO gap (≈ 0.2 eV). Squares and triangles denote majority α-spin and minority β-spin levels respectively. Filled data points represent occupied and white empty levels. There are six unpaired electrons for a spin $S = 3$ configuration. The protrusion connects to the two bordering edges seamlessly through the one ring high step edge. Green colored "parallel" zigzag C–C bonds (≈ 142.1 pm) form regions where the atomic spins are highest. Red "perpendicular" C–C bonds (≈ 145 pm, the longest in the system) separate perimeter atoms from the next zigzag row. The two edges bordering the protrusion have a spin pattern that is shifted towards the protrusion side and are enhanced (≈ 0.14) about 75% over the three remote edge spin patterns (≈ 0.08). Clearly, the interaction with the triangular protrusion dominates these edges. Partitioning the spin into edge and protrusion moieties is problematic because the spin on the hx09zg parent is weaker than the hx10zg parent as has been described earlier. We note that the HOMO–LUMO gap is reduced over that in the isolated tri-6-zg triangulene suggesting that any use of triangulenes as spin valves will require careful engineering of leads.

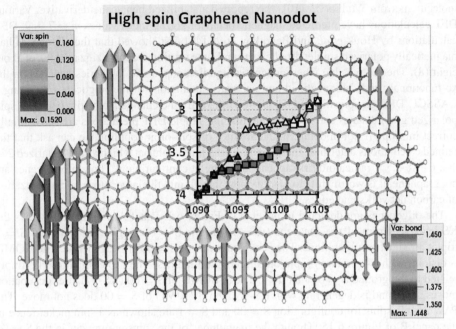

Figure 6.14 *Geometry and spin of a high spin graphene nanodot consisting of a parent hexangulene hx09zg with a large protrusion comprising six zigzag rows each contributing an unpaired spin. The inset panel plots KS level energy (eV) against the level order index on either side of the HOMO–LUMO gap (≈ 0.2 eV). Squares and triangles denote α-spin levels and β-levels. Filled data points are occupied and white are empty levels. See colour version in the colour plate section*

6.10 Magnetism of Oligocenes, Bisanthene-Homologs, Squares and Rectangles

The number of C_6-rings along an extended zigzag edge of a GND with equivalent sublattices needed to support one spin is an important question. Literature estimates vary from index $m = 5$ to values in the 10–15 range. The sensitivity to GND dimensions or GNR width has not been conclusively addressed. In this section we briefly describe DFT results on a variety of nano-graphenes with parallel zigzag edges that do not solve but do help delineate the problem. The width between zigzag edges is measured by n_a the number of C_6 rings measured perpendicular (armchair edge direction) to the zigzag edge. We start with the oligocenes ($n_a = 1$) since historically their magnetization [46] was discovered first and before graphene, and their electronics has been subject to the exacting scrutiny [47,48].

6.10.1 Oligocene Series: $C_{4m+2}H_{2m+4}$ ($n_a = 1; m = 2, 3, 4 \ldots$)

Panel B of Figure 6.1 shows the oligocene series $C_{4m+2}H_{2m+4}$ ($m = 1, 2, 3 \ldots$) a ladder-like chain of C_6-rings with zigzag edges, *cis*-butadiene like apex and no interior. Clar [8] has described partially conjugated derivatives as large as undecacene. The best conjugated molecule is from Wudl *et al.* [10] who reported a stabilized heptacene derivative. Various DFT calculations have predicted magnetic behavior zigzag index values $m = 7 - 9$. DFT calculations by Houk *et al.* and Bendikov *et al.* [46] first showed that the oligocenes had magnetically polarized singlet diradical ground state starting around a zigzag index of about eight (8). The index value where magnetization first appears in the series depends on the xc-functional used. More refined and accurate calculations of the oligocenes by using a CASSCF, DMRG, spin-flip and so on [46–48], verify the change from spin paired to spin polarized ground states and demonstrate that the simple DFT predictions are substantially correct in scope if not in detail. In the absence of spin-orbit coupling we can ask that the refined calculations determine the specific zigzag index value at which a spin-polarized $2-$, $4-$, $6-$, $8-$,... radicaloid first appears in a singlet configuration. Though oligocenes are not graphene, it is worthwhile to identify how they fit into the larger story of magnetization of carbon networks and edges.

The oligocenes are depicted in panel B of Figure 6.1. Panel A, Figure 6.15 shows the KS level spectrum of the decacene ($m = 10$) in the $S = 00$ singlet diradical and $S = 1$ triplet diradical configurations. The occupied levels of $S = 00$ are degenerate. The HOMO is disjoint with α-spin and β-spin on different sub-lattices. The triplet manifold is simply related to the ground $S = 00$ manifold. The α-LUMO of $S = 00$ (KS-97) after occupation moves down and is degenerate with KS-96. The β-LUMO of $S = 00$ does not move. The PAW atomic spins for configuration $S = 00$ and $S = 1$ are shown as a spin picket fence in the panel B of Figure 6.15. Though the magnitudes of the spins are greater in the $S = 00$

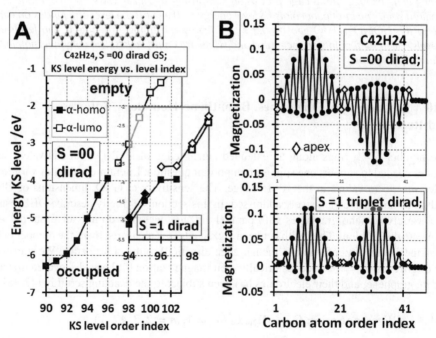

Figure 6.15 KS levels and spin of decacene $C_{42}H_{24}$. Panel A: KS level spectrum of decacene in the $S = 00$ singlet diradical and (inset) the $S = 1$ triplet diradical configuration. Panel B shows the PAW atomic spins for configuration $S = 00$ and $S = 1$ as a spin picket fence

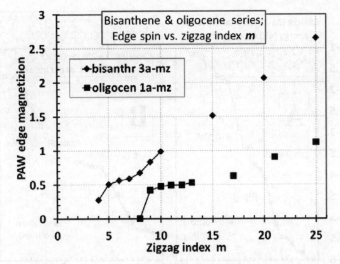

Figure 6.16 *Comparison of single edge PAW atomic magnetization of oligocene (data, -■- 1a-mz, m = 8, 9, 10...) and two rows of bisanthene homologous series (data, -♦- 3a-mz, m = 4, 5, 6...). The value 0.5 on the PAW scale corresponds to approximately unit spin*

configuration the difference in α-spin and β-spin are almost the same, implying that each edge of the S = 00 configuration is equivalent to one unpaired electron, thereby supporting the diradical assignment for the singlet ground state.

Figure 6.16 shows how the PAW atomic spin summed along one edge of a series of oligocenes varies with zigzag index (-■- 1a-mz). The value 0.5 on the PAW scale corresponds to approximately spin unit. Therefore, in the oligocene series we see that magnetization starts sharply at $m = 9$, plateaus through $m = 13$ and then rises roughly one spin unit (ΔPAW = 0.5) between $m = 17$ and 25. In other words the $\Delta m = 8$ zigzag units corresponds to space required to add the second spin on an edge.

6.10.2 Bisanthene Series: $C_{8m+4}H_{2m+8}$ ($n_a = 3; m = 2, 3, 4 ...$)

Stick bond models of two members of the bisanthene homologous series (3a-mz) are shown at the top of Figure 6.17. The series consists of two parallel oligocenes linked across every inside zigzag apex retaining D_{2h} symmetry. Molecules in this and wider ($n_a > 3$) series have been studied by Jiang et al. [39,54,55]. Some of their calculations for the $n_a = 3$ series have been repeated using the parameter set used in all the other DFT and the magnetization results are plotted in Figure 6.16 (-♦- 3a-mz, data) for comparison with the oligocene data. Figure 6.16 shows PAW magnetizations for the bisanthene homologs plotted on the same axes and scales as the oligocenes. We note that the threshold for first spin polarized ground configuration is lower at m = 4 – 5 and the slope is about twice. For the 3a-mz series it appears that five zigzag units supports one spin which is about twice the oligocene result. In other words the 3a-25z homolog is a possible 10-radical supporting five spins per long side.

Figure 6.17 compares the KS levels for bisanthenes with m = 10 and 25 respectively. The shorter system 3a-10z has a lower (HOMO) lip consisting of two non-degenerate

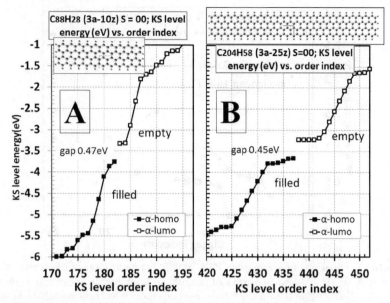

Figure 6.17 Comparison of KS energy levels for bisanthene homologs (3a-10z) and (3a-25z). Panel A, KS levels for (3a-10z) versus level order index. Panel B, KS levels for (3a-25z) versus KS level order index

levels separated by a near degenerate pair bordering the HOMO-LUMO gap (0.47 eV). The longer molecule 3a-25z has plateaus of slowly changing HOMOS and LUMOS bordering a similar gap (0.45 eV). The spin picket fence diagrams in Figure 6.18 look different from oligocene 1a-10z and hexangulene hx10zg. In the latter case the spin pattern of the two molecules has fewer high spin sites. Figure 6.18 shows the approximate linear dependence of edge (side with two rows) spin on length. So if 3a-10z has two spins per edge then 3a-25 has approximately 4–5 spins per side.

Figure 6.19 shows for bisanthene homologs, the energy difference curve Δ(ES-ES00) of the energy of spin configuration S relative to the S = 00, plotted against the zigzag index m. Curves are drawn through data points to guide the eye. All the curves show a "bounce" off the base line. Curve Δ(ES1-ES00) has a bounce m = 6, Δ(ES2-ES00) at m = 10, Δ(ES3-ES00) a bounce at m $\approx$ 17, Δ(ES4-ES00) at m $\approx$ 21, Δ(ES5-ES00) at m $\geq$ 25 and so on. The bounce corresponds to the zigzag index where the spin S configuration is optimized for the length of the molecule. So this provides a surrogate measure of zigzag index of the spin polarized singlet (S = 00) with edge spin $^1/_2$S. The numbers are consistent with the plot of PAW magnetization and the notion that one spin is added for approximately every four or five (4–5) zigzag units. The mechanism is simply the fitting of the high spin configuration into the coulomb box of the molecule. A given high spin configuration starts with high confinement energy that decreases with box length until there is an optimum fit. Thereafter its energy will increase and an even higher spin configuration takes it place. In this way the magnetization of the series starting with small molecules transforms into an extensive variable.

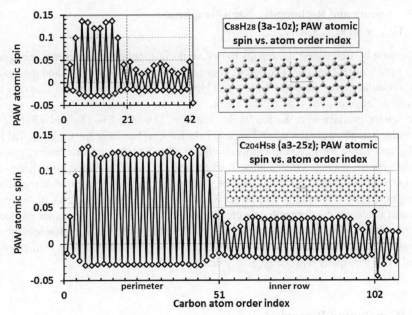

Figure 6.18 Comparison of the spin of bisanthene homologs (3a-10z) and (3a-25z). Top panel, the spin picket fence plotted for molecule (3a-10z) on an edge and penultimate row. Bottom panel, the spin picket fence plotted for (3a-25z) on an edge and penultimate row

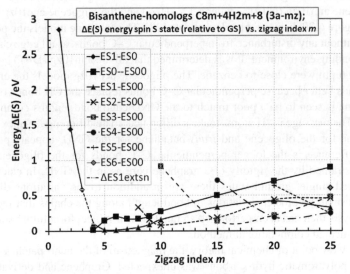

Figure 6.19 High spin configuration bounce of bisanthene homologs. Plot of energy difference $\Delta(ES\text{-}ES00)$ of spin S configuration relative to the singlet $S = 00$ versus zigzag index m. Curves are drawn through data points to guide the eye. Note the curves show a "bounce" off the base line

6.10.3 Square and Rectangular Nano-Graphenes: $C_{8m+4}H_{2m+8}$ ($m = 2, 3, 4 \ldots$)

We conclude by describing the results to two series of graphene patches that are approximately square or rectangular. The patches have low symmetry C_{2v} or C_{1h}, and the stoichiometric formula: $C_{2[(n+1)(m+1)-1]}H_{2(n+m+1)}$. Here $n = n_a$ is the number of C_6-rings along the armchair edge and m is the zigzag index.

The square patches were 8a-8z, 9a-9z, 10a-10z, 11a-11z, 13a-13z and 15a-15z. The rectangular patches were the $n_a = 7$ series: 7a7z for $C_{48}H_{30}$; 7a9z for $C_{158}H_{34}$; 7a17z for $C_{286}H_{50}$ and 7a27z for $C_{446}H_{70}$. Opposite zigzag edge has the opposite majority spin. In all these systems we noted that high spin levels also "crashed" as in the bisanthene series, and a corresponding spin polarized singlet ground configuration appear that replaced the existing one. Lower symmetry and larger widths created a confusing array of low energy levels that could not be unravelled as clearly and the existence of a sequence of bounces like those observed for the bisanthene series could not be confirmed. Nevertheless the crash behaviour of high spin configurations observed for the bisanthenes was confirmed as a general trend for square and rectangular patches.

6.11 Concluding Remarks

We have commented on geometry, magnetism and mention some recent experiments. First, the geometry of zigzag edges: It is of some value to compare singlet spin paired C–C bond lengths for the zigzag edge of the hexangulene hx10zg ($C_{600}H_{60}$) with the similar sized annulene ann10zg ([114]-annulene, $C_{114}H_{114}$) and the oligocene ace10zg (decacene, $C_{42}H_{24}$). Figure 6.20 displays the relevant data [28]. The annulene models the perimeter of the GND without any disturbance from perpendicular C–C bonds, as all edge carbons have the same bonding environment. It was determined that all the ann10zg (DFT) bond lengths C–C = 139.6 pm were close to benzene. The oligocene ace10zg models the environment of edge bonds better because perpendicular C–C bonds replace the CH bonds on one side. The annulene is seen to be a poor match to the GND edge bond lengths compared to the oligocene. The message is "H atoms make a difference". The difference in apex structures, *cis*-butadiene for the oligocene and *trans*-butadiene for the GND, appears perturbative. A major difference is the longer perpendicular C–C bonds for the oligocene as these are not constrained by the rigidity of a graphene core as in the GND. In calculations of spin-polarized singlet ground states there are in addition to the variations displayed in Figure 6.20 small picometer variations in bond length along the chains that require more careful analysis than has been performed so far. This is a subject for a future study using a more sophisticated computational method.

From the view point of chemical theory, the magnetism of the nano-patches considered as compact polyaromatic hydrocarbons was unexpected. Graphene and derivatives add a new category of magnetic organic material. Previously the magnetism in organics was dominated by linear polymeric poly-radical systems most of which lack the two-dimensional conjugation of the graphene network. Finally, some examples pertinent to the synthesis of structures with zigzag edges and GNDs: a review containing detailed images of graphene edges [56]; the synthetic patterning of hexagonal holes and triangular GNDs with zigzag edges [57]; and a close look at the C–C bonds of the smallest armchair "graphene", the

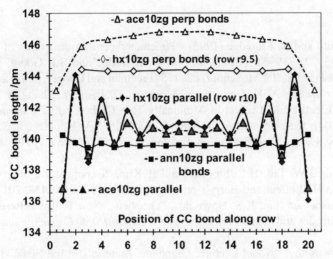

Figure 6.20 Comparison of the DFT calculated bond lengths for singlet spin paired state perimeters of hx10zg ($C_{600}H_{60}$), ace10zg (decacene $C_{42}H_{24}$) and ann10zg ([114]-annulene $C_{114}H_{114}$). Note the similarity of edge parallel C–C bonds for hexangulene and the acene but not the annulene. H atoms make a difference. The longest C–C bonds occur in ace10zg. The most uniform set of parallel CC bonds belong to the annulene. Adapted with permission from [28] © 2009 American Institute of Physics

fully benzenoid hydrocarbon s-hexabenzo-coronene ($C_{42}H_{18}$), using a CO functionalized AFM tip [58].

This article has reviewed aspects of the geometric and magnetic structures predicted for nanographenes using a modern *ab initio* DFT method. It is known that geometry is well predicted by the DFT but not so for HOMO-LUMO gaps and similar properties. So though we expect progress predicting edge geometry and properties resulting from addition or substitution reactions [59] there is no reason to believe the same for reaction paths and mechanisms. Whether the study of graphene reactivity will draw any benefit as benzene did from the electron pair gospel and mesomerism theory [60] remains to be seen.

Acknowledgment

I express sincere thanks to: Professor Y. Kawazoe for hospitality and financial support at the Center for Computational Materials Science during many visits to Japan; the IMR supercomputer staff for friendly and efficient help; the Japan Society for the Promotion of Science (JSPS) for long term fellowships during the period 2004–2012; Professor W.J. Lester Jr. for hospitality and support at the Kenneth S. Pitzer Center for Theoretical Chemistry, Department of Chemistry, University of California Berkeley where much of this report was researched and written.

Note added in proof. Recent publications by Deleuze *et al.* [61] have shown that the magnetism of zigzag edges on equivalent alternant PAH bilattices is due to spin contamination and the multi-radical ground state configurations are spurious.

References

[1] S.J. Gould and N. Eldredge (1993), "Punctuated equilibrium comes of age", *Nature* **366**, 223–227 (1993); original theory: N. Eldredge and S.J. Gould, in *Models in Paleobiology* (ed. T.J.M. Schopf) 82–115 (Freeman and Cooper, San Francisco 1972).

[2] K.S. Novoselov, A. K. Geim, S. V. Morozov, D. Jiang, Y. Zhang, S. V. Dubonos, et al., "Electric Field Effect in Atomically Thin Carbon Films", *Science* **306**, 666–669 (2004); K.S. Novoselov, D. Jiang, F. Schedlin, V.V. Khotkevich, S. V. Morozov and A.K. Geim, "Two dimensional atomic crystals", *PNAS*(US) **102**, 10451–10453 (2005).

[3] Y. Zhang, J.W. Tan, H.L. Stormer, and P. Kim, "Experimental observation of the quantum Hall effect and Berry's phase in graphene", *Nature* **438**, 201–204 (2005); M.I. Katsnelson and K.S. Novoselov, "Graphene: New bridge between condensed matter physics and quantum electrodynamics", *Solid State Communications* **143**, 3–13 (2007).

[4] K.S. Novoselov, "Nobel Lecture: Graphene: Materials in the Flatland", *Rev. Mod. Phys.* **83**, 837–848 (2011); A.K. Geim, "Nobel Lecture: Random walk to graphene", *Rev. Mod. Phys.* **83**, 851–862 (2011).

[5] E.H. Lieb, "Two theorems on the Hubbard model", *Phys. Rev. Lett.* **62**, 1201–1204 (1989).

[6] J.N. Murrell, *Theory of the Electronic Spectra of Organic Molecules* (Methuen, London 1963); L. Salem: *The Molecular Orbital Theory of Conjugated Systems* (Benjamin, New York, 1966).

[7] M.R. Philpott, F. Cimpoesu and Y. Kawazoe, "Geometry, bonding and magnetism in planar triangulene graphene molecules with D_{3h} symmetry: zigzag $C_{m**2+4m+1}H_{3m+3}$ (m = 2, .. 15)", *Chem. Phys.* **354**, 1–15 (2008) Figure 9.

[8] E. Clar: *Polycyclic Hydrocarbons*, Volume 1 and 2 (Academic Press, London, 1964); E. Clar, *The Aromatic Sextet*, (John Wiley & Sons, Inc., New York, 1972).

[9] J.C. Fetzer, *Large (C≥24) Polycyclic Aromatic Hydrocarbons: Chemistry and Analysis* (John Wiley & Sons, Inc., New York, 2000).

[10] D. Chun, Y. Cheng, and F. Wudl, "The most stable and fully characterized functionalized heptacene", *Angew. Chem. Inter. Ed.* **47**(44), 8380–8385 (2008).

[11] B. Beden, C. Lamy, A. Bewick and K. Kunimatsu, "Electrosorption of methanol on a platinum electrode. IR spectroscopic evidence for adsorbed CO species", *J. Electroanal. Chem.* **121**, 343—347 (1981).

[12] H. Wang, "Formation of nascent soot and other condensed-phase materials in flames", *Proc. Combust. Inst.* **33** 41–67 (2011); M. Frenklach and J. Ping, "On the role of surface migration in the growth and structure of graphene layers", *Carbon* **42**, 1209-1211 (2004); M. Frenklach, C.A. Schuetz, and J. Ping, "Migration Mechanism of Aromatic-edge Growth", *Proc. Combust. Inst.* **30**, 1389–1396 (2005); R. Whitesides, A.C. Kollias, D. Domin, W.A. Lester Jr. and M.Frenklach, "Graphene layer growth: collision of migrating five-member rings", *Proc. Combustion Inst.* **31**, 539–546 (2007).

[13] M. Müller, C. Kübel, and K. Müllen, "Giant polycyclic aromatic hydrocarbons", *Chem. Eur. J.*, **4**, 2099–2109 (1998).

[14] C.D. Simpson, J.D. Brand, A.J. Berresheim, L. Przybilla, H.J. Raeder and K. Müllen, "Synthesis of a giant 222 carbon graphite sheet", *Chem. Eur. J.* **8**, 1424–1429 (2002).

[15] K. Müllen and J.P. Rabe, "Nanographenes as active components of single molecule electronics and how a scanning tunneling microscope puts them to work", *Acc. Chem. Res.* **41**, 511–520 (2008).

[16] J. Cai, P. Ruffieux, R. Jaafar, M. Bieri, T. Braun, S. Blankenburg, et al., "Atomically precise bottom-up fabrication of graphene nanoribbons", *Nature* **466**, 470–473 (2010).

[17] M. Treier, C.A. Pignedoli, T. Laino, R. Rieger, K. Muellen, D. Passerone and R. Fasel, "Surface-assisted cyclodehydrogenation provides a synthetic route towards easily processable and chemically tailored nanographenes", *Nature* **3**, 61–67 (2011).

[18] C.O. Girit, J.C. Meyer, R. Erni, M.D. Rossell, C. Kisiellowski, L. Yang, et al., "Graphene at the edge: stability and dynamics", *Science* **323**, 1705–1708 (2009).

[19] VASP web site url: (www.vasp.at); manual (cms.mpi.univie.ac.at/vasp/vasp.html); recent workshop: (cms.mpi.univie.ac.at/vasp-workshop/slides/documentation.htm) Accessed April 30, 2013.

[20] P.E. Blöchl, "Projector augmented-wave method", *Phys. Rev. B* **50** 17953–17979 (1994); G. Kresse and J. Joubert, "From ultrasoft pseudopotentials to the projector augmented-wave method", *Phys. Rev. B* **59**, 1758–1775 (1999).

[21] J.P. Perdew, Unified theory of exchange and correlation beyond the local density approximation, in *Electronic Structure of Solids 1991*, (edS) P. Ziesche and H. Eschrig (Akademie-Verlag, Berlin, 1991); J.P. Perdew, J.A. Chevary, S.H. Vosko, K.A. Jackson, M.R. Pederson, D.J. Singh, and C. Fiolhais, *Phys. Rev. B* **46**, 6671–6687 (1992).

[22] Vaspview data viewer (http://vaspview.sourceforge.net, accessed April 30, 2013); GaussView 5 (www.gussian.com/g_prod/gv5.htm); VisIt at NERSC (http://www.nersc.gov/users/software/vis-analytics/visit/) accessed May 14, 2013.

[23] M.S. Schuurman, C. Pak and H. F. Schaefer, "What to do about unpaired electrons? A hydrocarbon hexaradical with three CIoss diradicals linked by 1,3,5-trimethylbenzene as ferromagnetic coupler", *J. Chem. Phys.* **117**, 7147–7152 (2002).

[24] H.A. Jahn and E. Teller, *Proc. Roy. Soc. (Lon)* **161A**, 220 (1937); W. Moffit and A.D. Liehr, *Phys. Rev.* **106**, 1195(1957); G. Herzberg, *Molecular Spectra and Molecular Structure III. Electronic Spectra and Electronic Structure of Polyatomic Molecules*, (Van Nostrand, Princeton, New Jersey 1967); T. Kato, K. Yoshihara and K. Hitao, "Electron–phonon coupling in negatively charged acene- and phenanthrene-edge-type hydrocarbon crystals" *J. Chem. Phys.* **116**, 3420–3429 (2002).

[25] S.E. Stein and R.L. Brown, "π-Electron properties of large condensed polyaromatic hydrocarbons", *J. Am. Chem. Soc.* **109**, 3721–3729 (1987).

[26] M.R. Philpott, F. Cimpoesu and Y. Kawazoe, "Bonding and magnetism in high symmetry nano-sized graphene molecules: linear acenes $C_{4m+2}H_{3m+3}$ (m = 2,... 25); zigzag hexangulenes $C_{6m**2}H_{6m}$ (m = 2,... 10); crenelated hexangulenes $C_{6(3m**2-3m+1)}H_{6(2m-1)}$ (m = 2,... 6); triangulenes $C_{m**2+4m+1}H_{6m}$ (m = 1, 2,... 15)", *Mat. Trans (Japan Instit. of Metals)* **49**, 2448–456 (2008).

[27] M.R. Philpott and Y. Kawazoe, "Edge versus interior in the chemical bonding of graphene materials", *Phys. Rev. B* **79**, 233303–233304 (2009).

[28] M.R. Philpott and Y. Kawazoe, "Bonding and magnetism in nanosized graphene molecules: Singlet states of zigzag edged hexangulenes", *J. Chem. Phys.* **131**, 214706–214712 (2009).

[29] M.R. Philpott and Y. Kawazoe, "Geometry and bonding in the ground and lowest triplet state of D_{6h} symmetric crenellated edged $C_{6[3m(m-1)+1]}H_{6(2m-1)}$ (m = 2,.. 6) graphene hydrocarbon molecules", *Chem. Phys.* **358**, 85–95 (2009).

[30] J.W. Armitt and R. Robinson, "Polynuclear hetrocyclic aromatic types. Part II. Some anhydronium bases." *J. Chem. Soc.* (London) Trans., 1925, 1604–1618.

[31] D.P. Craig, "Polar structures in the theory of conjugated molecules III. The energy levels of benzene.", *Proc. Roy. Soc.* (London) **A200**, 401–409 (1949).

[32] Y. Heejun, A.J. Mayne, M. Boucherit, G. Comtet, G. Dujardin and Y. Kuk, "Quantum interference channeling at graphene edges", *Nano Lett.* **10**, 943–947 (2010).

[33] T. Wassmann, A.P. Seitsonen, A.M. Saitta, M. Lazzeri and F. Mauri, "Clar's theory, π-electron distribution, and geometry of graphene nanoribbons", *J. Am. Chem. Soc.* **132**, 3440–3451 (2010).

[34] T. Wassmann, A.P. Seitsonen, A.M. Saitta, M. Lazzeri and F. Mauri, "The thermodynamic stability and simulated STM images of graphene nanoribbons", *physica status solidi(b)* **246**, 2586–2591 (2009).

[35] M.R. Philpott, S. Vukovic, Y. Kawazoe and W.A. Lester Jr., "Edge versus interior in the chemical bonding and magnetism of zigzag edged triangular graphene molecules", *J. Chem. Phys.* **133**, 44708–44708 (2010).

[36] M.R. Philpott and Y. Kawazoe, "Vestigial Clar fully benzenoid rings in armchair graphene nanodots", (unpublished calculations, July 2012).

[37] T.L. Makarova and F. Palacio, (eds) *Carbon-based Magnetism* (Elsevier, Amsterdam, 2006).

[38] J. Fernandez-Rossier, J.J. Palacio, "Magnetism in graphene nanoislands", *Phys. Rev. Lett.* **99**, 177204–177204 (2007).

[39] D.-E. Jiang, B.G. Sumpter, and S. Dai, First principles study of magnetism in nanographenes", *J. Chem. Phys.* **127**, 124703–124705 (2007).

[40] O. Hod, V. Barone, and G.E. Scuseria, "Half-metallic graphene nanodots: A comprehensive first-principles theoretical study", *Phys. Rev.* **B 77**, 35411–35416 (2008).

[41] W.L. Wang, S. Meng and E. Kaxiras, "Graphene nanoflakes with large spin", *Nano Lett.* **8**, 241–245 (2008).

[42] W.L. Wang, O.V. Yazyev, S. Meng, and E. Kaxiras, "Topological frustration in graphene nanoflakes: magnetic order and spin logic devices", *Phys. Rev. Lett.* **102**, 157201–157204 (2009).

[43] J.H. Wang, D.Y. Zubarev, M.R. Philpott, S. Vukovic, W.A. Lester Jr., T. Cui, and Y. Kawazoe, "Onset of diradical character in small nanosized graphene patches", *Phys. Chem. Chem. Phys.* **12**(33), 9839–9844 (2010).

[44] Y.-W. Son, M.L. Cohen and S.G. Louie, "Half-metallic graphene nanoribbons", *Nature* **444**, 347–349 (2006).

[45] V. Barone, O. Hod, and G.E. Scuseria, "Electronic structure of semiconducting graphene nanoribbons", *Nano Lett.* **6**, 2748–2754 (2006).

[46] K.N. Houk, P.S. Lee and M. Nendel, *J. Org. Chem.* **66**, 5107 (2001); M. Bendikov, H.M. Duong, K. Starkey, K.N. Houk, E.A. Carter and F. Wudl, "Oligoacenes: Theoretical prediction of open-shell singlet diradical ground states", *J. Amer. Chem. Soc. Comm.* **126** 7416–7417 (2004).

[47] J. Hachmann, J.J. Dorando, M. Aviles and G.K.-L. Chan, "The radical character of the acenes: A density matrix renormalization group study", *J. Chem. Phys.* **127**, 134309–134309 (2007).
[48] D. Casanova and M. Head-Gordon, "Restricted active space spin-flip configuration interaction approach: theory, implementation and examples", *Phys. Chem. Chem. Phys.* **11**, 9779–9790 (2009).
[49] M.R. Philpott, Prabhat and Y. Kawazoe, "Magnetism and bonding in Graphene Nanodots with H modified Interior, Edges and Apex", *J. Chem. Phys.* **135**, 84707–84711 (2011).
[50] M.R. Philpott and Y. Kawazoe, "Graphene nanodots with intrinsically magnetic protrusions", *J. Chem. Phys.* **136**, 64706–64708 (2012).
[51] M.R. Philpott and Y. Kawazoe, "Triplet States of Zigzag Edged Hexagonal Graphene Molecules $C_{6m**2}H_{6m}$ ($m = 2, 3, \ldots 10$) and Carbon Based Magnetism", *J. Chem. Phys.* **134**, 124706–124709 (2011).
[52] H. Sahin, R.T. Senger, and S. Ciraci, "Spintronic properties of zigzag-edged triangular graphene flakes", *J. Appl. Phys.* **108**, 074301 (2010).
[53] M.R. Philpott and Y. Kawazoe, Unpublished calculations of chimeric magnetic systems.
[54] D.-E. Jiang and S. Dai, "Electronic ground state of higher acenes.", *J. Phys. Chem. A* **112**, 332 (2008).
[55] D.-E. Jiang, X.-Q. Chen, W. Luo, and W.A. Shelton, "From transpolyaceylene to zigzag edged graphene nanoribbons", *Chem. Phys. Lett.* **483**, 120–123 (2009).
[56] M. Acik and Y.J. Chabal, "Nature of graphene edges: A review", *Japanese J. Appl. Phys.* **50**, 07010-16 (2011).
[57] Z. Shi, R. Yang, L. Zhang, Y. Wang, D. Liu, D. Shi, *et al.*, "Patterning graphene with zigzag edges by self-aligned anisotropic etching", *Adv. Mater.* **20**, 1–5 (2011).
[58] L. Gross, F. Mohn, N. Moll, B. Schuler, A. Criado, E. Guitián, *et al.*, "Bond-order discrimination by atomic force microscopy", *Science* **337**, 1326–1329 (2012).
[59] M.R. Philpott and Y. Kawazoe, "Magnetism and structure of graphene nanodots with interiors chemically modified by boron, nitrogen and charge", *J. Chem. Phys.* **137**, 054715-10 (2012).
[60] C.K. Ingold, *Structure and Mechanism in Organic Chemistry*, (Bell and Sons, London 1953).
[61] B. Hajgató and M. S. Deleuze, "Quenching of magnetism in hexagonal graphene nanoflakes by non-local electron correlation" *Chem. Phys. Lett.* **553**, 6–10 (2012).

7
Porous Graphene and Nanomeshes

Yan Jiao,[a,b] Marlies Hankel,[a] Aijun Du,[a] and Sean C. Smith[c]

[a] Centre for Computational Molecular Science, Australian Institute for Bioengineering and Nanotechnology, The University of Queensland, Australia
[b] School of Chemical Engineering, The University of Queensland, Australia
[c] Center for Nanophase Materials Sciences, Oak Ridge National Laboratory, USA

7.1 Introduction

The goal of this chapter is to briefly introduce concepts relating to the design and modeling of single atomic layer carbon membranes for highly efficient gas separation. The inspiration for this work is the active development in recent years of membrane technologies for gas separation characterized by facile operation, low energy consumption, and easy maintenance [1]. In principle, microporous materials can be used as membranes to separate a mixture of molecules based either on their size, shape, or differences in chemical affinity. For example the purification of hydrogen from CH_4 is typically based on the idea that CH_4 would be physically excluded from a sufficiently narrow pore while H_2 could freely pass through resulting in selective permeation of H_2. Another rather more challenging application of such "molecular sieving" is the separation of different isotopes of molecular hydrogen. In a classical description, different isotopic species of the same molecule have identical size, shape, and adsorption properties. However, in "quantum sieving" the separation is not based on size but rather on the differences between hindered rotational (or, vibrational) quantum energy levels of the lighter and heavier molecules when confined in a very narrow pore.

Since the permeance of a membrane is inversely proportional to its thickness [2], a single atomic layer membrane may potentially enable very high transmission flux. Several monolayer porous structures have been proposed previously and studied toward their potential as

---

Graphene Chemistry: Theoretical Perspectives, First Edition. Edited by De-en Jiang and Zhongfang Chen.
© 2013 John Wiley & Sons, Ltd. Published 2013 by John Wiley & Sons, Ltd.

gas separation membranes. This chapter provides an overview of recent work in this area and considers the evaluation of performance in gas separation using transition state theory (TST) [3]. Following the introduction, a summary is provided of experimentally available single atomic layer carbon membranes with their pros and cons. The second section outlines TST implementations and related aspects germane to such membrane applications (i.e., 4D molecular calculation and density functional theory calculation). The third part of this chapter illustrates three cases utilizing TST to derive the diffusion rate and selectivity of several nanomesh membranes based on our recent publications. The last part is dedicated to a conclusion including how to improve the performance and experimental realization of these ultrathin membranes in gas separation. We have aimed to make this chapter largely self-explanatory, with all the models and analysis methods given in sufficient detail to reproduce the calculations.

7.1.1 Graphene-Based Nanomeshes

Carbon atoms can form several distinct types of valence bonds by different hybridization [4]; hence single layer carbon assumes many structural forms. The first group of porous single atomic layer carbon membranes could be described as graphene-based nanomeshes whose synthetic processes all include as a first step manufacture of graphene. Graphene without defects is impermeable to gases as small as Helium with an 11.67 eV diffusion barrier [5, 6]; therefore to realize the goal of gas permeation and separation, pores need to be created in the graphene. The earliest proposal from such a perspective was a nitrogen-functionalized porous graphene [7] which was based on graphene and could in principle be prepared by traditional electron beam lithography [8]. The model was constructed by removing several rings from the graphene sheet and passivating unsaturated carbon atoms with hydrogen atoms and also replacing some of the carbon atoms by nitrogen atoms. One-sided decoration of such an ultrathin membrane with metal was subsequently proposed to create an asymmetric pathway to facilitate gas passage [9]. Other similar graphenes with designed pores were proposed for H_2/N_2 separation [10] and for He isotope separation [11, 12]. It should be noted that due to the restrictions of current technology, the precise manufacture of such structured graphene-based networks cannot be guaranteed.

Similar graphene nanomesh structures based on such a route have been successfully synthesized with variable periodicities from 27–39 nm [13]. The nanomeshes are prepared by punching a high-density array of nanoscale holes into a single or a few layers of graphene using self-assembled block copolymer thin films as the mask template. The precise morphology control of this structure could be guaranteed, however, the pore size is several tens of Angstroms, which is too large for small gas molecule separation. The molecular structures of the aforementioned graphene based nanomeshes are illustrated in Figure 7.1.

7.1.2 Graphene-Like Polymers

The first group of graphene-based nanomeshes for gas separation summarized earlier suffers from some intrinsic deficiencies: they are either difficult to realize with precise morphology control or the pore size is too large to enable selectivity in small gas molecule diffusion. Accompanying the exploratory wave of graphene research since its successful isolation [14], other single-atomic-layer carbon structures that could realize precise morphology

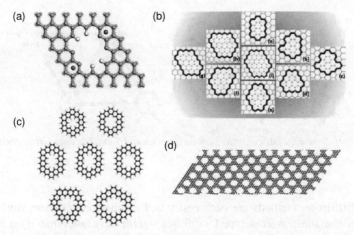

Figure 7.1 Structure of (a) nitrogen functionalized porous graphene with titanium. Dark orange: carbon, blue: nitrogen, white: hydrogen, pink: titanium. Reprinted with permission from [9] © 2012 American Chemical Society. (b) Graphene with designed pores for H_2/N_2 separation. Yellow hexagons represents graphene, the pore shape is colored red, and the removed carbon atoms are shown by transparent gray balls. For more detail please refer to Ref. [10]. Reprinted with permission from [10] © 2011 American Chemical Society. (c) Graphene with designed pore size for He isotope separation. Color code is same as (a). Reprinted with permission from [11] © 2012 American Chemical Society. (d) Free standing graphene nanomesh by block copolymer lithography. Gray: carbon. Reprinted with permission from [13] © 2010 Nature Publishing Group. See colour version in the colour plate section

control have also been proposed and synthesized [4,15–17]. This second group may be summarized as graphene-like polymers. Bieri *et al.* have for the first time succeeded in synthesizing a well-defined porous graphene [15], which was subsequently proposed for hydrogen purification [18, 19] and for extremely high H_2 selectivity with respect to other gases such as CH_4. Another work by quantum chemistry on a precise level showed that this type of porous graphene allows a very high selective separation for helium [20]. Subsequently graphdiyne, a member in the large family of single carbon layer allotropes, was successfully fabricated via a cross-linking reaction on a copper surface [21]. The framework of graphdiyne is constructed by hexagonal carbon rings cross-linked by diacetylene. The size of van der Waals (vdW) pores defined by the framework is in between that of hydrogen and methane/carbon monoxide. This makes graphdiyne potentially utilizable as a separation membrane [22] for hydrogen purification from syngas – a product of steam-methane reforming by partial oxidation of methane [23] that contains CH_4 and CO as undesirable gas molecules [24]. The molecular structures of these two polymeric structures are illustrated in Figure 7.2.

7.1.3 Other Relevant Subjects

7.1.3.1 Isotope Separation

The separation of isotopic mixtures, such as D_2 from H_2 and ^{3}He from ^{4}He, is a difficult and energy intensive process requiring special techniques. Existing centrifugation and

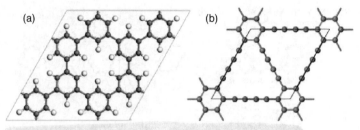

Figure 7.2 *Structure and lattice of (a) porous graphene (b) graphdiyne. Gray: carbon; white: hydrogen*

cryogenic distillation methods are both costly and cumbersome. Recent works [25–28] demonstrated that the heavier isotope D_2 diffuses significantly faster than H_2 at sufficiently low temperature in narrow pore nanomaterials, raising the possibility of kinetic separation as a competitive option for isotope separation. This concept was first proposed by Beenakker et al. [29], and its intriguing effects have since been the focus of several theoretical [25, 26,29–45] and experimental [27,46–48] studies. Molecular dynamics [25] and experimental [48] studies confirmed the effect of quantum sieving in isotope separation. Bhatia and coworkers [25] investigated the kinetic molecular sieving approach for isotope separation based on materials such as zeolites and carbon molecular sieves (CMS). The study showed that high transport selectivity for D_2 (exceeding 20) can be achieved at low temperatures (30–60 K) even when the equilibrium selectivity is significantly lower, opening the door for practical isotope separation using membranes. Indeed, independent support for this approach has subsequently been obtained by Chu et al. [28] and Zhao et al. [27]. Their results demonstrating faster desorption of D_2 compared to H_2 at 77 K are consistent with the prediction of ref. [39] of faster diffusion of D_2 compared to H_2 below 94 K in zeolite of essentially the same critical pore size.

Several studies by quantum tunneling on low temperature $^3He/^4He$ separation, which responds to the increasing need of fermionic ^3He in ultracold physics research, neutron detection or medical applications such as lung tomography, however reveals different phenomenon where the lighter isotope ^3He diffuse quicker through porous graphenes [11,12,20,49]. Due to the shape of the kinetic energy distribution of fermionic and bosonic Helium, a higher ^3He tunneling probability for kinetic energies below the barrier is obtained, and the transmission of the lighter isotope is preferred. Such situation is interpreted as a pressure-driven process using differences in the "quantum tunneling" properties, while the former approach is a temperature-driven process using the concept of "quantum sieving" as suggested by Ref. [11]. In this chapter, we demonstrate by transition state theory that the mono carbon layer porous structures exhibit excellent performance in thermally driven isotope separation.

7.1.3.2 Van der Waals Correction for Density Functional Theory

DFT suffers currently from inadequate incorporation of long-range dispersion, or van der Waals (vdW) interactions [50]. vdW-corrected DFT was tested to give a reasonable physisorption energy and equilibrium distance to be obtained for H_2 on several carbon structures including a graphite surface and a nanotube bundle. The vdW correction is a very effective method for implementing DFT calculations, allowing a reliable description of both short-range chemical bonding and long-range dispersive interactions.

The vdW correction method adopted in this chapter is the D2 set [51] of Grimme's schemes [51–54] as this one is readily incorporated in Dmol³. The dispersion correction for systems containing only carbon and hydrogen using the PBE [55] functional within such scheme is briefly introduced here as follows:

$$E_{disp} = -0.75 \sum_{i=1}^{N_{at}-1} \sum_{j=i+1}^{N_{at}} \frac{C_6^{ij}}{R_{ij}^6} f_{dmp}(R_{ij}) \quad (7.1)$$

where N_{at} is the number of atoms in the system, 0.75 is a global scaling factor that depends on the density functional. R_{ij} is the interatomic distance. C_6^{ij} is the dispersion coefficient for atom pair ij; for C_6(C-C) the value is 1.75 Jnm⁶mol⁻¹, C_6(C-H) is 0.495 Jnm⁶mol⁻¹, and C_6(H-H) is 0.14 Jnm⁶mol⁻¹: f_{dmp} is the damping function used to avoid near-singularities for small R_{ij} and is calculated to be:

$$f_{dmp}(R_{ij}) = \frac{1}{1 + e^{-d(R_{ij}/R_r - 1)}} \quad (7.2)$$

where R_r(C-C) = 2.904 Å, R_r(C-H) = 2.453 Å, R_r(H-H) = 2.002 Å.

The total energy of a system is then obtained by adding the E_{disp} component to the energy value derived by regular density functional theory calculation.

7.1.3.3 Potential Energy Surfaces for Hindered Molecular Motions Within the Narrow Pores

As will be seen later, to attempt a reasonable description of the kinetics of passage of H_2/D_2 through a narrow pore with the transition state formalism, we have implemented the calculation of partition functions for the coupled rotational and translational degrees of freedom of the molecule in the plane of the pore. The need to solve a coupled four-dimensional Schrödinger equation for these motions stems in principle from the fact that the spatial confinement effectively causes both hindering and coupling of the two rotational degrees of freedom of the diatomic molecule and its two translational degrees of freedom perpendicular to the axis of the pore. The potential energy of such a system could in principle be computed directly using the van der Waals-corrected DFT approach summarized previously. However, this becomes prohibitively expensive at the exploratory stages when many different systems are being trialed and a full potential energy surface would in principle be needed for each different system. In light of this, we have pursued two separate strategies: (1) for the full coupled four-dimensional partition function calculations we use Lennard–Jones pairwise interactions to generate the necessary potential energy surfaces; and (2) we also consider circumventing the four-dimensional calculations by using an approximate harmonic separation of the hindered motions of the molecules in the pore. This latter approach allows us to use simple harmonic partition functions with frequencies derived from the properties of the saddle-point computed via van der Waals-corrected DFT.

Where van der Waals pairwise interactions are used, these are taken to be a sum of empirical H−C and H−H potentials for each hydrogen atom (or deuterium) with each atom in the porous graphene structure. We use a Lennard–Jones (LJ) 6-12 interaction potential in which the interactions of the hydrogens with atoms of the porous graphene structures are given by:

$$V(r) = 4\varepsilon \left(\frac{\sigma^{12}}{r^{12}} - \frac{\sigma^6}{r^6} \right) \qquad (7.3)$$

In our study we employ structures containing hydrogen and carbon atoms. The parameters for the C—H interactions are chosen as $\varepsilon = 19.2$ cm^{-1} and $\sigma = 3.08$ Å [45] and $\varepsilon = 6.99$ cm^{-1} and $\sigma = 2.0$ Å for H—H and $\varepsilon = 19.94$ cm^{-1} and $\sigma = 2.75$ Å for N—H which we have taken from the AutoDock force fields [56].

7.2 Transition State Theory

Calculating the exact dynamics to get an exact diffusion rate of gas permeation through the aforementioned single layer structures is a very large computational job, hence we turn to alternative methods that could give a reasonably accurate prediction of the diffusion rate and hence the selectivity. Transition state theory [3] is well established as a cost effective approach to computing reaction rates and can also be readily applied in the present context. Consequently we use TST to describe the gas transmission behavior, and the following section is devoted to briefly introduce the theory in the framework of gas transmission.

7.2.1 A Brief Introduction of the Idea

For gas permeation through a membrane, if we neglect the role of surface adsorption, the transportation of gases through the membrane reduces to the analysis of the potential barrier a single molecule has to overcome when passing through the pore center. Due to the constraint of the pore size, the highest position on the energy profile of gas transmission across the single layer nanomesh usually is when it is right inside – or close to – the pore center (some exceptions are discussed with an example in Section 7.3.2). We take this highest position as the transition state (TS) – the "point of no return" in the TS approximation. As in Figure 7.3, this barrier represents a "rate determining" position on the potential surface, that is, a bottleneck area through which the flux of gas molecules passing is a minimum.

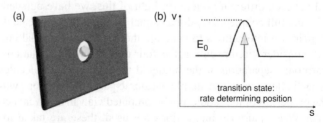

Figure 7.3 (a) Gas molecule in the pore center as the transition state; (b) schematic energy profile and the transition state in gas transmission through a single atomic layer membrane

In order to calculate the transmission flux we simply need to calculate the relative *concentration* of the gas species at the transition state (on top of the barrier) and multiply this by the average velocity of motion along the reaction coordinate (i.e., over the barrier).

$$\text{Flux of gas molecules through a membrane} = \text{concentration} \times <\text{velocity}> \quad (7.4)$$

Expressing energies relative to that of the well-separated gas molecule on the incoming side of the nanomesh, and borrowing the standard thermal probability of any given state from equilibrium statistical mechanics, the relative concentration of species at the transition state can be obtained from the partition function of the transition state Q^+:

$$\text{Relative concentration at transition state} = e^{-E_o/k_B T} Q^+ \quad (7.5)$$

where k_B is the Boltzmann constant, T the temperature and E_0 the energy difference between the well-separated gas molecule + mesh system and the transition state when the molecule is in the pore. This implies that Q^+ is referenced to the local minimum energy of the transition state, and this E_o is the energy difference between transition state and ground state.

Noting that the potential on top of the barrier is locally flat along the reaction coordinate and that this motion is unbound, we can approximate this motion as a one-dimensional translation of a particle with a characteristic reduced mass for the reaction coordinate, μ_s. It is convenient to write this one-dimensional translational partition function explicitly so we rewrite the Equation 7.5 as:

$$\text{Relative concentration at transition state} = e^{-E_o/k_B T} \times \frac{1}{2}\left(\frac{2\pi \mu_s k_B T}{h^2}\right)^{\frac{1}{2}} \times Q^{TS} \quad (7.6)$$

where h is the Planck constant. So, Q^{TS} is the partition function at the *TS* for all its motions *except* the pseudo-translational motion along the reaction coordinate. Again appealing to the locally 'flat' and unbound nature of the reaction coordinate potential energy profile at the barrier, we approximate the motion along the reaction coordinate at the barrier as a one-dimensional translational motion. Hence, the average velocity $\langle v_s \rangle = \langle p_s/\mu_s \rangle$ is given by:

$$\langle v_s \rangle = \frac{\int_0^\infty dp_s (p_s/\mu_s) e^{-p_s^2/2\mu_s k_B T}}{\int_0^\infty dp_s e^{-p_s^2/2\mu_s k_B T}} = 2\left(\frac{k_B T}{2\pi \mu_s}\right)^{1/2} \quad (7.7)$$

We can now evaluate the flux of species through the transition state by multiplying the relative concentration at the transition state by average velocity along the reaction coordinate:

$$\text{Flux through the pore center} = 2\left(\frac{k_B T}{2\pi \mu_s}\right)^{1/2} \times e^{-E_o/k_B T} \times \frac{1}{2}\left(\frac{2\pi \mu_s k_B T}{h^2}\right)^{1/2} \times Q^{TS}$$

$$= \frac{k_B T}{h} Q^{TS} e^{-E_o/k_B T} \quad (7.8)$$

Dividing this flux by the thermal probability of the asymptotic 'reactant' state (the partition function, Q^{react}) leads to the standard Transition State Theory expression for the rate coefficient of gas transmission across porous nanomesh:

$$k = \frac{k_B T}{h} \frac{Q^{TS}}{Q^{react}} e^{-E_o/k_B T} \tag{7.9}$$

The selectivity of a porous nanomesh to two gas molecules (gas 1 and gas 2) is then expressed by the ratio of the two rate coefficients:

$$S = k_1/k_2 = \frac{Q_1^{TS} \cdot Q_2^{react}}{Q_1^{react} \cdot Q_2^{TS}} e^{-(E_1 - E_2)/k_B T} \tag{7.10}$$

where E_1 and E_2 are the corresponding energy barrier for each gas to pass through.

7.2.2 Evaluating Partition Functions: The Well-Separated "Reactant" State

Some preliminary considerations allow us to significantly simplify the evaluation of the TST rate constant as in Equation 7.9. In the implementation of TST for the present application, we make the approximation that the two-dimensional membrane is not significantly affected by the approach of the gas molecule; hence the partition function for the membrane is unaltered and will cancel in the numerator and denominator of Equation 7.9. Furthermore, we assume the internal vibrational frequency or frequencies of the gas molecule are unperturbed upon approach to the membrane; hence the vibrational partition function for these internal vibrations also will cancel. In like manner, electronic degeneracies for the ground electronic state are assumed unchanged and the electronic partition function will cancel.

With the above approximations in mind, the partition functions remaining in Equation 7.9 for the well-separated "reactant" system are therefore just the three-dimensional translational partition function for the gas molecule,

$$Q_{trans} = \left(\frac{2\pi m k_B T}{h^2} \right)^{3/2} \tag{7.11}$$

and its rotational partition function,

$$Q_{rot} = \frac{1}{\sigma} \frac{k_B T}{B} \tag{7.12}$$

for a linear molecule and

$$Q_{rot} = \frac{\pi^{1/2}}{\sigma} \left(\frac{k_B T}{A} \right)^{\frac{1}{2}} \left(\frac{k_B T}{B} \right)^{\frac{1}{2}} \left(\frac{k_B T}{C} \right)^{\frac{1}{2}} \tag{7.13}$$

for a non-linear molecule. Such partition functions are in the high-temperature range and are valid when the rotational constants are much smaller than $k_B T$. In Equations 7.12 and 7.13, σ is the rotational symmetry number of the molecule, that is, the number of symmetrically equivalent orientations it has. Thus, the effective total number of degrees of freedom in the model is five (for a diatomic or linear molecule) or six (for a non-linear molecule).

7.2.3 Evaluating Partition Functions: The Fully Coupled 4D TS Calculation

For the case of the diatomics H_2 and D_2, the five degrees of freedom in the model equate to four degrees of freedom (i.e., less the reaction coordinate) in the calculation of the partition function for the *TS*. In this section we consider the direct quantum calculation of the *TS* eigenstates for the coupled 4D motions at the *TS*, comprising of 2D hindered rotation coupled with 2D hindered translation within the pore. The transition state is defined with respect to a fixed value of the z coordinate that runs through the center of the pore perpendicular to its plane (the reaction coordinate). To a first approximation, a fixed value of z at the middle of the pore might be envisaged as the bottleneck that the molecule has to overcome to enter the structure. The rate is calculated from Equation 7.9 as:

$$k = \frac{k_B T}{h} \frac{Q^{TS}}{Q^{react}} e^{-E_0/k_B T} = \frac{k_B T}{h} \frac{Q^{TS}}{Q_{rot}^{react} Q_{trans}^{react}} e^{-E_0/k_B T} \qquad (7.14)$$

where Q^{TS} is the partition function at the transition state. Q_{trans}^{react} is the translational partition function, Equation 7.11, and Q_{rot}^{react} is the quantum rotational partition function of the reactants as will be discussed.

At low temperatures where the separation between rotational levels of H_2 and D_2 with respect to $k_B T$ will be large, the rotational partition function is treated quantum mechanically. This situation is most germane for the molecular isotope separation of H_2 from D_2, in which case the quantum rotational partition function of the diatomic well separated from the nanomesh is given by:

$$Q_{rot}^{react} = g_e \sum_{j=even} (2j+1) \exp[-B(j(j+1))/k_B T]$$
$$+ g_o \sum_{j=odd} (2j+1) \exp[-B(j(j+1))/k_B T] \qquad (7.15)$$

where B, the rotational constant, is given by $\hbar^2/2\mu R_e^2$. μ is the reduced mass of the molecule and R_e is the equilibrium bond length of the diatom. g_e and g_o are the spin-statistical weighting factors. For hydrogen, they are 1 and 3 respectively. For deuterium, they are 6 and 3 respectively.

The partition function at the transition state Q^{TS} is computed using discrete energy levels E_n obtained from four-degrees-of freedom quantum mechanical calculations:

$$Q^{TS} = g_e \sum_n \exp[-E_n^e/(k_B T)] + g_o \sum_n \exp[-E_n^o/(k_B T)] \qquad (7.16)$$

In Equation 7.16 the first and second summations represent eigenstates of even and odd parity, respectively.

The coordinate system for our four-degree-of-freedom model is displayed in Figure 7.4. In our model the coordinates x, y, and z are Cartesian coordinates, where z represents the pore axis (perpendicular to the plane of the nanomesh), which is taken to be the reaction coordinate (fixed at $z = 0$ for the calculations). The x and y coordinates represent the translational motion of the system in the plane of the pore/nanomesh. θ is the angle between the molecular bond and the z axis. The azimuthal angle, ϕ, represents rotation about the z-axis.

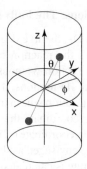

Figure 7.4 *Coordinates used in the 4D calculations*

For reasons discussed earlier, we fix the bond length of the diatomic in its equilibrium position and do not explicitly treat motion of the atoms in the nanomesh. The Hamiltonian for this system is given by:

$$\hat{H} = -\frac{\hbar^2}{2M}\left[\frac{\partial^2}{\partial x^2} + \frac{\partial^2}{\partial y^2}\right] - B\left[\left(\frac{1}{\sin\theta}\right)\frac{\partial}{\partial\theta}(\sin\theta)\frac{\partial}{\partial\theta} + \frac{1}{\sin^2\theta}\frac{\partial^2}{\partial\phi^2}\right]$$
$$+ V(x, y, \theta, \phi) \qquad (7.17)$$

where M is the total mass of the diatomic. The potential $V(x,y,\theta,\phi)$ includes all interactions between the atoms of the diatomic and the carbon atoms of the pore mouth. The wavefunction is expanded in a set of real eigenfunctions of the rotational part of the Hamiltonian:

$$\psi(x, y, \theta, \phi) = \sum_{j=0}^{j_{max}}\sum_{m=0}^{j} \bar{P}_j^m(\cos\theta)(C^{j,m}(x, y)\cos m\phi + D^{j,m}(x, y)\sin m\phi) \qquad (7.18)$$

where $\bar{P}_j^m$ is a normalized associated Legendre function and $C^{j,m}$ and $D^{j,m}$ are expansion coefficients. The Cartesian coordinates x, y are represented on evenly spaced grids and a high order finite difference method is used to compute the action of the kinetic energy operator. The homonuclear symmetry of the diatomic molecule results in a decoupling of the Hamiltonian matrix into even ($j = 0, 2, \ldots$) and odd ($j = 1, 3, \ldots$) blocks.

7.2.4 Evaluating Partition Functions: Harmonic Approximation for the TS Derived Directly from Density Functional Theory Calculations

When investigating molecular structures by quantum chemistry software such as *Dmol³* [57,58], it is not practical at the exploratory stages of an investigation to obtain exact energy levels of the coupled 4D or 5D effective Hamiltonian at the *TS* due to the expenses of computing the full potential energy surface. However, harmonic frequencies at a given position along the reaction coordinate can be readily computed. At sufficiently low temperatures, the separable harmonic approximation might be expected to be valid, in which case the *TS* partition functions can be computed from the simple analytic form of the quantum harmonic oscillator partition function – and in fact at this level of approximation all the molecular degrees of freedom, for example, translational, rotational and vibrational, associated with the gas molecule can equally readily be incorporated.

The partition functions were obtained in standard fashion as:

$$Q = \prod_{i=1}^{n} Q_i = \prod_{i=1}^{n} \frac{1}{1 - e^{-h\nu_i/k_B T}}, (n = 5 \text{ for } H_2, \, D_2 \text{ and } CO, \, n = 14 \text{ for } CH_4) \quad (7.19)$$

with Equation 7.19 as defined previously, E_0 in Equations 7.9 and 7.15 becomes more specifically the *zero-point energy difference* between reactant and transition states.

7.3 Gas and Isotope Separation

7.3.1 Gas Separation and Storage by Porous Graphene

7.3.1.1 Porous Graphene for Hydrogen Purification and Storage

Several studies have been reported concerning hydrogen purification by porous graphene. The first one was reported by Li et al. [19] that used density functional theory to obtain barriers for H_2, CH_4, CO_2, and CO diffusion. They claim in their paper that porous graphene exhibit premier performance in H_2 purification, and the selectivity of H_2/CO_2, H_2/CO and H_2/CH_4 are 10^{26}, 10^{29}, and 10^{76}, respectively. However, the barrier for H_2 diffusion was calculated to be 0.61 eV by PW91 functional [59] and 0.68 eV by BLYP functional [60]; these values are somewhat large and we couldn't expect a fair diffusion rate across the membrane under room temperature. Blankenburg et al. [18] claimed later in their work that the porous graphene model for H_2 purification should be larger, and van der Waals interaction should also be considered. In their work, the H_2 diffusion barrier was calculated to be 0.37 eV using PBE functional plus dispersion correction using the scheme as we discussed in Section 7.1.3; this value is reduced to about half the value Li at el. [19] reported and could lead to a higher diffusion rate. This type of porous graphene was also studied toward its ability for Helium separation whose passage barriers for He, Ne, and CH_4 are 0.523, 1.245, and 4.832 eV, respectively [20].

Graphene doped with Li atoms has been predicted to be a promising candidate for hydrogen storage in clean energy area [61–63]. Indeed, Li-doped graphene is expected to be superior to a Li-doped carbon nanotube because both sides might be readily utilized to ensure efficient hydrogen storage [61–65]. However, with a high coverage of Li atoms on graphene, the adsorption of hydrogen will be significantly weakened by the strong electrostatic interaction between Li cations. On porous graphene, with the insertion of holes of specific size and distribution into graphene sheets, the separation of Li dopants on porous graphene is expected to be large (4.23 Å by LDA [17]) compared to that on Li-doped graphene and hence the electrostatic interaction between Li cations should be lower, leading to enhanced adsorption of hydrogen molecules. They found the hexagonal center is the most energetically favorable adsorption site for a Li atom on porous graphene. The adsorption energy (-1.81 eV by LDA) for a Li atom onto porous graphene (7.45 Å × 7.45 Å) is stronger than that on graphene model about the same size [61]. Clearly, this effect may therefore be attributed to the natural separation of Li adsorption sites on porous graphene, which avoids strong electrostatic interactions between the Li cations. Moreover the adsorption of a second Li adsorbs preferentially on another hexagonal center, which potentially excludes the possibility of Li clustering. The H_2 storage capacity by this Li

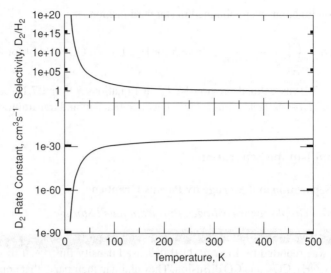

Figure 7.5 Top panel: D_2/H_2 selectivity of porous graphene. Bottom panel: Diffusion rate constant for D_2 for PG

doped porous graphene was subsequently calculated to be up to 12 wt.%, in significant contrast to the case of Li-doped graphene. Several other metal doped porous graphene were also investigated toward their ability of H_2 adsorption and Ca decorated porous graphene exhibits the highest storage capacity [66].

7.3.1.2 Porous Graphene for Isotope Separation

As discussed earlier, molecules with identical size and shape but different mass, such as H_2 and D_2, could also be separated by sufficiently small pores that allow utilization of the quantum effect of the differential zero-point energies. H_2/D_2 separation by porous graphene could be visualized in Figure 7.5.

While narrow pores lead to large selectivities they also restrict the actual flux through the pore due to the narrower configuration space. Due to the very small size of the pore on the porous graphene, we therefore also plot the rate constant for the passage of D_2 for porous graphene in the bottom panel of Figure 7.5. The figure shows that the rate constant for PG is negligible below 200 K. This comes as no surprise as the passage through the pore presents a barrier. The performance of such porous graphene for Helium isotope separation (^{3}He/^{4}He) was also evaluated and ^{3}He passage is strongly favored at low temperatures, however, with a very small diffusion rate [20].

7.3.2 Nitrogen Functionalized Porous Graphene for Hydrogen Purification/Storage and Isotope Separation

7.3.2.1 Introduction

Several works studied nitrogen functionalized porous graphene (NPG) for gas/isotope separation, and this structure exhibit good performance. The passage of H_2 and CH_4 into

NPG by first principles density functional theory calculations showed a high selectivity on the order of 10^8 for H_2/CH_4 with a high H_2 permeance [7]. The pore on NPG presents a formidable barrier (0.41 eV by PBE and 0.33 eV by vdW-DF [67,68]) for CH_4 but easily surmountable for H_2 (with barriers of 0.025 and 0.04 eV for PBE and vdW-DF, respectively). A related study based on such structures includes F–N functionalized pores and hydrogen passivated pores to investigate the passage of ions through porous graphene by Sint et al. [69]. NPG have also the capability to separate ^{3}He from ^{4}He efficiently by quantum tunneling through the pore [11,12,49]; calculations on size-reduced model systems of single-graphene pores reveal that a balanced choice of tunneling barrier height and low gas temperature boosts the selectivity and keeps the helium gas flux at an industrially exploitable level.

7.3.2.2 NPG and its Asymmetrically Doped Version for D_2/H_2 Separation – A Case Study

NPG is a possible membrane model for H_2/D_2. Our calculation shows that H_2/D_2 passage through the pore proceeds without a barrier and is in fact attractive. This should be attributed to the vdW correction scheme we used is different with that of Ref. [7]. Please see the NPG line in Figure 7.6.

Variational transition state theory (VTST) could be used in treating such a barrierless reaction pathway [70]. Within such an approach, the conventional transition state (*TS*) is replaced by a proper variation of location of the transition state on the minimum energy pathway (*TS**); corresponding partition function is then replaced by Q^{TS^*}. For the NPG system, such *TS** was chosen as the gas molecule in the pore center and we brief it by *middle state*. The frequencies for each state were calculated as in Table 7.1. The selectivities of D_2 over H_2 calculated by method as described in Section 7.2.4 from these frequencies

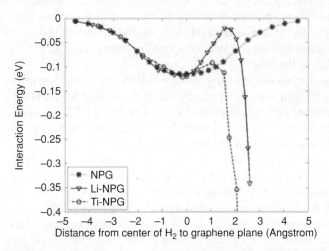

Figure 7.6 *Minimum energy pathway of hydrogen passing through NPG, Li-NPG, and Ti-NPG from DFT calculations with vdW correction. Adapted with permission from [9] © 2012 American Chemical Society*

Table 7.1 Frequencies calculated from vdW corrected DFT calculations for nitrogen functionalized porous grapheme (NPG) with the H_2/D_2 molecule located inside the pore.[a] Units are in cm^{-1}

H_2	D_2	Mode
28.61	27.12	Translation
121.91	96.48	Translation
293.49	210.72	Translation
430.40	305.19	Rotation
458.95	328.25	Rotation
4252.85	3008.64	Vibration

[a] There are no imaginary frequencies in the table which indicates this is a stable state rather than a transition state.

showed that below 125 K, D_2 is favored to pass through over H_2, above which the trend is inverted.

There is a fundamental problem with the symmetric membranes discussed above: although entrance into the membrane may be barrierless in some cases, with apparently good kinetics and selectivity for entrance into the pore, anything gained in terms of selectivity on the way in will be lost on the way out because of the symmetry in the interaction potential (see the NPG curve in Figure 7.6). As described in Section 7.3.1, hydrogen binds to the lithium sites on Li-decorated porous graphene with an apparently optimal exothermicity for reversible hydrogen storage applications. Additionally, the intrinsic porosity of the 2D structure tends to favor atomic dispersion of the metal onto the adsorption sites, potentially avoiding unwanted clumping of the metal atoms. This metal adsorption concept, if implemented asymmetrically on the exit side of the ultrathin membrane, offers a solution for this problem that has not previously been discussed. Hence, we now explore the effects on the minimum energy path of decorating the backside of the porous structure with Li or Ti. Since the H_2 (D_2) will bind favorably onto the metal, we would expect its exit energy to be lowered relative to undecorated NPG, creating effectively a downhill pathway all the way into and out of the pore.

The computed minimum energy pathways (MEPs) by firstly a transition state search by LST/QST [71] then a reaction pathway search by Nudged Elastic Band algorithm [72] from vdW corrected DFT calculations of Li- and Ti-doped NPG are shown in Figure 7.6. Using the MEP for the Li case as an example, the hydrogen isotope approaches the porous graphene without a barrier from the non-lithium-doped side. It will then reside just inside of the pore with adsorption energy of −0.12 eV. The orientation of the hydrogen at this state (*middle state*) is parallel to the graphene plane, and is pointing to the center of two adjacent nitrogen atoms. After passing through the pore center, the hydrogen diatom will go over a 0.10 eV barrier (*barrier state*) and finally reside near the lithium atom (*final state*). The energy line shows that once the hydrogen overcomes the energy barrier and adsorb onto the lithium atom, it will be very difficult to pass back through the pore which makes the reaction preferably going forward rather than going backward. Hence the inclusion of lithium doping on the exit side successfully modifies the symmetrical energy curve of the NPG to be energetically favorable in the forward direction and lowers the barrier value,

Table 7.2 Frequencies of H_2 and D_2 at different locations for the Li-NPG. The units are cm^{-1}. Middle state corresponds to the molecule being inside the pore and barrier state corresponds to the molecule at the top of the barrier before adsorption onto lithium

State	H_2	D_2	Mode
middle state	32.88	31.02	Translation
	124.83	97.19	Translation
	302.17	216.83	Translation
	453.92	325.34	Rotation
	467.43	331.34	Rotation
	4244.46	3002.70	Vibration
barrier state	319.02i	227.75i	Rotation[a]
	65.57	56.60	Translation
	81.41	67.23	Translation
	165.91	126.39	Rotation
	227.37	166.44	Translation
	4320.78	3056.70	Vibration

[a] Notice the imaginary frequency induced by the barrier here.

thereby enabling quantum induced kinetic sieving. The frequencies of H_2/D_2 at transition state and at the center of the pore are in Table 7.2.

However, we find that while the selectivity for passage into the pore and also for crossing the exit barrier is good, the exit barrier is the rate-determining step of this process, with the 0.1 eV barrier still leading to unfavorable kinetics at the low temperatures needed for the selectivity. The rate constant is decreasing with decreasing energy and is therefore smallest where the selectivity is highest.

To overcome this effect we also employed titanium doped nitrogen functionalized porous graphene (Ti-NPG) membrane. Figure 7.6 also shows the MEP for Ti-NPG where titanium atoms have been absorbed on the exit side of the membrane only. The approach from the undoped side is very similar to the lithium doped structure and the hydrogen absorbs with an energy of −0.12 eV. The barrier to exit from the pore is now just 0.03 eV. After surmounting the barrier the hydrogen finally absorbs onto the titanium at an energy of −0.57 eV.

Figure 7.7(a) shows the calculated selectivities of D_2 over H_2 for the passage through the pore and over the exit barrier calculated from the DFT frequencies as in Table 7.3. The selectivity shows that D_2 preferably passes through the pore for temperatures below 125 K while the exit barrier favors D_2 for temperatures below 100 K. Actually the selectivity over the barrier of Ti-NPG is much better than Li-NPG, exhibiting an increase on the selectivity.

As we now look at the corresponding rate constants, Figure 7.7(b), we find that rate constants for H_2 and D_2 for passage through the pore are still larger for temperatures below 200 K than the ones for passage over the barrier. We therefore find that for the titanium-doped NPG the small exit barrier is still the rate determining step for temperatures where the sieving process favors D_2. However, the rate constant at the exit barrier is much higher and now displays a favorable negative temperature dependence (it is increasing with decreasing energy and is therefore highest where the selectivity is also highest) — facilitating good kinetics at the low temperatures needed for the selectivity.

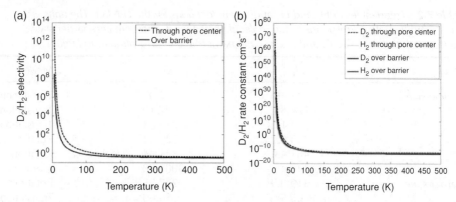

Figure 7.7 (a) Selectivities of and (b) rate constant for D_2 over H_2 for passage through the pore center and over the exit barrier for Ti-NPG. Adapted with permission from [9] © 2012 American Chemical Society

7.3.3 Graphdiyne for Hydrogen Purification

Unlike NPG we just discussed in Section 7.3.2, the pores on graphdiyne we are going to discuss in this section are naturally and uniformly distributed with the same shape and dimension. The side length of the triangle defined by vdW surface of the pore on graphdiyne is about 3.8 Å. It is worth noticing that the pore size is in-between the vdW dimension of H_2 and CH_4/CO. Due to these novel structure properties, graphdiyne could be an ideal H_2 purification membrane. We give in this chapter a precise description of the hydrogen diffusion rate and selectivity to CH_4/CO, firstly using dispersion corrected DFT by PBE functional then using TST on gas diffusion.

For hydrogen diffusion, the most stable configuration of H_2 in plane of graphdiyne is lying flat in the pore center with the hydrogen parallel to one side of the triangular pore

Table 7.3 Frequencies of H_2 and D_2 at different locations for the Ti-NPG. The units are cm^{-1}. Middle state corresponds to the molecule being inside the pore and barrier state corresponds to the molecule at the top of the barrier before adsorption onto titanium

State	H_2	D_2	Mode
middle state	51.27	44.43	Translation
	122.54	95.42	Translation
	285.31	204.55	Translation
	434.53	311.33	Rotation
	443.98	314.70	Rotation
	4250.61	3.008.59	Vibration
barrier state	262.44i	190.41i	Translation
	116.3	88.89	Translation
	182.00	133.65	Translation
	256.23	185.04	Rotation
	531.17	377.20	Rotation
	4123.24	2923.32	Vibration

Table 7.4 Adsorption energy E_{ad} and frequencies of H_2, CH_4 and CO at React and TS

Gas	Position	E_{ad}(eV)	Frequencies (cm^{-1})[a]					
H_2	React	−0.07	130.08	135.13	148.49	164.29	185.10	4328.13
	TS	0.03	189.78i	318.84	329.31	433.74	459.98	4310.08
CH_4	React	−0.14	70.73	72.67	90.18	204.10	238.68	297.10
			1306.27	1316.11	1324.35	1528.52	1530.82	2988.90
			3099.20	3108.04	3131.49			
	TS	0.58	148.45i	210.59	218.62	304.49	319.03	402.81
			1300.27	1318.84	1331.80	1536.03	1540.53	3043.01
			3139.19	3219.92	3255.56			
CO	React	−0.11	53.42	55.64	57.25	168.17	182.92	2129.10
	TS	0.22	64.26i	107.56	109.54	189.11	190.30	2089.64

[a] The imaginary frequencies are denoted by an i after values.

with an adsorption energy of 0.03 eV which means the interaction between H_2 and the framework is slightly repulsive. H_2 configuration outside graphdiyne plane is parallel to the surface with a height of 1.75 Å and an adsorption energy of −0.07 eV – this indicates a weak vdW interaction. The energy barrier of H_2 passing through the pores on graphdiyne is 0.10 eV.

Following Section 7.2.4, that is, harmonic approximation derived directly from DFT, we define the reactant state as the situation where hydrogen is adsorbed outside the graphdiyne plane and the transition state being hydrogen right in the pore center of graphdiyne. The adsorption energies and frequencies are summarized in Table 7.4.

The MEP is then searched using the identified stationary points on the energy surface including the most stable adsorption position of gas above the pore and the structure of gas center in the graphdiyne plane, and is shown in Figure 7.8. The barrier for H_2 diffusion is easily surmountable and is much smaller than that across the porous graphene as in Section 7.3.1. Using the Arrhenius equation [73], we estimate that graphdiyne exhibits more than 10^4 times faster rate for H_2 permeation than the porous graphene structures as in Section 7.3.1 investigated by Blankenburg et al. [18] which used the same functional and vdW correction.

The barriers for CH_4 and CO passing through the pore were computed to be 0.72 and 0.33 eV, respectively. The values indicate that in comparison with H_2, it is much more difficult for these molecules to diffuse across a graphdiyne-based nanomesh. The most stable adsorption configuration is that of a gas molecule sitting in the center of a graphdiyne pore. The corresponding MEPs are shown in Figure 7.8. It is worth noting that the graphdiyne framework slightly distorts when CH_4/CO is in the pore center; the diacetylene linking the benzene ring buckles outwards from the gas molecule. This distortion illustrates the stronger repulsion from graphdiyne to CH_4/CO hence a higher diffusion barrier comparing to that of H_2 on another perspective.

To quantitatively describe the hydrogen purification behavior, gas diffusion rate across graphdiyne and selectivities of H_2 to CH_4/CO were calculated using TST and are plotted

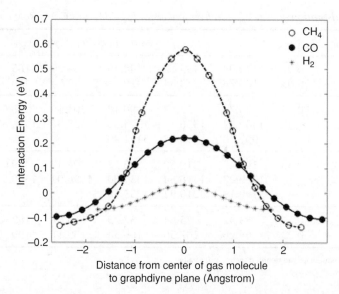

Figure 7.8 Minimum energy pathways for H_2, CO, and CH_4 diffusion through graphdiyne. Reference energy level is set to be gas and graphdiyne with infinite distance in-between. The line serves as a guide. From Ref. [22] Reproduced with permission from [22] © 2011 The Royal Society of Chemistry

in Figure 7.9 as (a) and (b). Figure 7.9(b) clearly indicates that the selectivity of H_2 to CH_4 is very high; at room temperature it is about 10^{10}. Comparing with silica and carbon membranes [74,75], which have the selectivity for H_2 to CH_4 in the order of 10 to 10^3, the performance of graphdiyne is hence predicted to be dramatically superior. The selectivity of H_2 to CO is lower, but the value of 10^3 at room temperature is still superior for effective separation.

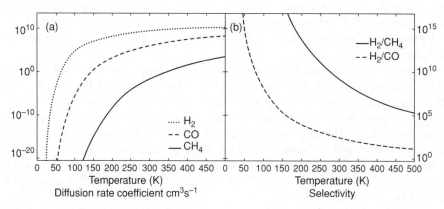

Figure 7.9 Diffusion rate coefficient (a) and selectivity (b) of gas molecules diffusing through the pores on graphdiyne as functions of temperature. From Ref. [22] Reproduced with permission from [22] © 2011 The Royal Society of Chemistry

7.4 Conclusion and Perspectives

With significant advances in recent years, graphene and other single atom layer carbon allotropes can now be synthesized. These porous single carbon atom layer membranes exhibit extraordinary performance in gas separation such as H_2 purification, isotope separation, and He purification. We presented within this chapter an affordable yet precise method to evaluate the performance of mono-atom layer porous membrane in gas/isotope separation; such method include firstly identifying the minimum energy pathway following by applying transition state theory to obtain the selectivity/diffusivity.

The following strategies could be explored to further increase the performance of such porous membranes for gas separation, including investigating alternative carbon allotropes [4,76], and the substitution of heterogeneous atoms (nitrogen, boron, etc.). Such strategies might lead to a larger difference in energy barriers for different gas molecules to pass through. For example, the performance for hydrogen purification would be better if the barrier for H_2 diffusion is reduced and the barrier for CH_4 diffusion is increased; this might be realized by applying alternative membrane framework or heteroatoms substitution. Another issue needing to be addressed is that due to the "softness" of these one atom thin layer structures, robust porous materials, for example, Al_2O_3 as proposed by Ref. [18], that exhibit superior flux rate could be included as a mechanical support for the membranes which could prevent the membrane from distortion under a pressure differential, while at the same time retaining the high permeability of the membrane.

Due to their intrinsic nature of being only one atom thick, these nanomeshes are expected to potentially serve as markedly superior membranes over traditional ones for several applications including H_2 purification – a vital step for the realization of a clean energy economy, isotope separation – whose process is very energy intensive, and N_2 separation in the context of natural gas purification, and so on. Since most of the studied single-atomic layer carbon membranes have been successfully synthesized, the separation scheme we summarized in this book chapter provides an interesting target for experimental validation studies.

Acknowledgement

Computational resources used in this work were provided by the University of Queensland (Centre for Computational Molecular Science), the Australian Research Council (LIEF grant LE0882357: A Computational Facility for Multiscale Modeling in Computational Bio and Nanotechnology), the Queensland Cyber Infrastructure Foundation (QCIF), and the NCI National Facility in Australia which is supported by the Australian Commonwealth Government. Use of the Center for Nanoscale Materials was supported by the US Department of Energy, Office of Science, Office of Basic Energy Sciences, under Contract No. DE-AC02-06CH11357. S.C.S. acknowledges support from the Center for Nanophase Materials Sciences, which is sponsored at the Oak Ridge National Laboratory by the Scientific User Facilities Division, US Department of Energy.

References

[1] P. Bernardo, E. Drioli and G. Golemme, Membrane gas separation: a review/state of the art. *Ind. Eng. Chem. Res.* 2009, **48**, 4638–4663.

[2] S. T. Oyama, D. Lee, P. Hacarlioglu and R. F. Saraf, Theory of hydrogen permeability in nonporous silica membranes. *J. Membr. Sci.*, 2004, **244**, 45–53.
[3] R. G. Gilbert and S. C. Smith, *Theory of Unimolecular and Recombination Reactions*, Blackwell Scientific Publications, Oxford, 1990.
[4] A. Hirsch, The era of carbon allotropes. *Nat. Mater.*, 2010, **9**, 868–871.
[5] J. S. Bunch, S. S. Verbridge, J. S. Alden, A. M. van der Zande, J. M. Parpia, H. G. Craighead and P. L. McEuen, Impermeable atomic membranes from graphene sheets. *Nano Lett.*, 2008, **8**, 2458–2462.
[6] O. Leenaerts, B. Partoens and F. M. Peeters, Graphene: A perfect nanoballoon. *Appl. Phys. Lett.*, 2008, **93**, 193107.
[7] D. E. Jiang, V. R. Copper and S. Dai, Porous graphene as the ultimate membrane for gas separation. *Nano Lett.*, 2009, **9**, 4019–4024.
[8] M. D. Fischbein and M. Drndic, Electron beam nanosculpting of suspended graphene sheets. *Appl. Phys. Lett.*, 2008, **93**, 113107.
[9] M. Hankel, Y. Jiao, A. Du, S. K. Gray and S. C. Smith, Asymmetrically decorated, doped porous graphene as an effective membrane for hydrogen isotope separation. *J. Phys. Chem. C*, 2012, **116**, 6672–6676.
[10] H. Du, J. Li, J. Zhang, G. Su, X. Li and Y. Zhao, Separation of hydrogen and nitrogen gases with porous graphene membrane. *J. Phys. Chem. C*, 2011, **115**, 23261–23266.
[11] A. W. Hauser, J. Schrier and P. Schwerdtfeger, Helium tunneling through nitrogen-functionalized graphene pores: pressure- and temperature-driven approaches to isotope separation. *J. Phys. Chem. C*, 2012, **116**, 10819–10827.
[12] A. W. Hauser and P. Schwerdtfeger, Nanoporous graphene membranes for efficient ^{3}He/^{4}He Separation. *J. Phys. Chem. Lett.*, 2012, **3**, 209–213.
[13] J. Bai, X. Zhong, S. Jiang, Y. Huang and X. Duan, Graphene nanomesh. *Nat. Nano.*, 2010, **5**, 190–194.
[14] K. S. Novoselov, A. K. Geim, S. V. Morozov, D. Jiang, Y. Zhang, S. V. Dubonos, *et al.*, Electric field effect in atomically thin carbon films. *Science*, 2004, **306**, 666–669.
[15] M. Bieri, M. Treier, J. M. Cai, K. Ait-Mansour, P. Ruffieus, O. Groning, *et al.*, Two-dimensional polymer synthesis with atomic precision. *Chem. Commun.*, 2009, **45**, 6919–6921.
[16] W. Pisula, M. Kastler, C. Yang, V. Enkelmann and K. Müllen, Columnar mesophase formation of cyclohexa-*m*-phenylene-based macrocycles. *Chem. Asian J.*, 2007, **2**, 51–56.
[17] A. Du, Z. Zhu and S. C. Smith, Multifunctional porous graphene for nanoelectronics and hydrogen storage: new properties revealed by first principle calculations. *J. Am. Chem. Soc.*, 2010, **132**, 2876–2877.
[18] S. Blankenburg, M. Bieri, R. Fasel, K. Müllen, C. A. Pignedoli and D. Passerone, Graphene as an atmospheric nanofilter. *Small*, 2010, **6**, 2266–2271.
[19] Y. Li, Z. Zhou, P. Shen and Z. Chen, Two-dimensional polyphenylene: experimentally available porous graphene as a hydrogen purification membrane. *Chem. Commun.*, 2010, **46**, 3672–3674.
[20] J. Schrier, Helium separation using porous graphene membranes. *J. Phys. Chem. Lett.*, 2010, **1**, 2284–2287.
[21] G. Li, Y. Li, H. Liu, Y. Guo, Y. Li and D. Zhu, Architecture of graphdiyne nanoscale films. *Chem. Commun.*, 2010, **46**, 3256–3258.

[22] Y. Jiao, A. Du, M. Hankel, Z. Zhu, V. Rudolph and S. C. Smith, Graphdiyne: A versatile nanomaterial for electronics and hydrogen purification. *Chem. Commun.*, 2011, **47**, 11843–11845.

[23] J. H. Lunsford, Catalytic conversion of methane to more useful chemicals and fuels: A challenge for the 21st century. *Catal. Today*, 2000, **63**, 165–174.

[24] N. W. Ockwig and T. M. Nenoff, Membranes for hydrogen separation. *Chem. Rev.*, 2007, **107**, 4078–4110.

[25] A. V. A. Kumar and S. K. Bhatia, Quantum effect induced reverse kinetic molecular sieving in microporous materials. *Phys. Rev. Lett.*, 2005, **95**, 245901.

[26] A. V. A. Kumar, H. Jobic and S. K. Bhatia, Quantum effects on adsorption and diffusion of hydrogen and deuterium in microporous materials. *J. Phys. Chem. B*, 2006, **110**, 16666–16671.

[27] X. Zhao, S. Villar-Rodil, A. J. Fletcher and K. M. Thomas, Kinetic isotope effect for H_2 and D_2 quantum molecular sieving in adsorption/desorption on porous carbon materials. *J. Phys. Chem. B*, 2006, **110**, 9947–9955.

[28] X. Z. Chu, Y. P. Zhou, Y. Z. Zhang, W. Su, Y. Sun and L. J. Zhou, Adsorption of hydrogen isotopes on micro- and mesoporous adsorbents with orderly structure. *J. Phys. Chem. B*, 2006, **110**, 22596–22600.

[29] J. J. M. Beenakker, V. D. Borman and S. Y. Krylov, Molecular transport in subnanometer pores: Zero-point energy, reduced dimensionality and quantum sieving. *Chem. Phys. Lett.*, 1995, **232**, 379–382.

[30] S. W. Rick, D. L. Lynch, J. D. Doll, The quantum dynamics of hydrogen and deuterium on the Pd(111) surface: a path integral transition state theory study. *J. Chem. Phys.*, 1993, **99**, 8183–8193.

[31] M. Pavese and G. A. Voth, Pseudopotentials for centroid molecular dynamics: application to self-diffusion in liquid para-hydrogen. *Chem. Phys. Lett.*, 1996, **249**, 231–236.

[32] S. Miura, S. Okazaki and K. Kinugawa, A path integral centroid molecular dynamics study of nonsuperfluid liquid helium-4. *J. Chem. Phys.*, 1999, **110**, 4523–4532.

[33] Q. Y. Wang, S. R. Challa, D. S. Sholl and J. K. Johnson, Quantum sieving in carbon nanotubes and zeolites. *Phys. Rev. Lett.*, 1999, **82**, 956–959.

[34] S. R. Challa, D. S. Sholl and J. K. Johnson, Light isotope separation in carbon nanotubes through quantum molecular sieving. *Phys. Rev. B*, 2001, **63**, 245419.

[35] B. C. Hathorn, B. G. Sumpter and D. W. Noid, Contribution of restricted rotors to quantum sieving of hydrogen isotopes. *Phys. Rev. A*, 2001, **64**, 022903.

[36] S. R. Challa, D. S. Sholl and J. K. Johnson, Adsorption and separation of hydrogen isotopes in carbon nanotubes: multicomponent grand canonical Monte Carlo simulations. *J. Chem. Phys.*, 2002, **116**, 814–824.

[37] T. Lu, E. M. Goldfield and S. K. Gray, The equilibrium constants for molecular hydrogen adsorption in carbon nanotubes based on iteratively determined nano-confined bound states. *J. Theor. Comput. Chem.*, 2003, **2**, 621–625.

[38] T. Lu, E. M. Goldfield and S. K. Gray, Quantum states of molecular hydrogen and its isotopes in single-walled carbon nanotubes. *J. Phys. Chem. B*, 2003, **107**, 12989–12995.

[39] T. D. Hone and G. A. Voth, A centroid molecular dynamics study of liquid *para*-hydrogen and *ortho*-deuterium. *J. Chem. Phys.*, 2004, **121**, 6412–6422.

[40] G. Garberoglio, M. M. DeKlavon and J. K. Johnson, Quantum sieving in single-walled carbon nanotubes: effect of interaction potential and rotational-translational coupling. *J. Phys. Chem. B*, 2006, **110**, 1733–1741.

[41] T. Lu, E. M. Goldfield and S. K. Gray, Quantum states of hydrogen and its isotopes confined in single-walled carbon nanotubes: dependence on interaction potential and extreme two-dimensional confinement. *J. Phys. Chem. B*, 2006, **110**, 1742–1751.

[42] P. Kowalczyk, P. A. Gauden, A. P. Terzyk and S. K. Bhatia, Thermodynamics of hydrogen adsorption in slit-like carbon nanopores at 77 K. Classical versus path-integral Monte Carlo simulations. *Langmuir*, 2007, **23**, 3666–3672.

[43] A. V. A. Kumar, H. Jobic, H. and S. K. Bhatia, Quantum effect induced kinetic molecular sieving of hydrogen and deuterium in microporous materials. *Adsorption*, 2007, **13**, 501–508.

[44] A. V. A. Kumar and S. L. Bhatia, Is kinetic molecular sieving of hydrogen isotopes feasible. *J. Phys. Chem. C*, 2008, **112**, 11421–11426.

[45] M. Hankel, H. Zhang, T. X. Nguyen, S. K. Bhatia, S. K. Gray and S. C. Smith, Kinetic modelling of molecular hydrogen transport in microporous carbon materials. *Phys. Chem. Chem. Phys.*, 2011, **13**, 7834–7844.

[46] H. Tanaka, H. Kanoh, M. Yudaska, S. Ijima and K. Kaneko, Quantum effects on hydrogen isotope adsorption on single-wall carbon nanohorns. *J. Am. Chem. Soc.*, 2005, **127**, 7511–7516.

[47] Y. Hattori, H. Tanaka, F. Okino, H. Touhara, Y. Nagahigashi, S. Utsumi, *et al.*, Quantum sieving effect of modified activated carbon fibers on H_2 and D_2 adsorption at 20 K. *Phys. Chem. B*, 2006, **110**, 9764–9767.

[48] T. X. Nguyen, H. Jobic and S. K. Bhatia, Microscopic observation of kinetic molecular sieving of hydrogen isotopes in a nanoporous material. *Phys. Rev. Lett.*, 2010, **105**, 085901.

[49] J. Schrier and J. McClain, Thermally-driven isotope separation across nanoporous graphene. *Chem. Phys. Lett.*, 2012, **521**, 118–124.

[50] A. Du and S. C. Smith, Van der Waals-corrected Density functional theory: benchmarking for hydrogen-nanotube and nanotube-nanotube interactions. *Nanotechnology*, 2005, **16**, 2118–2123.

[51] S. Grimme, Semiempirical GGA-type Density functional constructed with a long-range dispersion correction. *J. Comput. Chem.*, 2006, **27**, 1787–1799.

[52] S. Grimme, Accurate Description of van der Waals complexes by density functional theory including empirical corrections. *J. Comput. Chem.*, 2004, **25**, 1463–1473.

[53] S. Grimme, J. Antony, S. Ehrlich and H. Krieg, A consistent and accurate ab initio parameterization of density functional dispersion correction (DFT-D) for the 94 elements H-Pu. *J. Chem. Phys.*, 2010, **132**, 154104.

[54] S. Grimme, S. Ehrlich and L. Goerigk, Effect of the damping function in dispersion corrected density functional theory. *J. Comput. Chem.*, 2011, **32**, 1456–1465.

[55] J. P. Perdew, K. Burke, K. and M. Ernzerhof, Generalized gradient approximation made simple. *Phys. Rev. Lett.*, 1996, **77**, 3865–3868.

[56] G. M. Morris, D. S. Goodsell, R. S. Halliday, R. Huey, W. E. Hart, R. K. Belew and A. J. Olson, AutoDock. *J. Comput. Chem.*, 1998, **19**, 1639–1662.

[57] B. Delley, An all-electron numerical-method for solving the local density functional for polyatomic-molecules. *J. Chem. Phys.*, 1990, **92**, 508–517.

[58] B. Delley, From molecules to solids with the DMol(3) approach. *J. Chem. Phys.*, 2000, **113**, 7756−7764.

[59] J. P. Perdew and Y. Wang, Accurate and simple analytic representation of the electron-gas correlation energy. *Phys. Rev. B*, 1992, **45**, 13244−13249.

[60] C. Lee, W. Yang and R. G. Parr, Development of the Colle–Salvetti correlation energy formula into a functional of the electron density. *Phys. Rev. B*, 1988, **37**, 785−789.

[61] I. Cabria, M. J. Lopez and J. A. Alonso, Enhancement of hydrogen physisorption on graphene and carbon nanotubes by Li doping. *J. Chem. Phys.*, 2005, **123**, 204721.

[62] C. Ataca, E. Akturk, S. Ciraci and H. Ustunel, High-Capacity Hydrogen Storage By Metallized Graphene. *Appl. Phys. Lett.* 2008, **93**, 043123.

[63] W. Liu, Y. H. Zhao, J. Nguyen, Y. Li, Q. Jiang and E. J. Lavernia, Electric field induced reversible switch in hydrogen storage based on single-layer and bilayer graphenes. *Carbon*, 2009, **47**, 3452–3460.

[64] T. Yildirim and S. Ciraci, Titanium-decorated carbon nanotubes as a potential high-capacity hydrogen storage medium. *Phys. Rev. Lett.*, 2005, **94**, 175501.

[65] C. Ataca, E. Akturk and S. Ciraci, Hydrogen storage of calcium atoms adsorbed on graphene. first-principles plane wave calculations. *Phys. Rev. B*, 2009, **79**, 041406.

[66] P. Reunchan and S.-H. Jhi, Metal-dispersed porous graphene for hydrogen storage. *Appl. Phys. Lett.*, 2011, **98**, 093103.

[67] T. Thonhauser, V. R. Cooper, S. Li, A. Puzder, P. Hyldgaard and D. C. Langreth, van der Waals density functional: self-consistent potential and the nature of the van der Waals bond. *Phys. Rev. B*, 2007, **76**, 125112.

[68] M. Dion, H. Rydberg, E. Schröder, D. C. Langreth and B. I. Lundqvist, van der Waals density functional for general geometries. *Phys. Rev. Lett.*, 2004, **92**, 246401.

[69] K. Sint, B. Wang and P. Král, Selective ion passage through functionalized graphene nanopores. *J. Am. Chem. Soc.*, 2008, **130**, 16448−16449.

[70] D. G. Truhlar, B. C. Garrett and S. J. Klippenstein, Current status of transition state theory, *J. Phys. Chem.*, 1996, **100**, 12771−12800.

[71] T. A. Halgren, W. N. Lipscomb, Synchronous-transit method for determining reaction pathways and locating molecular transition-states. *Chem. Phys. Lett.*, 1977, **49**, 225–232.

[72] G. Henkelman, H. Jonsson, Improved tangent estimate in the nudged elastic band method for finding minimum energy paths and saddle points. *J. Chem. Phys.*, 2000, **113**, 9978–9985.

[73] Arrhenius equation: $(A = A_0\exp(-\Delta E/k_B T))$, for which A is the diffusion rate, A_0 is the diffusion prefactor, k_B the Boltzmann constant, and T the temperature.

[74] R. M. de Vos and H. Verweij, High-selectivity, high-flux silica membranes for gas separation. *Science*, 1998, **279**, 1710–1711.

[75] M. B. Shiflett and H. C. Foley, Ultrasonic deposition of high-selectivity nanoporous carbon membranes. *Science*, 1999, **285**, 1902–1905.

[76] A. N. Enyashin and A. L. Ivanovskii, Graphene allotropes, *Phys. Status Solidi B*, 2011, **248**, 1879–1883.

8

Graphene-Based Architecture and Assemblies

Hongyan Guo,[a] Rui Liu,[b] Xiao Cheng Zeng,[b] and Xiaojun Wu[a]

[a]*Department of Materials Science and Engineering, CAS Key Laboratory of Materials for Energy Conversion, and Hefei National Laboratory for Physical Science at the Microscale, University of Science and Technology of China, China.*
[b]*Department of Chemistry, University of Nebraska-Lincoln, USA*

8.1 Introduction

Carbon, the element that plays an important role for life on earth, possesses rich allotropes, partly due to its diverse electronic hybridization characteristics. Especially, sp^2-hybridized carbon atoms can form three distinct low-dimensional allotropic nanostructures, namely, buckminsterfullerene (C_{60}), carbon nanotubes, and graphene with fully conjugated π-electrons confined in either zero, quasi-one or two-dimensions [1–3]. The two-dimensional graphene is a single layer of graphite with sp^2-hybridized carbon atoms arranged in honeycomb structure, [4, 5] while the zero-dimensional C_{60} fullerene and quasi-one-dimensional carbon nanotube can be viewed as structural transformation of graphene pieces. The C_{60} is composed of 60 sp^2-hybridized carbon atoms that form a spherical cage, which can be viewed as a result of wrapping a finite-sized graphene piece containing 60 carbon atoms, whereas the one-dimensional carbon nanotube can be viewed as a result of rolling a graphene nanoribbon [6, 7].

Manifestation of quantum-confinement effect in these low-dimensional carbon nanostructures endows them with unique chemical and physical properties. For example, a single-walled carbon nanotube can be either metallic or semiconducting, depending on both their helix angles and diameters. A graphene monolayer is a semimetal with an exactly zero

---

Graphene Chemistry: Theoretical Perspectives, First Edition. Edited by De-en Jiang and Zhongfang Chen.
© 2013 John Wiley & Sons, Ltd. Published 2013 by John Wiley & Sons, Ltd.

band gap, and its charge carriers' velocity is constant throughout the band edge. As a consequence of their remarkable structures and properties, these sp^2-hybridized carbon nanostructures have attracted intense research interest for their great potential for applications in nanoelectronics, sensors, energy storage, drug delivery, field emission, composite materials, and so on.

Over past decades, scientists have performed extensive theoretical and experimental studies of these carbon nanostructures. Recently, many efforts have been made to design sp^2-carbon-based superarchitectures by physically merging units of fullerene, carbon nanotube, and graphene. One motivation for the design of sp^2-carbon-based super-architectures is the possibility to achieve new multi-functionality through combining novel properties of individual building units. Another motivation is to overcome deficiency of carbon-based materials in some applications, for example, energy storage. It is known that energy can be stored in chemical, electrochemical, and electrical forms, such as in hydrogen storage and supercapacitors. Among many energy storage materials, [8, 9] sp^2-hybridized carbon nanomaterials are preferred candidates due to their light mass, high specific surface area, and energy capacity, as well as excellent electrical conductivity. Compared with graphite (10–20 m^2 g^{-1}), [3, 10] a single-walled carbon nanotube, [11–13] possesses a unit surface area of up to 1315 m^2 g^{-1}, and graphene has the unit surface area of 2630 m^2 g^{-1}, which is suitable for energy storage applications. However, the van der Waals interactions between surfaces of sp^2-hybridized carbon nanostructures usually cause the bundling of carbon nanotubes or the restacking of graphenes [14–16]. One possible solution is to develop three-dimensional sp^2-hybridized carbon superstructures with large surface areas and high electrical conductivity, such as three-dimensional graphene and carbon nanotube superarchitectures [17–23].

The purpose of this chapter is to review properties of sp^2-hybridized carbon-based superstructures that have been theoretically studied. The graphene based super-architecture and some hybrid structures are presented. Also, boron-nitride-nanotube and monolayer-based super-architectures are also briefly discussed.

The organization of the chapter is as follows. Section 8.1 summarizes some progress on the polymerization of fullerenes. Section 8.2 provides discussions about the structures and electronic properties of carbon nanotube superarchitectures. Section 8.3 presents some theoretical designs of three-dimensional graphene structure. The theoretical design of fullerene, carbon nanotube and graphene based hybrid carbon nanostructures are the subjects of Section 8.4. In Section 8.5, some boron-nitride nanotube and monolayer-based superstructures are briefly discussed. Finally, we conclude with the discussion.

8.2 Fullerene Polymers

The fullerenes are the molecular carbon allotropes of which the most prominent one is C_{60} fullerene, first reported by Kroto *et al.* in 1985 [1]. C_{60} fullerene consists of 60 sp^2-hybridized carbon atoms in the shape of a soccer ball composed of 12 pentagons and 20 hexagons (I_h symmetry) [24]. What makes fullerene so attractive is not only its unique structure, but also its vast potential by generating various fullerene derivatives. Over the past two decades, several possible derivations of fullerene have been reported, such as fullerene salts, exohedral adducts, endohedral fullerene, heterofullerenes. Applications of

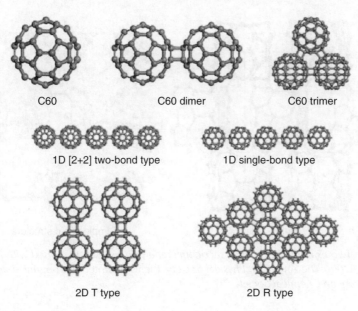

Figure 8.1 The structures of fullerene molecule, dimer, trimer, 1D, and 2D polymeric fullerene

these fullerene derivatives in materials science include the field of superconductors and optoelectronic devices.

From the viewpoint of bond structure, the sp^2-hybridized carbon atoms form two types of carbon–carbon covalent bonds in C_{60}, which are either between pentagon and hexagon (neighboring [5,6]-bond) or two hexagons (neighboring [6,6]-bond). Compared with the neighboring [5,6]-bonds, the [6,6] bonds have more double-bond character. Thus, it is proposed that C_{60} molecules can be merged to form fullerene polymers, from one-dimensional (1D) to three-dimensional fullerene polymer, via [2 + 2] cycloaddition reactions between two neighboring [6,6]-bonds. Some theoretical designs of C_{60}-polymers are illustrated in Figure 8.1. The simplest case is that two C_{60} molecules are linked together to form a C_{60} dimer [27, 28]. Also, an interesting model of a C_{60} trimer is obtained by covalently bonding three C_{60} molecules via a [2 + 2] cycloaddition reaction [29].

Similarly, the 1D, 2D, and 3D fullerene polymer structures are proposed by merging C_{60} molecules in three directions via a covalent carbon–carbon bond. Theoretical study suggested that a 1D fullerene polymer is a semiconductor with a direct band gap of about 2.18 eV, which is smaller than the HOMO-LUMO gap of C_{60} molecule [30, 31]. Upon the 2D fullerene polymers, two types of structure were reported, including tetragonal (T) and rhombohedral (R) [32, 33]. The T-type fullerite is a semiconductor with an indirect band gap of 1.852 eV [30]. The most stable R type fullerite ($C_{60(66)}$) is a semiconductor with an indirect band gap of 0.35∼0.577 eV, which is considered a major phase in the rhombohedral C_{60} fullerite [30, 34, 35]. Upon 3D fullerene polymers, more structures are investigated mainly for the purpose of searching for super-hard materials [25, 36–41]. Besides some simple 3D structures of simple-cubic, body-centered-orthorhombic, and body-centered-tetragonal phases, more 3D structures are proposed theoretically, such as the (60–0), (52–8), and

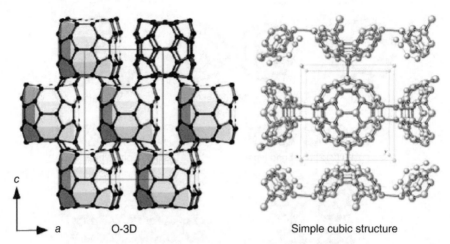

Figure 8.2 Two examples of 3D polymeric fullerene [25, 26]. Left Reprinted with permission from [25] © 2006 The American Physical Society. Right Reprinted with permission from [26] © 2006 American Chemical Society

(52–8) with the bulk moduli 295–300 GPa, [38] the (32–28) with a bulk modulus of 304 GPa simulated at 13 GPa and 820K,[39] the metallic O-3D (Figure 8.2) simulated under 15 GPa with a conductivity of about 10^{-2} S cm^{-1} at 600°C, [25] the rhombohedral structure, R-3, with a Micro-Vickers Hardness (MVH) of 4500 kg/mm^2 (10000 kg/mm^2 for diamond), [40] and a series of designed 3D polymeric C_{60} fullerene with bulk moduli 156–248 GPa [41].

In experiments, several techniques have been developed to realize fullerene polymers, such as solid state polymerization and photo induced polymerization [42–46]. The fullerene dimer, trimer, 1D, 2D, and 3D fullerene polymers have been synthesized [32, 33, 36, 37, 42–52]. Importantly, these fullerene polymers present novel properties, for example, the rhombohedral C_{60} polymer with ferromagnetism up to 500 K and some 3D structures with bulk moduli from 156 to 300 GPa, [52] implying the great potential of fullerene polymers in the applications of magnetic and mechanical materials.

8.3 Carbon Nanotube Superarchitecture

Carbon nanotubes, first experimentally reported by Iijima in 1991, display a combination of remarkable properties such as an ultrahigh Young's modulus and tensile modulus, high thermal conductivity, ballistic electron transport, and high aspect ratio [11–13]. These properties render carbon nanotubes a unique 1D material for applications in nanoelectronics. Using carbon nanotubes as building blocks in tailoring materials functionality, many unique properties of carbon nanotubes can be captured.

Like fullerene polymerization, previous theoretical and experimental studies have indicated that sp^2-carbon based carbon nanotubes can be merged covalently to form polymerized nanotube structures under high pressure, or form molecular junctions by using electron beam welding [53–55]. Such carbon nanotube superarchitectures can not only retain some of the extraordinary properties of 1D carbon nanotubes in 3D structures but also offer a class

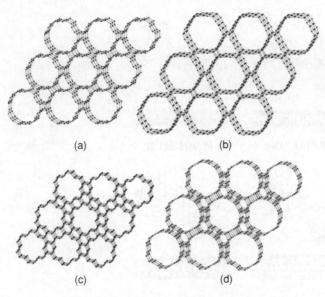

Figure 8.3 Various polymeric solids formed from (a) (6,6), (b) (12,0), (c) (9,0) and, (d) another (6,6) carbon nanotube via 2+2 and/or 2+4 cycloadditions. While (a), (b), and (c) contain geometries with mixed sp^2- and sp^3-bonded atoms, the structure in (d) has all sp^3-bonded atoms and forms a nanotube clathrate [54]. Reprinted with permission from [54] © 2002 The American Physical Society

of 2D and 3D truss-like materials with unprecedented architectures and open-ended materials functionality and tunability. For examples, Popov et al. have synthesized superhard materials of polymerized nanotubes under high-pressure treatment (up to 29 GPa) [56]. A theoretical study suggested that the fully relaxed structures of polymerized nanotubes show inter-nanotube connectivity via the 2+2 and 2+4 cycloaddition process, as shown in Figure 8.3. A generalized tight-binding molecular-dynamics simulation predicted that the polymeric solid are less stable than graphite and nanotube bundles [54]. The polymerized carbon nanotubes are semiconductors or insulators, depending on the nature of inter-tube bonds. In particular, their band gaps increase with the number of sp^3-hybridized carbon atoms in the materials.

Alternately, to retain metallic properties of carbon nanotubes in the super-architecture, it is important to build a carbon nanotube super-architecture containing only sp^2-hybridized carbon atoms. Note that carbon nanotube junctions with T-, Y-, and X-shaped structures have been realized in the laboratory [58]. By introducing some polygons (other than hexagons) in the carbon nanotube network, only sp^2-hybridized carbon atoms can be involved in these junctions, as shown in Figures 8.4 and 8.5. It was also shown that CNTs can be aligned into arrays or periodic superarchitectures [59]. With successful realization of aligned CNT arrays and T-, Y- and X-shaped CNT junctions, covalent assembly of CNT superarchitectures could be realized by welding arrays of nanocomponents together at high temperature [55, 60–62].

Several theoretical designs of 3D SWCNT super-architectures have been proposed, as shown in Figures 8.4 and 8.5 [19–21, 57]. Interestingly, the first-principles calculations

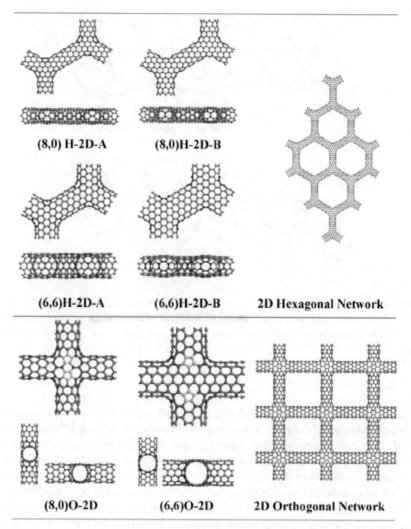

Figure 8.4 Two representative planar hexagonal and orthogonal single-walled carbon nanotubes superarchitectures (top panel) and optimized structural units based on the (6,6) and (8,0) Y- and T-junctions, respectively (middle and bottom panel). Heptagonal and pentagonal rings in the joint regions are highlighted in gray shading and light shading, respectively, and octagonal rings are dark shading [57]. Reprinted with permission from [57] © 2011 American Chemical Society

suggest that 2D hexagonal SWCNT superarchitectures are universally semiconducting, while 3D SWCNT superarchitectures are mostly metallic, regardless of whether the constituent SWCNTs are metallic or semiconducting [57]. This finding is informative to experimentalists for synthesizing metallic 3D carbon nanotube networks, since their metallic properties are only decided by the 3D structure and are independent of the building block. On the mechanical properties, it has been shown theoretically that SWCNT superarchitectures

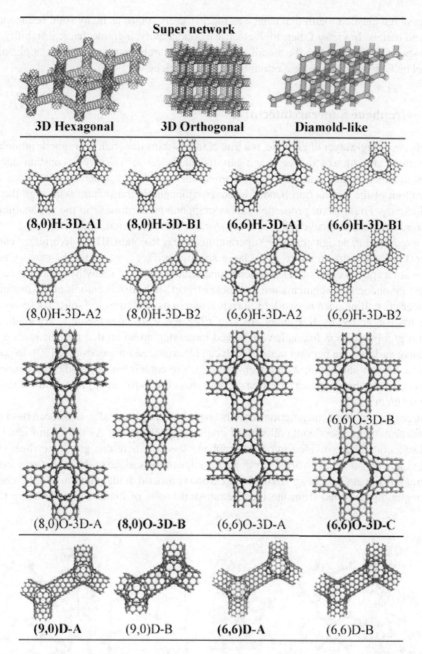

Figure 8.5 Three representative 3D hexagonal, orthogonal (simple cubic), and diamond-like single-walled carbon nanotube superarchitectures (top panel) and optimized structural units (in a supercell) of 16 3D SWCNT superarchitectures. Heptagonal and pentagonal rings at the joint regions are highlighted in gray and light shading, respectively, and octagonal rings are highlighted in dark shading. Two different views are displayed for three 3D orthogonal superarchitectures [57]. Reprinted with permission from [57] © 2011 American Chemical Society

possess high elastic moduli that can be as mechanically robust as many solid semiconductors and metals. In view of their high electrical conductivity, high mechanical stability, and large specific surface areas, the metallic SWCNT superarchitectures may have applications in fuel cells, battery electrodes, or nanoelectronic devices.

8.4 Graphene Superarchitectures

Graphene, a single layer of graphite, is a true 2D atom-thick material. A graphene monolayer is a semimetal with exactly zero band gap. Its charge carrier velocity is constant through the band edge, including the Dirac point where the top of the valence band overlaps with the bottom of the conduction band. Moreover, graphene is transparent with high thermal conductivity. These novel properties endow graphene great promise in the applications of electronic devices, optical devices, energy storage and conversion, and so on.

A major goal to design graphene superarchitecture is to obtain 3D sp^2-hybridized carbon network with porous structure, very large specific surface area and light mass. Carbon foams are a special class of 3D graphene superarchitectures. In many hypothetical carbon foams, graphene like segments are connected by sp^3-hybridized carbon atoms, resulting in either crystalline open network or porous amorphous structures [63–69]. These systems were first suggested by Karfunkel et al. in 1992 and were later theoretically studied by several groups. Carbon foams have received increasing attention due to their low density and large surface area per unit mass, which can be exploited for gas storage [70]. In particular, the carbon nanofoam, with a web-like structure containing randomly interconnected carbon clusters, has large surface area per unit mass (300–400 m^2/g) and ultralow density (< 0.01 g/cm^3) [71].

Three types of graphene junctions have been proposed theoretically, which can be viewed as the result of intersectional collision of graphene patches [71]. As shown in Figure 8.6, each junction can be divided into two graphene sheets, where one graphene sheet (light shaded spheres) attaches an armchair or zigzag edge to the surface of the other (gray shaded spheres). A linear chain of sp^3-hybridized carbon is formed at the junction and their bonds are negligibly distorted from the ideal tetrahedral bonds of diamond. In structure C, an

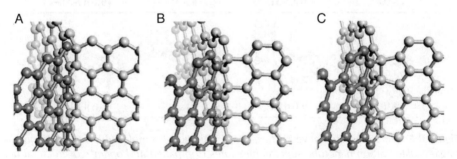

Figure 8.6 Schematic view of structure A, B, and C which are three-dimensional junction structures of sp^2 network, where light and darker spheres represent carbon atoms [71]. Reprinted with permission from [71] © 2005 The American Physical Society

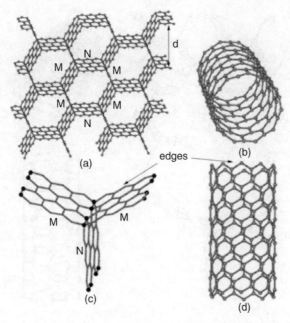

Figure 8.7 Zigzag (3,3) carbon foam structure (a) and the junction (c) in comparison with the zigzag carbon nanotube (10,0) (b) and (d). Foams are built using AA stacking of graphite planes [69]. Reprinted with permission from [69] © 2006 The American Physical Society

array of 5- and 8-membered ring defects is introduced. As the computational investigation shows, the reaction barrier height involved for the formation of these junctions is almost zero, and the binding energies per bond for each structure are about 1 eV. Furthermore, the high thermal stability and high formation probabilities for these junction structures were also shown.

Two types of hexagonal carbon foam can be built depending on the pattern of open edges, which are called armchair or zigzag foams in analogy to the nomenclature of carbon nanotubes. Their pore size can be defined by two numbers N and M indicating the number of hexagonal units between the junctions, as shown in Figures 8.7 and 8.8. The zigzag foam was the first theoretically proposed carbon foam model, as shown in Figure 8.7 [63]. According to simulation results, their electronic structure has similar size dependence to zigzag carbon nanotubes, as they are metallic if the distance between two junctions is a multiple of three hexagonal units, that is (n,M) = [3m, M] and/or (N,M) = [N,3m], and otherwise semiconducting with a wider range of the band gap than for carbon nanotubes. The metallic character is due to graphene-like stripes, where the sp^3-hybridized carbon chains at the junction are insulating [66,69]. The armchair foam is another type of carbon foam with sp^2-sp^3 hybridization, as shown in Figure 8.8(a), and its junction shown in Figure 8.8(c) is on the armchair edge of graphite segments, which is different from Figure 8.7(c). The armchair foams are metallic, independent of size, which is similar to armchair carbon nanotubes [67,69].

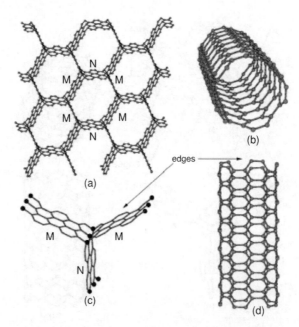

Figure 8.8 Armchair (3,5) carbon foam structure (a) and the junction (c) in comparison with the armchair carbon nanotube (5,5) (b) and (d). Foams are built using AA stacking of graphite planes [69]. Reprinted with permission from [69] © 2006 The American Physical Society

The same number of N and M result in the symmetrical carbon foam. When the number of hexagonal units between the junctions is not the same, more complicated structures like those shown in Figure 8.9 can be built with various electronic properties [63].

Recently, another kind of armchair carbon foams has been proposed, as shown in Figure 8.10, where the junctions are composed of carbon–carbon dimers along the center [72]. In addition, a kind of triangular foam has been designed, as shown in Figure 8.10(c), where graphene-like stripes are connected by hexagonal carbon rings at each joint, and this foam contains only sp^2–sp^2 carbon bonds. According to calculations, these two carbon foams are both metallic with low mass density (0.80 and 1.67 g ml^{-1}, respectively) compared to graphite and diamond ($\sim$2.16 and $\sim$3.15g ml^{-1}). Their bulk moduli are 72 and 202 GPa, respectively, and their cohesive energies are 0.11 and 0.23 eV/atom lower than that of graphite, respectively.

In experiment, the synthesis process of 3D carbon foam is rather simple. The mesophase pitch precursor is molten at high temperatures resulting in so-called graphitic foams [73,74]. Other methods, such as a template-directed CVD approach, ethanol-CVD over Ni foam and so on, have been reported [75–77]. Moreover, great efforts have been made to combine various nanomaterials with graphene composed of nanocomposites and explore their applications. For example, Wang *et al.* have synthesized the 3D architecture of Sn/graphene nanocomposites in experiment [78]. Li *et al.* have synthesized 3D hierarchical Fe_3O_4/graphene composites using a simple in situ hydrothermal method [79]. Recently, Tao *et al.* reported the 3D hierarchical NiO/graphene nanosheet composites through a simple

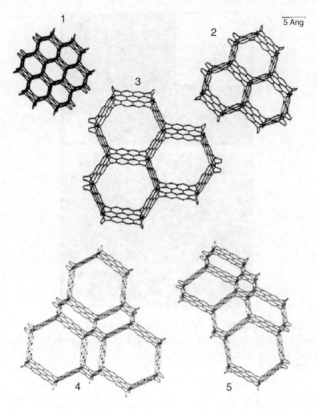

Figure 8.9 *Hexagonal crystal structures. In structures 1, 2, and 3 exactly zero, one, and two fused benzene rings are located between the junctions, respectively. Structures 4 and 5 are more complex structures of carbon foams [63]. Reprinted with permission from [63] © 1992 American Chemical Society*

ultrasonic method [80]. In addition, the 3D porous architecture of Si/graphene nanocomposites has also been studied [81].

8.5 C_{60}/Carbon Nanotube/Graphene Hybrid Superarchitectures

Besides the carbon superarchitectures constructed solely with either fullerenes, or carbon nanotubes, or graphene, it is desirable to build hybrid carbon superarchitecture by assembling C_{60}, carbon nanotubes, and graphene, that is, C_{60}/carbon nanotubes and C_{60}/graphene hybrid carbon superarchitectures and so on. The representative includes the carbon nanopeapod, nanobud, nanosieve, nanofunnel structures, and so on.

8.5.1 Nanopeapods   Erbsenschote

Carbon nanotube with quasi-one-dimensional space can be used as a nanometer-scale container for atoms or molecules [82–85]. The first attempt to build hybrid carbon

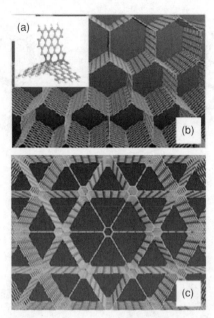

Figure 8.10 Geometric structures of (a) the building block of armchair foam, (b) armchair foam (honeycomb) and (c) triangular carbon foam [72]. Reproduced with permission from [72] © 2011 Royal Society of Chemistry

superarchitectures is to encapsulate C_{60} in carbon nanotubes by Smith *et al.* [86]. The hybrid nanostructure is known as a *peapod*. The van der Waals forces and the good geometrical match between fullerenes and SWCNTs make the encapsulation of fullerenes into SWCNTs a permanent and spontaneous process. It was theoretically predicted that the encapsulating process is exothermic when the radius of the nanotube for the encapsulation is larger than 6.4 Å (Figure 8.11) [87]. The carbon nanopeapod exhibits metallic property with multicarriers, which distributes either along the carbon nanotube or on the C_{60} chain, as shown in Figure 8.12. For the DWCNTs, the diameter of the internal tube regulates the structures of the encapsulated C_{60} crystal to maximize the van der Waals interactions between the tube and C_{60}. In addition, C_{60} molecules also interact with the outer layer of DWCNT as the diameter of the internal tube is smaller than 12 Å [88]. Using classical molecular dynamics simulations coupled with tight-binding calculations, Troche *et al.* have investigated the structural and electronic properties of zigzag carbon nanotubes filled with small fullerenes. They have observed that the electronic properties of the peapod have been significantly changed by the encapsulation of C_{20} and C_{30} fullerenes, implying the prospects in the design of nanoelectronic devices [89]. Recently, theoretical simulation has revealed the mechanism of carbon peapod fusion into double-walled nanotubes [90]. Furthermore, Duan *et al.* indicates that C_{60} cannot be encapsulated into a (16, 0) carbon nanotube due to a much larger energy barrier with the relaxation of the open end [91]. However, a C_{60} molecule could spontaneously insert into the SWCNT in supercritical CO_2, which is in good agreement with experimental observation [92].

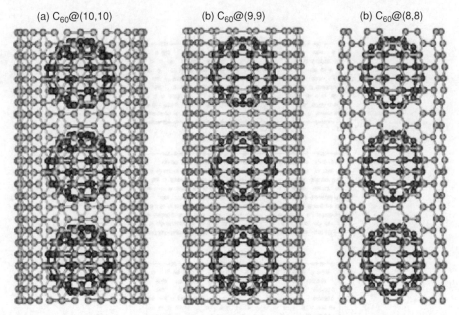

Figure 8.11 Total-energy optimized geometries of C_{60} encapsulated in (a) the (10, 10), (b) the (9, 9), and (c) the (8, 8) nanotubes. Distances between C_{60}s are 3.14, 3.15, and 2.45 Å, for (10, 10), (9, 9), and (8, 8) peapods, respectively [87]. Reprinted with permission from [87] © 2001 The American Physical Society

In experiment, the high-resolution transmission electron microscopy (HRTEM) image shows that the inner circle diameter of carbon nanotube is about 0.7 nanometers [86]. Burteaux et al. synthesized these hybrid nanomaterials in large quantities by pulsed laser vaporization of graphite, and C_{60} fullerenes are observed in ordered chains with about 10 Å separation [93]. Since then, several kinds of fullerenes have been encapsulated into carbon nanotubes to assemble the hybrid structures between these two allotropes of carbon, [94–110] such as C_{70}-encapsulated nanopeapod, DWCNT-based peapod, and metallofullerenes-encapsulated nanopeapod and so on [111–119]. Interestingly, metallofullerene-based nanopeapod can be used in the engineering of field effect transistors (FET) by varying the encapsulated metallofullerenes in SWCNTs [117–119].

8.5.2 Carbon Nanobuds

Different to peapods, carbon nanobuds (CNBs) are new hybrid carbon structures in which fullerenes are covalently bonded to the outer surface of carbon nanotubes, as shown in Figure 8.13. There are two types of CNBs, which are named the embedding and attaching configurations, respectively. In the embedding configuration, C_{60} molecule or part of it is attached to an imperfect carbon nanotube, whereas in the attaching configuration, C_{60} molecule is attached to the wall of perfect carbon nanotube [120, 121]. The attaching configuration can be viewed as the result of cycloaddition reaction between the carbon–carbon double bonds of C_{60} molecule and carbon nanotubes, respectively.

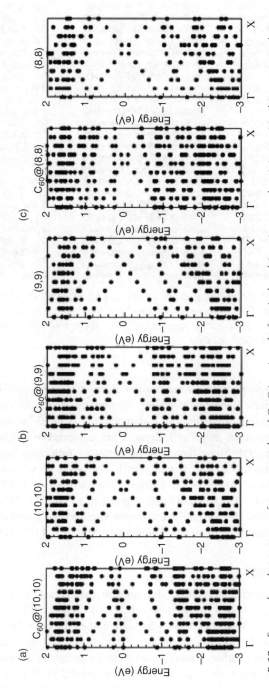

Figure 8.12 Energy band structures of an encapsulated C_{60}@(n, n) and of an isolated (n, n) nanotube. (a) $n = 10$, (b) $n = 9$, and (c) $n = 8$. Energies are measured from the Fermi level energy ε_F [87]. Reprinted with permission from [87] © 2001 The American Physical Society

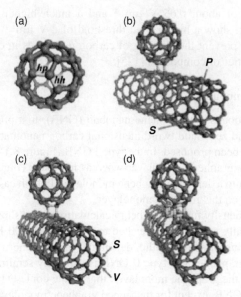

Figure 8.13 Carbon nanobud structures for different connections of fullerene to SWCNTs. (a) Structure of C_{60} molecule. (b)–(d) are three CNBs with the attaching configurations [121]. Reprinted with permission from [121] © 2008 American Chemical Society

The electronic properties of CNBs strongly depend on their structures. For the attaching configuration, all CNBs are semiconducting, regardless of whether the original SWCNT base is metallic or semiconducting. Chemical attachment of C_{60} to SWCNTs can either open up the band gap (e.g., for armchair SWCNT) or introduce impurity states within the band gap, thereby reducing the band gap (for semiconducting SWCNT). In addition, the band gap of CNBs can be modified by changing the density of C_{60} attached to the sidewall of the SWCNT [121]. For the embedding configuration, it was theoretically predicted that the carbon nanobuds based on zigzag SWCNTs are semiconducting, while those based on armchair SWCNTs are metallic [120]. Due to the coupling and charge transfer between fullerenes and carbon nanotubes, the electronic and optical properties of the hybrid materials can be tuned [120–122]. Also, substantial amounts of magnetic moments are predicted to arise in CNBs due to the presence of carbon radicals introduced by the geometry-induced electronic frustration [123]. Furthermore, the electronic structure and transport properties of carbon nanobuds can be affected by chemical modification such as F and Li adatoms, and the changes in the conductance of semiconducting carbon nanobuds are systematic, meaning that the CNBs are hopeful contenders for molecular sensor applications [124].

In experiment, the CNBs can be synthesized through two different one-step continuous methods. It was first seen in HRTEM images and then confirmed by Raman spectroscopy [125, 126]. The calculated Raman spectrotrum of CNBs with attaching configurations is very consistent with the experiment [127]. The CNBs present more advantageous properties than pristine SWCNTs and fullerenes. Compared with pristine SWNTs synthesized in the same condition but without water vapor during the synthesis process, nanobuds exhibit a low field threshold of about 0.65 V μm^{-1} and a much higher current density than pristine SWNTs which show a higher field threshold of 2 V μm^{-1} [125]. Moreover, the attached fullerene can prevent the slipping of carbon nanotubes in composite to improve the mechanical properties of composites [125].

8.5.3 Graphene Nanobuds

Similar to carbon nanobuds, the graphene nanobud (GNB), first proposed by theoretical calculations, is a hybrid zero- and two-dimensional carbon nanomaterial [128]. Two prototypes of GNBs have been proposed. In the type I GNB (Figure 8.14), C_{60} buckyballs are covalently bonded to a graphene monolayer whereas in type II (Figure 8.15), fragmented buckyballs are fused onto a defective graphene monolayer. In both cases, the C_{60} molecules form a periodic lattice on the graphene monolayer.

The first-principles density functional theory calculations have shown that various forms of GNB are all thermally stable at an elevated temperature (~800 K). Type I GNBs can be either semiconducting or semimetallic, depending on the pattern of chemical bonding between the C_{60} and graphene. The type II GNB is generally semimetallic. The hallmark electronic structure of the graphene monolayer, that is, the conic Dirac points, is still preserved in the type II PGNB, except for the ripped graphene monolayer. Finally, multilayer GNBs are a porous network structure and may be exploited for gas storage [128]. Theoretical calculations also predicted that graphene nanobuds may present better field emission properties than graphene and fullerenes themselves, suggesting the possibility of producing planar filed emission devices with periodic graphene nanobuds. Recently, it was predicted that some graphene nanobuds may show magnetic behavior [129]. The magnetic graphene nanobuds may play an important role in spintronics. Moreover, graphene nanobuds may have potential applications in gas storage, such as hydrogen, methane, and carbon dioxide storage [130, 131].

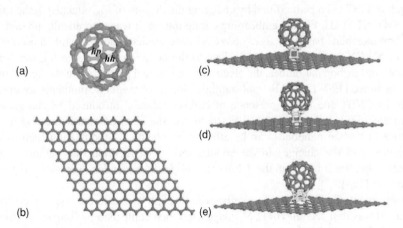

Figure 8.14 Type I Graphene nanobud structure. (a) The optimized structure of a C_{60} molecule. (b) The optimized structure of graphene. (c)–(d) The optimized structures of graphene nanobuds, where C_{60} is attached with graphene through a [2 + 2] or [6 + 6] cycloaddition reaction [128]. Reprintcd with permission from [128] © 2009 American Chemical Society

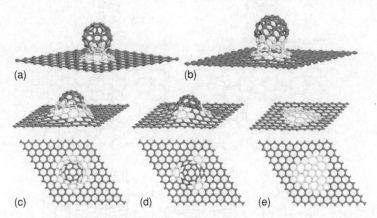

Figure 8.15 (a)–(e) The optimized structures of type II graphene nanobud structures. Fullerene segments are merged with graphene. The contact regions are lighter-shaded [128]. Reprinted with permission from [128] © 2009 American Chemical Society

The design of graphene nanobuds presents an interesting solution to avoid the restacking of graphene in experiments. Multilayer GNBs are a porous network structure with large surface area and may be exploited for gas storage [128]. Moreover, considering the electron mobility in graphene, the graphene part in the graphene nanobud can be used as a kind of electronic transport channel. Experimentally, graphene nanobuds can be obtained through a lithiation reaction [132]. The synthesized graphene nanobuds can be used as electron acceptors in poly-(3-hexylthiophene)-based bulk heterojunction solar cells and significantly improve the power conversion efficiency.

8.5.4 Nanosieves and Nanofunnels

Assembling carbon nanotube and graphene can form hybrid 1D/2D carbon superarchitectures, that is, the carbon nanosieve and nanofunnel, which have not been produced in experiments. A carbon nanofunnel can be created by seamlessly joining one end of a SWCNT with an aperture-containing graphene. Figure 8.16 displays two designed nanofunnels whose necks are (9,0) or (6,0) SWCNT, respectively. In the joint region, carbon atoms of the SWCNT are covalently bonded with atoms of the graphene at edge of the aperture, where six heptagonal defects (shown in Figure 8.16) are introduced to reduce local stress due to bending. The calculated electronic band structures and density of states suggest that the well-known Dirac point that cross the Fermi level in perfect graphene is slightly shifted away from the Fermi level in periodic nanofunnel structures. Meanwhile, a localized state with nearly no dispersion is introduced at the Fermi level. Both (9,0) and (6,0) SWCNT-based nanofunnels are metallic.

A carbon nanosieve structure can be constructed by joining two identical graphene-based nanofunnels together at the open end of the SWCNTs part (necks), as shown in Figure 8.17. Two graphene nanosieves, of which the neck is either (9,0) or (6,0) SWCNT, are displayed. Unlike the (9,0) nanofunnel which has a zero band gap, the (9,0) nanosieve becomes a semiconductor. The DOS near the Fermi level is mainly contributed by carbon atoms in the graphene layer, while the contribution to the DOS from the carbon atoms of the necks can be seen in some lower energy bands with nearly no dispersion.

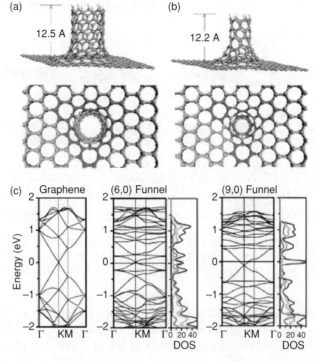

Figure 8.16 Side and bottom views of the optimized structures of nanofunnels whose necks are (a) (9,0) and (b) (6,0) finite-size SWCNTs. The open end of the neck is passivated by hydrogen atoms. The heptagons in the joint region are highlighted in light shading. (c) The band structures and density of states (DOS) of the perfect graphene, (6,0) nanofunnel, and (9,0) nanofunnel. The Fermi level is set as zero. The total DOS, PDOS projected on the SWCNT neck and the graphene are plotted with black, light gray, and dark gray solid lines, respectively. $\Gamma = (0, 0, 0)$, $K = (-1/3, 2/3, 0)$, and $M = (0, 0.5, 0)$ in the first Brillouin zone.

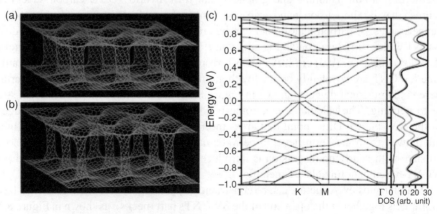

Figure 8.17 Periodic nanosieve structures whose necks are finite-sized (a) (9,0) and (b) (6,0) SWCNTs. (c) The band structure, total and PDOS of the nanosieve with the (9,0) SWCNT necks. The Fermi level is set as zero, and marked by a dotted line. The total DOS, PDOS projected on the necks, and PDOS of the remaining regions are plotted with black, light gray, and dark gray solid lines, respectively

Besides the large surface area in nanofunnel and nanosieve structures, these two designs can also be used as artificial semipermeable membrane, which allows water molecules passing through but blocking larger ionic species in aqueous solutions. Molecular dynamic simulation suggests that the nanosieve could be an artificial semipermeable membrane to achieve short-time unidirectional water flow within the SWCNT neck [133]. The diversity in CNT-based morphologies combined with their robust mechanical properties and tunable electronic properties can be exploited for future applications in nanotechnology.

8.6 Boron-Nitride Nanotubes and Monolayer Superarchitectures

As structural analogues to CNTs, boron nitride nanotubes (BNNTs) possess many structural and mechanical properties similar to their CNT counterparts. Moreover, BNNTs possess some distinctive properties, such as size- and chirality-independent wide band gap, exceptional resistance to oxidation, and high thermal conductivity [134–136]. Hence, BNNTs have also received growing attention over the past decade. Like CNTs, BNNTs can be joined together as well to form a variety of nanojunctions. To date, BNNT nanojunctions fabricated in the laboratory are predominantly T-type and Y-type [137–139]. However, few theoretical studies have reported on BNNT nanojunctions and superarchitectures. Note that due to the existence of two elements in BNNTs, it is expected that local structures of BNNT nanojunctions, especially in the nodal region, can be substantially more complex than the CNT counterparts. The differences in local structures between BNNT and CNT nanojunctions can be exploited for different applications.

In Figure 8.18, four prototype nanojunctions (T-, L-, X- and Y-) built from (6,6) or (8,0) BNNTs are displayed. Octagonal rings are found to be more favored than heptagonal rings in

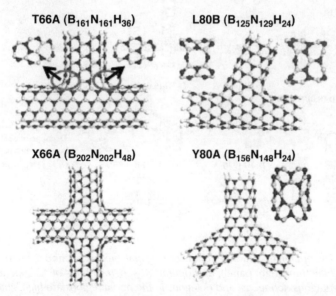

Figure 8.18 *The optimized structures of four prototype nanojunctions (T-, L-, X- and Y-) build from (6,6) or (8,0) BNNT*

172 Graphene Chemistry

the nodal region for most BNNT nanojunctions due to smaller number of polygonal defects needed to smoothly merge multi-terminal junctions together, as well as a smaller number of B–B and N–N bonds. 2D or 3D superarchitectures of BNNT can be constructed by covalently connecting the BNNT branches of these nanojuctions. For 3D superarchitectures, as shown in Figure 8.19, the diamond-like structures are energetically preferred to the other two, while the orthogonal structures are the least favorable. In particular, the hexagonal

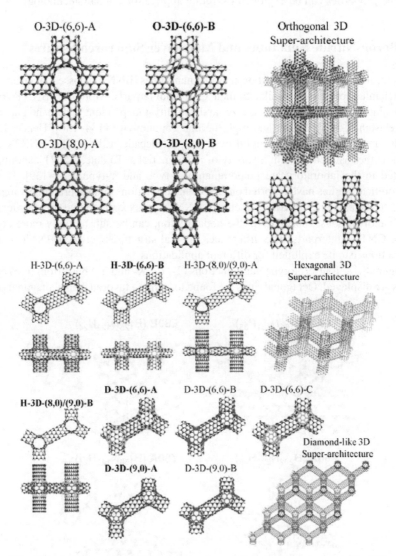

Figure 8.19 *The optimized structures of the orthogonal, hexagonal, and diamond-like 3D SW-BNT superarchitectures (right panel), and optimized supercell units of 13 superarchitectures. Pentagons, heptagons, tetragons, and octagons in the nodal regions are highlighted in yellow, red, green, and cyan, respectively. Bold type highlights the superarchitectures in each family with the greatest cohesive energy. See colour version in the colour plate section*

(6,6)-based 2D BNNT superarchitecture and diamond-like (6,6)-based 3D superarchitecture possess the greatest cohesive energy and the least strain energy mainly due to the smallest number of polygonal defects entailed in the nodal region without forming any B–B and N–N bonds. The band gaps of all superarchitectures are somewhat reduced compared to the constituent BNNTs due to the nodal-defect induced impurity states within the band gap. Again, among all the 2D and 3D BNNT superarchitectures considered, the hexagonal (6,6)-based 2D BNNT superarchitecture and the diamond-like (6,6)-based 3D superarchitecture possess the largest band gaps. The architecture-dependent electronic properties (e.g., band gaps) may be exploited for designing BNNT-based assemblies with a tunable functionality.

8.7 Conclusion

In conclusion, several sp^2-hybridized carbon based superarchitectures are proposed by assembling fullerenes, carbon nanotubes, and graphene sheets together. These first-principles calculation-based designs present a serial of theoretical prototypes of three-dimensional sp^2-hybridized carbon structures with large specific surface areas and light masses, possibly suitable for applications such as gas storage, parts in nanodevices, and so on. Moreover, using fullerenes, carbon nanotubes, and graphene as building blocks, the design of hybrid carbon nanostructures presents a possible way to examine a variety of multifunctional carbon nanostructures. Experimentally, some proposed carbon structures have been achieved and exhibit great potential in applications for energy storage and conversion, nanoelectronics, and so on.

However, there still remain many challenges, but also opportunities, for theoretical designs of sp^2-hybridized carbon-based superarchitectures. For example, the theoretically proposed structures are ideal models, which may not be easily obtained in experimentation. Thus, it is important to explore the defect structures in these 3D superarchitectures and their distribution to obtain both a large specific surface area and high electrical conductivity in carbon superarchitectures. The existence of defects may also alter chemical reactivity of carbon materials and enhance their energy storage capability. Finally, embedding some functional nanomaterials within carbon superarchitectures will pose new opportunities to produce a hybrid carbon-based complex, which may find applications in catalysis, electronic and optical devices, and information storage.

Acknowledgments

We acknowledge Keke Mao, Xiuling Li, Zhiwen Zhuo, Na Zhang for the contribution to this work. USTC group is supported by the National Basic Research Program of China (Nos. 2011CB921400, 2012CB922001), NSFC (grant no. 11004180, 51172223), One Hundred Person Project of CAS, the Shanghai Supercomputer Center, and the Hefei Supercomputer Center. UNL group is supported by NSF (grant No. DMR-0820521 and CBET-1036171), ARL (grant No. W911NF1020099), the Nebraska Research Initiative, the University of Nebraska's Holland Computing Center, and a travel grant from USTC International Center for Quantum Design of Functional Materials.

References

[1] H. W. Kroto, J. R. Heath, S. C. O'Brien, R. F. Curl, and R. E. Smalley, C_{60}: Buckminsterfullerene, *Nature*, **318**(6042), 162–163 (1985).

[2] O. Stephan, P. M. Ajayan, C. Colliex, P. Redlich, J. M. Lambert, P. Bernier, and P. Lefin, Doping graphitic and carbon nanotube structures with boron and nitrogen, *Science*, **266**(5191), 1683–1685 (1994).

[3] K. S. Novoselov, A. K. Geim, S. V. Morozov, D. Jiang, Y. Zhang, S. V. Dubonos, *et al.*, Electric field effect in atomically thin carbon films, *Science*, **306**(5696), 666–669 (2004).

[4] B. A. McKinnon and T. C. Choy, A tight binding model for the density of states of graphite-like structures, calculated using Green's functions, *Aust. J. Phys.*, **46**(5), 601–612 (1993).

[5] M. S. Dresselhaus, G. Dresselhaus, and R. Saito, Physics of carbon nanotubes, *Carbon*, **33**(7), 883–891 (1995).

[6] D. Ugarte, HREM characterization of graphitic nanotubes, *Microsc. Microanal. Microstruct.*, **4**, 505–512 (1993).

[7] A. K. Geim and K. S. Novoselov, The rise of graphene, *Nat. Mater.*, **6**, 183–191 (2007).

[8] C. Liu, F. Li, L.-P. Ma and H.-M. Cheng, Advanced materials for energy storage, *Adv. Mater.*, **22**(8), E28–E62 (2010).

[9] J. Liu, G. Cao, Z. Yang, D. Wang, D. Dubois, X. Zhou, *et al.*, Oriented nanostructures for energy conversion and storage, *ChemSusChem*, **1**(8–9), 676–697 (2008).

[10] M. Pumera, B. Smid, and K. Veltruska, Influence of nitric acid treatment of carbon nanotubes on their physico-chemical properties, *J. Nanosci. Nanotechno.*, **9**, 2671–2676 (2009).

[11] A. Hirsch, The era of carbon allotropes, *Nat. Mater.*, **9**, 868–871 (2010).

[12] S. Iijima, Helical microtubules of graphitic carbon, *Nature*, **354**, 56–58 (1991).

[13] A. Jorio, G. Dresselhaus, and M. S. Dresselhaus, Carbon nanotubes: advanced topics in the synthesis, structure, properties and applications, in Claus E. Ascheron (ed.), *Topics in Applied Physics, Volume 111*, Springer-Verlag GmbH, Heidelberg, Germany (2008).

[14] A. Thess, R. Lee, P. Nikolaev, H. J. Dai, P. Petit, J. Robert, *et al.*, Crystalline *ropes of metallic carbon nanotubes*, *Science*, **273**(5274), 483–487 (1996).

[15] X. Yang, J. Zhu, L. Qiu, and D. Li, Bioinspired effective prevention of restacking in multilayered graphene films: towards the next generation of high-performance supercapacitors, *Adv. Mater.*, **23**(25), 2833–2838 (2011).

[16] C. H. Lui, Z. Li, K. FaiMak, E. Cappelluti, and T. F. Heinz, Observation of an electrically tunable band gap in trilayer graphene, *Nat. Phys.*, **7**, 944–947 (2011).

[17] V. R. Coluci, D. S. Galvão, and A. Jorio, Geometric and electronic structure of carbon nanotube networks: 'super'-carbon nanotubes, *Nanotechnology*, **17**, 617–621 (2006).

[18] V. R. Coluci, N M Pugno, S. O. Dantas, D. S. Galvão and A. Jorio, Atomistic simulations of the mechanical properties of 'super' carbon nanotubes, *Nanotechnology*, **18**, 335702 (2007).

[19] J. M. Romo-Herrera, M. Terrones, H. Terrones, S. Dag, and V. Meunier, Covalent 2D and 3D networks from 1D nanostructures: designing new materials, *Nano Lett.*, **7**(3), 570–576 (2007).

[20] J. M. Romo-Herrera, M. Terrones, H. Terrones, and V. Meunier, Guiding electrical current in nanotube circuits using structural defects: a step forward in nanoelectronics, *ACS Nano*, **2**(12), 2585–2591 (2008).

[21] J. M. Romo-Herrera, M. Terrones, H. Terrones, and V. Meunier, Electron transport properties of ordered networks using carbon nanotubes, *Nanotechnology*, **19**, 315704 (2008).

[22] V. V. Ivanovskaya and A. L. Ivanovskii, Simulation of novel superhard carbon materials based on fullerenes and nanotubes, *J. Superhard Mater.*, **32**, 67–87 (2010).

[23] E. Hernández, V. Meunier, B. W. Smith, R. Rurali, H. Terrones, M. Buongiorno Nardelli, et al., Fullerene coalescence in nanopeapods: a path to novel tubular carbon, *Nano Lett.*, **3**(8), 1037–1042 (2003).

[24] R. D. Bendale, J. D. Baker, and M. C. Zerner, Calculations on the electronic structure and spectroscopy of C_{60} and C_{70} cage structures, *Int. J. Quantum Chem.*, **40**, 557–568 (1991).

[25] S. Yamanaka, A. Kubo, K. Inumaru, K. Komaguchi, N. S. Kini, T. Inoue, and T. Irifune, Electron conductive three-dimensional polymer of cuboidal C_{60}, *Phys. Rev. Lett.*, **96**, 076602 (2006).

[26] S. Ueda, K. Ohno, Y. Noguchi, S. Ishii, and J. Onoe, Dimensional dependence of electronic structure of fullerene polymers, *J. Phys. Chem. B*, **110**, 22374–22381 (2006).

[27] C. Yeretzian, K. Hansen, F. Diederich, and R. L. Whetten, Coalescence reactions of fullerenes, *Nature*, **359**(6390), 44–47 (1992).

[28] S. Tsukamoto and T. Nakayama, First-principles electronic structure calculations for peanut-shaped C_{120} molecules, *Sci. Technol. Adv. Mater.*, **5**(5–6), 617–620 (2004).

[29] M. Fujitsuka, K. Fujiwara, Y. Murata, S. Uemura, M. Kunitake, O. Ito, and K. Komatsu, Properties of photoexcited states of C_{180}, a triangle trimer of C_{60}, *Chem. Lett.*, **30**, 384–385 (2001).

[30] V. V. Belavin, L. G. Bulusheva, A. V. Okotrub, D. Tomanek, Stability, electronic structure and reactivity of the polymerized fullerite forms, *J. Phys. Chem. Sol.*, **61**(12), 1901–1911 (2000).

[31] T. A. Beu, J. Onoe, and A. Hida, First-principles calculations of the electronic structure of one-dimensional C_{60} polymers, *Phys. Rev. B*, **72**, 155416 (2005).

[32] O. Bethoux, M. Nuñez-Regueiro, L. Marques, J-L. Hodeau, and M. Perroux, Abstracts of contributed Papers (Material Research Society, Pittsburgh, 1993), Abstract No. G2.9, in *Proceedings of the Materials Research Society, Boston, 1993*, p. 202.

[33] Y. Iwasa, T. Arima, R. M. Fleming, T. Siegrist, O. Zhou, R. C. Haddon, et al., New phases of C_{60} synthesized at high pressure, *Science*, **264**(5165), 1570–1572 (1994).

[34] S. Okada and A. Oshiyama, Electronic structure of metallic rhombohedral C_{60} polymers, *Phys. Rev. B*, **68**, 235402 (2003).

[35] S. Okada and S. Saito, Rhombohedral C_{60} polymer: mA semiconducting solid carbon structure, *Phys. Rev. B*, **55**, 4039–4041 (1997).

[36] S. Okada, S. Saito, and A. Oshiyama, New metallic crystalline carbon: three dimensionally polymerized C_{60} fullerite, *Phys. Rev. Lett.*, **83**, 1986–1989 (1999).

[37] S. G. Buga, V. D. Blank, N. R. Serebryanaya, A. Dzwilewski, T. Makarova, and B. Sundqvist, Electrical properties of 3D-polymeric crystalline and disordered C_{60} and C_{70} fullerites, *Diamond Relat. Mater.*, **14**(3-7), 896–901 (2005).

[38] E. Burgos, E. Halac, R. Weht, H. Bonadeo, E. Artacho, and P. Ordejón, New superhard phases for three-dimensional C_{60}-based fullerites, *Phys. Rev. Lett.*, **85**, 112328–112331 (2000).

[39] C. A. Perottoni and J. A. H. da Jornada, First-principles calculation of the structure and elastic properties of a 3D-polymerized fullerite, *Phys. Rev. B*, **65**(22), 224208 (2002).

[40] S. Yamanaka, N. S. Kini, A. Kubo, S. Jida, and H. Kuramoto, Topochemical 3D polymerization of C_{60} under high pressure at elevated temperatures, *J. Am. Chem. Soc.*, **130**(13), 4303–4309 (2008).

[41] F. Zipoli and M. Bernasconi, First principles study of three-dimensional polymers of C_{60}: Structure, electronic properties, and Raman spectra, *Phys. Rev. B*, **77**(11), 115432 (2008).

[42] S. Pekker, L. Forró, L. Mihály, and A. Jánossy, Orthorhombic A_1C_{60}: A conducting linear alkali fulleride polymer?, *Solid State Commun.*, **90**(6), 349–352 (1994).

[43] A. M. Rao, P. Zhou, K. A. Wang, G. T. Hager, J. M. Holden, Y. Wang, *et al.*, Photoinduced polymerization of solid C_{60} films, *Science*, **259**(5097), 955–957 (1993).

[44] L. J. Barbour, G. W. Orr, and J. L. Atwood, Supramolecular assembly of well-separated, linear columns of closely-spaced C_{60} molecules facilitated by dipole induction, *Chem. Commun.*, **17**, 1901–1902 (1998).

[45] Y. Iwasa, K. Tanoue, T. Mitani, A. Izuoka, T. Sugawara, and T. Yagi, Supramolecular assembly of well-separated, linear columns of closely-spaced C_{60} molecules facilitated by dipole induction, *Chem. Commun.*, **13**, 1411–1412 (1998).

[46] G. W. Wang, Y. Murata, and M. Shiro, Synthesis and X-ray structure of dumb-bell-shaped C_{120}, *Nature*, **387**, 583–586 (1997).

[47] K. P. Meletov, J. Arvanitidis, S. Ves, and G. A. Kourouklis, High-pressure phase in tetragonal two-dimensional polymeric C_{60}, *Phys. Stat. Sol. (b)*, **223**(2), 489–493 (2001).

[48] P. Zhou, Z. H. Dong, A. M. Rao, and P. C. Eklund, Reaction mechanism for the photopolymerization of solid fullerene C_{60}, *Chem. Phys. Lett.*, **211**(4–5), 337–340 (1993).

[49] P. W. Stephens, G. Bortel, G. Faigel, M. Tegze, A. Jánossy, S. Pekker, *et al.*, Polymeric fullerene chains in RbC_{60} and KC_{60}, *Nature*, **370**, 636–639 (1994).

[50] G. M. Bendele, P. W. Stephens, K. Prassides, K. Vavekis, K. Kordatos, and K. Tanigaki, Effect of charge state on polymeric bonding geometry: the ground state of Na_2RbC_{60}, *Phys. Rev. Lett.*, **80**(4), 736–739 (1998).

[51] L. Marques, M. Mezouar, J.-L. Hodeau, M. Núñez-Regueiro, N. R. Serebryanaya, V. A. Ivdenko, *et al.*, "Debye-Scherrer Ellipses" from 3D fullerene polymers: an anisotropic pressure memory signature, *Science*, **283**, 1720–1723 (1999).

[52] T. L. Makarova, B. Sundqvist, R. Höhne, P. Esquinazi, Y. Kopelevichk, P. Scharff, *et al.*, Magnetic carbon, *Nature*, **413**, 716–718 (2001).

[53] L. A. Chernozatonskii, Polymerized nanotube structures – new zeolites?, *Chem. Phys. Lett.*, **297**, 257–260 (1998).

[54] L. Chernozatonskii, E. Richter, and M. Menon, Crystals of covalently bonded carbon nanotubes: Energetics and electronic structures, *Phys. Rev. B*, **65**, 241404(R) (2002).

[55] M. Terrones, F. Banhart, N. Grobert, J.-C. Charlier, H. Terrones, and P. M. Ajayan, Molecular junctions by joining single-walled carbon nanotubes, *Phys. Rev. Lett.*, **89**, 075505 (2002).

[56] M. Popov, M. Kyotani, Y. Koga, and R. J. Nemanich, Superhard phase composed of single-wall carbon nanotubes, *Phys. Rev. B*, **65**, 033408 (2002).

[57] R. L. Zhou, R. Liu, L. Li, X. J. Wu, and X. C. Zeng, Carbon nanotube superarchitectures: an ab initio study, *J. Phys. Chem. C*, **115**(37), 18174–18185 (2011).

[58] D. Wei and Y. Liu, The intramolecular junctions of carbon nanotubes, *Adv. Mater.*, **20**(15), 2815–2841 (2008).

[59] A. Ismach, D. Kantorovich, and E. Joselevich, Carbon nanotube graphoepitaxy: highly oriented growth by faceted nanosteps, *J. Am. Chem. Soc.*, **127**(33), 11554–11555 (2005).

[60] M. Terrones, H. Terrones, F. Banhart, J.-C. Charlier, and P. M. Ajayan, Coalescence of single-walled carbon nanotubes, *Science*, **288**(5469), 1226–1229 (2000).

[61] A. V. Krasheninnikov, K. Nordlund, J. Keinonen, and F. Banhar, Making junctions between carbon nanotubes using an ion beam, *Nucl. Instrum. Meth. B*, **202**, 224–229 (2003).

[62] C. H. Jin, K. Suenaga, and S. Iijima, Plumbing carbon nanotubes, *Nature Nanotech.*, **3**, 17–21 (2008).

[63] H. R. Karfunkel and T. Dressler, New hypothetical carbon allotropes of remarkable stability estimated by MNDO solid-state SCF computations, *J. Am. Chem. Soc.*, **114**(7), 2285–2288 (1992).

[64] A. T. Balaban, D. J. Klein, and C. A. Folden, Diamond-graphite hybrids, *Chem. Phys. Lett.*, **217**(3), 266–270 (1994).

[65] A. V. Rode, E. G. Gamaly, and B. Luther-Davies, Formation of cluster-assembled carbon nano-foam by high-repetition-rate laser ablation, *Appl. Phys. A*, **70**(2), 135–144 (2000).

[66] N. Park and J. Ihm, Electronic structure and mechanical stability of the graphitic honeycomb lattice, *Phys. Rev. B*, **62**(11), 7614–7618 (2000).

[67] K. Umemoto, S. Saito, S. Berber, and D. Tománe, Carbon foam: Spanning the phase space between graphite and diamond, *Phys. Rev. B*, **64**, 193409 (2000).

[68] F. J. Ribeiro, P. Tangney, S. G. Louie, and M. L. Cohen, Structural and electronic properties of carbon in hybrid diamond-graphite structures, *Phys. Rev. B*, **72**(21), 214109 (2005).

[69] A. Kuc and G. Seifert, Hexagon-preserving carbon foams: Properties of hypothetical carbon allotropes, *Phys. Rev. B* **74**, 214104 (2006).

[70] N. Park, S. Hong, G. Kim, and S. H. Jhi, Computational study of hydrogen storage characteristics of covalent-bonded graphenes, *J. Am. Chem. Soc.*, **129**(29), 8999–9003 (2007).

[71] T. Kawai, S. Okada, Y. Miyamoto, and A. Oshiyama, Carbon three-dimensional architecture formed by intersectional collision of graphene patches, *Phys. Rev. B*, **72**, 035428 (2005).

[72] M. H. Wu, X. J. Wu, Y. Pei, Y. Wang, and X. C. Zeng, Three-dimensional network model of carbon containing only sp^2-carbon bonds and boron nitride analogues, *Chem. Commun.*, **47**(15), 4406–4408 (2011).

[73] J. W. Klett, A. D. McMillan, N. C. Gallego, and C. A. Walls, The role of structure on the thermal properties of graphitic foams, *J. Mater. Sci.*, **39**(11), 3659–3676 (2004).

[74] J. Klett, R. Hardy, E. Romine, C. Walls, and T. Burchell, High-thermal-conductivity, mesophase-pitch-derived carbon foams: effect of precursor on structure and properties, *Carbon*, **38**(7), 953–973 (2000).

[75] Z. Chen, W. Ren, L. Gao, S. Pei, and H. M. Cheng, Three-dimensional flexible and conductive interconnected graphene networks grown by chemical vapour deposition, *Nat. Mater.*, **10**, 424–428 (2011).

[76] X. H. Cao, Y. M. Shi, W. H. Shi, G. Lu, X. Huang, Q. Y. Yan, et al., Preparation of novel 3D graphene networks for supercapacitor applications, *Small*, **7**(22), 3163–3168 (2011).

[77] X. Y. Xiao, T. E. Beechem, M. T. Brumbach, T. N. Lambert, D. J. Davis, J. R. Michael, et al., Lithographically defined three-dimensional graphene structures, *ACS Nano*, **6**(4), 3573–3579 (2012).

[78] G. Wang, B. Wang, X. L. Wang, J. Park, S. Dou, H. Ahn, and K. Kim, Sn/graphene nanocomposite with 3D architecture for enhanced reversible lithium storage in lithium ion batteries, *J. Mater. Chem.*, **19**(44), 8378–8384 (2009).

[79] X. Li, X. Huang, D. Liu, X. Wang, S. Song, L. Zhou, and H. Zhang, Synthesis of 3D hierarchical Fe_3O_4/graphene composites with high lithium storage capacity and for controlled drug delivery, *J. Phys. Chem. C*, **115** (44), 21567–21573 (2011).

[80] L. Q. Tao, J. T. Zai, K. X. Wang, Y. H. Wan, H. J. Zhang, C. Yu, et al., 3D-hierarchical NiO–graphene nanosheet composites as anodes for lithium ion batteries with improved reversible capacity and cycle stability, *RSC Adv.*, **2**(8), 3410–3415 (2012).

[81] X. Xin, X. F. Zhou, F. Wang, X. Y. Yao, X. X. Xu, Y. M. Zhu, and Z. P. Liu, A 3D porous architecture of Si/graphene nanocomposite as high-performance anode materials for Li-ion batteries, *J. Mater. Chem.*, **22**(16), 7724–7730 (2012).

[82] P. M. Ajayan and S. Iijima, Capillarity-induced filling of carbon nanotubes, *Nature*, **361**, 333–334 (1993).

[83] R. S. Lee, H. J. Kim, J. E. Fisher, A. Thess, and R. E. Smalley, Conductivity enhancement in single-walled carbon nanotube bundles doped with K and Br, *Nature*, **388**(6639), 255–257 (1997).

[84] C. Bower, S. Suzuki, K. Tanigaki, and O. Zhou, Synthesis and structure of pristine and alkali-metal-intercalated single-walled carbon nanotubes, *Appl. Phys. A: Mater. Sci. Process.*, **67**(1), 47–52 (1998).

[85] C. Bower, A. Kleinhammes, Y. Wu, and O. Zhou, Intercalation and partial exfoliation of single-walled carbon nanotubes by nitric acid, *Chem. Phys. Lett.*, **288**(2–4), 481–486 (1998).

[86] B. W. Smith, M. Monthioux, and D. E. Luzzi, Encapsulated C_{60} in carbon nanotubes, *Nature*, **396**(6709), 323–324 (1998).

[87] S. Okada, S. Saito, and A. Oshiyama, Energetics and electronic structures of encapsulated C_{60} in a carbon nanotube, *Phys. Rev. Lett.*, **86**(17), 3835–3838 (2001).

[88] A. N. Khlobystov, D. A. Britz, A. Ardavan, and G. A. D. Briggs, Observation of ordered phases of fullerenes in carbon nanotubes, *Phys. Rev. Lett.*, **92**(24), 245507 (2000).

[89] K. S. Troche, V. R. Coluci, R. Rurali, and D. S. Galvão, Structural and electronic properties of zigzag carbon nanotubes filled with small fullerenes, *J. Phys.: Condens. Matter*, **19**(23), 236222 (2007).

[90] F. Ding, Z. W. Xu, B. I. Yakobson, R. J. Young, I. A. Kinloch, S. Cui, *et al.*, Formation mechanism of peapod-derived double-walled carbon nanotubes, *Phys. Rev. B*, **82**(4), 041403(R) (2010).

[91] H. Duan, X. Gao, G. J. Fu, and J. M. Li, Theoretical investigation of encapsulation processes of C_{60} into single-wall carbon nanotubes, *Phys. Lett. A*, **375**(11), 1412–1416 (2011).

[92] L. J. Tang and X. N. Yan, Molecular dynamics simulation of C_{60} encapsulation into single-walled carbon nanotube in solvent conditions, *J. Phys. Chem. C*, **116**(21), 11783–11791 (2012).

[93] B. Burteaux, A. Claye, B. W. Smith, M. Monthioux, D. E. Luzzi, and J. E. Fischer, Abundance of encapsulated C_{60} in single-wall carbon nanotubes, *Chem. Phys. Lett.*, **310**(1–2), 21–24 (1999).

[94] B. W. Smith, M. Monthioux, and D. E. Luzzi, Carbon nanotube encapsulated fullerenes: A unique class of hybrid materials, *Chem. Phys. Lett.*, **315**(1–2), 31–36 (1999).

[95] J. Sloan, R. E. Dunin-Borkowski, J. L. Hutchison, K. S. Coleman, V. C. Williams, J. B. Claridge, *et al.*, The size distribution, imaging and obstructing properties of C_{60} and higher fullerenes formed within arc-grown single walled carbon nanotubes, *Chem. Phys. Lett.*, **316**(3–4), 191–198 (2000).

[96] K. Hirahara1, K. Suenaga1, S. Bandow, H. Kato, T. Okazaki, H. Shinohara, and S. Iijima, One-Dimensional Metallofullerene Crystal generated inside single-walled carbon nanotubes, *Phys. Rev. Lett.*, **85**(25), 5384–5387 (2000).

[97] R. Kitaura and H. Shinohara, Carbon-nanotube-based hybrid materials: nanopeapods, *Chem. Asian J.*, **1**(5), 646–655 (2006).

[98] D. Eder, Carbon nanotube–inorganic hybrids, *Chem. Rev.*, **110**(3), 1348–1385 (2010).

[99] D. E. Luzzi and S. W. Smith, Carbon cage structures in single wall carbon nanotubes: a new class of materials, *Carbon*, **38**(11–12), 1751–1756 (2000).

[100] S. Bandow, M. Takizawa, K. Hirahara, M. Yudasaka, and S. Iijima, Raman scattering study of double-wall carbon nanotubes derived from the chains of fullerenes in single-wall carbon nanotubes, *Chem. Phys. Lett.*, **337**(1–3), 48–54 (2001).

[101] S. Berber, Y. K. Kwon, and D. Tománek, Microscopic formation mechanism of nanotube peapods, *Phys. Rev. Lett.*, **88**, 185502 (2002).

[102] H. Ulbricht and T. Hertel, Dynamics of C_{60} encapsulation into single-wall carbon nanotubes, *J. Phys. Chem. B*, **107**(51), 14185–14190 (2003).

[103] F. Simon, H. Kuzmany, H. Rauf, T. Pichler, J. Bernardi, H. Peterlik, *et al.*, Low temperature fullerene encapsulation in single wall carbon nanotubes: Synthesis of N@C_{60}@SWCNT, *Chem. Phys. Lett.*, **383**(3–4), 362–367 (2004).

[104] M. Yudasaka, K. Ajima, K. Suenaga, T. Ichihashi, A. Hashimoto, and S. Iijima, Nano-extraction and nano-condensation for C_{60} incorporation into single-wall carbon nanotubes in liquid phases, *Chem. Phys. Lett.*, **380**(1–2), 42–46 (2003).

[105] A. N. Khlobystov, D. A. Britz, A. S. O'Neil, J. Wang, M. Poliakoff, and G. A. D. Briggs, Low temperature assembly of fullerene arrays in single-walled carbon nanotubes using supercritical fluids, *J. Mater. Chem.*, **14**(19), 2852–2857 (2004).

[106] T. W. Chamberlain, A. M. Popov, A. A. Knizhnik, G. E. Samoilov, and A. N. Khlobystov, The role of molecular clusters in the filling of carbon nanotubes, *ACS Nano*, **4**(9), 5203–5210 (2010).

[107] A. Botos, A. N. Khlobystov, B. Botka, R. Hackl, E. Szekely, B. Simandi, and K. Kamaras, Investigation of fullerene encapsulation in carbon nanotubes using a complex approach based on vibrational spectroscopy, *Phys. Stat. Sol. B*, **247**(11–12), 2743–2745 (2010).

[108] J. Fan, T. W. Chamberlain, Y. Wang, S. Yang, A. J. Blake, M. Schröder, and A. N. Khlobystov, Encapsulation of transition metal atoms into carbon nanotubes: a supramolecular approach, *Chem. Commun.*, **47**(20), 5696–5698 (2011).

[109] D. Kocsis, D. Kaptás, Á. Botos, Á. Pekker, and K. Kamarás, Ferrocene encapsulation in carbon nanotubes: Various methods of filling and investigation, *phys. Stat. sol. (b)*, **248**(11), 2512–2515 (2011).

[110] A. L. Torre, G. A. Rance, H. E. Jaouad, J. N. Li, D. J. Irvine, P. D. Brown, and A. N. Khlobystov, Transport and encapsulation of gold nanoparticles in carbon nanotubes, *Nanoscale*, **2**(6), 1006–1010 (2010).

[111] H. Kataura, Y. Maniwa, M. Abe, A. Fujiwara, T. Kodama, K. Kikuchi, *et al.*, Optical properties of fullerene and non-fullerene peapods, *Appl. Phys. A: Mater. Sci. Process*, **74**(3), 349–354 (2002).

[112] G. Q. Ning, N. Kishi, H. Okimoto, M. Shiraishi, Y. Kato, R. Kitaura, *et al.*, Synthesis, enhanced stability and structural imaging of C_{60} and C_{70} double-wall carbon nanotube peapods, *Chem. Phys. Lett.*, **441**(1–3), 94–99 (2007).

[113] K. Hirahara, K. Suenaga, S. Bandow, H. Kato, T. Okazaki, H. Shinohara, and S. Iijima, One-dimensional metallofullerene crystal generated inside single-walled carbon nanotubes, *Phys. Rev. Lett.*, **85**(25), 5384–5387 (2000).

[114] A. Debarre, R. Jaffiol, C. Julien, A. Richard, D. Naturelli, and P. Tchenio, Specific raman signatures of a dimetallofullerene peapod, *Phys. Rev. Lett.*, **91**(8), 085501 (2003).

[115] J. H. Warner, A. A. R. Watt, L. Ge, K. Porfyrakis, T. Akachi, H. Okimoto, *et al.*, Dynamics of paramagnetic metallofullerenes in carbon nanotube peapods, *Nano Lett.*, **8**(4), 1005–1010 (2008).

[116] K. Suenaga, T. Okazaki, K. Hirahara, S. Bandow, H. Kato, A. Taninaka, *et al.*, High-resolution electron microscopy of individual metallofullerene molecules on the dipole orientations in peapods, *Appl. Phys. A: Mater. Sci. Process.*, **76**(4), 445–447 (2003).

[117] Y. Kurokawa, Y. Ohno, T. Shimada, M. Ishida, S. Kishimoto, T. Okazaki, *et al.*, "Fabrication and characterization of peapod field-effect transistors using peapods synthesized directly on Si substrate", *Jpn. J. Appl. Phys.*, **44**(2005), 1341–1343.

[118] A. Guo, Y. Y. Fu, L. H. Guan, J. Liu, Z. J. Shi, Z. N. Gu, et al., Thermally assisted tunnelling in ambipolar field-effect transistors based on fullerene peapod bundles, *Nanotechnology*, **17**(10), 2655–2660 (2006).

[119] T. Shimada, Y. Ohno, K. Suenaga, T. Okazaki, S. Kishimoto, T. Mizutani, et al., Tunable field-effect transistor device with metallofullerene nanopeapods, *Jpn. J. Appl. Phys.*, **44**(1A), 469–472 (2005).

[120] T. Meng, C. Y. Wang, and S. Y. Wang, First-principles study of a hybrid carbon material: Imperfect fullerenes covalently bonded to defective single-walled carbon nanotubes, *Phys. Rev. B*, **77**(3), 033415 (2008).

[121] X. J. Wu and X. C. Zeng, First-principles study of a carbon nanobud, *ACS Nano*, **2**(7), 1459–1465 (2008).

[122] Y. Tian, D. Chassaing, A. G. Nasibulin, P. Ayala, H. Jiang, A. S. Anisimov, et al., The local study of a nanobud structure, *Phys. Stat. Sol. (B)*, **245**(10), 2047–2050 (2008).

[123] X. Zhu and H. B. Su, Magnetism in hybrid carbon nanostructures: Nanobuds, *Phys. Rev. B*, **79**(16), 165401 (2009).

[124] P. Havu, A. Sillanpää, N. Runeberg, J. Tarus, E. T. Seppälä, and R. M. Nieminen, Effects of chemical functionalization on electronic transport in carbon nanobuds, *Phys. Rev. B*, **85**(11), 115446 (2012).

[125] A. G. Nasibulin, P. V. Pikhitsa, H. Jiang, D. P. Brown, A. V. Krasheninnikov, A. S. Anisimov, et al., A novel hybrid carbon material, *Nature Nanotech.*, **2**, 156–161 (2007).

[126] Y. Tian, D. Chassaing, A. G. Nasibulin, P. Ayala, H. Jiang, A. S. Anisimov, and E. I. Kauppinen, Combined Raman spectroscopy and transmission electron microscopy studies of a nanobud structure, *J. Am. Chem. Soc.*, **130**(23), 7188–7189 (2008).

[127] H. Y. He and B. C. Pan, Electronic structures and Raman features of a carbon nanobud, *J. Phys. Chem. C*, **113**(49), 20822–20826 (2009).

[128] X. J. Wu and X. C. Zeng, Periodic graphene nanobuds, *Nano Lett.*, **9**(1), 250–256 (2009).

[129] M. Wang, C. M. Li, Magnetic properties of all-carbon graphene-fullerene nanobuds, *Phys. Chem. Chem. Phys.*, **13**, 5945–5951 (2011).

[130] X. Wang, C. Ma, K. Chen, H. Li, and P. Wang, Interaction between nanobuds and hydrogen molecules: A first-principles study, *Phys. Lett. A*, **374**(1), 87–90 (2009).

[131] A. P. Terzyk, S. Furmaniak, P. A. Gauden, and P. Kowalczyk, Fullerene-intercalated graphene nano-containers – mechanism of argon adsorption and high-pressure CH_4 and CO_2 storage capacities, *Adsorpt. Sci. Technol.*, **27**(3), 281–296 (2009).

[132] D. Yu, K. Park, M. Durstock, and L. M. Dai, Fullerene-grafted graphene for efficient bulk heterojunction polymer photovoltaic devices, *J. Phys. Chem. Lett.*, **2**(10), 1113–1118 (2011).

[133] R. Z. Wan, H. J. Lu, J. Y. Li, J. D. Bao, J. Hu, and H. P. Fang, Concerted orientation induced unidirectional water transport through nanochannels, *Phys. Chem. Chem. Phys.*, **11**(42), 9898–9902 (2009).

[134] D. Golberg, Y. Bando, C. C. Tang, and C. Y. Zhi, Boron nitride nanotubes, *Adv. Mater.*, **19**(18), 2413–2432 (2007).

[135] C. H. Sun, H. X. Yu, L. Q. Xu, Q. Ma, and Y. T. Qian, Recent development of the synthesis and engineering applications of one-dimensional boron nitride nanomaterials, *J. Nanomater.*, **2010**, 163561 (2010).

[136] M. L. Cohen and A. Zettl, The physics of boron nitride nanotubes, *Phys. Today*, **63**(11), 34–38 (2010).

[137] L. M. Cao, X. Y. Zhang, H. Tian, Z. Zhang, and W. K. Wang, Boron nitride nanotube branched nanojunctions, *Nanotechnology*, **18**(15), 155605 (2007).

[138] S. K. Singhal, A. K. Srivastava, N. Dilawar, and A. K. Gupta, Growth and characterization of boron nitride nanotubes having novel morphologies using mechanothermal process, *J. Nanopart. Res.*, **12**(6), 2201–2210 (2010).

[139] J. Zhang, Z. Q. Li, and J. Xu, Formation and structure of boron nitride nanotubes, *J. Mater. Sci. Technol.*, **21**(1), 128–130 (2005).

9

Doped Graphene: Theory, Synthesis, Characterization, and Applications

Florentino López-Urías,[a,b] Ruitao Lv,[a] Humberto Terrones,[a,c] and Mauricio Terrones[a,d]

[a]*Department of Physics, The Pennsylvania State University, USA*
[b]*Advanced Materials Department, IPICYT, México*
[c]*Departamento de Física, Universidade Federal do Ceará, Brazil*
[d]*Department of Materials Science and Engineering and Materials Research Institute,
The Pennsylvania State University, USA and
Research Center for Exotic Nanocarbons (JST), Shinshu University, Japan*

9.1 Introduction

Two-dimensional (2D) monolayer sp^2 hybridized carbon, known as *graphene*, was firstly isolated by Andre Geim, Konstantin Novoselov and co-workers [1–4], using the so-called "scotch tape method". These authors also developed an ingenious technique for graphene identification using an optical microscope. In particular, Novoselov, Geim, and collaborators fabricated a field-effect transistor (FET) using a single sheet of graphene [1]. In 2010, Andre Geim and Konstantin Novoselov were awarded the Nobel Prize in Physics, and the fascinating properties of graphene kept stimulating extensive experimental and theoretical studies. Graphene is a semi-metal material which exhibits unprecedented physico-chemical properties, such as high electron mobility, excellent thermal conductivity, the integer quantum Hall effect, quantum blockades, enhanced Raman scattering, and so on [5–9].

Soon after the successful isolation of graphene, various groups worldwide started to explore different synthetic approaches to produce single and few-layered graphenes. In addition, other groups started to investigate the effects of doping in the electronic properties of graphene [10–24]. From this, substitutional doping of graphene with other atoms (e.g.,

Graphene Chemistry: Theoretical Perspectives, First Edition. Edited by De-en Jiang and Zhongfang Chen.
© 2013 John Wiley & Sons, Ltd. Published 2013 by John Wiley & Sons, Ltd.

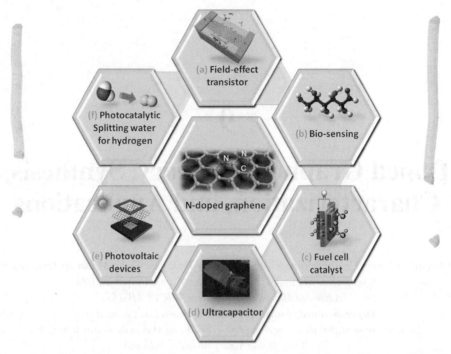

Figure 9.1 *Sketch showing different applications of N-doped graphene:* **(a)** *n-type field-effect transistor (FET) (Reprinted with permission from [41] © 2009 AAAS);* **(b)** *Bio-sensing for glucose molecules [73];* **(c)** *Metal-free fuel-cell catalyst [13] or catalyst support materials [69];* **(d)** *Flexible ultracapacitor (Reprinted with permission from [21] © 2011 American Chemical Society);* **(e)** *Transparent conducting film for photovoltaic devices [70], and* **(f)** *Catalyst for photocatalytic splitting water to generate hydrogen [74]*

nitrogen) will lead to new interesting properties and applications as illustrated in Figure 9.1. The use of several characterization techniques such as scanning electron microscopy (SEM), high-resolution transmission electron microscopy (HRTEM), Raman spectroscopy (RS), X-ray diffraction (XRD), X-Photoelectron spectroscopy (XPS), and atomic force microscopy (AFM), has resulted in a better understanding of the chemical and physical properties of graphene. The graphene structure belongs to one of the five 2D Bravais lattices called the hexagonal (triangular or honeycomb) lattice. It is noteworthy that, by piling up individual graphene layers in an ordered way with ABAB... stacking, one could form hexagonal 3D Bernal graphite, which belongs to the $P6_3/mmc$ (194) space group. It is noteworthy that a single layer of sp^2 hybridized carbon has also been used to describe the structure of carbon nanotubes (rolled graphene sheets).

9.2 Substitutional Doping of Graphene Sheets

Graphene possess a unique fingerprint: an electronic structure with a linear E(k) dispersion close to the Fermi level at the K and K' points in the Brillouin Zone. Figure 9.2(a) depicts

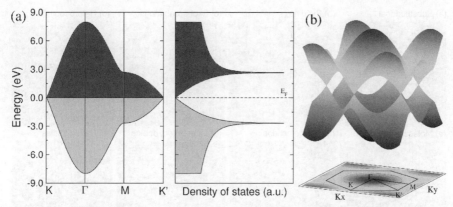

Figure 9.2 Graphene sheet electronic properties obtained using a nearest-neighbors tight binding model. **(a)** Band-structure and electronic density of states (DOS) calculations. In both cases, the Fermi level is set at zero (dashed line). The paler shaded regions correspond to the valence bands (occupied states) and the darker shaded regions depict the conduction band (unoccupied states); note that the system exhibits a zero electronic band-gap. **(b)** Energy surfaces for both occupied and unoccupied states. Note that the zero gap behavior occurs at energies corresponding to K and K' points which are degenerate. Therefore, K and K' points are the principal communication channels between the valence and conduction bands. In our calculations, the electronic hopping t_0 was set to 2.66 eV

the density of states (DOS) and the band-structure of graphene using a tight binding model considering one orbital per atom and nearest neighbor electronic hopping. Negative (positive) energy values in the DOS represent the occupied (unoccupied) states. From the DOS plot, it is possible to observe the semi-metallic behavior of graphene with a zero electronic band-gap. The band structure in conjunction with the most symmetrical points of the Brillouin zone K-Γ-M-K' reveal that the zero band-gap arises from the K and K' points. In addition, around these symmetry points, the energy exhibits a linear behavior. It is important to mention that different paths can be found in the literature such as Γ-K-M-Γ, but the physics is essentially the same. Figure 9.2(b) shows the energy as a function of k_x and k_y wave vectors. The low (top) surface energy corresponds to the valence (conduction) band. In this case, we have two energy surfaces since the calculation was carried out with two atoms in the unit cell. Note that at K and K' symmetry points, both valence and conduction energy plots exhibit a conical behavior; the contact points at K and K' are called Dirac points.

Figure 9.3 depicts different ways of doping graphene. Substitutional doping (Figure 9.3a) occurs when a carbon atoms is removed and replaced by a different atom. This type of doping is found experimentally for boron (B) and nitrogen (N), since the interatomic distance of B–C or N–C is similarly to the C–C bond in sp^2 hybridized carbon structures (∼1.42 Å). Note that both chemical elements (B and N) are C neighbors in the periodic table, the B $1s^2 2s^2 p^1$ (N $1s^2 2s^2 p^3$) with one electron less (plus) when compared to the C atom $1s^2 2s^2 p^2$. Therefore, the B (N) atoms introduce holes (electrons) into the graphene sheet. B and N atoms have been widely studied and have played a key role when understanding the doping effects within carbon nanostructures [25,26]. In addition to B and N, other elements

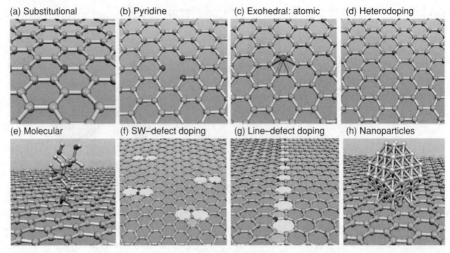

Figure 9.3 *Molecular models representing different types of doped graphenes.* **(a)** *Substitutional doping, in which a carbon atom is removed and replaced by the dopant;* **(b)** *pyridine-like doping, in which a single vacancy is generated and the low coordinated carbon atoms (three carbons around the vacancy) are replaced by dopants;* **(c)** *Exohedral atomic doping where a dopant is hosted on the surface of graphene;* **(d)** *Heterodoping which consists in doping simultaneously the graphene with two or more types atoms different from carbon;* **(e)** *Molecular doping or chemical functionalization of the graphene sheet;* **(f)** *55–77 Thrower–Stone–Wales defects in which 5–7–5–7 rings are generated by rotating a C–C bond by 90 degrees (dopants are also included in some defects),* **(g)** *5–7 line-defect in combination with substitutional doping, and* **(h)** *Nanoparticles or clusters anchored on the graphene surface. See colour version in the colour plate section*

could be substitutionally doping the graphene sheet, however, since the distance between C and dopant atom exceeds the C–C distance (∼1.42 Å), the dopant atoms remain out of the planar structure of graphene. Figure 9.3(b) shows the pyridine-like doping which consists of removing a C atom from the hexagonal lattice (creation of a vacancy in the graphene lattice), and the remaining two-coordinated C atoms surrounding the vacancy are replaced by the dopant. This type of doping is generally applied to N. In Figure 9.3(c) the exohedral doping is described, and corresponds to adatoms located above (or below) the sheet (e.g., phosphorus, silicon, sulfur, metal atoms, etc.). Figure 9.3(d) depicts the heterodoping of graphene, which corresponds to the addition of two or more types of atoms (besides carbon). The synthesis of heterodoped carbon nanostructures was first introduced by Cruz-Silva et al. [27] in carbon nanotubes. Graphene could be also be functionalized with molecules as shown in Figure 9.3(e). Another type of doping is the combination of substitutional doping together with defects embedded into the graphene lattice. The dopants could stabilize the formation of isolated Thrower–Stone–Wales 55–77 defects [28,29] (see Figure 9.3f) or the formation of line of defects in which two different graphene orientations are joined by several 5–7 defects (see Figure 9.3g). Graphene could be also doped with clusters (nanoparticles), thus increasing its surface area (see Figure 9.3h). There are other configurations of doped graphene and they are out of the scope of this chapter, however,

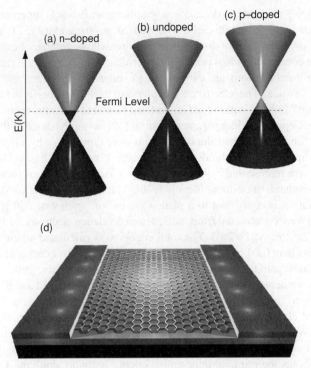

Figure 9.4 Schematic representation of the electronic energy E(k) of graphene sheets around of K and K' points (see Figure 9.2). The different shades depict the occupied states (dark shading) and unoccupied states (lighter shading) and the Fermi level is indicated by the dashed line. (**a**) Electron doping (n-type doping); (**b**) undoped or pristine, and (**c**) hole-doping (p-type doping). Note that in this representation, both types of doping (n and p) keep the zero band-gap behavior, and (**d**) Schematic representation of a field effect transistor (FET) based on graphene

the most important issue about doping consists of controlling the type of doping, as well as the location and concentration of dopants.

The graphene sheet could be doped with electrons (n-type doping) or holes (p-type doping). The value n (p) means negative (positive) doping. A common way to represent doping in graphene is depicted in Figure 9.4(a–c). Here, the systems retain the zero band-gap and after doping, the effect of a hole or an electron shifts the Fermi level. The FET could then modify the charge carriers concentration by the presence of an external electric field, thus electrons could be replaced by holes and vice versa (see Figure 9.4d). However, this is an ideal situation that can be modified with the doping concentration and the type of dopant. For example, a single B or N atom hosted in graphene could open an electron band-gap, and the zero band-gap will be retained only if the dopant atoms are located in specific atomic positions. From a theoretical standpoint (tight-binding models or first-principles density functional theory), slightly doped graphene systems represent more computational effort since it is necessary to consider a relatively large doped supercell. The number of ways to accommodate n_d dopant atoms in a supercell with N sites is $N!/[(N-n_d)!$

$n_d!$]. For example, in order to dope 3% a graphene with 6 × 6 supercell with N = 98 atoms (3 dopants ($n_d = 3$) and 95 carbons) dopant atoms there exits 152 096 permutations. Therefore, to consider all dopant distributions requires a large number of calculations, a task that can be time consuming if first principles electronic structure methods are used. Density functional theory calculations are carried out in order to understand the role of doping in the graphene lattice. The electronic calculations have been performed using Density Functional Theory [30, 31] within the Generalized Gradient Density Approximation (DFT-GGA) using the Ceperley–Alder parameterization [32] as implemented in the SIESTA code [33]. The wave functions for the valence electrons were represented by a linear combination of pseudo-atomic numerical orbitals using a double-ζ polarized basis (DZP) [34], while core electrons were represented by norm-conserving Troullier–Martins pseudopotentials in the Kleynman–Bylander non-local form [35, 36]. The real-space grid used for charge and potential integration is equivalent to a planewave cut-off energy of 250 Ry. The pseudopotentials (pps) were constructed from the 3, 4, and 5 valence electrons for B, C, and N ions (B: $2s^2 2p^1$, C: $2s^2 2p^2$, N: $2s^2 2p^3$). The total energy was calculated when the forces were converged to less than 0.04 eV/Å. In our calculations, all atoms contained in the supercell (6 × 6) are relaxed using conjugated gradient method with variable cell.

In Figure 9.5, we depict the structure and band structure of N-doped and B-N-heterodoped graphene considering different doping concentrations (3.06, 4.08, and 7.14%). Results reveal important changes around the Fermi level since the nitrogens introduce electrons to the system. We have found that the system with 4.08% of doping exhibit an indirect electronic band of 0.16 eV, the maximal occupied energy (valence band) is located in the Γ-point whereas the minimal unoccupied energy is found along the Γ–K path points (see Figure 9.5a). However for 3.06 and 7.14% doping, the band structure exhibits metallic features (see Figures 9.5b–c). Also, results shown in Figure 9.5 reveal the opening of an electronic band gap at K-point which is identified around an energy of −1eV, for undoped graphene systems, the band structure exhibit zero gap and energy linear behavior around of the K-point. Interesting behavior is found when the combination of boron and nitrogen doping is performed. The corresponding band structures for the different BN-doping concentrations are shown in Figure 9.5(d–f). The results show that a balanced doping (same B and N concentrations, 4.08% doping in our case), the band structure exhibits a centered electronic band-gap at a K-point of 0.4 eV approximately (see Figure 9.5d). For low and large concentrations, 3.6 and 7.14% respectively, the system exhibits a metallic behavior and the band-gaps are shifted to conduction or valence bands (see Figure 9.5e–f).

Results shown previously reveal the sensibility of the electronic properties of doped graphenes to the doping concentration. Now, we explore the electronic properties dependence of the position of dopants. We focus on the 4.08% doping concentration cases and the sublattices' division in honeycomb structures. Figure 9.6(a) shows the graphene supercell which has been divided in two sublattices (see the shaded atoms in Figure 9.6b). Figure 9.6(c–d), show the structures and the band structure calculation for N– and –BN-doping graphene. Note that the difference in structure with respect to the Figures 9.5(a) and 9.5(d) is the position of two dopants as is shown by the squares in Figure 9.6(c–d), here both dopants have been changed to an opposite sublattice with respect to the remaining dopants. In contrast to the case shown in Figure 9.5(a), the N-doped case exhibits a metallic behavior (see Figure 9.6c) whereas the BN case exhibits a reduction of the electronic band

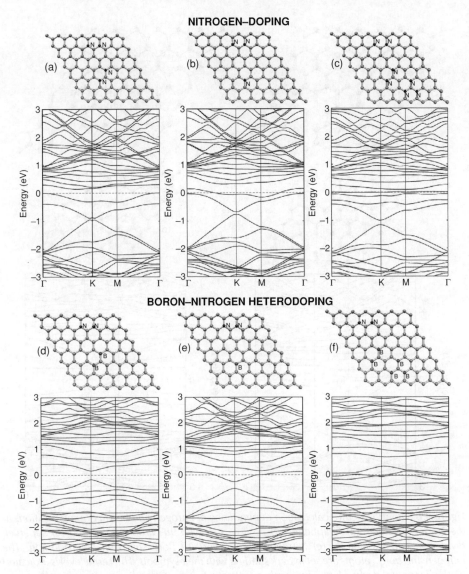

Figure 9.5 Band-structure calculations dependence of nitrogen and boron doping concentration in graphene lattice. (a)–(c) Nitrogen doping and (d)–(f) Boron-nitrogen heterodoping cases. The nitrogen and boron atoms are set in dark shades with their symbols. The Fermi level is indicated by the horizontal dashed line. All graphene structures were relaxed using conjugated gradient method. The corresponding formation energies of doping for cases (a)–(f) are 7.53, 5.55, 13.8, 9.92, 6.69, and 21.34 eV respectively, calculated as Ref. [27]

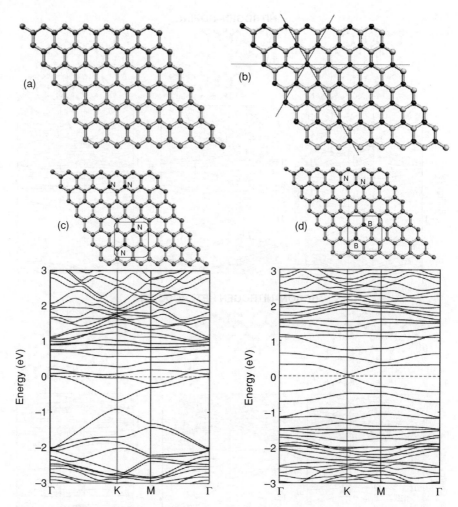

Figure 9.6 Band-structure calculation dependence of nitrogen and boron doping positions in the graphene lattice. **(a)** Graphene supercell 6 × 6 with 98-carbon atoms and **(b)** supercell divided in two sublattices. **(c)** Nitrogen doping and **(d)** Boron-nitrogen heterodoping. The nitrogen and boron atoms are set in dark shading with their symbols. The Fermi level is indicated by the horizontal dashed line. The graphene structures were relaxed using conjugated gradient method. The dopants indicated by squares are in different sublattices related with the two remaining dopants. Note that both cases depicted in **(a)** and **(b)** were calculated in Figures 9.5(a) and 9.5(d) respectively with all dopants in the same sublattice. The corresponding formation energies of doping for cases **(c)** and **(d)** are 7.85 and 10.22 eV respectively, calculated as Ref. [27]

gap (compare Figure 7.5d and Figure 9.6d). The two cases shown in Figure 9.6 present larger formation energies than when the doping is set only in one sublattice, these results are in according with a recent report by Lv et al. [37]. Additional theoretical and experimental research in this direction is needed in order to understand the physics of doped graphenes more clearly.

Recently, several investigations have demonstrated that heterodoping could improve the chemical reactivity and the electronic properties of these materials. In this context, Cruz-Silva et al. [27, 38] reported the doping of carbon nanotubes and graphene sheets. The authors demonstrated that low concentration of substitutional N doping does not cause major distortions of the graphene sheet (Figure 9.7). The C–N bond length is less than 2% shorter than the C–C bond, and first and second neighbors only experience small displacements (less than 0.7% change in their bond lengths) in order to compensate this shortening. This effect is also true for N3 pyridine-like doping of the graphene layer (Figure 9.7). However, when doping with P (with or without N), the graphene lattice experiences significant distortions (Figure 9.7c,d). The P atom preserves its sp^3 character, and bonds with tetrahedral-like configurations appear. The P–C bond length is 1.79 Å, larger when compared to 1.42 Å for C–C sp^2 bonds. The 26% increase in the bond length combined with the difference in bond angles, forces P to protrude from the graphene plane, thus displacing the positions of the first, second, and third neighbors out of the plane (see Figure 9.7c). This could be interpreted as a corrugation induced by the presence of P

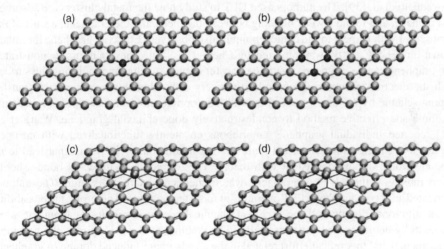

Figure 9.7 *First-principles density functional calculations of doped and heterodoped graphene on a supercell with 98 atoms. (a) substitutional nitrogen doping; (b) pyridine-like nitrogen doping; (c) phosphorus doping, and (d) nitrogen-phosphorous heterodoping. The pristine carbon positions are represented with a black hexagonal network. It can be observed that the presence of phosphorus induces great stress and corrugation within graphene. The formation energy of the different cases can be seen in Table 9.1. The most stable doping system corresponds to the substitutional doping followed by the pyridine-like doping. Note that the PN-heterodoped system is more stable that when only the system is doped with P (Reprinted with permission from [27] © 2009 American Chemical Society)*

Table 9.1 Formation energies associated with the doping of a graphene layer, as well as the changes in bond length for the doping atom to its first neighbors. The energies are expressed in eV and lengths in Å. The defect formation energy was calculated in the following way. First, the binding energy of the complete supercell was calculated as defined $E_{B\text{-}SC} = (N_C E_C + N_N E_N + N_P E_P) - E_{total}$ where E_{B_SC} is the binding energy of the supercell, E_{Total} is the total energy of the supercell, N_X is the number of atoms of the element X and E_X is the total energy of an isolated atom of the element X. The defect energy is calculated using the expression, $E_{defect} = E_{B\text{-}pristine} - E_{B\text{-}doped}$ [27]

System	Defect energy	1st nn bond A	1st nn bond B
Pristine	–	1.42	1.42
N-doped	0.732	1.39	1.39
N_3-doped	4.423	1.33	1.33
P-doped	6.637	1.78	1.78
PN-doped	6.180	1.78	1.80

atoms. For PN-doped graphene (see Figure 9.7d), the presence of N helps to reduce the displacements caused by the introduction of P, thus resulting in a "damping" effect on the structural strain, especially within first and second neighbors. This effect can be quantified by analyzing the formation energy total energy calculations (see Table 9.1).

Graphene doped with metals such as Al, Ag, Cu, Au, and Pt has been reported by Giovannetti et al. [39]. The authors used DFT to study how the metal clusters are adsorbed on graphene, and found a weak bonding between the metal and the graphene; a shift of the Fermi level with respect to the conical point is also observed. Epitaxial graphene thermally grown on 6H-SiC(0001) was reported by Chen et al. [40]. These authors demonstrated that graphene can be p-type via a surface transfer doping scheme by modifying the surface with the electron acceptor. Chen and co-workers claim that this novel surface transfer doping scheme by surface modification with appropriate molecular acceptors, represents a simple and effective method to nondestructively dope epitaxial graphene. Wang et al. [41] reported individual graphene nanoribbons covalently functionalized with nitrogen species using high-power electrical joule heating in ammonia (NH_3) gas, that lead to n-type electronic doping consistent with theory. The formation of the C–N bonds should occur mostly at the edges of graphene where chemical reactivity is high. The authors fabricated an n-type graphene FET operating at room temperature. Therefore, the possibility of substitutional doping and the excellent chemical sensor properties of graphene were successfully demonstrated experimentally [42]. Wehling et al. [43] reported the first joint experimental and theoretical effort related to the an adsorbate-induced doping of graphene. A general relationship between the doping performance and whether the adsorbates are open- or closed-shell systems, is demonstrated with NO_2: The single, open shell NO_2 molecule is found to be a strong acceptor, whereas its closed shell dimer N_2O_4 causes only weak doping. Xiaolin et al. [44] reported a simple chemical method to obtain bulk quantities of N-doped, reduced graphene oxide (rGO) sheets via thermal annealing of GO in NH_3. The N-doped rGO clearly shows n-type behavior with the Dirac point at negative gate voltages in three terminal devices. These authors claim that their method could lead to the synthesis of

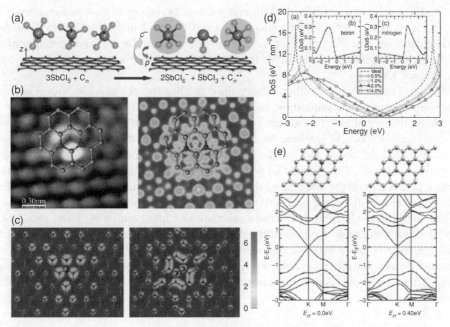

Figure 9.8 *(a) Closed shell molecular dopant destabilized by partial carrier transfer within the graphene surface, p-type doping SbCl$_5$ complex (Reprinted with permission from [45] © 2011 American Chemical Society); (b) High resolution images with N defects arranged in different configurations in a graphene sheet and simulated STM image (Reprinted with permission from [47] © 2011 American Chemical Society); (c) Graphene acting as a metal surface, real space properties of impurity states for single and double impurities, note that the impurity state due to the single impurity in one sublattice is almost entirely localized in the other sublattice (Reprinted with permission from [48] © 2009 Elsevier); (d) DOS of B- and N-doped 2D graphene (Reprinted with permission from [49] © 2008 American Physical Society); and (e) First-principles calculations of the electronic properties of B and N codoping graphene sheet, results obtained by X. Deng et al. (Reprinted with permission from [50] © 2009 Elsevier). The atomic structures (top panel) and the electronic band structures (bottom panel) of B/N codoped graphene. Note that for the codoping case a band-gap is enhanced*

bulk amounts of N-doped, rGO sheets useful for various practical applications. Nistor et al. [45] demonstrated that the graphene surface acts as a catalytic reducing/oxidizing agent, driving the chemical disproportionation of adsorbed dopant layers into charge-transfer complexes which inject majority carriers into the 2D carbon lattice. The authors studied the molecular SbCl$_5$ (see Figure 9.8a) and HNO$_3$ intercalates, and the solid compound AlCl$_3$. Raman spectroscopy data related to doped graphene was reported by Andrea Ferrari [46]. He investigated the evolution of the electronic properties and electron-phonon interaction of doped graphene by applying a gate-voltage in order to tune the Fermi level. He observed a stiffening of the Raman G-band for holes and electrons. The author claims that Raman spectroscopy could be efficiently used to monitor the doping level and confinement. Deng et al. [47] developed a novel method for one-pot direct synthesis of N-doped graphene (in gram scale) via the reaction of tetrachloromethane (CCl_4) with lithium nitride (Li_3N) under mild conditions. The electronic structure perturbation induced by the incorporation of N in the graphene network was observed for the first time by scanning tunneling microscopy (see Figure 9.8b).

Wehling et al. [48] studied different interaction mechanisms of impurities with graphene. The authors studied the impurity states in the vicinity of the Fermi level, and compared graphene to normal metals and semiconductors (see Figure 9.8c). Electronic transport in chemically doped 2D graphene materials was reported by Lherbier et al. [49]. They used *ab initio* calculations and found charge mobilities and conductivities of systems with impurities (B or N). They found that for a doping concentration as large as 4.0 wt%, the conduction is marginally affected by quantum interference effects. They also found that electron-hole mobilities and conductivities become asymmetric with respect to the Dirac point (see Figure 9.8d). First principles calculations carried out by X. Deng et al. [50] revealed that there exists the possibility of opening an electronic band-gap with B and N codoping (see Figure 9.8e). The authors studied different N and B doping configurations and found that when these dopants are located far away from each other, the band-gap was large. They also studied the chemical reactivity of the doped graphene by setting a Li-atom around the dopants.

9.3 Substitutional Doping of Graphene Nanoribbons

An extended line of defects (ELD) in graphene has been found using nickel [111] as substrate [51]. The authors propose that the defects involved are a combination of two pentagons and one octagon (5–8–5) forming a nanowire embedded in the graphene matrix (see Figure 9.9a), with strong metallicity. This experimental result has motivated the study of different possibilities of the ELD in graphene, and other layered materials such as BN and metal dichalcogenides. For example, first principles calculations of the ELD including two pentagons and two heptagons (D5D7), and three pentagons and three heptagons (T5T7) (see Figure 9.9a), have been studied using LDA and GGA [52]. The theoretical results indicate that the ELD are stable and could be produced experimentally. In addition, first principles STM simulations confirm the structure of the 5–8–5 ELD [52]. Since the ELD is more reactive than graphene, it might be interesting to study how the ELD respond to different dopants, in particular B and N since these are close to C in the Periodic table. As we have discussed earlier, larger atoms embedded in graphene break the planarity of the graphene structure. In this context, Kan and colleagues have reported the doping a 5–8–5 ELD with B and N, and have only analyzed one site for dopants, which was found to be magnetic in the case of B [53]. In order to understand the most suitable site for dopants in the ELD, different sites need to be tested (see Figure 9.9a). According to planewave calculations as implemented in the CASTEP code using GGA-PBE exchange-functional, we have found that N and B are more stable in different sites [54] (see Table 9.2). Figure 9.9(b) depicts the non-polarized density of states (DOS) of the undoped and doped nanoribbons. DOS of the doped cases corresponds to the sites where the dopant is most stable.

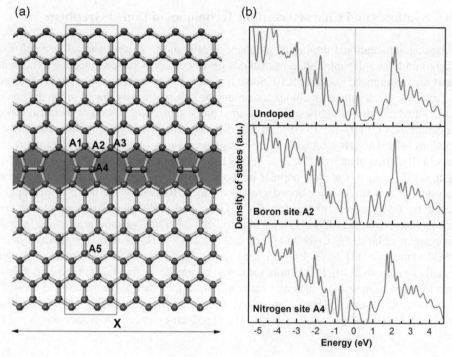

Figure 9.9 (a) Molecular model of zigzag GNRs with an extended line of defects (ELD) with 5–8 carbon rings. The ribbon is periodic in the x-direction and the unit cell is marked by the rectangle. The B or N atom could be set in sites labeled by A1, A2, A3, A4, and A5. (b) Electronic density of states (DOS) for undoped, B-, and N-doped GNRs. Interestingly, the DOS for the doped systems corresponds to the most stable configurations

Table 9.2 Doped energy (E_{dop}) of zigzag GNRs with an extended line of defects (ELD) containing 5–8 carbon rings. Note that the site A2 (A4) is the most stable configuration for B (N) atom (the corresponding site labels can be seen in Figure 9.9a). The doping energy has been calculated from the following expression: $E_{dop} = E_{tot} - E_{prist} - N_d E_{dc} + N_d E_{gc}$ as is calculated in Ref. [81]. E_{tot} refers to the total energy of the doped structure, N_d to the number of dopant atoms, E_{dc} to the energy per atom of the dopant atom in its crystalline phase, E_{gc} is the energy per atom of graphene. It is interesting to note that N at the sites A1, A2 and A4 is more stable than B and even C at the same sites (negative value of the doping energy). These sites involve pentagonal rings of carbon

Structure	Site A1	Site A2	Site A3	Site A4	Site A5
B-doping	1.0567	0.3717	0.4448	0.5155	1.0781
N-doping	−0.2041	−0.0354	0.6222	−0.3936	0.9074

9.4 Synthesis and Characterization Techniques of Doped Graphene

Regarding substitutional doping of graphene sheets, most of experimental research has focused on N as a dopant. Different methods developed for N-doped graphene (NG) synthesis are summarized in Figure 9.10. Some typical experimental parameters are also listed in Table 9.3. The synthetic methods for producing substitutionally doped graphene could be classified into two categories. One is *in situ (single step) doping*, such as chemical vapor deposition (CVD) [10–13, 15–19], the arc discharge method [20, 26], and solvothermal reactions [47]. The other category is *post-treatment doping*, such as electrothermal reactions [41], plasma treatment [14, 21, 22], and thermal treatment [23, 55–57]. Among all of them, CVD is one of the most popular methods due to its ability of achieving large-area, high-quality and single-layer doped graphene sheets. Regarding the nitrogen precursors used in the CVD process, except for NH_3 gas, some other N-containing *liquid* precursors, such as acetonitrile (CH_3CN) [12], pyridine (C_5H_5N) [17, 26] or *solid* precursors, including cyanuric chloride ($C_3Cl_3N_3$) [47] and melamine ($C_3N_6H_6$)/poly(methyl methacrylate) (PMMA) mixture [11], have also been used for the synthesis of NG sheets with few layers (usually no more than 10). In addition, some post treatments on few-layer pristine graphene, such as Joule heating [41], plasma treatment [14, 21], or thermal annealing [23, 55–57] in N-containing gases (NH_3, N_2, etc.) have been used for synthesizing NG. These post treatments have some unique advantages when producing devices simultaneously.

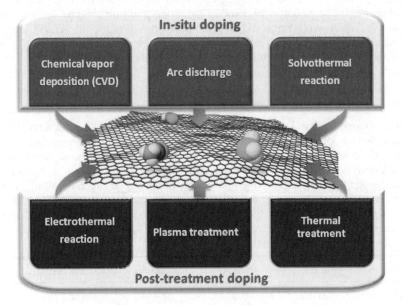

Figure 9.10 Diagram showing different experimental methods used to synthesize a heteroatom- (e.g., N) doped graphene. The methods could be classified into two categories. One is in situ (single step) doping. The other technique corresponds to post-treatment doping, in which the doping was achieved after the growth of graphene

Table 9.3 Summary of some experimental work related to N-doped graphene (NG) synthesis. In particular, chemical vapor deposition (CVD) is one of the popular methods due to its advantages when achieving large-area, high-quality and single-layer doped graphene sheets

Layers	Method	Substrate	Precursors	Growth parameters	Reference
2–6	CVD	Cu (25 nm) film on silicon wafer	Methane (CH_4) and Ammonia (NH_3)	800°C, 10 min	[10]
≥1	CVD	Cu foil (25 μm thick)	Acetonitrile (CH_3CN) and ammonia (NH_3)	950°C, 500 mTorr, 3–15 min	[12]
2–8	CVD	Ni (300 nm) film on silicon wafer	Methane (CH_4) and ammonia (NH_3)	1000°C, 5 min	[13]
≥1	CVD	Ni (300 nm) or Cu (300 nm) film on silicon wafer	Methane (CH_4) and Ammonia (NH_3)	980°C, 3 min (for Ni) or 20 min (for Cu)	[16]
1	CVD	Cu foil (25 μm thick, 99.999%)	Pyridine vapor	1000°C, 10 min at ~7 Torr	[17]
1–2	CVD	Cu foil (34 μm thick, 99.95%)	Ethylene (C_2H_4) and ammonia (NH_3)	900°C, 4.6 Torr, 30 min	[19]
>2	Arc discharge	No substrates	Graphite rod, pyridine vapor or ammonia (NH_3)	Flowing H_2 (200 Torr) and He (500 Torr) through a pyridine bubbler	[26]
1–6	Solvothermal reaction	No substrates	Lithium nitride, tetrachloromethane and cyanuric chloride	250–350°C, 6–10 h	[47]
1–3	Electrothermal reactions	Graphene nanoribbons on silicon wafer	Ammonia (NH_3)	e-annealed graphene nanoribbons in NH_3	[41]
1–2	Plasma post treatment	Pristine graphene	Ammonia (NH_3)	NH_3 plasma at a dose of 3×10^{14} cm^{-2}	[14]
>2	Plasma post treatment	No substrates	Reduced graphene oxide and N_2	Treatment in nitrogen plasma (500 W power, 14 torr N_2)	[21]
>2	Thermal treatment	No substrates	Ammonia (NH_3)	Heat treating graphite oxide (GO) at 800°C for 2 h in NH_3 atmosphere	[57]

Characterization of these doped layered materials is crucial because it includes the identification of the dopant, its concentration, its bonding environment, and its precise location. Different spectroscopy and microscopy techniques have been widely used when characterizing the structure and morphology of dopants in graphenes. For example, X-ray photoelectron spectroscopy (XPS) [10], Raman spectroscopy [58, 59], and Auger electron spectroscopy (AES) [60] have been employed to determine the doping level, doping type, and dopant species. High-resolution transmission electron microscopy (HRTEM) [16] and scanning tunneling microscopy (STM) [15] are used to visualize individual atomic dopants within the graphene lattice. Some examples related to NG characterization techniques are shown in Figure 9.11.

In particular, XPS is a quantitative spectroscopic technique, which could be used to determine the atomic species, chemical state of the dopants and doping level. Figure 9.11(a) shows XPS spectra of NG and pristine (undoped) graphene sheets [10]. From the survey scan, one could identify the dopants by analyzing the binding energies and the doping levels

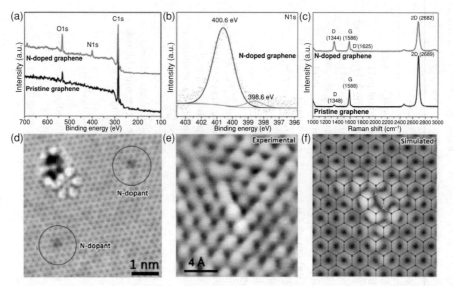

Figure 9.11 Different techniques for characterizing N-doped graphene. **(a)** X-ray photoelectron spectroscopy (XPS) survey scan (Reprinted with permission from [10] © 2009 American Chemical Society); **(b)** N1s line scan [37]. The XPS techniques could be used to identify the atomic species, chemical state of the dopants and doping level (Reprinted with permission from [37] © 2012 Nature Publishing Group); **(c)** Raman spectra of N-doped and pristine graphene sheets [37]. Raman spectroscopy could be used to distinguish n-type or p-type doping and even provide information about the dopant concentration. It is also powerful in identifying the number of graphene layers (Reprinted with permission from [37] © 2012 Nature Publishing Group); **(d)** HRTEM image of N-doped graphene (Reprinted with permission from [16] © 2011 Nature Publishing Group); **(e)** Experimental [37] and **(f)** Simulated scanning tunneling microscopy (STM) images and a ball-and-stick structural model of the N_2^{AA} dopant in graphene. The C and N atoms are depicted in dark and gray balls, respectively [37]. (Parts e and f Reprinted with permission from [37] © 2012 Nature Publishing Group)

of the corresponding peak areas of the other elements present within samples. In this case, the peaks located at around 284.8 eV, 401.6 eV, and 531.9 eV correspond to the C1s (sp^2 hybridized) of carbon, the N1s of nitrogen, and the O1s of oxygen, respectively. One could further obtain information of the atomic percentage of N by integrating the corresponding peak in survey scan. Moreover, the chemical states of each element could be analyzed by high-resolution XPS scans. Figure 9.11(b) demonstrates the N1s line scan spectrum which could be fitted with two components corresponding to graphitic (400.6 eV) and pyridine-like (398.6 eV) nitrogen dopants [37]. However, it is worth mentioning that XPS constitutes a surface chemical analysis technique with a depth range of 1~10 nm. Therefore, for monolayer or bi-layer graphene sheets, the materials irradiated underneath the substrate might influence the final XPS spectrum because of the atomic thickness of graphene samples. This effect will become negligible as the number of stacking graphene layers increases. In addition to XPS, Raman spectroscopy is another important and widely used technique when studying doped graphenes [61]. The technique could be used to distinguish n-type or p-type doping, and even provide information about the dopant concentration [62]. In particular, it is powerful in identifying the number of graphene layers and it is nondestructive [63]. Moreover, the peak position and the full-width at half-maximum (FWHM) of the G ($\sim$1584 cm^{-1}) and 2D ($\sim$2700 cm^{-1}) peaks of graphene will vary with different doping levels. Figure 9.11(c) shows typical Raman spectra of pristine and N-doped graphene on SiO$_2$/Si substrates using 514 nm laser excitation [37]. The symmetric line-shape of the 2D-band and the large ratio of the 2D-band over G-band intensities ($I_{2D}/I_G > 2$) confirms the growth of mostly a single layer in both NG and pristine graphene samples [63]. The spectra of NG reveal a more intense D-band ($I_D/I_G \sim 1.00$), which can be explained by the elastically scattered photo-excited electron by the large number of N atoms embedded in the graphene lattice before emitting a phonon. The Raman spectra of the NG also exhibit an obvious D'-band, which is believed to be originated from intravalley double resonance scattering processes [64]. When compared to pristine graphene, the Raman spectrum of NG shows a downshift of the 2D-band (7 cm^{-1}), which is an indicative of n-type doping [37].

HRTEM [16] and STM [15] have been recently used to visualize individual atomic dopants within doped graphene lattices. By combining HRTEM measurements with first-principle electronic structure calculations, Meyer et al. investigated the electronic configurations of nitrogen-substitution point defects in graphene sheets. The N substitutions exhibit a weak dark contrast in the larger defocus HRTEM images, as shown in Figure 9.11(d) [16]. Zhao et al. prepared monolayer NG sheets using a low-pressure (high vacuum) chemical vapor deposition (LP-CVD) method and visualized the *single substitution* of carbon atoms by N-dopants (i.e., an individual sp^2-like N-dopant, N$_1$) within the graphene lattice; this N1 defect was notably abundant when studying all regions by scanning tunneling microscopy/spectroscopy (STM/STS) [15]. Recently, we developed a simple and efficient method to synthesize large-area and highly-crystalline monolayer NG sheets on copper (Cu) foils by atmospheric-pressure chemical vapor deposition (AP-CVD) [37]. Figures 9.11(e) and (f) show the experimental and density functional theory (DFT)-simulated STM images of N dopant in our samples. It can be observed that two N atoms are separated by one C and sit on the same A-sub-lattice, thus forming an N$_2^{AA}$ configuration [37]. Besides the aforementioned characterization techniques, other methods, such as X-ray absorption near-edge structure (XANES) spectroscopy [65] and transport measurements [17, 66, 67], have been used to detect dopants in graphene. We believe that in the near future other alternative

200 Graphene Chemistry

characterization techniques will be developed in order to identify and quantify more precisely the dopant concentrations and their local structure within the graphene lattice.

9.5 Applications of Doped Graphene Sheets and Nanoribbons

As mentioned previously, the properties of graphene sheets could be remarkably modified by substitutional doping. Based on their unique structures and properties, various novel and exciting applications have been explored, including high-performance FET devices [10, 41, 68], energy conversion and storage components [12, 21, 69–72], biosensors [73], photocatalysts [74]. Figure 9.1 depicts some possible applications of N-doped graphene. When compared to pristine graphene, N-doped graphene FET devices exhibit an n-type semiconductor behavior, which leads to a decreased electrical conductivity and an improved on/off ratio [10]. By using N-doped graphene nanoribbons (GNRs) with widths of around 5 nm, a very high on/off ratio of $\sim 10^5$ could be achieved. The N-doped GNR devices also exhibit an n-type behavior [41]. These findings are meaningful for future graphene-based semiconductor electronics. In the fuel cell area, Qu et al. have demonstrated the use of NG as metal-free catalyst for oxygen reduction [13]. Furthermore, one could also use doped graphene sheets as a catalytic support. Recent work has revealed that an enhanced performance of a methanol fuel cell (DMFC) could be achieved when using N-doped carbon nanotube-graphene hybrid nanostructures (NCNT-GHN) as PtRu catalyst support [69]. Well-dispersed PtRu nanoparticles with diameters of 2$\sim$4 nm could be immobilized onto these NCNT-GHN supports without using any pretreatment. This could be attributed to a synergistic effect of the hierarchical structure (graphene-CNT hybrid) and electronic modulation (N-doping) during the methanol electro-oxidation reaction [69]. In addition, heteroatom-doped graphene could be also used in the area of ultracapacitors. By treating pristine graphene sheets with a N plasma, Jeong et al. developed NG-based ultracapacitors with capacitances of $\sim$280 F/g, excellent cycle life ($>$200 000) and high power capabilities. It is worth noting that N-doping could significantly increase the specific capacitance by about four times when compared to that of pristine graphene. In this context, it has been demonstrated that NG could be used to fabricate an ultracapacitor on commercial conductive carbon textiles and be wrapped onto a human arm. This flexible ultracapacitor could then turn on a light emitting diode (LED) device. This improved capacitance has been attributed to the N-doping effect, especially due to the N-doped sites embedded in the hexagonal lattice [21].

With the increasing energy shortage of fossil fuels, research related to clean and renewable energy, such as solar cells, has attracted the attention of scientists and government agencies worldwide. When using acetonitrile (CH$_3$CN) as a liquid precursor, it is possible to grow nanometer-thick N-doped carbon thin films (N-CFMs) on Cu foils via CVD [70]. Interestingly, as-synthesized N-CFMs are highly-transparent and electrically conducting polycrystalline carbon films. These materials could be used as transparent electrodes and active layers for electron-hole pair separation and transport in solar cells. The photovoltaic device based on 200 µl acetonitrile shows the highest power conversion efficiency up to $\sim$1.55% [70]. The mechanism of this thin-film solar cell device could be described as follows: Firstly, a built-in electric field forms between N-CFM and n-type Si due to the difference in work functions. Upon light illumination, electron-hole pairs are generated when the heterojunction of N-CFM, and Si responds to incident light. Subsequently, the photogenerated holes and electrons are separated and driven into the N-CFM and n-Si layer by the built-in electric field. As a result, a conversion from solar energy to electrical energy is achieved [70]. The N-CFM used in these studies is just a polycrystalline film. However, if high-quality NG sheets with a relatively high transparency and good electrical conductivity could be used, we believe that much higher power conversion efficiency could be achieved.

For N-doped CNTs, it has been demonstrated that nitrogen-doping could enhance the biocompatibility of CNTs for biosensing [75, 76]. Inspired on these results, Wang et al. demonstrated the electrochemical biosensing of NG sheets [73]. In particular, NG sheets exhibited excellent selectivity and sensitivity for glucose with concentrations as low as 0.01 mM in the presence of interferences (e.g., uric acid, ascorbic acid). This result could be attributed to the good biocompatibility and fast electron transfer kinetics of NG sheets for glucose molecules [73]. In addition, doped graphene could also be used to produce hydrogen from water. In this context, the best way to produce H_2 from renewable sources is water splitting under solar irradiation via photocatalysts. Interestingly, NG has demonstrated to be an excellent matrix material for the fabrication of CdS semiconductor photocatalysts [74]. By producing NG/CdS nanocomposites, high photocatalytic activity could be achieved under visible irradiation. In this case, NG sheets work as a co-catalyst, which could protect CdS from photocorrosion under light irradiation, and could also promote the separation and transfer of photogenerated carriers [74].

Many other applications of doped graphenes in superconductors [77], nanogenerators [78], and lithium ion batteries [12, 79], have also been recently explored. It is therefore expected that other fascinating properties and phenomena will be unveiled in the near future.

9.6 Future Work

Graphene doping is a very recent and fascinating field which opens up the possibility of fine-tuning the electronic and chemical properties of pristine graphene. For instance, from XPS characterization, NG reveals the presence of nitrogen substitutional sites and weak traces of pyridine-like nitrogen atoms. Therefore, in the future it might be possible to control the type, sublattice location, and concentration of nitrogen doping. Based on theoretical and experimental calculations, nitrogen substitutional doping should be better for fabricating electronic devices when compared to prydine-like or pyrrolic-like nitrogen doping. However, the latter is more chemically active and could be used for anchoring molecules or polymers that could facilitate the fabrication of composites. In addition, graphene could be an excellent material for spintronic applications due to the large spin-relaxation times, thus the spin information can be transported and manipulated long distances [80]. It has been recently found that lines of defects (e.g., 5–7 or 5–8–5) could be inserted into graphene layers. These lines of defects could act as metal wires embedded in the graphene lattice and structurally behaving as 2D grain boundaries. In the future, it is important to understand the role of different dopants within these lines of defects. It might be possible that the presence of dopants during growth could induce the formation of different lines of defects, and these could be used as molecular sensors, photoluminescent emission sources or 1D nanowires. It is clear that further experimental and theoretical investigations are necessary to clearly understand the role of dopants in graphene. A real challenge in the near future consists of controlling the doping of graphene layers, in which the doping type, sublattice location of dopants, and the dopant concentration, are carefully tuned. In this way different devices with a wide range of possible applications could be constructed.

Acknowledgments

This work was supported in part by CONACYT-México grants: 60218-F1 (FLU). MT thanks JST-Japan for funding the Research Center for Exotic NanoCarbons, under the Japanese regional Innovation Strategy Program by the Excellence. H.T. acknowledges the support of CAPES Brazil through funding from the Programa Professor Visitante do Exterior – PVE as "Bolsista CAPES/BRASIL". MT also thanks the Penn State Center for Nanoscale Science for seed grant entitled on 2D Layered Materials.

References

[1] K. S. Novoselov, A. K. Geim, S. V. Morozov, D. Jiang, Y. Zhang, S. V. Dubonos, et al., Electric field effect in atomically thin carbon films, Science, **306**(5696), 666–669 (2004).

[2] K. S. Novoselov, A. K. Geim, S. V. Morozov, D. Jiang, M. I. Katsnelson, I. V. Grigorieva, et al., Two-dimensional gas of massless Dirac fermions in graphene, Nature, **438**(7065), 197–200 (2005).

[3] K. S. Novoselov, E. McCann, S. V. Morozov, V. I. Fal'ko, M. I. Katsnelson, U. Zeitler, et al., Unconventional quantum Hall effect and Berry's phase of 2 pi in bilayer graphene, Nature Physics, **2**(3), 177–180 (2006).

[4] K. S. Novoselov, Z. Jiang, Y. Zhang, S. V. Morozov, H. L. Stormer, U. Zeitler, et al., Room-temperature quantum hall effect in graphene, Science, **315**(5817), 1379–1379 (2007).

[5] A. H. Castro Neto, F. Guinea, N. M. R. Peres, K. S. Novoselov and A. K. Geim, The electronic properties of graphene, Reviews of Modern Physics, **81**(1), 109–162 (2009).

[6] A. K. Geim and K. S. Novoselov, The rise of graphene, Nature Materials, **6**(3), 183–191 (2007).

[7] M. I. Katsnelson, K. S. Novoselov and A. K. Geim, Chiral tunnelling and the Klein paradox in graphene, Nature Physics, **2**(9), 620–625 (2006).

[8] L. A. Ponomarenko, F. Schedin, M. I. Katsnelson, R. Yang, E. W. Hill, K. S. Novoselov and A. K. Geim, Chaotic dirac billiard in graphene quantum dots, Science, **320**(5874), 356–358 (2008).

[9] L. M. Xie, X. Ling, Y. Fang, J. Zhang and Z. F. Liu, Graphene as a substrate to suppress fluorescence in resonance Raman spectroscopy, Journal of the American Chemical Society, **131**(29), 9890–9891 (2009).

[10] D. C. Wei, Y. Q. Liu, Y. Wang, H. L. Zhang, L. P. Huang and G. Yu, Synthesis of N-doped graphene by chemical vapor deposition and its electrical properties, Nano Letters, **9**(5), 1752–1758 (2009).

[11] Z. Z. Sun, Z. Yan, J. Yao, E. Beitler, Y. Zhu and J. M. Tour, Growth of graphene from solid carbon sources, *Nature*, **468**(7323), 549–552 (2010).

[12] A. L. M. Reddy, A. Srivastava, S. R. Gowda, H. Gullapalli, M. Dubey and P. M. Ajayan, Synthesis of nitrogen-doped graphene films for lithium battery application, *ACS Nano*, **4**(11), 6337–6342 (2010).

[13] L. T. Qu, Y. Liu, J. B. Baek and L. M. Dai, Nitrogen-doped graphene as efficient metal-free electrocatalyst for oxygen reduction in fuel cells, *ACS Nano*, **4**(3), 1321–1326 (2010).

[14] Y. C. Lin, C. Y. Lin and P. W. Chiu, Controllable graphene N-doping with ammonia plasma, *Applied Physics Letters*, **96**(13), 133110 (2010).

[15] L. Y. Zhao, R. He, K. T. Rim, T. Schiros, K. S. Kim, H. Zhou, *et al.*, Visualizing individual nitrogen dopants in monolayer graphene, *Science*, **333**(6045), 999–1003 (2011).

[16] J. C. Meyer, S. Kurasch, H. J. Park, V. Skakalova, D. Kunzel, A. Gross, et al., Experimental analysis of charge redistribution due to chemical bonding by high-resolution transmission electron microscopy, *Nature Materials*, **10**(3), 209–215 (2011).

[17] Z. Jin, J. Yao, C. Kittrell and J. M. Tour, Large-scale growth and characterizations of nitrogen-doped monolayer graphene sheets, *ACS Nano*, **5**(5), 4112v4117 (2011).

[18] G. Imamura and K. Saiki, Synthesis of nitrogen-doped graphene on pt(111) by chemical vapor deposition, *Journal of Physical Chemistry C*, **115**(20), 10000–10005 (2011).

[19] Z. Q. Luo, S. H. Lim, Z. Q. Tian, J. Z. Shang, L. F. Lai, B. MacDonald, et al., Pyridinic N doped graphene: synthesis, electronic structure, and electrocatalytic property, *Journal of Materials Chemistry*, **21**(22), 8038–8044 (2011).

[20] L. Guan, L. Cui, K. Lin, Y. Y. Wang, X. T. Wang, F. M. Jin, et al., Preparation of few-layer nitrogen-doped graphene nanosheets by DC arc discharge under nitrogen atmosphere of high temperature, *Applied Physics A-Materials Science & Processing*, **102**(2), 289–294 (2011).

[21] H. M. Jeong, J. W. Lee, W. H. Shin, Y. J. Choi, H. J. Shin, J. K. Kang and J. W. Choi, Nitrogen-doped graphene for high-performance ultracapacitors and the importance of nitrogen-doped sites at basal planes, *Nano Letters*, **11**(6), 2472–2477 (2011).

[22] T. Kato, L. Y. Jiao, X. R. Wang, H. L. Wang, X. L. Li, L. Zhang, R. Hatakeyama and H. J. Dai, Room-temperature edge functionalization and doping of graphene by mild plasma, *Small*, **7**(5), 574–577 (2011).

[23] K. Brenner and R. Murali, In situ doping of graphene by exfoliation in a nitrogen ambient, *Applied Physics Letters*, **98**(11), 113115 (2011).

[24] H. Wang, T. Maiyalagan and X. Wang, Review on recent progress in nitrogen-doped graphene: synthesis, characterization, and its potential applications, *ACS Catalysis*, **2**(5), 781–794 (2012).

[25] R. Czerw, M. Terrones, J. C. Charlier, X. Blase, B. Foley, R. Kamalakaran, et al., Identification of electron donor states in N-doped carbon nanotubes, *Nano Letters*, **1**(9), 457–460 (2001).

[26] L. S. Panchokarla, K. S. Subrahmanyam, S. K. Saha, A. Govindaraj, H. R. Krishnamurthy, U. V. Waghmare and C. N. R. Rao, Synthesis, structure, and properties of boron- and nitrogen-doped graphene, *Advanced Materials*, **21**(46), 4726–4730 (2009).

[27] E. Cruz-Silva, F. Lopez-Urias, E. Munoz-Sandoval, B. G. Sumpter, H. Terrones, J. C. Charlier, et al., Electronic transport and mechanical properties of phosphorus- and phosphorus-nitrogen-doped carbon nanotubes, *ACS Nano*, **3**(7), 1913–1921 (2009).

[28] P. A. Thrower, The study of defects in graphite by transmission electron microscopy, *Chemistry and Physics of Carbon*, **5** 217–319 (1969).

[29] A. J. Stone and D. J. Wales, Theoretical studies of icosahedral C_{60} and some related species, *Chemical Physics Letters*, **128**(5–6), 501–503 (1986).

[30] P. Hohenberg and W. Kohn, Inhomogeneous electron gas, *Physical Review B*, **136**(3B), B864 (1964).

[31] W. Kohn and L. J. Sham, Self-consistent equations including exchange and correlation effects, *Physical Review*, **140**(4), A1133 (1965).

[32] D. M. Ceperley and B. J. Alder, Ground-state of the electron-gas by a stochastic method, *Physical Review Letters*, **45**(7), 566–569 (1980).

[33] J. M. Soler, E. Artacho, J. D. Gale, A. Garcia, J. Junquera, P. Ordejon and D. Sanchez-Portal, The SIESTA method for ab initio order-N materials simulation, *Journal of Physics-Condensed Matter*, **14**(11), 2745–2779 (2002).

[34] J. Junquera, O. Paz, D. Sanchez-Portal and E. Artacho, Numerical atomic orbitals for linear-scaling calculations, *Physical Review B*, **64**(23), 235111 (2001).

[35] N. Troullier and J. L. Martins, Efficient pseudopotentials for plane-wave calculations, *Physical Review B*, **43**(3), 1993–2006 (1991).

[36] L. Kleinman and D. M. Bylander, Efficaious form for model pseudopotentials, *Physical Review Letters*, **48**(20), 1425–1428 (1982).

[37] R. Lv, Q. Li, A. R. Botello-Méndez, T. Hayashi, B. Wang, A. Berkdemir, et al., Nitrogen-doped graphene: beyond single substitution and enhanced molecular sensing, *Scientific Reports*, **2**, 586 (2012).

[38] E. Cruz-Silva, F. Lopez-Urias, E. Munoz-Sandoval, B. G. Sumpter, H. Terrones, J. C. Charlier, et al., Phosphorus and phosphorus-nitrogen doped carbon nanotubes for ultrasensitive and selective molecular detection, *Nanoscale*, **3**(3), 1008–1013 (2011).

[39] G. Giovannetti, P. A. Khomyakov, G. Brocks, V. M. Karpan, J. van den Brink and P. J. Kelly, Doping graphene with metal contacts, *Physical Review Letters*, **101**(2), 026803 (2008).

[40] W. Chen, S. Chen, D. C. Qi, X. Y. Gao and A. T. S. Wee, Surface transfer p-type doping of epitaxial graphene, *Journal of the American Chemical Society*, **129**(34), 10418–10422 (2007).

[41] X. R. Wang, X. L. Li, L. Zhang, Y. Yoon, P. K. Weber, H. L. Wang, J. Guo and H. J. Dai, N-doping of graphene through electrothermal reactions with ammonia, *Science*, **324**(5928), 768–771 (2009).

[42] F. Schedin, A. K. Geim, S. V. Morozov, E. W. Hill, P. Blake, M. I. Katsnelson and K. S. Novoselov, Detection of individual gas molecules adsorbed on graphene, *Nature Materials*, **6**(9), 652–655 (2007).

[43] T. O. Wehling, K. S. Novoselov, S. V. Morozov, E. E. Vdovin, M. I. Katsnelson, A. K. Geim and A. I. Lichtenstein, Molecular doping of graphene, *Nano Letters*, **8**(1), 173–177 (2008).

[44] X. L. Li, H. L. Wang, J. T. Robinson, H. Sanchez, G. Diankov and H. J. Dai, Simultaneous nitrogen doping and reduction of graphene oxide, *Journal of the American Chemical Society*, **131**(43), 15939–15944 (2009).

[45] R. A. Nistor, D. M. Newns and G. J. Martyna, The role of chemistry in graphene doping for carbon-based electronics, *ACS Nano*, **5**(4), 3096–3103 (2011).

[46] A. C. Ferrari, Raman spectroscopy of graphene and graphite: Disorder, electron-phonon coupling, doping and nonadiabatic effects, *Solid State Communications*, **143**(1–2), 47-57 (2007).

[47] D. H. Deng, X. L. Pan, L. A. Yu, Y. Cui, Y. P. Jiang, J. Qi, et al., Toward N-Doped graphene via solvothermal synthesis, *Chemistry of Materials*, **23**(5), 1188–1193 (2011).

[48] T. O. Wehling, M. I. Katsnelson and A. I. Lichtenstein, Adsorbates on graphene: Impurity states and electron scattering, *Chemical Physics Letters*, **476**(4–6), 125–134 (2009).

[49] A. Lherbier, X. Blase, Y. M. Niquet, F. Triozon and S. Roche, Charge transport in chemically doped 2D graphene, *Physical Review Letters*, **101**(3), 036808 (2008).

[50] X. H. Deng, Y. Q. Wu, J. Y. Dai, D. D. Kang and D. Y. Zhang, Electronic structure tuning and band gap opening of graphene by hole/electron codoping, *Physics Letters A*, **375**(44), 3890–3894 (2011).

[51] J. Lahiri, Y. Lin, P. Bozkurt, Oleynik, II and M. Batzill, An extended defect in graphene as a metallic wire, *Nature Nanotechnology*, **5**(5), 326–329 (2010).

[52] A. R. Botello-Mendez, X. Declerck, M. Terrones, H. Terrones and J. C. Charlier, One-dimensional extended lines of divacancy defects in graphene, *Nanoscale*, **3**(7), 2868–2872 (2011).

[53] M. Kan, J. Zhou, Q. Sun, Q. Wang, Y. Kawazoe and P. Jena, Tuning magnetic properties of graphene nanoribbons with topological line defects: From antiferromagnetic to ferromagnetic, *Physical Review B*, **85**(15), 155450 (2012).

[54] S. J. Clark, M. D. Segall, C. J. Pickard, P. J. Hasnip, M. J. Probert, K. Refson and M. C. Payne, First principles methods using CASTEP, *Zeitschrift Fur Kristallographie*, **220**(5–6), 567–570 (2005).

[55] D. S. Geng, Y. Chen, Y. G. Chen, Y. L. Li, R. Y. Li, X. L. Sun, et al., High oxygen-reduction activity and durability of nitrogen-doped graphene, *Energy & Environmental Science*, **4**(3), 760–764 (2011).

[56] Z. H. Sheng, L. Shao, J. J. Chen, W. J. Bao, F. B. Wang and X. H. Xia, Catalyst-free synthesis of nitrogen-doped graphene via thermal annealing graphite oxide with melamine and its excellent electrocatalysis, *ACS Nano*, **5**(6), 4350–4358 (2011).

[57] H. B. Wang, C. J. Zhang, Z. H. Liu, L. Wang, P. X. Han, H. X. Xu, et al., Nitrogen-doped graphene nanosheets with excellent lithium storage properties, *Journal of Materials Chemistry*, **21**(14), 5430–5434 (2011).

[58] A. Das, S. Pisana, B. Chakraborty, S. Piscanec, S. K. Saha, U. V. Waghmare, et al., Monitoring dopants by Raman scattering in an electrochemically top-gated graphene transistor, *Nature Nanotechnology*, **3**(4), 210–215 (2008).

[59] M. Bruna and S. Borini, Raman signature of electron-electron correlation in chemically doped few-layer graphene, *Physical Review B*, **83**(24), 241401 (2011).

[60] B. D. Guo, Q. A. Liu, E. D. Chen, H. W. Zhu, L. A. Fang and J. R. Gong, Controllable N-doping of graphene, *Nano Letters*, **10**(12), 4975–4980 (2010).

[61] P. T. Araujo, M. Terrones and M. S. Dresselhaus, Defects and impurities in graphene-like materials, *Materials Today*, **15**(3), 98–109 (2012).

[62] I. O. Maciel, N. Anderson, M. A. Pimenta, A. Hartschuh, H. H. Qian, M. Terrones, et al., Electron and phonon renormalization near charged defects in carbon nanotubes, *Nature Materials*, **7**(11), 878–883 (2008).

[63] A. C. Ferrari, J. C. Meyer, V. Scardaci, C. Casiraghi, M. Lazzeri, F. Mauri, et al., Raman spectrum of graphene and graphene layers, *Physical Review Letters*, **97**(18), 187401 (2006).

[64] M. S. Dresselhaus, G. Dresselhaus and M. Hofmann, Raman spectroscopy as a probe of graphene and carbon nanotubes, *Philosophical Transactions of the Royal Society A – Mathematical Physical and Engineering Sciences*, **366**(1863), 231–236 (2008).

[65] J. Robinson, X. J. Weng, K. Trumbull, R. Cavalero, M. Wetherington, E. Frantz, et al., Nucleation of epitaxial graphene on SiC(0001), *ACS Nano*, **4**(1), 153–158 (2010).

[66] H. Medina, Y. C. Lin, D. Obergfell and P. W. Chiu, Tuning of charge densities in graphene by molecule doping, *Advanced Functional Materials*, **21**(14), 2687–2692 (2011).

[67] W. Qian, X. Cui, R. Hao, Y. L. Hou and Z. Y. Zhang, Facile preparation of nitrogen-doped few-layer graphene via supercritical reaction, *ACS Applied Materials & Interfaces*, **3**(7), 2259–2264 (2011).

[68] B. N. Szafranek, D. Schall, M. Otto, D. Neumaier and H. Kurz, High on/off ratios in bilayer graphene field effect transistors realized by surface dopants, *Nano Letters*, **11**(7), 2640–2643 (2011).

[69] R. T. Lv, T. X. Cui, M. S. Jun, Q. Zhang, A. Y. Cao, D. S. Su, et al., Open-ended, N-doped carbon nanotube-graphene hybrid nanostructures as high-performance catalyst support, *Advanced Functional Materials*, **21**(5), 999–1006 (2011).

[70] T. X. Cui, R. T. Lv, Z. H. Huang, H. W. Zhu, J. Zhang, Z. Li, et al., Synthesis of nitrogen-doped carbon thin films and their applications in solar cells, *Carbon*, **49**(15), 5022–5028 (2011).

[71] Y. C. Qiu, X. F. Zhang and S. H. Yang, High performance supercapacitors based on highly conductive nitrogen-doped graphene sheets, *Physical Chemistry Chemical Physics*, **13**(27), 12554–12558 (2011).

[72] T. Q. Lin, F. Q. Huang, J. Liang and Y. X. Wang, A facile preparation route for boron-doped graphene, and its CdTe solar cell application, *Energy & Environmental Science*, **4**(3), 862–865 (2011).

[73] Y. Wang, Y. Y. Shao, D. W. Matson, J. H. Li and Y. H. Lin, Nitrogen-doped graphene and its application in electrochemical biosensing, *ACS Nano*, **4**(4), 1790–1798 (2010).

[74] L. Jia, D. H. Wang, Y. X. Huang, A. W. Xu and H. Q. Yu, Highly durable N-doped graphene/CdS nanocomposites with enhanced photocatalytic hydrogen evolution from water under visible light irradiation, *Journal of Physical Chemistry C*, **115**(23), 11466–11473 (2011).

[75] J. C. Carrero-Sanchez, A. L. Elias, R. Mancilla, G. Arrellin, H. Terrones, J. P. Laclette and M. Terrones, Biocompatibility and toxicological studies of carbon nanotubes doped with nitrogen, *Nano Letters*, **6**(8), 1609–1616 (2006).

[76] S. Y. Deng, G. Q. Jian, J. P. Lei, Z. Hu and H. X. Ju, A glucose biosensor based on direct electrochemistry of glucose oxidase immobilized on nitrogen-doped carbon nanotubes, *Biosensors & Bioelectronics*, **25**(2), 373–377 (2009).

[77] B. Uchoa and A. H. C. Neto, Superconducting states of pure and doped graphene, *Physical Review Letters*, **98**(14), 146801 (2007).

[78] H. J. Shin, W. M. Choi, D. Choi, G. H. Han, S. M. Yoon, H. K. Park, et al., Control of electronic structure of graphene by various dopants and their effects on a nanogenerator, *Journal of the American Chemical Society*, **132**(44), 15603–15609 (2010).

[79] X. F. Li, D. S. Geng, Y. Zhang, X. B. Meng, R. Y. Li and X. L. Sun, Superior cycle stability of nitrogen-doped graphene nanosheets as anodes for lithium ion batteries, *Electrochemistry Communications*, **13**(8), 822–825 (2011).

[80] M. Popinciuc, C. Jozsa, P. J. Zomer, N. Tombros, A. Veligura, H. T. Jonkman and B. J. van Wees, Electronic spin transport in graphene field-effect transistors, *Physical Review B*, **80**(21), 214427 (2009).

[81] B. G. Sumpter, J. S. Huang, V. Meunier, J. M. Romo-Herrera, E. Cruz-Silva, H. Terrones and M. Terrones, A theoretical and experimental study on manipulating the structure and properties of carbon nanotubes using substitutional dopants, *International Journal of Quantum Chemistry*, **109**(1), 97–118 (2009).

10
Adsorption of Molecules on Graphene

O. Leenaerts, B. Partoens, and F. M. Peeters
Department of Physics, University of Antwerp, Belgium

10.1 Introduction

Molecules in the environment of graphene devices can have an important impact on their performance. They can change the doping level and the transport characteristics of the graphene sample which causes different observable phenomena. Notable examples of this are the hysteretic behavior of field effect devices and the asymmetries in electron and hole transport which have both been ascribed to adsorbed molecules from ambient air [1]. The influence of such molecules is not always a nuisance that one should get rid of, but can sometimes be exploited to deliberately change the electronic properties of graphene devices. The transport properties and conductivity of graphene are determined by two important factors, namely the concentration of charge carriers, n, and the carrier mobility, μ. Both can be altered by the adsorption of molecules, and their relative importance has been a matter of continuing debate over the last five years. For example, Schedin *et al.* found that they could dope a graphene sample by NO_2 adsorption without any significant loss of carrier mobility [1]. This has been attributed to an increased screening of scattering centers [1,2]. On the other hand, Chen *et al.* observed a reduction of the mobility with increasing density of K on graphene caused by charged impurity scattering [3]. Another application for molecular functionalization of graphene is to increase the solubility of graphene samples. This can for instance be achieved by non-covalent functionalization with organic molecules [4].

From another point of view, it is also possible to use the electronic changes that result from molecular adsorption to detect particular molecules in sensor applications [1,5,6]. This has proven to be an important application: In 2007 a graphene-based gas sensor was

---

Graphene Chemistry: Theoretical Perspectives, First Edition. Edited by De-en Jiang and Zhongfang Chen.
© 2013 John Wiley & Sons, Ltd. Published 2013 by John Wiley & Sons, Ltd.

used to detect, for the first time, single molecular adsorption events in a strongly diluted environment [1]. The superior gas sensing properties of graphene have been attributed to the following characteristics: (1) Graphene is a 2D material and has, consequently, its whole volume exposed to possible adsorbates, (2) graphene is highly conductive and has therefore low Johnson-noise even in the absence of charge carriers, (3) in its neutral state, there are no charge carriers in graphene so that few charge carriers can cause notable changes in the carrier concentration and hence the conductivity, and (4) graphene crystals are almost defect-free which ensures decreased noise due to thermal movement of defects. Combining all these features, graphene forms a unique material for sensing applications.

But, whether one aims to decrease the impact of molecules from the ambient on the properties of graphene, or use their influence to modify these properties or make graphene into a sensitive gas sensor, it is essential to have a good understanding of the interaction of molecules with the graphene surface.

In this chapter, we take a closer look at this interaction. Our focus is on physisorption (see next), while chemisorption and dissociative adsorption are treated more superficially. A variety of different doping mechanisms is examined and also the effect of external (e.g., an electric field) and internal (e.g., defects) modifications on the interaction is investigated.

10.2 Physisorption *versus* Chemisorption

Adsorption processes can be divided into two large classes, namely *physisorption* and *chemisorption*. Generally, one speaks of chemisorption when a strong bond (> 0.5 eV), with a length from about 1 to 2.5 Å, is formed between adsorbate and adsorbent. Physisorption, on the other hand, is characterized by weak bonding (10–100 meV) and larger bond lengths (> 2.5 Å). Not all kinds of adsorption can be fitted unambiguously into this scheme: ionic bonds, for example, have features of both chemisorption and physisorption. In the case of graphene, however, it is possible to make a rather clear distinction between the two. Chemisorption on graphene is characterized by a change of orbital hybridization of the graphene carbon atoms involved in the bonding. The carbon atoms of graphene are sp^2-hybridized which makes it possible to have an efficient bonding with their three closest neighbors. If an adsorbate is to make a chemical bond with graphene, one or more C atoms have to change their hybridization to sp^3 to make this additional binding feasible. This change in hybridization costs a lot of energy because the C atom bumps out of the graphene plane in the process. Therefore, there is a substantial barrier involved in the chemisorption process which only the strongest bonds are able to overcome. The presence of this barrier lies at the origin of this rather clear separation between chemisorbed and physisorbed adsorbates.

Another apparent difference among adsorbates on graphene is the height of their migration barriers (MB) [7]. Physisorbed molecules or atoms are almost free to wander around the graphene surface (MB < 0.1 eV) while chemisorbed ones are often strongly anchored to their adsorption site (MB > 0.5 eV).

In the following sections, we mainly study physisorbed molecules on graphene. The interaction between these molecules and graphene is, by definition, rather weak, and therefore it is important to take into account dispersion forces.

Most of the theoretical work on the adsorption of molecules on graphene has been performed through *ab initio* density functional theory (DFT) simulations. But it is only in recent years that widely-used DFT packages, such as VASP, have implemented the possibility of taking the dispersion forces into account [8]. Consequently, most of the theoretical research has been performed without incorporating these dispersion forces and one should be aware of this when considering their results. In order to estimate the importance of these inaccuracies and to put restrictions on the applicability of the calculated properties, we give a brief description of the van der Waals interaction (vdW) and point out where the DFT description usually fails.

The van der Waals interaction between two neutral systems is generally considered to consist of three kinds of interaction:

- *dipole–dipole*: The permanent dipoles of the two systems interact with each other through an attractive or repulsive force that decreases rapidly for large distances. This interaction usually dominates when permanent dipoles are present.
- *dipole–induced dipole*: The permanent dipole of one of the systems induces a charge polarization (i.e., a dipole) in the other subsystem which leads to an attractive force.
- *induced dipole–induced dipole*: This interaction, also called the London dispersion interaction, is the result of correlation effects between the charge distributions of the two systems. An instantaneous charge fluctuation in one system results in a small dipole which induces a dipole in the other system and vice versa. The resulting interaction is attractive and can be very important in cases where other (stronger) interactions are absent as is the case for, as an example, noble gases at low temperatures.

The first and the second interaction are present in all DFT calculations, but the third, which is the actual dispersion force and which is usually the smallest, is rarely included.

In our case, the interactions that are examined are molecule-graphene interactions. The dipole–dipole interaction is thus absent, but, because many molecules are polar, the second interaction (dipole–induced dipole) is usually present. When the adsorbed molecules are non-polar, an appreciable underestimation of the adsorption energy can be expected. This has, for example, been demonstrated recently (2010) by vdW density functional studies on the adsorption of diatomic halogen molecules on graphene [9]. In this study, the dispersion force was explicitly incorporated and its results are shown in Figure 10.1.

In this figure the total energy is plotted as a function of the distance between the molecules and graphene for DFT-GGA calculations with (blue curves) and without (dark curves) vdW-interaction. It can be seen that PBE-GGA, which is one of the most widely used exchange-correlation functionals [10], would give more or less the correct adsorption distance, but underestimates seriously the adsorption energy in comparison with the vdW-corrected functional. Although this might suggest that most DFT studies, namely those that do not incorporate dispersion forces, are too crude approximations to give reliable results, this is not necessarily the case because they do give some insights: The adsorption energies are not well reproduced, but more qualitative aspects, such as adsorption sites and doping, are usually accurately predicted.

One should also note another problem with different approaches in DFT calculations: The two most widely used approximations for the exchange correlation functional in DFT, namely the local density approximation (LDA) and the generalized gradient approximation (GGA), tend to overestimate and underestimate binding energies, respectively. In cases

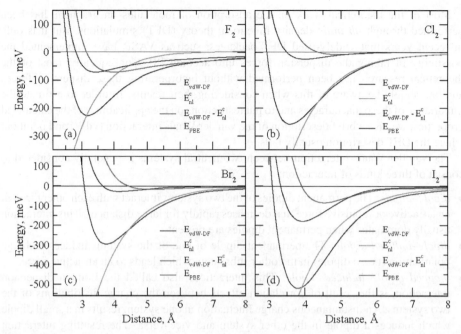

Figure 10.1 *Adsorption of different halogen molecules on graphene. The binding energy as a function of the distance between the molecule and graphene is shown for the vdW-corrected density functional (blue curve) and the PBE-GGA functional (brown curve). The other two curves show partial contributions to the vdW-corrected functional and are of no importance here. See colour version in the colour plate section. Reprinted with permission from [9] © 2010 American Physical Society*

where the difference between chemisorption and physisorption is small, this can lead to contrary predictions (see following).

10.3 General Aspects of Adsorption of Molecules on Graphene

In this section we take a closer look at the adsorption process of molecules on graphene. We do not give an exhaustive treatment of all molecule-graphene systems that have been studied so far, but rather indicate some general trends and use specific molecules to exemplify these.

First, let us state the most important aspects involved in molecular adsorption on graphene. The most obvious characteristic is the adsorption energy, E_a. This energy is usually defined as the energy difference between the total graphene-molecule system and the sum of the energies of the separate graphene sample and the molecule:

$$E_a = E_{graph+mol} - (E_{graph} + E_{mol}) \tag{10.1}$$

E_a is thus defined per adsorbed molecule and a negative adsorption energy indicates that the molecule will be attached to the graphene surface.

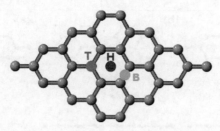

Figure 10.2 *Different adsorption sites for physisorption on graphene*

If a small molecule or atom adsorbs on graphene, it can do so at various inequivalent sites of the surface. It is common in the literature to distinguish between three adsorption sites (see Figure 10.2): (1) on top of a particular carbon atom of graphene (T), (2) on top of a nearest-neighbor C–C bond (B), and (3) at the center of a C hexagon (H). There appears to be some distinction in the preferred adsorption sites of chemisorbed and physisorbed adsorbates. Chemisorption processes always happen at the T (e.g. for H atoms [11]) or B (e.g. for O atoms [12]) site, while physisorption usually occurs at the H site [13, 14].

Molecules can also bind to the graphene surface with different orientations. For example, an ammonia molecule can adsorb with its nitrogen atom or hydrogen atoms closer to the surface [13]. The adsorption energy and other properties (see later) are highly dependent on the orientation of the molecule, so this is an important factor.

As an example, let us look at a typical physisorption process as studied in Ref. [13], namely the adsorption of a water molecule on graphene. In Table 10.1, the adsorption energies and distances are given for various adsorption sites (T, B, H) and orientations (u, d, n, v). Beware that these calculations were performed without dispersion forces, so the effective adsorption energies are probably considerably higher. The explanation of the different orientations is given in Ref. [13], but is of no importance here. It is important, however, that the adsorption distance, d, and energy, E_a, depend on the orientation of the molecule but appear to be rather independent of the adsorption site. Consequently, the water molecule is able to move around a (pristine) graphene surface at low temperatures.

These results are not restricted to water molecules, but can be seen to be generally applicable to loosely bound molecules on pure graphene [7,13]. Therefore, it is expected that physisorbed adsorbates on pure graphene tend to cluster if long-range attractive forces between adsorbates are present. Leenaerts *et al.* showed that this is, for example, the case for water molecules on graphene [15]. The attracting force between the water molecules

Table 10.1 *Adsorption of H_2O on graphene: the adsorption energy, E_a, and distance, d, are shown for different orientations and adsorption sites*

orientation	u			d			n			v
position	T	B	H	T	B	H	T	B	H	H
E_a (meV)	19	18	20	24	24	27	19	18	19	47
d (Å)	3.70	3.70	3.69	3.56	3.55	3.55	4.05	4.05	4.02	3.50

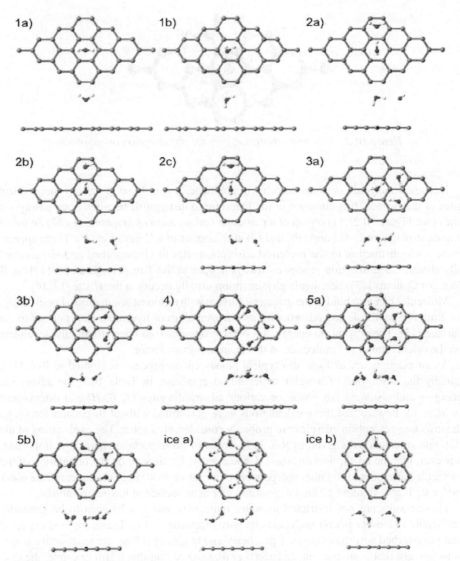

Figure 10.3 *Adsorption of water clusters on graphene: (1–5) Water clusters with 1–5 molecules with different orientations, and (6–7) ice structures on graphene. Reprinted with permission from [15] © 2009 American Physical Society*

results from dipole–dipole interaction at large distances and hydrogen bonding at shorter distances. When water molecules are added one by one on a graphene surface, they readily group together to form water clusters (see Figure 10.3). At still larger concentrations, ice layers can be formed [15]. Water molecules are strongly polar molecules and in small water clusters, the dipoles tend to compensate each other. In ice layers, on the other hand, the different dipoles add together and form a large dipole layer. Since the orientation of the dipoles of polar molecules can have an important impact on the doping of graphene (see later), the doping can depend on the concentration of adsorbates. This is only one example in which the adsorbate–adsorbate interaction can have a substantial impact on the physics of the adsorption process. Another example is given by the reduced impurity scattering and subsequent increase in carrier mobility after clustering of the adsorbates [16].

However, most *ab initio* calculations do not take this into account because it is rather hard to simulate systems that are large enough to obtain realistic adsorbate–adsorbate interactions. On the contrary, most studies try to investigate single adsorbate–graphene systems and, when the simulation is done with periodic boundary conditions, try to reduce the (artificial) adsorbate–adsorbate interaction from periodic images as much as possible. Consequently, there is still not much known about the interadsorbate interaction from *ab initio* calculations, and we will not examine this any further in the following.

To summarize, it appears that most properties of physisorbed molecules on pure graphene are only slightly dependent on the adsorption site (T, B, or H), but they do depend considerably on the orientation of the molecules with respect to the graphene surface. The molecules are mobile and tend to cluster on the surface when attractive forces between different adsorbates are present. The clustering of adsorbates can have important consequences on the graphene–adsorbate interaction, so that the concentration of adsorbates can be an important factor in the physisorption process.

10.4 Various Ways of Doping Graphene with Molecules

One of the most important aspects of physisorption on graphene is the doping induced by the molecules. In semiconductors, doping is usually achieved by incorporating foreign atoms into the bulk which introduce a hole or an extra electron in the semiconducting material. But doping can also be realized through electron exchange with dopants situated at the surface [17, 18]. This kind of doping is referred to as *surface transfer doping* and it plays a particular important role in graphene. Substitutional doping (i.e., doping in the bulk) can also be realized in graphene (by doping the graphene crystal with B and N atoms) but this introduces important structural and electronic changes because of the low dimensionality of graphene [19]. Therefore, surface transfer doping might be a more convenient way of achieving doping in graphene.

There are many different mechanisms that can lead to surface transfer doping and we will investigate some of them in this section. An important distinction in our description is made between inert and open-shell adsorbates [13, 14, 20]. Adsorbates with an open-shell structure, such as alkali metals and paramagnetic molecules, are usually more reactive than closed-shell atoms or molecules. These two classes of adsorbates lead to different charge transfer mechanisms and are, therefore, treated separately.

10.4.1 Open-Shell Adsorbates

Strong chemical doping of graphene can only be realized when the adsorbates induce impurity states close to the Fermi-level in graphene. Wehling *et al.* have argued for two realistic ways of achieving this [14]. First, by the formation of a covalent bond with a C atom of graphene so that an electron is effectively removed from the resonant bonding structure of graphene and universal mid-gap states are created in the electronic band structure of graphene. This is a clear case of chemisorption and the graphene layer will undergo substantial structural changes in the adsorption process. Another possibility is weakly bonded open-shell systems which have, by definition, partly occupied valence orbitals. The occupied and unoccupied electron states of these adsorbates are separated by a Hund exchange energy of the order of 1 eV so that there are always molecular levels in the vicinity of the Dirac point of graphene. If the highest occupied molecular orbital (HOMO) level is higher in energy than the Dirac level, an electron will be transferred from the molecule to the graphene layer. If, on the other hand, the lowest unoccupied molecular orbital (LUMO) level is below the Dirac point, an electron is transferred from graphene to the molecule. Both cases can induce a large charge transfer and effectively dope the graphene sample. To illustrate this charge transfer mechanism, consider the well-studied case of an NO_2 molecule on graphene [13,20,21,22].

NO_2 is a paramagnetic molecule and is therefore expected to have molecular orbitals in the vicinity of the Dirac point. The total spin-polarized density of states (DOS) of this molecule adsorbed on a graphene surface is shown in Figure 10.4. The molecular orbitals (MO) are visible as peaks in the DOS and are named according to their symmetry. The LUMO ($6_{a1,\downarrow}$) of NO_2 appears to be located at approximately 0.3 eV below the Dirac point. This induces a large charge transfer of about one electron to the molecule [21] so that NO_2 is a suitable molecule to induce p-doping in graphene [1,23,24].

Note, however, that an open-shell structure is not a sufficient criterion for an adsorbate to have good doping characteristics. Consider, for example, the adsorption of an NO molecule on graphene. Although NO is paramagnetic and has, consequently, MO levels close to the Dirac point, it does not induce any appreciable doping and leaves the graphene sample

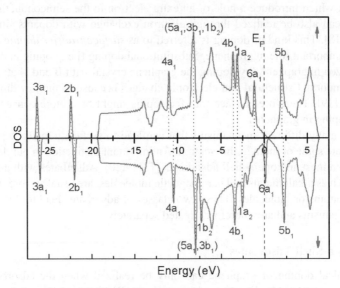

Figure 10.4 Spin-polarized DOS of NO_2 on graphene. The Fermi-level is set at zero energy and the position of the MOs of NO_2 are indicated with dotted lines. Reprinted with permission from [13] © 2008 American Physical Society

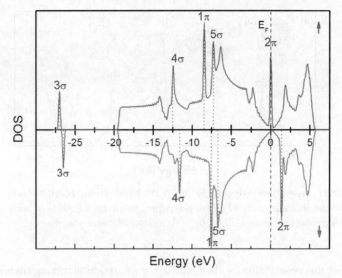

Figure 10.5 *Spin-polarized DOS of NO on graphene. The Fermi-level is set at zero energy and the position of the MOs of NO are indicated with dotted lines. Reprinted with permission from [13] © 2008 American Physical Society*

almost unaffected [13]. The DOS for this molecule on graphene is shown in Figure 10.5 and it can be seen that the LUMO ($2\pi_\uparrow$) of this molecule almost coincides with the Dirac point of graphene. Therefore, no substantial charge transfer is expected.

The effect of most efficient graphene dopants can be explained by this charge transfer mechanism. Some notable p-dopants are NO_2 [13,20], CrO_3 [25], and the organic molecules F4-TCNQ [26,27], DDQ [28], and TCNE [29]. Some well-studied n-dopants are the alkali metals Li, Na, K, and Cs [30] and organic molecules such as TTF [28].

10.4.2 Inert Adsorbates

Most molecules are closed-shell and are therefore relatively inert. Their occupied and unoccupied molecular orbitals have energy differences of about 5–10 eV and are thus lying far away from the Dirac point. This means that any direct charge transfer between these molecules and graphene is necessarily small. However, if a molecule adsorbs on graphene, there will always be some interaction and this can cause a small charge transfer. As an example, consider the physisorption of ammonia on graphene as studied in Ref. [13]. The total DOS of the system is shown in Figure 10.6. It is seen that the HOMO is located around 2.5 eV below the Dirac point whereas the LUMO is 3 eV above the Dirac point. This indicates that there is no direct charge transfer as in the case of the NO_2 molecule. But when the charge on the molecule is calculated, there appears to have been a small charge transfer with a size that depends on the orientation of the molecule.

The two orientations of the ammonia molecule that were investigated are those with the H atoms pointing away from the surface (u) and towards the surface (d). A small charge transfer from the molecule to the graphene surface of approximately 0.03 e was calculated for the u orientation and there is (almost) no charge transfer for the d orientation [13]. The

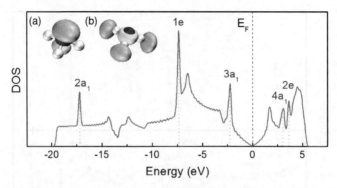

Figure 10.6 NH$_3$ on graphene. Inset: (a) the HOMO and (b) the LUMO of NH$_3$ (the N atom is the darkest central one and the H atoms are white). Main panel: DOS of NH$_3$ on graphene. Reprinted with permission from [13] © 2008 American Physical Society

explanation for this observation can be found in the small orbital mixing (hybridization) of graphene states with the molecular orbitals. We can see how this comes about by looking at the HOMO (3a$_1$) and LUMO (4a$_1$) of the NH$_3$ molecule (Figure 10.6a and b). In the *u* orientation, the HOMO is the only orbital that can have a significant overlap with the graphene orbitals and thus can cause charge transfer. As a consequence the NH$_3$ molecule will act as a donor. In the *d* orientation, both HOMO and LUMO can cause charge transfers which are similar in magnitude but in opposite directions and the net charge transfer is therefore close to zero. But the charge transfer is small in any case and, therefore, we do not expect strong doping from inert molecules in general. The fact that graphene reacts intrinsically very little with inert molecules such as NH$_3$ has been observed in experiment [31], but should be contrasted with those experiments in which ammonia is found to be a strong n-dopant [1,32,33]. This gives an indication that there is another charge transfer mechanism present in these experiments. Some experiments suggest that the polar nature of molecules such as NH$_3$ and H$_2$O should have an important influence on the doping [34] and causes the hysteretic behavior of the electric field effect [35].

On the theoretical side, Wehling *et al.* proposed two possible doping mechanisms that are present in the vicinity of strong dipole layers [36]. The first system that they investigated consists of a freestanding graphene layer covered with high concentrations of water molecules that are described by specific model structures such as ice layers or unrelaxed water molecules with fixed orientations. These water molecules form a dipole layer which appears to shift the nearly free electron bands in the graphene spectrum. When the dipole layer is strong enough, that is, when enough uniformly oriented water layers are present, the free energy band can be shifted below the Fermi-level of graphene and cause doping. As indicated by Wehling *et al.*, the problem with this explanation, is that the dipole layer should be much stronger than what is observed experimentally [34].

A more realistic system that was treated by Wehling *et al.* is a graphene layer on top of a SiO$_2$ substrate with an adsorbed layer of water molecules [36]. The SiO$_2$ substrate was simulated by taking a slab of a SiO$_2$ crystal, in which the dangling bonds of the Si at the surface are saturated with hydroxyl groups so that a silanol layer is formed at the interface. Defects were introduced into this idealized substrate to get a better approximation of a

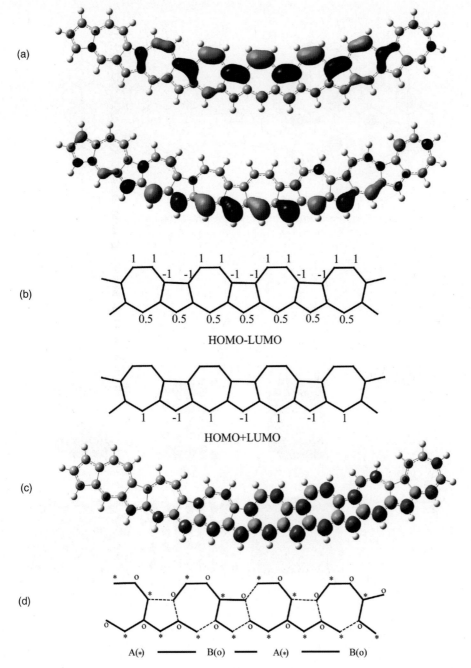

Figure 2.4 *Two near-degenerate FMOs (a) calculated with UB3LYP/6-31G(d), and (b) determined by molecular graph theoretical methods. Ground state spin distribution of [6]-azulene (c) calculated with EVB. (d) Mapping of fused-azulene into a spin 1/2 chain. Reprinted with permission from [89] © 2011 American Institute of Physics*

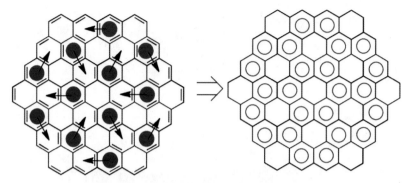

Figure 3.4 $C_{96}H_{24}$ Kekulé structure (left): Clar sextet (solid dot) migration, which is responsible for the NICS aromaticity pattern on the right, is indicated by the arrows. Redrawn from Ref. [34]

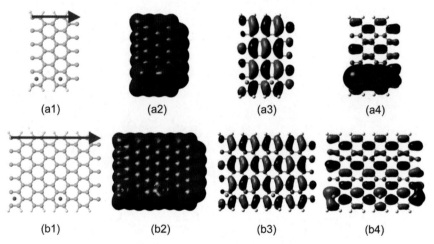

Figure 4.7 HSE charge densities (a2 and b2), HOCOs (a3 and b3), and LUCOs (a4 and b4) at the singlet spin state of the lithium doped 4×11 (a1) and 8×11 (b1) armchair graphene nanoribbon unit-cells, respectively

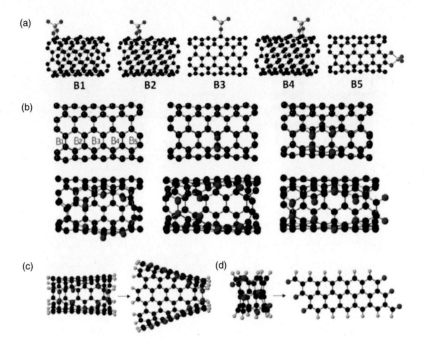

Figure 5.10 (a) Adsorption sites of MnO_4^- anions on an open (5, 5) carbon nanotube; (b) Adsorption of oxygen pairs on the nanotube; (c) Schematic diagram of the unzipping of the nanotube; (d) Schematic diagram of the unzipping of a hydrogen passivated (6, 0) carbon nanotube. The adsorption sites are labeled in (b). Reprinted with permission from [31] © 2009 American Institute of Physics

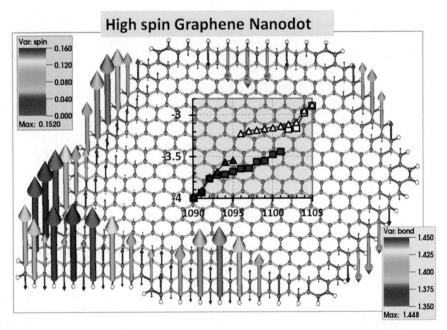

Figure 6.14 Geometry and spin of a high spin graphene nanodot consisting of a parent hexangulene hx09zg with a large protrusion comprising six zigzag rows each contributing an unpaired spin. The inset panel plots KS level energy (eV) against the level order index on either side of the HOMO–LUMO gap (≈ 0.2 eV). Squares and triangles denote α-spin levels and β-levels. Filled data points are occupied and white are empty levels

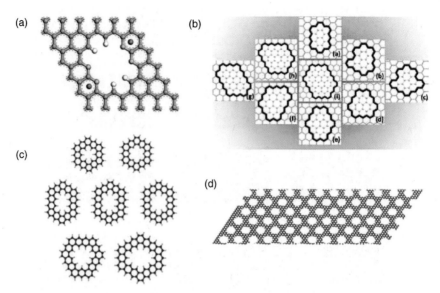

Figure 7.1 Structure of (a) nitrogen functionalized porous graphene with titanium. Dark orange: carbon, blue: nitrogen, white: hydrogen, pink: titanium. Reprinted with permission from [9] © 2012 American Chemical Society. (b) Graphene with designed pores for H_2/N_2 separation. Yellow hexagons represents graphene, the pore shape is colored red, and the removed carbon atoms are shown by transparent gray balls. For more detail please refer to Ref. [10]. Reprinted with permission from [10] © 2011 American Chemical Society. (c) Graphene with designed pore size for He isotope separation. Color code is same as (a). Reprinted with permission from [11] © 2012 American Chemical Society. (d) Free standing graphene nanomesh by block copolymer lithography. Gray: carbon. Reprinted with permission from [13] © 2010 Nature Publishing Group

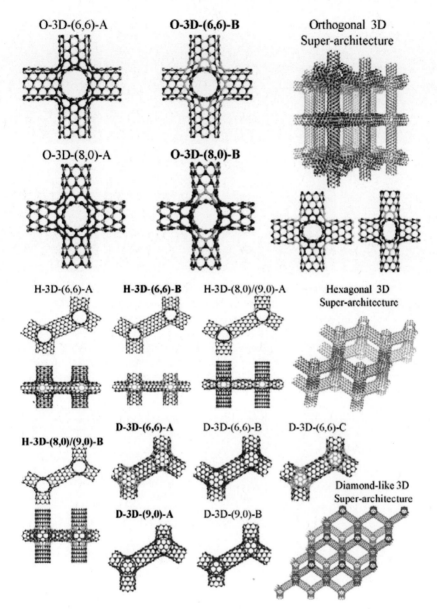

Figure 8.19 The optimized structures of the orthogonal, hexagonal, and diamond-like 3D SW-BNT superarchitectures (right panel), and optimized supercell units of 13 superarchitectures. Pentagons, heptagons, tetragons, and octagons in the nodal regions are highlighted in yellow, red, green, and cyan, respectively. Bold type highlights the superarchitectures in each family with the greatest cohesive energy

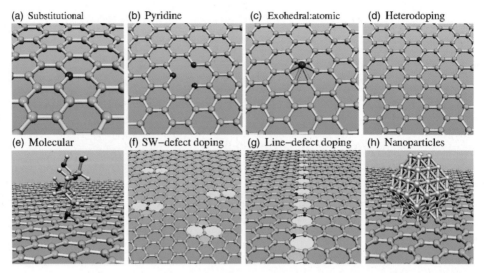

Figure 9.3 Molecular models representing different types of doped graphenes. *(a)* Substitutional *doping, in which a carbon atom is removed and replaced by the dopant;* *(b)* pyridine-like *doping, in which a single vacancy is generated and the low coordinated carbon atoms (three carbons around the vacancy) are replaced by dopants;* *(c)* Exohedral atomic *doping where a dopant is hosted on the surface of graphene;* *(d)* Heterodoping *which consists in doping simultaneously the graphene with two or more types atoms different from carbon;* *(e)* Molecular *doping or chemical functionalization of the graphene sheet;* *(f)* **55–77** Thrower–Stone–Wales *defects in which 5-7-5-7 rings are generated by rotating a C–C bond by 90 degrees (dopants are also included in some defects),* *(g)* 5–7 line-defect *in combination with substitutional doping, and* *(h)* Nanoparticles *or clusters anchored on the graphene surface*

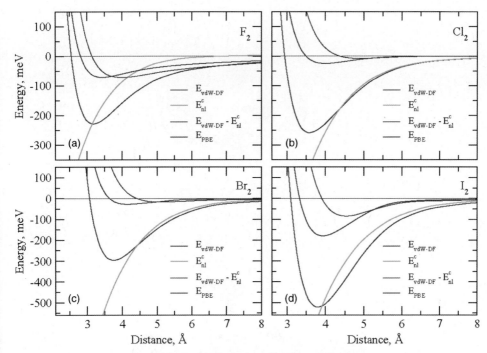

Figure 10.1 Adsorption of different halogen molecules on graphene. The binding energy as a function of the distance between the molecule and graphene is shown for the vdW-corrected density functional (blue curve) and the PBE-GGA functional (brown curve). The other two curves show partial contributions to the vdW-corrected functional and are of no importance here. Reprinted with permission from [9] © 2010 American Physical Society

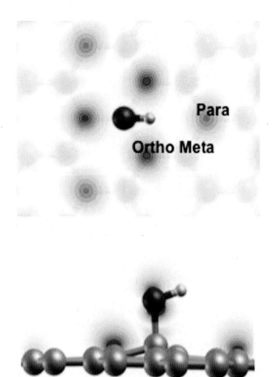

Figure 11.6 Ball-and-stick model of OH adsorbed on pristine graphene (gray: carbon atoms, small light blue: hydrogen atoms, red: oxygen atoms) and spin density distribution, calculated as the difference between the majority and minority spin densities (three-dimensional and projection onto the pristine graphene plane: red/blue indicate positive/negative values)

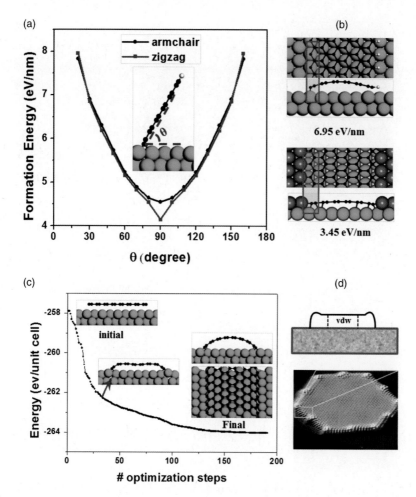

Figure 12.11 (a) The preference of an upright standing graphene nanoribbon (GNR) on an Ni(111) surface. (b) Optimized zigzag (ZZ) graphene edges on a Ni(111) terrace and near a metal step, along with the corresponding formation energies for each structure. (c) Formation of a curved GNR on Ni(111) surface and the total energy change during the structural optimization. (d) The scheme of graphene formation on metal substrate and the experimentally observed STM for graphene on Ir(111) surface. (Parts (a–c) Reprinted with permission from [111,116] © 2009 American Chemical Society; (d) Reprinted with permission from [117] © 2012 American Physical Society)

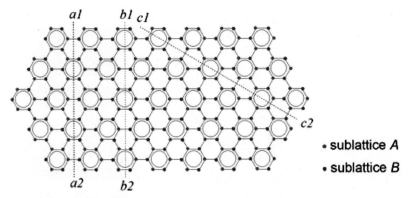

Figure 13.7 Oxidation-unzipping of graphene fragments along dashed lines a1−a2, b1−a2 and c1−c2. Reprinted with permission from [31] © 2009 American Chemical Society

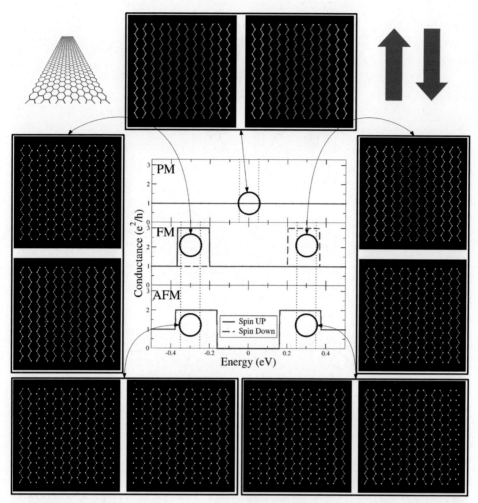

Figure 14.8 Quantum conductance as a function of energy (center) for the PM, FM and AFM states in a 12-ZGNR (full blue curves correspond to spin-up and red dashed lines to spin-down). For each state we present local current plots corresponding to $\mu_1 = -0.35$ eV and $\mu_2 = -0.25$ eV and to $\mu_1 = +0.25$ eV and $\mu_2 = +0.35$ eV for both the AFM and FM states and $\mu_{1/2} = +/-0.05$ eV for the PM case (chemical potential windows marked by vertical black dotted lines), both to spin-up (blue) and -down (red)

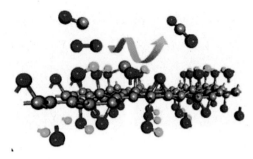

Figure 15.14 *Scheme of CO oxidation on an Fe-anchored GO in Ref. [71] Reprinted with permission from [71] © 2012 American Chemical Society*

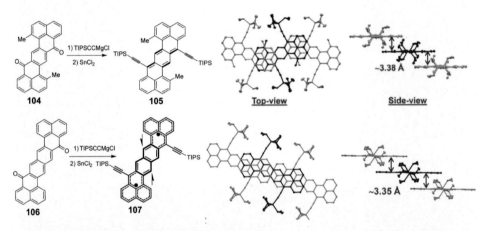

Figure 17.23 *Synthesis and crystal structures of heptazethrene and octazethrene derivatives*

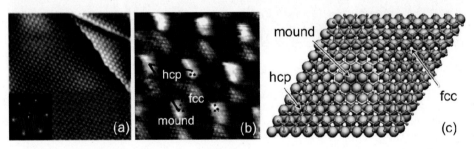

Figure 18.1 *(a) A typical STM image of graphene on Ru(0001) (100 × 100 nm^2, V_{sample} = +0.1 V, $I_{tunneling}$ = 0.2 nA); a typical LEED pattern is shown as an inset. Reprinted with permission from [27] © 2012 the American Physical Society. (b) A close-up STM image (8 × 8 nm^2, V_{sample} = −0.3 V, $I_{tunneling}$ = 1.0 nA). These and the STM images below are taken at room temperature. Reprinted with permission from [28] © Royal Society of Chemistry. (c) The optimized structural model for (12 × 12)–C on (11 × 11)-Ru(0001). Gray and green spheres represent lower- and top-layer Ru atoms, respectively. Carbon atoms are shown in ball-and-stick model and colored by height (red = lowest; blue = highest) for clarity. The three different regions in the moiré unit cell are labeled in (b) and (c)*

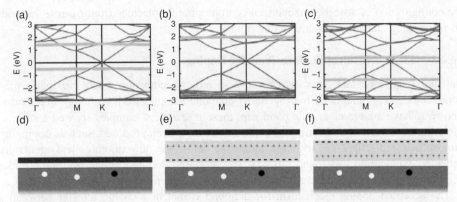

Figure 10.7 Shifting of the defect bands by water layers. The lower figures (d–f) show schematic pictures of the different investigated systems and the upper figures: The white and black dots indicate defects in the SiO$_2$ substrate. Figures (a–c) show the corresponding electronic band structures. The impurity bands are indicated in light shading and the bands from the water layer in darker shading. Reprinted with permission from [36] © 2008 American Institute of Physics

real SiO$_2$ substrate. The silanol layer is very hydrophilic and a water layer can be readily adsorbed. Wehling *et al.* used two different orientations of the water layer on top of the hydrophilic substrate. A schematic picture of two of these two models is shown in Figure 10.7(e) and (f). Again, the water molecules form a strong dipole layer that can shift the energy levels of the system. This time, the levels of the defects in the SiO$_2$ layer are shifted up and down with respect to the Dirac point, depending on the orientation of the dipole layer. When an impurity band crosses the Fermi-level, doping is induced in the graphene layer. This picture suggests a possible route to reduce the doping level and diminish the hysteresis in the electric field effect by depositing a hydrophobic layer on top of the SiO$_2$ substrate to prevent the formation of a water layer from ambient air. This has indeed been demonstrated by Lafkioti *et al.* in an experiment in which a self-assembled hydrophobic hexamethyldisilazane (HMDS) layer was created on top of a SiO$_2$ substrate before depositing the graphene layer [35].

It should be noted that this doping mechanism is only present when the dipole layer is placed between graphene and the substrate. When the water layer is placed on top of the graphene–substrate system, no influence of the dipole layer is expected. This can be understood from the fact that the dipole layer divides space into two regions with a different potential. When both systems (i.e., graphene and the substrate) experience the same potential shift, no changes are expected. But if the two systems are placed on different sides of the dipole layer, a relative shift of their respective band structures follows and a charge transfer is possible. This shift, and therefore also the charge transfer, depends on the orientation of the dipole layer. This makes it interesting to search for ways of modifying the direction of the molecular dipoles by some external effect as a way to change the doping in graphene. Some routes to achieve this have already been demonstrated: Chen *et al.* used large bias pulses to align NH$_3$ molecules on graphene in the direction of the bias [37], and Kim *et al.* obtained light-driven doping in graphene by reversible switching of the configuration of adsorbed azobenzene chromophore molecules from *trans* to *cis* with corresponding changes in the molecular dipoles [38].

10.4.3 Electrochemical Surface Transfer Doping

In 1989 Landstrass and Ravi discovered a strange phenomenon when they performed conductance experiments on hydrogenated diamond samples [39]. Although diamond is known to have a large electronic band gap, these researchers' samples showed a surprisingly high conductivity. A variety of explanations have been proposed, such as doping by atmospheric molecules, but it took a long time before the now (more or less) generally accepted explanation was found [40, 41]. The proposed doping mechanism might, under certain circumstances, be relevant for the doping of graphene samples as well [26,42].

The accepted doping mechanism for diamond is that of a charge transfer between a hydrogenated diamond surface on one side and an aqueous redox couple on the other side. This mechanism is only possible when there is an aqueous phase present near the diamond surface and if the redox couple has the right electrochemical potential. A hydrogenated diamond surface is hydrophobic but the charge transfer enhances the mutual attraction so that the hydrophobicity is reduced. It was suggested that the following reducing half reactions appear to underlie the doping mechanism for diamond [40, 41]:

$$O_2(g) + 4H^+(aq) + 4e^- \rightarrow 2H_2O(l) \tag{10.2}$$

$$O_2(g) + 2H_2O(l) + 4e^- \rightarrow 4OH^- \tag{10.3}$$

It depends on the acidity of the aqueous solution as to which one is dominant. The electrochemical potentials for these reactions under ambient conditions are -5.66 eV and -4.83 eV, respectively. In other words, energies from 4.83–5.66 eV (depending on the acidity of the solution) are gained when an electron is brought from vacuum into the system to trigger these reactions. When the electron is not coming from vacuum but from the hydrogenated diamond, one should subtract the energy that is needed to remove the electron from the hydrogenated diamond into vacuum. So there will be a spontaneous electron transfer from the hydrogenated diamond to the aqueous solution if the valence band maximum of diamond (-5.2 eV) is higher in energy than the electrochemical potentials for the half reactions. For equation (10.2) this is indeed the case so that it can cause hole doping in hydrogenated diamond (see Figure 10.8).

In the case of graphene, the electrochemical potential should be compared to the work function of graphene, which is about 4.5 eV, so that the same mechanism as for diamond might induce hole doping in graphene (as can be seen from Figure 10.6). Note that for this doping mechanism to be present, an aqueous phase is needed near the graphene layer and furthermore O_2 gas molecules should be available from the atmosphere. As mentioned earlier, such an aqueous phase can be present between graphene and a SiO_2 substrate due to the hydrophilic silanol layer at the SiO_2 surface. The observations of Ryu et al. of hole doping through atmospheric oxygen in deformed graphene on a SiO_2 substrate can be explained in this way [43].

One can also imagine other redox couples, that is, different from those relevant for doping hydrogenated diamond, to induce doping in graphene under certain circumstances. Pinto et al. have suggested similar redox reactions to cause doping with NO_2, NH_3, and toluene molecules [26].

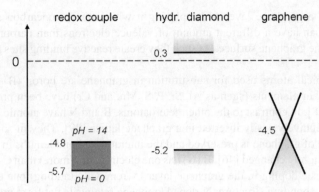

Figure 10.8 *The energy of the redox couple (for different pH) in comparison with the valence and conduction band of hydrogenated diamond and graphene. The energies are in eV*

In the transfer doping mechanisms discussed, the work function (WF) of graphene plays an important role because it determines the relative alignment of the electronic band structure of graphene and the molecular orbital levels. Consequently, the interaction and doping can be changed by modifying the WF of graphene. Actually, the doping that results from a water layer between graphene and a SiO_2 substrate can also be regarded as a change of the WF due to the dipole layer on the surface. But this is only one way to change the WF and other possibilities are readily found in the literature. It is, for instance, possible to change the work function of graphene by introducing strain in the sample [44], or by changing the number of layers in few layer graphene on a substrate [45].

10.5 Enhancing the Graphene-Molecule Interaction

We have described previously how the interaction of open-shell molecules and dipolar molecules with graphene can lead to a change of the electronic properties of graphene by causing n or p-type doping in graphene. Inert molecules with small electric dipoles, such as CO molecules, interact very weakly with graphene and in general are not able to induce noticeable changes. Therefore, it would be helpful to modify the graphene sample to enhance the interaction with these kinds of molecules. Several approaches to achieve this have been investigated in the literature, such as substitutional doping of graphene (e.g., B or N-doping), attaching more reactive adatoms on graphene to mediate the graphene-molecule interaction, the introduction of defects (e.g., vacancies) to make the graphene sample more reactive, and the application of a perpendicular electric field over the graphene layer. Furthermore, these modifications to the graphene sample can lead to an improved selectivity in the adsorption process. Some molecules attach more easily on the optimized graphene surface, while others do not bind at all. These are very useful properties for gas sensor applications. In this section, we give a short overview of some of the results that were obtained using various modification techniques.

10.5.1 Substitutional Doping

One of the most studied ways to increase the interaction between relatively inert molecules and graphene is to dope the graphene layer with heteroatoms. This is done by removing a C atom from a graphene layer and replacing it with various noncarbon atoms. These foreign atoms can have a different amount of valence electrons than carbon and usually protrude from the graphene surface. As such they create reactive binding sites for molecular adsorption.

The archetypical atoms used for substitution in graphene are boron (B) and nitrogen (N), but also other elements (such as Al, Si, P, S, Mn, and Cr) have been proposed in the literature [46–48]. In contrast to the other heteroatoms, B and N have atomic sizes close to C and can be relatively easily inserted in a graphene lattice [49]. They fit so well that the planar structure of graphene is preserved and the interatomic bond lengths in the graphene layer are only slightly changed [46]. B (N) has one electron less (more) than C and therefore it induces intrinsic doping in the graphene layer by removing (adding) one electron from (to) the aromatic bonding structure. N-doped graphene retains its flat form upon molecular adsorption and no chemical bond is made between the nitrogen atom and the molecule. But DFT calculations show that both adsorption energy and charge transfer are increased for typical molecules such as CO, NO, NO_2, and NH_3 [48] which indicates an increased interaction. This has been confirmed by experiments in which enhanced oxygen reduction was achieved with N-doped graphene devices [50, 51].

B-doped graphene, on the other hand, has been demonstrated to allow for chemical bond formation with NO_2 molecules [46]. The boron atom bulges out from the graphene layer and binds strongly with an O atom of the NO_2 molecule (see Figure 10.9). This reaction is rather similar to the chemisorption process of, for example, H atoms on pristine graphene [11]. There are some contradicting claims in the literature about the adsorption of NH_3: Depending on the kind of approximation for the exchange-correlation functional, LDA or GGA, NH_3 is found to chemisorb [48] or physisorb [46] on B-doped graphene. This reflects the tendency of LDA and GGA to overestimate and underestimate the binding energies, respectively.

Other dopants, such as Al and S, are found to protrude from the graphene plane [46, 48]. This indicates that their molecular orbitals are more sp^3 hybridized and that these atoms will try to bond with four partners. Three of these bonds are formed with carbon atoms in the graphene layer, while the fourth one is unsaturated. In this way, a very reactive adsorption site is created which causes most molecules to chemisorb on the dopant [46]. Ao et al. were the first to propose Al doped graphene as a means for sensing CO molecules [52]. Pristine graphene and B, N, or S doped graphene interact very weakly with CO molecules, so Al-doped graphene appears to be a useful candidate. However, the Al atom is so reactive that it binds to almost every molecule, even the very inert N_2 molecule [46], which renders the practicality of Al-doped graphene sensors rather doubtful.

10.5.2 Adatoms and Adlayers

A related technique to substitutional doping, is the coverage of the graphene surface with an additional layer of adsorbates. These extra adsorbates can enhance the interaction between molecules and graphene in a way similar to substitutional dopants. But adsorbates allow for much more variety because they do not need to be fitted into the graphene lattice. An often-made choice is to cover the graphene surface with a layer of metal adatoms such as Li atoms [30,53–55]. Metal adatoms from groups I–III of the Periodic table usually form ionic bonds with graphene and induce a substantial n-doping in graphene [30]. The bonding

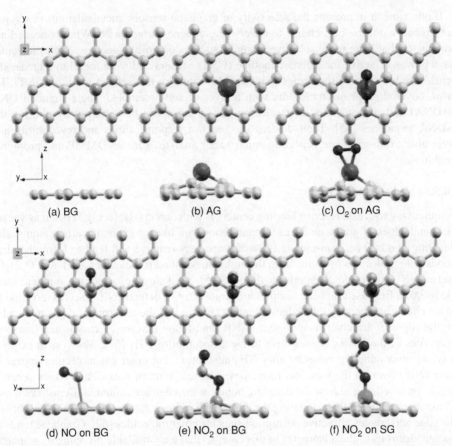

Figure 10.9 The most stable configurations for (a) B-doped graphene, (b) Al-doped graphene, (c) an O_2 molecule on Al-doped graphene, (d) an NO molecule on B-doped graphene, (e) an NO_2 molecule on B-doped graphene, and (f) an NO_2 molecule on S-doped graphene. Reprinted with permission from [46] © 2009 American Institute of Physics

of transition, noble, and group IV metals are more covalent-like and are characterized by strong hybridization between adatom and graphene orbitals [30]. The latter can also induce p-doping in graphene [56]. Zhang et al. investigated the effect of metallic adatoms on the adsorption of organic molecules [53]. Their *ab initio* study revealed that the adatoms can deform the organic molecules upon adsorption and that large charge transfers could be induced. The strongest effect was obtained with Fe atoms which were shown by their transport studies to increase the sensitivity for organic molecules with two orders of magnitude. Similar as with carbon nanotubes [57], one can think of increasing the selectivity by attaching small metal clusters that provide a broad range of reactivity with different molecules.

Chemically bonded adatoms, such as oxygen, transform graphene into a semiconducting material, called graphene oxide (GO), which, in reduced form, also shows some promise to be used in gas sensor applications [58–60]. In this case, the residual oxygen adsorbates provide the binding sites for the molecules.

If one aims at improving the selectivity of graphene sensors, metal adatoms or oxygen are probably not the best choice to cover the graphene surface with. What is needed are adsorbates that have more selective features, such as organic molecules. Such molecules have proven effective for carbon nanotube (CNT) sensors and were found to increase the sensitivity for specific molecules [61–64]. Motivated by this related work on CNTs, Lu et al. covered graphene transistors with a layer of self-assembled single-stranded DNA (ssDNA) and demonstrated that selective sensitivity could be achieved by altering the ssDNA sequences [65]. Their devices showed fast response times and reversibility and were able to discriminate between similar vapor analytes such as DMMP and propionic acid [65].

10.5.3 Edges and Defects

Another way to create reactive binding centers is with defects. Defect sites (such as vacancies) and edges of graphene flakes or nanoribbons are likely to contain carbon atoms with dangling bonds. These unsaturated C atoms are very reactive and tend to form chemical bonds with molecules or atoms from the ambient. Most inert molecules, such as CO, H_2O, and even N_2, are covalently bonded to the defect sites and are able to dope the graphene sample [66,67]. Because graphene samples are almost free of defects [68], it seems more likely to have dangling bonds at the edges. Huang et al. studied the adsorption of gas molecules on the edges of armchair nanoribbons (ANR) and found that the electronic and transport properties of the ANRs are sensitive to the adsorption of NH_3 [67]. The system exhibits n-type semiconducting behavior after NH_3 adsorption, but other gas molecules appear to have little effect on the transport properties. The reactivity of nanoribbon edges depends highly on whether and how the dangling bonds at the edge are saturated. Berashevich and Chakraborty showed that saturating the edges with oxygen atoms enhances the stability of the edge and creates attractive adsorption sites for polar molecules [69]. Graphene tends to donate electrons to the adsorbates in this case, so that a controllable p-doping of graphene with a variety of molecules can be achieved.

But, since it is rather difficult to control nanoribbon edges with good precision, it seems hard to derive a clear understanding of the interaction of these edges with adsorbates directly from *ab initio* simulations. Therefore, Saffarzadeh studied molecular adsorption on ANRs by applying the results of first-principles calculations to a single-band tight-binding approximation [70]. ANR is semiconducting and if the gas molecules are assumed to adsorb on the edges with a uniform probability distribution, the adsorbed molecules were shown to induce localized states near the center of the band gap which suggests that it is difficult to change considerably the charge transport in such systems.

10.5.4 External Electric Fields

An important way to control the physical properties of hybrid systems is by application of an external electric field. The electric field can cause substantial changes in the structure of, for example, an adsorbed layer on the substrate and can therefore alter the interaction between them. In recent years, a lot of research has been done on the use of electric fields to enhance the graphene–adsorbate interaction. A notable example is the electric field controlled hydrogen uptake of pristine and doped graphene systems. Without any external electric field, the interaction between graphene and H_2 molecules is extremely weak. Even in the case of strongly reacting substitutional heteroatoms such as Al, the doped graphene layer remains almost inert to hydrogen adsorption [46].

Ao and Peeters studied the hydrogenation of pristine graphene in the presence of an electric field and found a reduction of the energy barrier for molecular dissociative adsorption [71]. Shi *et al.* have experimentally verified an increased hydrogen adsorption on platinum-supported carbon samples due to an applied electric field [72]. The barrier can be decreased even further by doping the graphene layer. Ao *et al.* studied the diffusion of H atoms on N-doped graphene in the presence of a perpendicularly applied electric field [73]. They showed that an electric field facilitates the dissociative binding of hydrogen molecules on N-doped graphene. The process appears to be reversible and the absorbed H atoms can be released when the field is removed. Also doping with adatoms, such as Li, can be used to attach H_2 molecules in the presence of an electric field [74,75]. The interaction between the H_2 molecule and the adatoms is enhanced by the electric field and the molecules physisorb on the doped graphene layer.

But the usefulness of an electric field to enhance the adsorbate-adsorbent interaction is not restricted to the hydrogen-graphene systems. It was shown by Ao *et al.* that the interaction between a CO molecule and Al-doped graphene can be reduced or increased depending on the sign of the electric field [76]. It is thus possible to continuously tune the interaction. This has very interesting consequences as was demonstrated by a DFT study of Lu *et al.* [77]. In this study, the adsorption of gold atoms and NO_2 molecules was examined under varying electric fields. The doping was shown to depend almost linearly on the applied electric field and could even be altered from p to n-type doping. This is rather similar to the ambipolar electric field effect of graphene on a substrate [1]. But in the case under study, the magnetic moment of the adsorbates is changed according to the change in charge and therefore magnetic centers with a controllable magnetic moment are obtained [77]. The adsorbates can also be removed from the graphene surface when the electric field is further increased. This can be explained by the separation of the electric charges under the applied electric field: when the force on the charges becomes larger than the strength of the bond between the adsorbate and graphene, the adsorbate is pulled away from the surface [77].

As already mentioned before, it is also possible to align the electric dipoles of the polar adsorbates by the application of a perpendicular electric field over the graphene sample. Chen *et al.* did an experiment in which they changed the direction of the dipoles of adsorbed NH_3 molecules [37]. They obtained this NH_3 "flipping" by a sequence of high back-gate bias pulses of $+100$ and -100 V. After a bias pulse, they subsequently observed a change in the doping from the NH_3 molecules which they explained by a change in the dipole orientation of the molecules (see Section 10.4).

10.5.5 Surface Bending

Graphene samples that are attached to a substrate often take the corrugated form of that substrate [68,78] and even suspended graphene samples are not completely flat but contain elongated ripples. As with chemisorption [79], it has been suggested that these corrugation effects have an important influence on the physisorption of molecules on graphene [80]. Boukhvalov performed an *ab initio* simulation of physisorbed molecules on a corrugated graphene surface and observed two tendencies in his study [80]: (1) Large molecules are more likely to adsorb on top of ripples and small molecules in the valleys, and (2) electron donors tend to be adsorbed on the top of ripples because of a decreased charge density there, while electron acceptors are more likely to adsorb in the valleys between the ripples [80]. The charge transfers between molecules and the corrugated graphene sample are increased because of the stronger interaction. The enhanced interaction can be understood by a small change of the sp^2 and p_z orbitals in graphene towards sp^3 hybridization as a consequence of the bending (similar as with small carbon nanotubes). This results in an increased tendency for bond formation.

10.6 Conclusion

The interaction of graphene with molecules in its environment is an important research topic in the field of applied graphene physics. Adsorbed molecules can be used to dope graphene samples with electrons or holes and have a significant influence on the carrier mobility which is not yet completely understood. The doping can result from substitutional doping or surface transfer doping. Various charge transfer mechanisms can be distinguished in the last case, such as direct doping with paramagnetic molecules or electrochemical surface transfer doping in the presence of an aqueous phase. An important distinction can be made between open-shell and inert adsorbates of which the first usually induce large n or p-doping. The latter do not interact strongly with graphene, but can cause a substantial doping level in the case of polar molecules. The dipoles of such molecules can give rise to relative shifts of the energy states of graphene and, for example, impurity bands and can lead to significant charge transfers. However, many inert molecules interact very weakly with graphene and are therefore undetectable with a graphene sensor. To increase the interaction with these molecules, one can follow different strategies. A first one is to create more reactive binding centers in graphene by substitutional doping, adatoms, or adlayers, and defects sites with unsaturated carbon atoms. This has been demonstrated to enhance the interaction with some orders of magnitude in the case of very inert molecules such as CO and N_2. The interaction can also be tuned, that is, increased or decreased, by the application of a perpendicular electric field or the curvature of the graphene surface.

References

[1] F. Schedin, A. K. Geim, S. V. Morozov, E. W. Hill, P. Blake, M. I. Katsnelson, and K. S. Novoselov, Detection of individual gas molecules adsorbed on graphene, *Nature Materials*, **6**(9), 652–655 (2007).

[2] C. Hummel, F. Schwierz, A. Hanisch, and J. Pezoldt, Ambient and temperature dependent electric properties of backgate graphene transistors, *Physica Status Solidi (B)*, **247**(4), 903–906 (2010).

[3] J.-H. Chen, C. Jang, S. Adam, M. Fuhrer, E. D. Williams, and M. Ishigami, Charged-impurity scattering in graphene, *Nature Physics*, **4**(5), 377–381 (2008).

[4] A. Ghosh, K. V. Rao, S. J. George, and C. N. R. Rao, Noncovalent functionalization, exfoliation, and solubilization of graphene in water by employing a fluorescent coronene carboxylate, *Chemistry – A European Journal*, **16**(9), 2700–2704 (2010).

[5] M. Qazi, T. Vogt, and G. Koley, Trace gas detection using nanostructured graphite layers, *Applied Physics Letters*, **91**(23), 233101 (2007).

[6] M. Qazi, T. Vogt, and G. Koley, Two-dimensional signatures for molecular identification, *Applied Physics Letters*, **92**(10), 103120 (2008).

[7] T. O. Wehling, M. I. Katsnelson, and A. I. Lichtenstein, Impurities on graphene: Midgap states and migration barriers, *Physical Review B*, **80**(8), 085428 (2009).

[8] J. Klimeš, D. R. Bowler, and A. Michaelides, Van der Waals density functional applied to solids, *Physical Review B*, **83**, 195131 (2011).

[9] A. N. Rudenko, F. I. Keil, M. I. Katsnelson, and A. I. Lichtenstein, Adsorption of diatomic halogen molecules on graphene: A van der Waals density functional study, *Physical Review B*, **82**(3), 035427–035433 (2010).

[10] J. P. Perdew, K. Burke, and M. Ernzerhof, Generalized gradient approximation made simple, *Physical Review Letters*, **77**(18), 3865–3868 (1996).

[11] D. W. Boukhvalov, M. I. Katsnelson, and A. I. Lichtenstein, Hydrogen on graphene: Electronic structure, total energy, structural distortions, and magnetism from first-principles calculation, *Physical Review B*, **77**, 035427 (2008).

[12] J. Ito, J. Nakamura, and A. Nator, Semiconducting nature of the oxygen-adsorbed graphene sheet, *Journal of Applied Physics*, **103**, 113712 (2008).

[13] O. Leenaerts, B. Partoens, and F. M. Peeters, Adsorption of H_2O, NH_3, CO, NO_2, and NO on graphene: A first-principle study, *Physical Review B*, **77**, 125416 (2008).

[14] T. O. Wehling, M. I. Katsnelson, and A. I. Lichtenstein, Adsorbates on graphene: Impurity states and electron scattering, *Chemical Physics Letters*, **476**(4–6), 125–134 (2009).

[15] O. Leenaerts, B. Partoens, and F. M. Peeters, Water on graphene: Hydrophobicity and dipole moment using density functional theory, *Physical Review B*, **79**, 235440 (2009).

[16] K. M. McCreary, K. Pi, A. G. Swartz, W. Han, W. Bao, C. N. Lau, et al., Effect of cluster formation on graphene mobility. *Physical Review B*, **81**(11), 115453 (2010).

[17] J. Ristein, Surface Transfer doping of semiconductors, *Science*, **313**, 1057 (2006).

[18] W. Chen, D. Qi, X. Gao, and A. T. S. Wee, Surface transfer doping of semiconductors, *Progress in Surface Science*, **84**, 279–321 (2009).

[19] L. Zhao, R. He, K. T. Rim, T. Schiros, K. S. Kim, H. Zhou, et al., Visualizing individual nitrogen dopants in monolayer graphene, *Science*, **333**, 999 (2011).

[20] T. O. Wehling, K. S. Novoselov, S. V. Morozov, E. E. Vdovin, M. I. Katsnelson, A. K. Geim, and A. I. Lichtenstein, Molecular doping of graphene, *Nano Letters*, **8**(1), 173–177 (2008).

[21] O. Leenaerts, B. Partoens, and F. M. Peeters, Paramagnetic adsorbates on graphene: a charge transfer analysis, *Applied Physics Letters*, **92**, 243125 (2008).

[22] O. Leenaerts, B. Partoens, and F. M. Peeters, Adsorption of small molecules on graphene, *Microelectronics Journal*, **40**(4–5), 860–862 (2009).

[23] A. C. Crowther, A. Ghassaei, N. Jung, and L. E. Brus, Strong charge-transfer doping of 1 to 10 layer graphene by NO(2), *ACS Nano*, **6**(2), 1865–1875 (2012).

[24] G. Ko, H.-Y. Kim, J. Ahn, Y.-M. Park, K.-Y. Lee, and J. Kim, Graphene-based nitrogen dioxide gas sensors, *Current Applied Physics*, **10**(4), 1002–1004 (2010).

[25] Zanella, S. Guerini, S. B. Fagan, J. Mendes Filho, and A. G. Souza Filho, Chemical doping-induced gap opening and spin polarization in graphene, *Physical Review B*, **77**, 073404 (2008).

[26] H. Pinto, R. Jones, J. P. Goss, and P. R. Briddon, Mechanisms of doping graphene, *Physica Status Solidi*, **207**(9), 2131–2136 (2010).

[27] C. Coletti, C. Riedl, D. S. Lee, B. Krauss, L. Patthey, K. Von Klitzing, et al., Band structure engineering of epitaxial graphene on SiC by molecular doping, *Physical Review B*, **81**(23), 235401 (2009).

[28] Y.-H. Zhang, K.-G. Zhou, K. F. Xie, J. Zeng, H.-L. Zhang, and Y. Peng, Tuning the electronic structure and transport properties of graphene by noncovalent functionalization: effects of organic donor, acceptor and metal atoms, *Nanotechnology*, **21**, 065201 (2010).

[29] Y. H. Lu, W. Chen, Y. P. Feng, and P. M. He, Tuning the electronic structure of graphene by an organic molecule, *Journal of Physical Chemistry B*, **113**(1), 2–5 (2009).

[30] K. T. Chan, B. Neaton, and M. L. Cohen, First-principles study of metal adatom adsorption on graphene, *Physical Review B*, **77**, 235430 (2008).

[31] Y. Dan, Y. Lu, N. J. Kybert, Z. Luo, and A. T. C. Johnson, Intrinsic response of graphene vapor sensors, *Nano letters*, **9**(4), 1472–1475 (2009).

[32] H. E. Romero, P. Joshi, A. K. Gupta, H. R. Gutierrez, M. W. Cole, S. A. Tadigadapa, and P. C. Eklund, Adsorption of ammonia on graphene, *Nanotechnology*, **20**(24), 245501 (2009).

[33] Y.-C. Lin, C.-Y. Lin, and P.-W. Chiu, Controllable graphene N-doping with ammonia plasma. *Applied Physics Letters*, **96**(13), 133110 (2010).

[34] J. Moser, A. Verdaguer, D. Jiménez, A. Barreiro, and A. Bachtold, The environment of graphene probed by electrostatic force microscopy, *Applied Physics Letters*, **92**, 123507 (2008).

[35] M. Lafkioti, B. Krauss, T. Lohmann, U. Zschieschang, H. Klauk, K. von Klizing, and J. H. Smet, Graphene on a hydrophobic substrate: Doping reduction and hysteresis suppression under ambient conditions, *Nano Letters*, **10**, 1149–1153 (2010).

[36] T. O. Wehling, A. I. Lichtenstein, and M. I. Katsnelson, First-principles study of water adsorption on graphene: the role of the substrate, *Applied Physics Letters*, **93**(20), 202110 (2008).

[37] S. Chen, W. Cai, D. Chen, Y. Ren, X. Li, Y. Zhu, J. Kang, and R. S. Ruoff, Adsorption/desorption and electrically controlled flipping of ammonia molecules on graphene, *New Journal of Physics*, **12**(12), 125011 (2010).

[38] M., Kim, N. S. Safron, C. Huang, M. S. Arnold, and P. Gopalan, Light-driven reversible modulation of doping in graphene, *Nano Letters*, **12**(1), 182–187 (2011).

[39] M. I. Landstrass, K. V. Ravi, Resistivity of chemical vapor deposited diamond films, *Applied Physics Letters*, **55**, 975 (1989).

[40] J. Ristein, Surface transfer doping of diamond, *Journal of Applied Physics D: Applied Physics*, **39**, R71–R81 (2006).

[41] V. Chakrapani, J. C. Angus, A. B. Anderson, S. D. Wolter, B. R. Stoner and G. U. Sumanasekera, Charge transfer equilibria between diamond and an aqueous oxygen electrochemical redox couple, *Science*, **318**, 1424–1430 (2007).

[42] S. J. Sque, R. Jones, and P. R. Briddon, The transfer doping of graphite and graphene, *Physica Status Solidi*, **204**(9), 3078–3084 (2007).

[43] S. Ryu, L. Liu, S. Berciaud, Y. Yu, H. Liu, and P. Kim, Atmospheric oxygen binding and hole doping in deformed graphene on a SiO_2 substrate, *Nano Letters*, **10**(12), 4944–4951 (2010).

[44] S.-M. Choi, S.-H. Jhi, and Y.-W. Son, Effects of strain on electronic properties of graphene, *Physical Review B*, **81**, 081407 (2010).

[45] H. Hibino, H. Kageshima, M. Kotsugi, F. Maeda, F.-Z. Guo, and Y. Watanabe, Dependence of electronic properties of epitaxial few-layer graphene on the number of layers investigated by photoelectron emission microscopy, *Physical Review B*, **79**, 125437 (2009).

[46] J. Dai, J. Yuan, and P. Giannozzi, Gas adsorption on graphene doped with B, N, Al, and S: A theoretical study, *Applied Physics Letters*, **95**, 232105–232107 (2009).

[47] J. Dai and J.Yuan, Adsorption of molecular oxygen on doped graphene: Atomic, electronic, and magnetic properties, *Physical Review B*, **81**(16), 165414–165420 (2010).

[48] Y.-H. Zhang, Y.-B. Chen, K.-G. Zhou, C.-H Liu, J. Zeng, H.-L. Zhang, and Y. Peng, Improving gas sensing properties of graphene by introducing dopants and defects: a first-principles study, *Nanotechnology*, **20**(18), 185504–185511 (2009).

[49] R. B. Pontes, A. Fazzio, and G. M. Dalpian, Barrier-free substitutional doping of graphene sheets with boron atoms: Ab initio calculations, *Physical Review B*, **79**(3), 033412 (2009).

[50] Y. Shao, S. Zhang, M. H. Engelhard, G. Li, G. Shao, Y. Wang, *et al.*, Nitrogen-doped graphene and its electrochemical applications, *Journal of Materials Chemistry*, **20**(35), 7491–7496 (2010).

[51] L. Qu, Y. Liu, and J. Baek, Nitrogen-doped graphene as efficient metal-free electrocatalyst for oxygen reduction in fuel cells. *ACS nano*, **4**(3), 1321–1326 (2010).

[52] Z. M. Ao, J. Yang, S. Li, and Q. Jiang, Enhancement of CO detection in Al doped graphene. *Chemical Physics Letters*, **461**(4–6), 276–279 (2008).

[53] Y.-H. Zhang, K.-G. Zhou, K. F. Xie, J. Zeng, H.-L. Zhang, and Y. Peng, Tuning the electronic structure and transport properties of graphene by noncovalent functionalization: effects of organic donor, acceptor and metal atoms, *Nanotechnology*, **21**, 065201 (2010).

[54] I. Cabria, M. J. López, and J. A. Alonso, Enhancement of hydrogen physisorption on graphene and carbon nanotubes by Li doping, *The Journal of Chemical Physics*, **123**(20), 204721 (2005).

[55] E. Rangel, G. Ruiz-Chavarria, and L. F. Magana, Water molecule adsorption on a titanium-graphene system with high metal coverage, *Carbon*, **47**(2), 531–533 (2009).

[56] I. Gierz, C. Riedl, U. Starke, C. R. Ast, and K. Kern, Atomic hole doping of graphene. *Nano Letters*, **8**(12), 4603–4607 (2008).

[57] Q. Zhao, M. B. Nardelli, W. Lu, and J. Bernholc, Carbon nanotubes-metal cluster composites: a new route to chemical sensors?, *Nano Letters*, **5**(5), 847–851 (2005).

[58] J. T. Robinson, F. K. Perkins, E. S. Snow, Z. Wei, and P. E. Sheehan, Reduced graphene oxide molecular sensors, *Nano Letters*, **8**(10), 3137–3140 (2008).

[59] J. D. Fowler, M. J. Allen, V. C. Tung, Y. Yang, R. B. Kaner, and B. H. Weiller, Practical chemical sensors from chemically derived graphene. *ACS Nano*, **3**(2), 301–306 (2009).

[60] G. Lu, L. E. Ocola, and J. Chen, Gas detection using low-temperature reduced graphene oxide sheets, *Applied Physics Letters*, **94**(8), 083111 (2009).

[61] P. Qi, O. Vermesh, M. Grecu, A. Javey, Q. Wang, and H. Dai, Toward large arrays of multiplex functionalized carbon nanotube sensors for highly sensitive and selective molecular detection, *Nano Letters*, **3**(3), 347–351 (2003).

[62] Y. Lin, F. Lu, Y. Tu, and Z. Ren, Glucose biosensors based on carbon nanotube nanoelectrode ensembles, *Nano Letters*, **4**(2), 191–195 (2004).

[63] C. Staii and A. T. Johnson Jr., DNA-decorated carbon nanotubes for chemical sensing, *Nano Letters*, **5**(5), 1774–1778 (2005).

[64] O. Kuzmych, B. L. Allen, and A. Star, Carbon nanotubes sensors for exhaled breath components, *Nanotechnology*, **18**, 375502 (2007).

[65] Y. Lu, B. R. Goldsmith, N. J. Kybert, and A. T. C. Johnson, DNA-decorated graphene chemical sensors, *Applied Physics Letters*, **97**, 083107 (2010).

[66] H. Ren, Q. Li, H. Su, Q. W. Shi, J. Chen, and J. Yang, Edge effects on the electronic structures of chemically modified armchair graphene nanoribbons, doi: arXiv:0711.1700v1.

[67] B. Huang, Z. Li, Z. Liu, G. Zhou, S. Hao, J. Wu, et al., Adsorption of gas molecules on graphene nanoribbons and its implication for nano-scale molecule sensor, *Journal of Physical Chemistry C*, **112**, 13442 (2008).

[68] E. Stolyarova, K.T. Rim, S. Ryu, J. Maultzsch, P. Kim, L.E. Brus, et al., High-resolution scanning tunneling microscopy imaging of mesoscopic graphene sheets on an insulating surface, *Proceedings of the National Academy of Sciences of the U.S.A.*, **104**, 9209–9212 (2007).

[69] J. Berashevich and T. Chakraborty, Doping graphene by adsorption of polar molecules at the oxidized zigzag edges, *Physical Review B*, **81**, 205431 (2010).

[70] A. Saffarzadeh, Modeling of gas adsorption on graphene nanoribbons, *Journal of Applied Physics*, **107**, 114309 (2010).

[71] Z. M. Ao and F. M. Peeters, Electric field: A catalyst for hydrogenation of graphene, *Applied Physics Letters*, **96**(25), 253106 (2010).

[72] S. Shi, J.-Y. Hwang, X. Li, X. Sun, and B. I. Lee, Enhanced hydrogen sorption on carbonaceous sorbents under electric field, *International Journal of Hydrogen Energy*, **35**(2), 629–631 (2010).

[73] Z. M. Ao, A. D. Hernández-Nieves, F. M. Peeters, and S. Li, The electric field as a novel switch for uptake/release of hydrogen for storage in nitrogen doped graphene, *Physical Chemistry Chemical Physics*, **14**(4), 1463–1467 (2012).

[74] Z.-W. Zhang, J.-C. Li, and Q. Jiang, Density functional theory calculations of the metal-doped carbon nanostructures as hydrogen storage systems under electric fields: A review, *Frontiers of Physics*, **6**(2), 162–176 (2011).

[75] W. Liu, Y. H. Zhao, J. Nguyen, Y. Li, Q. Jiang, and E. J. Lavernia, Electric field induced reversible switch in hydrogen storage based on single-layer and bilayer graphenes, *Carbon*, **47**(15), 3452–3460 (2009).

[76] Z. M. Ao, S. Li, and Q. Jiang, Correlation of the applied electrical field and CO adsorption/desorption behavior on Al-doped graphene, *Solid State Communications*, **150**(13–14), 680–683 (2010).

[77] Y.-H. Lu, L. Shi, C. Zhang, and Y.-P. Feng, Electric field control of magnetic states, charge transfer, and patterning of adatoms on graphene: First-principles density functional theory calculations, *Physical Review B*, **80**, 233410 (2009).

[78] M. Ishigami, J. H. Chen, W. G. Cullen, M. S. Fuhrer, and E. D. Williams, Atomic structure of graphene on SiO_2, *Nano Letters*, **7**(6), 1643–1648 (2007).

[79] D. W. Boukhvalov and M. I. Katsnelson, Enhancement of chemical activity in corrugated graphene, *Journal of Physical Chemistry C*, **113**(32), 14176–14178 (2009).

[80] D. W. Boukhvalov, Tuneable molecular doping of corrugated graphene, *Surface Science*, **604**(23–24), 2190–2193 (2010).

11
Surface Functionalization of Graphene

Maria Peressi
Department of Physics, University of Trieste, Trieste, Italy and
CNR-IOM DEMOCRITOS National Simulation Center, Italy

11.1 Introduction

Graphene is of enormous interest due to its unique physical properties, which make it a promising candidate for a wide range of applications in next-generation nanotechnologies overcoming the limits of current materials. However, there are some problems preventing graphene in reaching its full potential. Pure, ideal graphene (pristine) has an intrinsic zero band-gap which must be properly tuned to combine graphene with the well assessed semiconductor-based electronic technology. In addition, graphene is difficult to produce and process on large scale. Pure graphene is insoluble in organic solvents and tends to agglomerate into graphite in aqueous solutions. Modification of graphene sheets seems, therefore, necessary not only to tune its properties at will, but also to overcome the above mentioned problems, driving graphene towards unique applications in electronics and other physics and chemistry based technologies.

Without introducing substitutional impurities that could destroy the carbon network and affect its mechanical stability and performance, an ideal way of modifying the physical and chemical properties of graphene is through chemical adsorption of atoms or molecules. To our knowledge, the terms *chemically modified graphene* (CMG) and *functionalized graphene* (FG) were first used only a few years ago [1, 2]. Since then, the number of publications on the subject has rapidly increased, as shown in Figure 11.1, and up to today this is a very active research field. A few recent reviews [3–7], perspectives [8], collections

---

Graphene Chemistry: Theoretical Perspectives, First Edition. Edited by De-en Jiang and Zhongfang Chen.
© 2013 John Wiley & Sons, Ltd. Published 2013 by John Wiley & Sons, Ltd.

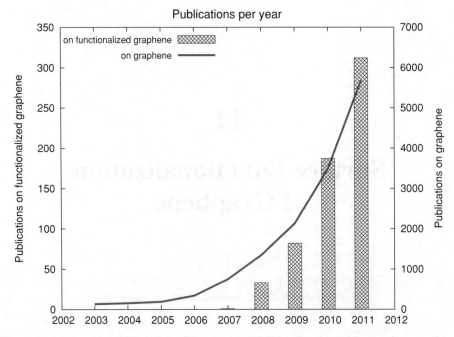

Figure 11.1 Number of scientific publications on functionalized graphene and on graphene. Data are obtained from the most widely used scientific databases with keywords "functionalized graphene" or "chemically modified graphene", and "graphene" in the title, respectively

of relevant papers and special issues on the subject [9–12] are already available, and updates on current research can be found on web sites of the major active research groups in the field (see e.g., the one at the Manchester University, where graphene was isolated by A. Geim and K. Novoselov in 2003 [13]).

Similar to the surface of graphite, graphene can adsorb and desorb various atoms and molecules through non-covalent and covalent bonding. Graphene is characterized by the sp^2 hybridization which leads to a strong covalent bonding in the 2D carbon network, accompanied by delocalization of π electrons. The interaction of the surface with atoms or molecules modifies the π-π interaction, and thus electron density distribution and physical and chemical properties. The nature of the chemical bonding with the surface (i.e., non-covalent versus covalent) determines the kind of functionalization. Weakly attached adsorbates (non-covalent bonding) often act as donors or acceptors and lead to changes in the carrier concentration, making graphene highly conductive. Other than weakly attached adsorbates, graphene can be modified by several chemical groups forming intermediate or strong bonds (covalent).

Functionalization of graphene sheets is strongly related with the production processes. Most work to date has focused on graphene oxide (GO) as one of the simplest covalently chemically modified forms of graphene that can be used to produce and process graphene-based materials on large scales [14–17]. A method of producing single FG sheets in bulk quantities is through simultaneous thermal exfoliation and reduction of GO [2, 18]. GO is

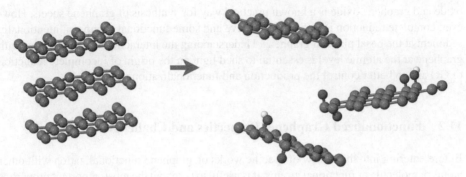

Figure 11.2 Sketch illustrating the fabrication process of functionalized graphene sheet by exfoliation from graphite

electrically insulating and typically highly defective, thus far from successful in recovering the electronic properties of pristine graphene, even upon reduction. Nevertheless, it is used in many applications, mainly as a precursor for further functionalization. It is worth mentioning that, in general, many forms of CMG can be further functionalized with the desirable physical and chemical properties. In all cases, surface modification of graphene can prevent agglomeration and facilitates the formation of stable dispersions of single-layer graphene sheets.

Several studies have been also reported on the preparation of FG directly from graphite, for instance with liquid-phase exfoliation [19, 20] (Figure 11.2). Recently, wet chemical bulk functionalization from pristine graphite was reported, which allows to avoid the initial oxidative damage of graphene [21]. Theoretical investigations suggested that functionalization of graphite by phenyl groups ($-C_6H_5$) reduces the interlayer binding energy by a factor of approximately 10 which makes exfoliation of the upper layer relatively simple [22].

For many applications, nanostructures with large active surface areas are desirable. Graphene, being the "thinnest material in the world", offers a large surface area (a very large surface area to mass ratio) for functionalization and consequent applications. *Surface functionalization* is therefore extremely important and is the main focus of this chapter. However, graphene can be functionalized both at the surface and edges. The latter are highly reactive to bind atoms or molecules. Edge functionalization has been widely studied in graphene nanoribbons [23–27].

Graphene functionalization constitutes a challenging topic not only for experimental, but also for theoretical and computational investigations. Accurate and highly predictive *ab initio* calculations can provide detailed information about the electronic structure, charge transfer and redistribution in graphene upon functionalization. It is therefore possible to study systematically the effective change of its electronic properties, obtaining insights into the graphene physics and chemistry.

In this chapter, we provide a short overview of the advances in the understanding of graphene functionalization from theoretical/computational perspectives. We focus in particular on the adsorption and reaction of hydrogen and oxygen-containing functional groups such as hydroxyl, epoxy, and carboxyl. Intentional functionalization of graphene with such groups is of great interest. On the other hand, as already mentioned, reduction of graphite oxide and graphene oxide is a known practical way for synthesis of graphene sheets. However, complete reduction is difficult to achieve and some functional groups unintentionally sediment at the basal plane of graphene. Understanding the interaction of these groups with graphene at the atomic level is essential to shed light on the origin of incomplete reduction of GO and to better control the production and functionalization processes.

11.2 Functionalized Graphene: Properties and Challenges

Before entering into the review of specific works on graphene functionalization with other atoms or molecule or functional groups, it is useful to point out the most important properties discovered till now, and some desired possible applications of functionalized graphene.

- *Electronics, band-gap engineering, spin devices:*
  Possible applications of graphene in electronic nanodevices (super-capacitor devices, transistor devices...) make it desirable to render graphene, semiconducting. In fact, traditional microelectronic devices such as those based on the well assessed silicon technology require a semiconductor material with an electronic band-gap. Monolayer pristine graphene does not possess such an electronic band-gap, and bi-layer graphene needs an applied gate voltage to have an induced band-gap. Various methods to control the graphene band-structure have been proposed. Chemical functionalization seems to be a promising way to reach this goal.

  For instance, CMG made from one-atom thick sheets of carbon, functionalized as needed, demonstrates its performance in an ultracapacitor cell [28]. Changes in the magnetic properties of graphene upon functionalization open the way for its use in spintronic devices.
- *Sensing, solar cells, selectivity/specificity for catalytic reactions:*
  CMG has the potentiality of detecting gas molecules with single-atom sensitivity. This makes graphene one of the most exciting materials for sensing applications. Commercial realization of graphene sensors will require the graphene to be selectively sensitive to the target molecule, while being unaffected by other interfering molecules. This requires special functionalization of the graphene, which should be achieved without destroying its conductivity and integrity. Various applications require different functionalization.

  FG electrodes were tested for catalytic performance in dye-sensitized solar cells giving promising results [29].
- *Biosystems:*
  Various graphene-based nanomaterials have been used to fabricate functionalized biosystems integrated with nucleic acids, peptides, proteins, and even cells. CMG started to reveal a great potential for applications in biological studies, and biotechnology. Graphene, graphene oxide (GO), CMG and in general graphene derivatives are promising candidates for biotechnology development [30].
- *Nanodevices with optimal mechanical performance:*
  Functionalization of graphene can improve its mechanical performance. For instance, the bending characteristics of functionalized graphene sheets have been proved with the tip of an atomic force microscope. Individual sheets were transformed from a flat into a

folded configuration. Sheets can be reversibly folded and unfolded multiple times, and the folding always occurs at the same location, suggesting that the folding and bending behavior of the sheets is dominated by pre-existing kink (or even fault) lines consisting of defects and/or functional groups [31].

- *Nanocomposites:*
Polymer-based composites have been considered since the 1960s as a new paradigm for strong, durable, multifunctional materials. By dispersing strong, highly stiff fibers in a polymer matrix, high-performance lightweight composites could be developed and tailored to individual applications. The creation of polymer nanocomposites with functionalized graphene sheets, which provide excellent polymer-particle interactions, has been recently reported [32].

11.3 Theoretical Approach

Graphene is one of the most exciting and challenging subjects of theoretical and computational works. Theoretical works focus on its interaction with atoms and molecules, diffusion of the adsorbed species, graphene's chemical derivatives such as *graphane* (hydrogenated graphene), graphene oxide, graphene with defects, and dopants. Numerical simulations can predict equilibrium geometries, structural and vibrational parameters, charge distribution, electronic structure, magnetic properties, adsorption energies, and energy barriers. Furthermore, they can predict how the adsorbed species can be detected by various spectroscopy [33,34] and microscopy tools [35]. Simulation can guide experiments in the optimal design of CMG.

Most of theoretical work on graphene based systems is within the framework of the spin polarized density functional theory (DFT) [36]. Both periodic (slab models) and molecular approaches (cluster models) [37] are used, as well as different basis set, including local orbitals and plane waves, and different treatment of the electrons, including pseudopotential and all-electron methods. Among the different functionals widely used to study graphene and CMG, we mention the Local Density Approximation (LDA) and the Generalized Gradient Corrected approximation (GGA) [36].

Standard DFT methods fail to describe long-range dispersion (van der Waals) interactions. They have to be taken into account to obtain reliable results for atoms and molecules physisorbed on graphene or non-covalently adsorbed, such as neutral (poly-)aromatic, anti-aromatic, and more generally π-conjugated systems on graphene [38,39]. Several correction methods to standard DFT are used to this purpose: one is an empirical r^{-6} correction to the DFT total energy (DFT-D) [40,41]. Another method is based on a nonlocal functional proposed by Dion et al. (vdW-DF) [42]; other more refined methods are also used, such as "double-hybrid" density functionals including second order perturbation theory corrections to the correlation energy, and others are under development [43].

Beyond standard DFT calculations, advanced techniques are used for the treatment of excited states and the calculations of optical properties, that require inclusion of many-body corrections (GW quasi-particle calculations).

Exploring potential energy surfaces is extremely important for the calculation of minimum-energy paths, transition states, activation energies and barriers in adsorption processes and other reactions. To this purpose, the Nudged Elastic Band (NEB) method

[44] is a widely used approach, implemented in many codes for periodic DFT calculations. Potential energy surfaces can be studied using metadynamics by including additional degrees of freedom, or using suitably chosen collective variables, to describe the evolution of the system towards the reactant state [45].

Several computer codes designed to perform electronic structure calculations are available (see e.g., Ref. [46] for a representative list). Among those based on slab models for periodic DFT calculations, which are widely used by the computational physics community, we mention the Quantum-ESPRESSO package [47] which allows for the numerical simulation of the electronic, structural, and dynamical properties of materials, including first-principles molecular dynamics and Car–Parrinello codes. It is distributed under the GNU General Public Licence. Cluster models are used in state-of-the-art quantum chemical codes, such as Gaussian [48], released and updated mainly by the computational chemistry community. Although in this chapter we mainly focus on results obtained by slab models, we note that in most cases cluster models, combined with high-level theory approaches, have been proved to be a successful alternative in modeling graphene surfaces and can often serve as a benchmark for the development of potentials aiming to describe graphene-based systems (for a recent example see Ref. [49]).

Ab initio methods are preferred since they adequately describe the chemical bonding, at the price of a rather limited size of the systems studied. For larger scale simulations, instead, classical molecular dynamics (CMD) methods are more suitable, although they can capture only qualitatively properties and mechanisms. A recent example of application is the study of the interaction of interfacial water with graphitic carbon at the atomic scale, which is studied as a function of the hydrophobicity of epitaxial graphene [50]. Another example is the evolution of randomly distributed hydroxyl and epoxy groups on GO sheets during high temperature annealing using molecular dynamics simulations where the atomic interactions are described by the so-called Reactive Force Field potential [51].

11.4 Interaction of Graphene with Specific Atoms and Functional Groups

11.4.1 Interaction with Hydrogen

H atom adsorption is known to induce structural, electronic, and chemical modifications at the graphene surface. The mechanism has been widely studied from first principles [49, 52–59]. An individual hydrogen chemisorbs on top of a carbon atom with a threefold symmetry configuration characterized by a C–H distance of about 1.5 Å and a slight out-of-plane protrusion of the hydrogenated carbon atom of about 0.4 Å. The binding energy reported by the literature is in the range of 0.3–0.7 eV. Itinerant magnetism due to the defect-induced extended states is induced, with a calculated magnetic moment equal to 1 μ_B per chemisorbed hydrogen. This makes the energy barrier for a single hydrogen migration very delicate to calculate and sensitive to the hydrogen coverage; an estimate similar to the adsorption energy is reported, for instance, in Ref. [57].

Experiments and theory point towards the existence of a physisorbed state also, with a much lower binding energy, allowing the physisorbed atoms to diffuse relatively rapidly. DFT calculations with various exchange-correlation functionals, including van der Waals

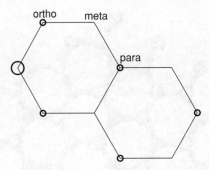

Figure 11.3 Labeling of different neighboring adsorption sites of the graphene lattice, relative to the one indicated as the reference (large circle). One sublattice includes the reference and meta position and the other sublattice includes para and ortho positions

corrected functionals, predict a wide range of binding energies for physisorbed hydrogen on graphene, going from 5 meV to about 0.1 eV [59]. The spreading of these results indicate that this system presents significant challenges to DFT methods, and even recent state-of-the-art van der Waals corrected DFT methods may not be accurate enough. Quantum Monte Carlo calculations using the diffusion Monte Carlo (DMC) method are in agreement with the lowest DFT value of the adsorption energy [59].

A hydrogen dimer can adsorb in different configurations, according to the relative position of the hydrogenated carbons (see Figure 11.3). Dimers adsorbed on graphene can be in ferromagnetic, antiferromagnetic, or nonmagnetic states, depending on the adsorption sites of the two hydrogen atoms and in general whether they correspond to the same or to different hexagonal sublattices of the graphene lattice. There is consensus on the fact that so-called *ortho* and *para* dimers are nonmagnetic and exhibit an enhanced stability with respect to all the other configurations. Further details and the mechanism through which stabilization occurs is analyzed and discussed in Refs. [53 and 54], with some discrepancies. The *ortho* and *para* dimers are important for the molecular hydrogen recombination mechanism. DFT calculations have also shown that adsorbed hydrogen atoms tend to aggregate, with a high probability of cluster formation in high adatom concentration [58]. The same occurs with fluorine atoms and hydroxyl groups functionalization, as we will discuss later on, but not with larger species (methyl and phenyl groups) due to the steric effect which renders cluster formation unfavored.

Full hydrogenation of graphene, giving rise to the so called *graphane*, has been first theoretically predicted in 2007 [60]. Shortly after, experimental evidence of its synthesis was achieved, together with the reversibility of the hydrogenation process upon annealing [61,62]. At full coverage, hydrogen adsorbs on graphene sublattices from the two opposite sides and carbon atoms of the two sublattices move out of the plane in opposite directions, producing a buckling of the carbon sheet as shown in Figure 11.4. Full hydrogenation implies a change of hybridization from sp^2 (graphene) to sp^3 (graphane), accompanied by a metal-insulator transition [62]. A widening of the C–C bond length has been reported from 1.42 Å (graphene) to 1.54 Å (graphane). The value of 1.54 Å corresponds to the C–C distance in bulk diamond, consistently with the sp^3 hybridization, whereas the value 1.42 Å is also the C–C bond length in graphite, consistently with the sp^2 hybridization.

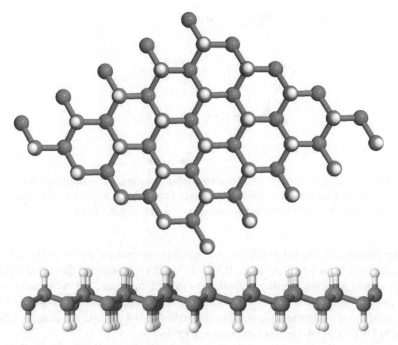

Figure 11.4 *Ball-and-stick model of graphane, fully hydrogenated graphene, in top and side view. Gray: carbon atoms, white: hydrogen atoms*

DFT-GGA *ab-initio* calculations of graphane have been performed to characterize the various properties of graphane, and in particular the metal-insulator transition [63]. The electronic band structure is characterized by the opening of a direct electronic gap at Γ. DFT value for the gap is reported to be about 3.5 eV. Many-body corrections, included within the GW approximation, strongly increase the fundamental gap of graphane giving a quasi-particle gap of about 5.7 eV. Dramatic changes in the optical absorption spectrum are reported with respect to perfect graphene [63].

11.4.2 Interaction with Oxygen

As mentioned in the Introduction, probably the most common route to graphene and CMG involves the production of GO, a highly exfoliated material that is dispersible in water, yielding colloidal suspensions. Therefore, we mention here for completeness the most relevant properties of GO but for further details we address the reader to Chapter 13 of the book specifically devoted to this.

The atomic and electronic structure of GO has been the subject of numerous experimental and theoretical studies, as reported in some recent reviews [14–17]. The structural model of GO has been the subject of considerable debate for many years, and even today no definite answer exists. The most recent models of GO point to a non-stoichiometric and non-regular pattern of adsorbed O atoms. GO is known to contain epoxide functional groups (C–O–C) as well as hydroxyl (–OH) and carboxyl (–COOH) along the basal plane.

DFT calculations predict that partial oxidation is thermodynamically favored over complete oxidation and that the precise kind and distribution of oxide functional groups depend strongly on the oxygen coverage [64].

GO is electrically insulating, and must be converted by chemical reduction to restore the electronic properties of graphene. The chemical and structural characteristics of GO may change during the various stages of reduction [65]. The properties of reduced GO (r–GO) do not approach those of intrinsic graphene obtained by mechanical cleavage for two reasons: (1) oxidation to GO and deoxydation introduce defects; (2) chemical reduction of GO is never fully achieved: the material remains significantly oxidized and the graphitic structure is not fully restored. Concerning (1), it is worth mentioning that numerical studies on the adsorption of atomic oxygen on graphene surface have shown that formation of CO_2 and CO leaves vacancy defects at the graphene surface saturated by oxygen groups [66]. Concerning (2), we will deeply address the problem in the section devoted to the interaction of graphene with hydroxyl groups.

Surface modification of GO followed by reduction has been carried out to obtain functionalized graphene. The addition of other groups to GO using various chemical reactions providing for either covalent or non-covalent attachment to the resulting CMG, add functionality to groups already present on GO and make it a more versatile precursor for a wide range of applications. Some reactions are reported in a recent review [14].

The production of GO without significant defects, the control of the degree of reduction and the understanding of the related mechanisms constitute the big challenges, still open, in the end applications of graphene.

11.4.3 Interaction with Hydroxyl Groups

Systematic studies from first-principles calculations have been carried out for the adsorption of hydroxyl on perfect and defected graphene, individually and in the presence of other coadsorbed functional groups. The aim was to shed light on the r-GO configuration, where epoxy and hydroxyl groups coexist at the graphene surface (Figure 11.5), and to clarify the role of topological defects (pentagon-heptagon, known as Stone–Wales defects) and holes which have been experimentally detected at r-GO and which may influence the adsorption and the reactivity of functional groups.

OH adsorbs on pristine graphene with O preferentially on top of carbon atoms and H pointing in the direction of the center of a six-fold carbon ring. The adsorption is rather weak, with an estimated energy of the order of 0.5 eV, with values in literature ranging from 0.21 eV to 0.70 eV [67–72]. Similarly to the adsorption of hydrogen on pristine graphene, the individually adsorbed hydroxyl group breaks one of the C–C π bonds and transforms the sp^2 hybridization to sp^3. One unpaired π electron is distributed on neighboring carbon atoms and gives a magnetic character to the system, with a total magnetization of about 0.75 μ_B [72]. Carbon atoms in *ortho* and *para* positions with respect to the one bonded to OH have the largest positive magnetic moments (Figure 11.6).

Individual OH can diffuse on graphene surface with energy barriers of about 0.3 eV [72]. Much higher values are reported in Ref. [57], probably related to a different, unfavored path. Dissociation of a single OH molecule on pristine graphene is highly unfavored, since the final possible configuration (epoxy with one coadsorbed hydrogen) has higher energy

242 Graphene Chemistry

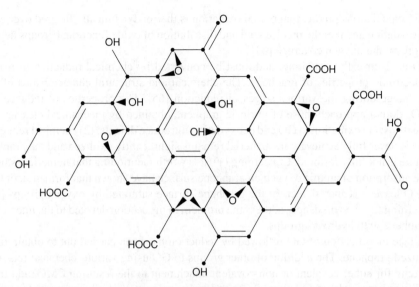

Figure 11.5 *Sketch of a model of r–GO with residual oxygen, hydroxyl and carboxylic groups. See Ref. [14] for a review of various models proposed in the literature*

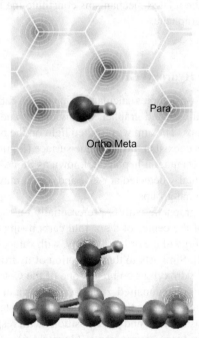

Figure 11.6 *Ball-and-stick model of OH adsorbed on pristine graphene (gray: carbon atoms, small light blue: hydrogen atoms, red: oxygen atoms) and spin density distribution, calculated as the difference between the majority and minority spin densities (three-dimensional and projection onto the pristine graphene plane: red/blue indicate positive/negative values). See colour version in the colour plate section*

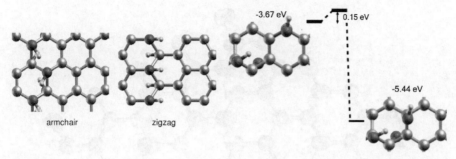

Figure 11.7 Optimized structural models and energies for OH groups upon aggregation. Left panel: OH groups adsorbed along a line in an armchair and zigzag configuration. Right panel: aggregation process of OH groups in presence of a coadsorbed epoxy. The total adsorption energies of the aggregates (initial and final configurations) are reported, together with the energy barrier for the aggregation. Gray: carbon atoms, small light gray: hydrogen atoms, black: oxygen atoms. Data from Ref. [72]

by about 2 eV with respect to the undissociated adsorbed OH. This suggests that, in the presence of epoxy groups, hydrogen prefers to bond to oxygen rather than to carbon.

Hydroxyl groups can adsorb on graphene either on the same side of the sheet (one-side adsorption) or on both sides (two-side adsorption). For OH pairs, two-side adsorption at the *ortho* position is the most favored, with an adsorption energy of about 1.3 eV per hydroxyl molecule and a correspondent energy gain of about 0.75 eV per each adsorbed OH. In general, it has been found that aggregation of hydroxyl groups, which is characterized by attractive hydrogen bonding interaction O–H···O between hydroxyl groups, is largely favored, and, in particular, configurations with adsorbates on alternate sides of graphene minimize the total energy of the system. Adsorption along a line in zigzag or armchair configuration (Figure 11.7, left hand side) form a very stable aggregate. At variance with single hydroxyl adsorbed on graphene which presents magnetic properties, the preferred configuration of hydroxyl pairs is non-magnetic. Configurations with an adsorbed aggregate containing an odd number of hydroxyl molecules exhibit a magnetic character [72].

Pristine graphene purely functionalized by hydroxyl has been studied in Ref. [73]. It is found that OH groups prefer to adsorb on graphene in pairs close to each other from both sides and the interaction between OH pairs is coverage-dependent. The most stable fully hydroxyl-functionalized graphene or *graphene hydroxide* (GOH) is half-covered by hydroxyl pairs along zigzag chains with alternating sp^2 and sp^3 hybridization between carbon atoms (Figure 11.8). This structure is semiconducting with a band-gap of about 1 eV in GW approximation, and it is very stable at room temperature. The mobility of charge carriers through such GOH is highly anisotropic, which is very useful for controlling current in electronics. These properties open the way for graphene-based electronic circuits from graphene hydroxide [73].

Concerning the coadsorption of hydroxyl and epoxy groups, calculations indicate that OH binds to the graphene epoxide with an adsorption energy considerably stronger than on pristine graphene (about 1.35 eV, to be compared with 0.5 eV) [72]. The energy gain is even larger in aggregates with more hydroxyl or epoxy groups adsorbed. The barriers for aggregation depend on the specific configuration, but the general conclusion is that

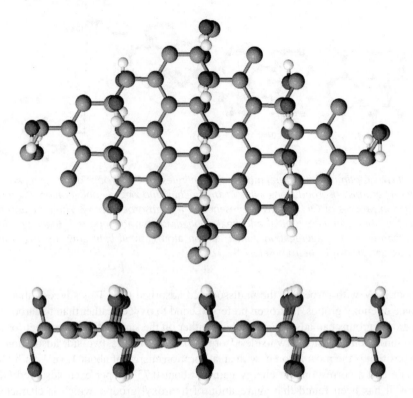

Figure 11.8 Ball-and-stick model of the most stable structure of graphene hydroxide: it corresponds at half OH coverage, with linear adsorption structures along zigzag chains. The model is proposed in Ref. [73]

aggregation of hydroxyl groups with or without coadsorbed epoxy groups is largely favored (Figure 11.7, right panel). Aggregation opens the way for the reaction of water formation (disproportionation reaction OH + OH → O + H$_2$O), which actually can occur rather easily with a barrier of about 0.5 eV on pristine graphene (Figure 11.9, top panel).

Defects in graphene crystal lattice, as well as the edges of graphene, drastically change the scenario of its chemical functionalization [74, 75]. Defect sites are more reactive for OH adsorption but play different roles. The adsorption of hydroxyl on defected graphene is much stronger than on pristine, with adsorption energy of 1.80 eV at Stone–Wales (SW) defects and larger than 4 eV at single vacancies. At variance with pristine and SW defects, hydroxyl adsorption at a single vacancy is highly dissociative, with a low barrier of about 0.2 eV, leading to the formation of stable ether groups with a strong magnetic character (Figure 11.10). The large adsorption energy at SW defects, together with the high barrier for the disproportionation reaction with water formation (about 1 eV, see Figure 11.9, bottom panel), suggests that SW defects could be responsible for the stabilization of the hydroxyl groups which are present at graphene oxide also after reduction.

The migration energy barrier for the hydroxyl group changes in the presence of water. Water weakens the covalent bonds and makes the hydroxyl groups more fluid. Increasing the number of water molecules provides an increase of the migration energy [57].

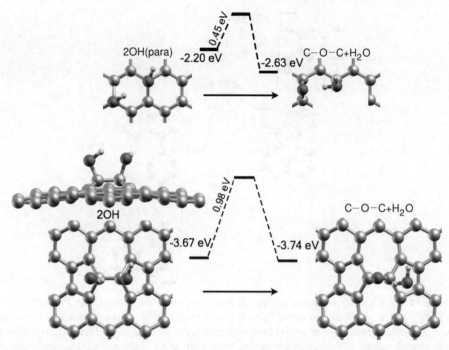

Figure 11.9 *Optimized ball-and-stick structural models and energies in the process of water formation from OH adsorbed on pristine graphene and on a Stone–Wales defect. Energies of the initial and final systems are relative to the energy of the isolated reactants; the energy of graphene with a Stone–Wales defect has been considered. Gray: carbon atoms, small light gray: hydrogen atoms, black: oxygen atoms*

More generally, an understanding of the fundamental interactions at water/graphene interface is essential to use graphene and CMG in electrochemical energy storage systems. A recent joint experimental/theoretical work focusing on the role of hydroxyl on the graphene sheet at the water/graphene interface is reported in Ref. [50]. The interaction of water with graphene epitaxially grown on SiC, multilayer and free-standing graphene is investigated. The epitaxial layer is more hydrophilic than subsequent layers owing to the increased perturbation by the substrate, surface defects, and functional groups. The macroscopic wettability of multilayer graphene is controlled by the local structure and in particular by the presence of functionalized surface groups such as hydroxyls, leading to lateral heterogeneity of interfacial water [50].

11.4.4 Interaction with Other Atoms, Molecules, and Functional Groups

Modification of graphene with different atoms, molecules, and functional groups is used to engineer different physical and chemical properties. Beside the effects that we have reviewed in the previous subsections concerning interaction with H, O, and OH groups, we now list a few other examples focused on engineering magnetic properties, chemical sensitivity, and reactivity.

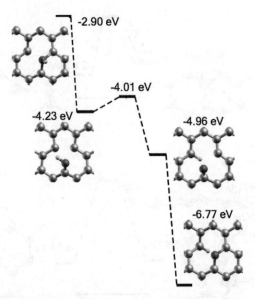

Figure 11.10 Optimized structural models and energies in the dissociation process of individual OH on a single vacancy and formation of a stable ether group. Energies of the initial, intermediate and final systems are relative to the energy of an isolated OH group and graphene with a single vacancy. Gray: carbon atoms, small light gray: hydrogen atoms, black: oxygen atoms. Data from Ref. [72]

Remarkably, it is possible to extract some general trends concerning the effects on the magnetic properties of different functional groups on graphene. In Ref. [76], for instance, it is reported that the induced magnetic properties are somehow universal, in the sense that they are largely independent of the particular adsorbates considered. When a weakly polar single covalent bond is established with the graphene basal plane, with the creation of a sp^3 type defect, DFT calculations predict a local spin moment of 1.0 μ_B. This effect is similar to that already discussed for H and OH adsorption. The magnetic couplings between the adsorbates strongly depend on whether the atoms or functional groups are adsorbed on the same carbon sublattice or not. Adsorbates at the same sublattice show a ferromagnetic coupling, with an exchange interaction that decays very slowly with distance, while no magnetism is found for adsorbates at different sublattices. Similar magnetic properties are obtained if several p_z orbitals are saturated simultaneously by the adsorption of a large molecule. These results are interesting in view of engineering the magnetic properties of graphene derivatives by chemical means.

Concerning again the magnetic properties, it is worth mentioning the possibility of using functionalized graphene as a high-performance two-dimensional spintronics device. First-principles calculations have shown that functionalizing graphene with O on one side and H on the other side in the chair conformation gives a structure which is a ferromagnetic metal with high spin-filter efficiency, whereas functionalizing graphene with F on one side in the chair conformation allows to obtain an antiferromagnetic semiconductor which is predicted to give a giant room-temperature magnetoresistance when used in a spin-valve [77].

A joint experimental and theoretical investigation of NO_2 and N_2O_4 adsorption on graphene reported in Ref. [78], treating on the same ground an open- and a closed-shell system, points out a general relation between the adsorption strength and the kind of the adsorbates. The single, open shell NO_2 molecule is found to be a strong acceptor, whereas its closed shell dimer N_2O_4 is only weakly adsorbed. The peculiar density of states of graphene is ideal for chemical sensor applications and explain the observed NO_2 single molecule detection. FG offers the potential for dramatically improving selectivity and efficiency. NO_2 and other molecules of toxic gases, such as CO, cannot be readily adsorbed onto perfect pristine graphene surface. They can be better detected on graphene functionalized with a thin coating layer of certain polymers. The thin polymer layer acts like a concentrator that absorbs gaseous molecules, making FG an efficient chemical sensor.

In this short review of CMG it is worth mentioning some examples of further functionalization of GO or r–GO or at least reporting on some works concerning its interaction with molecules. For instance, the interactions of GO with ammonia (NH_3) has been studied by DFT. The adsorption of NH_3 on GO is generally stronger than on graphene because of the presence of diverse active defect sites, such as the hydroxyl and epoxy functional groups and their neighboring carbon atoms. OH$\cdots$N and O$\cdots$HN hydrogen bonds can form between NH_3 and O and a large charge transfer is induced from NH_3 to GO. The adsorbed NH_3 can dissociate into chemisorbed NH_2 or NH species, accompanied by the hydroxyl group hydrogenation and ring-opening of epoxy group. The reactions of NH_3 with the hydroxyl and epoxy groups are predicted to be exothermic with energy barriers which depend on the oxidation species and the atomic arrangement of these groups. The hydroxyl group exhibits relatively higher reactivity toward hydrogen abstraction from the adsorbed NH_3 than the epoxy group in GO with a single oxygen group. Coadsorbed OH groups enhance the activation of the oxygen groups. The calculated density of states of the adsorbed systems also reveals strong interactions between GO and NH_3. Experimental data are also available on this system [79].

11.5 Surface Functionalization of Graphene Nanoribbons

As mentioned in the Introduction, graphene nanoribbons (GNRs) can be functionalized both at the edges and at the surface. We briefly mention here some examples of the effects of surface functionalization on nanoribbons. One of the most interesting applications of GNRs concerns their transport properties.

For instance, the impact of epoxide adsorbates on the transport properties of GNRs with width varying from a few nanometers to 15 nm has been addressed in Ref. [80]. For the wider ribbons, a scaling analysis of conductance properties is performed for low adsorbate density. Oxygen atoms introduce a large electron-hole transport asymmetry with mean free paths changing by up to one order of magnitude, depending on the hole or electron nature of charge carriers. The effect of the adsorbates in narrower GNRs is also investigated by full *ab initio* calculations to explore the limit of ultimate downsized systems. In this case, the inhomogeneous distribution of adsorbates and their interplay with the ribbon edge are found to play an important role [80].

GNRs chemically functionalized with coadsorption of hydroxyl and hydrogen groups on the surface have been also investigated by DFT as a function of defect location and coverage density. The chemical bonding of a single defect pair (C–OH and C–H) is shown to considerably alter the conduction capability of ribbon channels, similarly to an sp^3 type of defect. Transport properties are found to be severely damaged by the functionalization, indicating a strong tendency toward an insulating regime [81].

Another DFT study on GNRs afford both monovalent and divalent ligands, hydrogenated defects, and vacancies. The investigation reports that, at variance with the edge metallic states that are preserved under a variety of chemical environments, bulk (surface) conducting channels can be easily destroyed by either hydrogenation or ion or electron beams, resulting in devices that can exhibit spin conductance polarization close to unity [82].

A last contribution that we review concerning GNRs surface functionalization, compares the effects of surface with respect to edge functionalization. Molecular dynamics simulations have been performed to study the mechanical properties of methyl (CH_3) functionalized FG. It is found that the mechanical properties of FG greatly depend on the location, distribution and coverage of the adsorbed CH_3 radicals. Remarkably, surface functionalization exhibits a much stronger influence on the mechanical properties than edge functionalization [83].

11.6 Conclusions

The dramatic increase in the scientific production on the subject and the variety of possible ingredients for graphene functionalization made it impossible to give an exhaustive overview here. For the sake of definiteness and simplicity, we mainly focused on the interaction of graphene with the simplest functional groups, and small environmental abundant atoms and molecules, such as hydrogen, oxygen, hydroxyl, epoxy, water, and few others. In our opinion, these few case studies give a significant overview of the wider scenario for possible functionalization of graphene to tune its electronic and magnetic properties.

Considerable progress in understanding and controlling the functionalization of graphene surfaces has been achieved in recent years by combining experimental and theoretical efforts. Advances in computational physics have made possible accurate *ab initio* calculations of the structure and of many properties of different kinds of FG. These computations can be used to study the effect of various microscopic features in numerical experiments and to understand the underlying atomic-scale mechanisms. The complexity of the systems which can be examined is steadily increasing, ranging from graphene with simple adsorbates to multifunctional complex systems. Although we have mainly reviewed cases of covalent functionalization of graphene, it is worth mentioning that recent developments in electronic structure calculations allow us to account for dispersive forces and to also give a reliable description of systems characterized by non-covalent interactions.

Although the domain of application of the *ab initio* methods becomes wider and wider thanks to increasing computing power, when the number of atoms to be simulated becomes large (thousands of atoms), alternative approaches or novel algorithms are required.

Other approaches such as molecular dynamics with some empirical or semiempirical interaction potentials are in fact more appropriate for large systems and for the purpose of studying the evolution of functional groups on graphene basal plane, such as in the case of GO during high-temperature thermal reduction. Semiempirical approaches are doubtless a powerful resource; *ab initio* calculations can be used to test and validate the results in some particular cases.

Finally, we have to mention developments and progress towards large-scale atomistic simulations with fully quantum mechanical description of electrons, such as *ab initio* based multiscale methods and real-space approaches, which include the possibility of introducing local mesh refinements and are of immediate application to nonperiodic systems.

There remains a number of open issues which still limit the power of theoretical schemes. One concerns the mechanisms responsible for the actual atomic–scale arrangement of the functional groups at the graphene surface. As we mentioned, a definitive picture of the structure of GO is still lacking. Advances in this direction and the possibility of predicting the morphology of the functional groups that are actually established, would clearly improve our ability to engineer the functionality of graphene.

References

[1] S. Stankovich, D. A. Dikin, G. H. B. Dommett, K. M. Kohlhaas, E. J. Zimney, E. A. Stach, et al., Graphene-based composite materials, *Nature*, **442**(7100), 282–286 (2006).

[2] H. C. Schniepp, J.-L. Li, M. J. McAllister, H. Sai, M. Herrera-Alonson, D. H. Adamson, et al., Functionalized single graphene sheets derived from splitting graphite oxide, *Journal of Physical Chemistry B*, **110**(17), 8535–8539 (2006).

[3] T. Kuila, S. Bose, A. K. Mishra, P. Khanra, N. H. Kim, J. H. Lee, Chemical functionalization of graphene and its applications, *Progress in Materials Science*, **57**(7), 1061–1105 (2012).

[4] L. Yan, Y. B. Zheng, F. Zhao, S. Li, X. Gao, B. Xu, et al., Chemistry and physics of a single atomic layer: Strategies and challenges for functionalization of graphene and graphene-based materials, *Chemical Society Reviews*, **41**(1), 97–114 (2012).

[5] V. Georgakilas, M. Otyepka, A. B. Bourlinos, V. Chandra, N. Kim, K. C. Kemp, et al., Functionalization of graphene: Covalent and non-covalent approaches, derivatives and applications, *Chemical Reviews*, **112**(11), 6156–6214 (2012).

[6] W. Wei and X. Qu, Extraordinary physical properties of functionalized graphene (review), *Small*, **8**(14), 2138-2151 (2012).

[7] S. Park and R. S. Ruoff, Chemical methods for the production of graphenes, *Nature Nanotechnology*, **4**, 217–224 (2009).

[8] C. N. R. Rao, A. K. Sood, R. Voggu, and K. S. Subrahmanyam, Some novel attributes of graphene, *J. Phys. Chem. Lett.*, **1**, 572–580 (2010).

[9] Focus on Chemically Modified Graphene: Focus issue of new journal of physics: http://iopscience.iop.org/1367-2630/focus/Focus%20on%20Chemically%20Modified%20Graphene (last accessed: June 11, 2013).

[10] O. V. Prezhdo, P. V. Kamat, G. C. Schatz, Virtual Issue: Graphene and functionalized graphene, *J. Phys. Chem. C*, **115**, 3195–3197 (2011).

[11] *Nature*, supplement, **483**(7389) (15 March 2012), Graphene, available on: http://www.nature.com/nature/outlook/graphene/ (last accessed: June 11, 2013).

[12] F. Bonaccorso, J. Coraux, C. Ewels, G. Fiori, A. C. Ferrari, J-C. Gabriel, et al., Graphene Position Paper, *E-Nano Newsletter Special Issue*, (2012), available at: http://www.phantomsnet.net/Foundation/Enano_newsletterSG.php (last accessed: June 11, 2013).

[13] http://www.graphene.manchester.ac.uk/ (last accessed: June 11, 2013).

[14] D. R. Dreyer, S. Park, C. W. Bielawski and R. S. Ruoff, The chemistry of graphene oxide, *Chem. Soc. Rev.*, **39**, 228–240 (2010).
[15] R. S. Ruoff, Calling all chemists, *Nature Nanotechnology*, **3**, 10–11 (2008).
[16] Y. Zhu, S. Murali, W. Cai, X. Li, J. W. Suk, J. R. Potts, and R. S. Ruoff, Graphene and graphene oxide: synthesis, properties, and applications, *Adv. Mater.*, **22**(35), 3906–3924 (2010).
[17] G. Eda and M. Chhowalla, Chemically derived graphene oxide: towards large-area thin-film electronics and optoelectronics, *Adv. Mater.*, **22**, 2392–2415 (2010).
[18] M. J. McAllister, D. H. Adamson, H. C. Schniepp, A. A. Abdala, J. Liu, M. Herrera-Alonson, et al., Single sheet functionalized graphene by oxidation and thermal expansion of graphite, *Chem. Mater.*, **19**, 4396–4404 (2007).
[19] X. Cui, C. Zhang, R. Hao and Y. Hou, Liquid-phase exfoliation, functionalization and applications of graphene, *Nanoscale*, **3**, 2118–2126 (2011).
[20] M. Quintana, K. Spyrou, M. Grzelczak, W. R. Browne, P. Rudolf, and M. Prato, Functionalization of graphene via 1,3-dipolar cycloaddition, *ACS Nano*, **4**, 3527–3533 (2010).
[21] J. M. Englert, C. Dotzer, G. Yang, M. Schmid, C. Papp, J. M. Gottfried, et al., Covalent bulk functionalization of graphene, *Nature Chemistry*, **3**, 279–286 (2011).
[22] D. W. Boukhvalov and M. I. Katsnelson, A new route towards uniformly functionalized single-layer graphene, *J. Phys. D*, **43**(17), 175302 (5pp) (2010).
[23] X. Zhu and H. Su, Excitons of edge and surface functionalized graphene nanoribbons, *J. Phys. Chem. C*, **114**(41), 17257–17262 (2010).
[24] C. Cocchi, D. Prezzi, A. Ruini, M. J. Caldas, and E. Molinari, Optical properties and charge-transfer excitations in Edge-functionalized all-graphene nanojunctions, *J. Phys. Chem. Lett.*, **2**(11), 1315–1319 (2011).
[25] C. Cocchi, A. Ruini, D. Prezzi, M. J. Caldas, E. Molinari, Designing all-graphene nanojunctions by covalent functionalization, *J. Phys. Chem. C*, **115**, 2969–2973 (2011).
[26] O. Hod, V. Barone, J. E Peralta, G. E. Scuseria, Enhanced half-metallicity in edge-oxidized zigzag graphene nanoribbons, *Nano Letters*, **7**(8), 2295–2299 (2007).
[27] N. Gorjizadeh and Y. Kawazoe, Chemical functionalization of graphene nanoribbons, *Journal of Nanomaterials*, **2010**, 513501 (7pp) (2010).
[28] M. D. Stoller, S. Park, Z. Yanwu, J. An, R. S. Ruoff, Graphene-based ultracapacitors, *Nano Letters*, **8** 3498–3502 (2008).
[29] J. D. Roy-Mayhew, G. Boschloo, A. Hagfeldt, I. A. Aksay, Functionalized graphene sheets as a versatile replacement for platinum in dye-sensitized solar cells, *ACS Appl. Mater. Interfaces*, **4**(5), 2794–2800 (2012).
[30] Y. Wang, Z. Li, J. Wang, J. Li, Y. Lin, Graphene and graphene oxide: biofunctionalization and applications in biotechnology, *Trends in Biotechnology*, **29**(5), 205–212 (2011).
[31] H. C. Schniepp, K. N. Kudin, J.-L. Li, R. K. Prudhomme, R. Car, et al., Bending properties of single functionalized graphene sheets probed by atomic force microscopy, *ACS Nano*, **2**(12), 2577–2584 (2008).
[32] T. Ramanathan, A. A. Abdala, S. Stankovich, D. A. Dikin, M. Herrera-Alonso, R. D. Piner, et al., Functionalized graphene sheets for polymer nanocomposites, *Nature Nantechnology*, **3**, 327–331 (2008).

[33] K. N. Kudin, B. Ozbas, H. C. Schniepp, R. K. Prud'homme, I. A. Aksay, and R. Car, Raman spectra of graphite oxide and functionalized graphene sheets, *Nano Letters*, **8**(1), 36–41 (2008).

[34] R. Larciprete, S. Fabris, T. Sun, P. Lacovig, A. Baraldi, and S. Lizzit, Dual path mechanism in the thermal reduction of graphene oxide, *J. Am. Chem. Soc.*, **113**, 17315–17321 (2011).

[35] Md. Z. Hossain, J. E. Johns, K. H. Bevan, H. J. Karmel, Y. Teng Liang, S. Yoshimoto, et al., Chemically homogeneous and thermally reversible oxidation of epitaxial graphene, *Nature Chemistry*, **4**, 305–309 (2012).

[36] For a review see for instance: R. Martin, *Electronic Structure*, Cambridge University Press, and references therein.

[37] D. Umadevi and G. N. Sastry, Molecular and ionic interaction with graphene nanoflakes: a computational investigation of CO_2, H_2O, Li, Mg, Li^+, and Mg2+ interaction with polycyclic aromatic hydrocarbons, *J. Phys. Chem. C*, **115**, 9656–9667 (2011).

[38] J. Björk, F. Hanke, C.-A. Palma, P. Samori, M. Cecchini, and M. Persson, Adsorption of aromatic and anti-aromatic systems on graphene through $\pi-\pi$ stacking, *J. Phys. Chem. Lett.*, **1**, 3407–3412 (2010).

[39] C.-H. Chang, X. Fan, L.-J. Li, J.-L. Kuo, Band gap tuning of graphene by adsorption of aromatic molecules, *Nature Chemistry*, **4**, 305–309 (2012).

[40] M. Elstner, P. Hobza, T. Frauenheim, S. Suhai, E. Kaxiras, Hydrogen bonding and stacking interactions of nucleic acid base pairs: A density-functional-theory based treatment, *J. Chem. Phys.*, **114**, 5149–5155 (2001).

[41] S. Grimme, Accurate description of van der Waals complexes by density, *J. Comput. Chem.*, **25**, 1463–1473 (2004).

[42] M. Dion, H. Rydberg, E. Schroder, D. C. Langreth, B. I. Lundqvist, Van der Waals density functional for general geometries, *Phys. Rev. Lett.*, **92**(24), 246401 (4pp) (2004).

[43] C. David Sherrill, Frontiers in electronic structure theory, *J. Chem. Phys.*, **132**, 110902 (7pp) (2010).

[44] H. Jónsson, G. Mills, K. W. Jacobsen, Nudged elastic band method for finding minimum energy paths of transitions, in *Classical and Quantum Dynamics in Condensed Phase Simulations*, Eds., B. J. Berne, G. Ciccotti and D. F. Coker (World Scientific, 1998), pp 385–404; G. Henkelman and H. Jónsson, A dimer method for finding saddle points on high dimensional potential surfaces using only first derivatives, *J. Chem. Phys.*, **111**, 7010 (1999).

[45] C. Micheletti, A. Laio, M. Parrinello, Reconstructing the density of states by history-dependent metadynamics, *Phys. Rev. Lett.*, **92**, 170601 (2004).

[46] A list with related links is available for instance on http://www.psi-k.org/codes.shtml (last accessed: August 30, 2012).

[47] http://www.quantum-espresso.org (last accessed: June 11, 2013) and P. Giannozzi, S. Baroni, N. Bonini, M. Calandra, R. Car, C. Cavazzoni, D. Ceresoli, et al., QUANTUM ESPRESSO: a modular and open-source software project for quantum simulations of materials, *J. Phys. Condens. Matter*, **21**, 395502 (19pp) (2009).

[48] http://www.gaussian.com/ (last accessed: June 11, 2013); reference for current version: M. J. Frisch et al., Gaussian 09; Gaussian, Inc.: Wallingford, CT (2009).

[49] Y. Wang, H.-J. Qian, K. Morokuma, S. Irle, Coupled cluster and density functional theory calculations of atomic hydrogen chemisorption on pyrene and coronene as model systems for graphene hydrogenation, *J. Phys. Chem. A*, **116**, 7154–7160 (2012).

[50] H. Zhou, P. Ganesh, V. Presser, M. C. F. Wander, P. Fenter, P. R. C. Kent, et al., Understanding controls on interfacial wetting at epitaxial graphene: Experiment and theory, *Phys. Rev. B*, **85**(3), 035406 (11pp) (2012).

[51] A. Bagri, R. Grantab, N. V. Medhekar, V. B. Shenoy, Stability and formation mechanisms of carbonyl- and hydroxyl-decorated holes in graphene oxide, *J. Phys. Chem. C*, **114**, 12053–12061 (2010).

[52] O. V. Yazyev and L. Helm, Defect-induced magnetism in graphene, *Phys. Rev. B*, **75**(12), 125408 (5pp) (2007).

[53] D. W. Boukhvalov, M. I. Katsnelson, A. I. Lichtenstein, Hydrogen on graphene: Electronic structure, total energy, structural distortions and magnetism from first-principles calculations, *Phys. Rev. B*, **77**(3), 035427 (7pp) (2008).

[54] Y. Ferro, D. Teillet-Billy, N. Rougeau, V. Sidis, S. Morisset, A. Allouche, Stability and magnetism of hydrogen dimers on graphene, *Phys. Rev. B*, **78**(8), 085417 (8pp) (2008).

[55] S. Casolo, O. M. Lovvik, R. Martinazzo, G. F. Tantardini, Understanding adsorption of hydrogen atoms on graphene, *J. Chem. Phys.*, **130**(5), 054704 (10pp) (2009).

[56] D. W. Boukhvalov and M. I. Katsnelson, Chemical functionalization of graphene, *J. Phys. Cond. Matt.*, **21**(34), 344205 (12pp) (2009).

[57] D. W. Boukhvalov, Modeling of hydrogen and hydroxyl group migration on graphene, *Phys. Chem. Chem. Phys.*, **12**, 15367–15371 (2010).

[58] D. W. Boukhvalov, Modeling of epitaxial graphene functionalization, *Nanotechnology*, **22**(5), 055708 (5pp) (2011).

[59] J. Ma, A. Michaelides, and D. Alfè, Binding of hydrogen on benzene, coronene, and graphene from quantum Monte Carlo calculations, *J. Chem Phys.*, **134**, 134701 (6pp) (2011).

[60] J. O. Sofo, A. S. Chaudhari, and G. D. Barber, Graphane: A two-dimensional hydrocarbon, *Phys. Rev. B*, **75**(15), 153401 (4pp) (2007).

[61] S. Ryu, M. Y. Han, J. Maultzsch, T. F. Heinz, P. Kim, M. L. Steigerwald, L. E. Brus, Reversible basal plane hydrogenation of graphene, *Nano Letters*, **8**(12), 4597–4602 (2008).

[62] D. C. Elias, R. R. Nair, T. M. G. Mohiuddin, S. V. Morozov, P. Blake, M. P. Halsall, et al., Control of graphene's properties by reversible hydrogenation: evidence for graphane, *Science*, **323**(5914), 610–613 (2009).

[63] O. Pulci, P. Gori, M. Marsili, V. Garbuio, A. P. Seitsonen, F. Bechstedt, et al., Electronic and optical properties of group IV two-dimensional materials, *Phys. Status Solidi A*, **207**(2), 291–299 (2010).

[64] D. W. Boukhvalov and M. I. Katsnelson, Modeling of graphite oxide, *J. Am. Chem. Soc.*, **130**, 10697–10701 (2008).

[65] C. Mattevi, G. Eda, S. Agnoli, S. Miller, K. A. Mkhoyan, O. Celik, et al., Evolution of electrical, chemical, and structural properties of transparent and conducting chemically derived graphene thin films, *Adv. Funct. Mater.*, **19**, 2577–2583 (2009).

[66] T. Sun and S. Fabris, Mechanisms for oxidative unzipping and cutting of graphene, *Nano Lett.*, **12**(1), 17–21 (2012).

[67] A. Jelea, F. Marinelli, Y. Ferro, A. Allouche, C. Brosset, Quantum study of hydrogen-oxygen-graphite interactions, *Carbon*, **42**, 3189–3198 (2004).

[68] R. J. W. E. Lahaye, H. K. Jeong, C. Y. Park, Y. H. Lee, Density functional theory study of graphite oxide for different oxidation, *Phys. Rev. B*, **79**(12), 125435 (8pp) (2009).

[69] M. C. Kim, G. S. Hwang, R. S. Ruoff, Epoxide reduction with hydrazine on graphene: A first principles study, *J. Chem. Phys.*, **131**, 064704 (5pp) (2009).

[70] S. C. Xu, S. Irle, D. G. Musaev, M. C. Lin, Quantum chemical study of the dissociative adsorption of OH and H_2O on pristine and defective graphite surfaces: reaction mechanisms and kinetics, *J. Phys. Chem. C*, **111**, 1355–1365 (2007).

[71] P. A. Denis, Density functional investigation of thioepoxidated and thiolated graphene, *J. Phys. Chem. C*, **113**, 5612–5619 (2009).

[72] N. Ghaderi and M. Peressi, First-principle study of hydroxyl functional groups on pristine, defected graphene, and graphene epoxide, *J. Phys. Chem. C*, **114**, 21625–21630 (2010).

[73] Q. Zhu, Y. H. Lu, and J. Z. Jiang, Stability and properties of two-dimensional graphene hydroxide, *J. Phys. Chem. Lett.*, **2**, 1310–1314 (2011).

[74] D. W. Boukhvalov and M. I. Katsnelson, Chemical functionalization of graphene with defects, *Nano Lett.*, **8**(12), 4373–4379 (2008).

[75] F. Banhart, J. Kotakoski, A. V. Krasheninnikov, Structural defects in graphene, *ACS Nano*, **5**(1), 26–41 (2011).

[76] J. G. Santos Elton, A. Ayuela, D. Sanchez-Portal, Universal magnetic properties of sp(3)-type defects in covalently functionalized graphene, *New Journal of Physics*, **14**, 043022 (13pp) (2012).

[77] L. Linze, Q. Rui, L. Hong, et al., Functionalized graphene for high-performance two-dimensional spintronics devices, *ACS Nano*, **5**, 2601–2610 (2011).

[78] T. O. Wehling, K. S. Novoselov, S. V. Morozov, E. E. Vdovin, M. I. Katsnelson, A. K. Geim, A. I. Lichtenstein, Molecular doping of graphene with NO_2 and N_2O_4, *Nano Lett.*, **8**(1), 173–177 (2008).

[79] S. Tang and Z. Cao, Adsorption and dissociation of ammonia on graphene oxides: a first-principles study, *J. Phys. Chem. C*, **116**(15), 8778–8791 (2012).

[80] A. Cresti, A. Lopez-Bezanilla, P. Ordejon Pablo, et al., Oxygen surface functionalization of graphene nanoribbons for transport gap engineering, *ACS Nano*, **5**, 9271–9277 (2011).

[81] A. Lopez-Bezanilla, F. Triozon, S. Roche, Chemical functionalization effects on armchair graphene nanoribbon transport, *Nanoletters*, **9**, 2537–2541 (2009).

[82] G. Cantele, Y.-S. Lee, D. Ninno and N. Marzari, Spin channels in functionalized graphene nanoribbons, *Nano Lett.*, **9**(10), 3425–3429 (2009).

[83] Q.-X. Pei, Y.-W. Zhang and V. B. Shenoy, Mechanical properties of methyl functionalized graphene: a molecular dynamics study, *Nanotechnology*, **21**, 115709 (8pp) (2010).

12

Mechanisms of Graphene Chemical Vapor Deposition (CVD) Growth

Xiuyun Zhang, Qinghong Yuan, Haibo Shu, and Feng Ding
Institute of Textiles and Clothing, Hong Kong Polytechnic University, Hong Kong

12.1 Background

12.1.1 Graphene and Defects in Graphene

Research into graphene, a single atomic layer material with the symmetrical honeycomb lattice, is very active these days because of its intriguing properties and immense potential applications in nano-electronics, solar cells, sensors, and many others [1–19]. Numerous experiments and theoretical calculations have shown that the graphene has an extremely high thermal conductivity, 3000–5000W/mK, which is close to or exceeds that of a previously studied material, diamond [1–6]. This ensures a fast heat dissipation in large scale integrated electronics. Graphene also has a recorded ultra-high carrier mobility [1,7–9]. For example, Bolotin et al. [7] achieved a mobility of 200 000 cm^2 V^{-1} s^{-1} and electron density of $\sim$2 $\times$ 10^{11} cm^{-2} by suspending single layer graphene. Furthermore, graphene has been proved to be one of the strongest materials ever encountered in Nature [10–15], with a Young modulus of $\sim$ 1 TPa and tensile strength of $\sim$100 GPa, which is very similar to those of high quality carbon nanotubes (CNTs) [10].

It has been noted that these exceptional electronic, mechanical, and thermal properties are only achievable in graphene of very high quality and could be greatly downgraded if structural defects are introduced to the honeycomb lattice [19–33]. Due to the reduced dimensionality, zero-dimensional (0D) point defects and one-dimensional (1D) linear defects are two major types of defects in graphene. Figure 12.1(a–c) show a few types of point defects, a

---

Graphene Chemistry: Theoretical Perspectives, First Edition. Edited by De-en Jiang and Zhongfang Chen.
© 2013 John Wiley & Sons, Ltd. Published 2013 by John Wiley & Sons, Ltd.

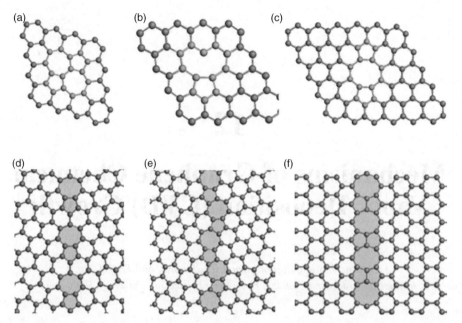

Figure 12.1 Various types of defects in graphene: (a–c) are three types of point defects and (d–f) are linear defects in graphene. (a) Stone–Wales defect SW. (b) Single vacancy, (c) Double vacancy. (d–e) Two linear grain boundaries (GB) in graphene, (f) a specific type of grain boundary without lattice orientation mismatching on both sides

Stone–Wales (SW) defect (Figure 12.1a) [19–23,30, 31], a single vacancy with one missing carbon atom from the lattice (Figure 12.1b) [19,23–27,32], a double vacancy with two missing atoms (Figure 12.1c) [19,22,26–33]. Figure 12.1(d–f) shows three different linear defects, each of which splits the graphene lattice into two independent parts. In Figure 12.1(d–e), the linear defects are in the form of a linear combination of pentagon-heptagon (5–7) pairs, by which the graphene is divided into two domains with different orientations and the linear concentration of 5–7 pairs is proportional to the tilt angle of the two domains [34–39]. Figure 12.1(f) is a symmetric boundary which links the zigzag edges of two graphene domains with same crystal orientation. It's a metastable structure but can be formed during the graphene CVD growth on catalyst surface due to the symmetry difference between the catalyst surface and graphene [19,40].

In graphene, both point defects and linear defects serve as carrier trapping centers and scattering centers and these may greatly affect its properties. For example, a linear defect interferes with the carrier/heat flow as a dam in a river and thus a large concentration of linear defects must dramatically downgrade both the carrier mobility and thermal conductivity of graphene [34]. Beyond this, recent theoretical studies showed that a linear defect also reduces the mechanical strength of graphene radically [37,38]. Comparing the linear defects, a point defect only affects a very small area around it in the graphene and is thus expected to be less effective than the linear defects [19,21,33].

As discussed previously, the key issue of synthesizing high quality graphene is to greatly suppress the concentrations of linear defects and point defects.

Table 12.1 Methods of graphene synthesis and their comparisons

	mechanical peeling	graphite oxide reduction	SiC sublimation	chemical intercalation	CVD growth
C source	graphite	graphite oxide	SiC	graphite oxide	hydrocarbon
temperature	ambient	373 K	1900–2300 K	ambient	~1300 K
thickness	MLG[1] & FLG[2]	1–5 layers	MLG and FLG	MLG (>90%)	MLG and FLG
size	~100 μm	~10 μm	~1 μm	~20 μm	30 inch
domain size	~10–100 μm	small	<1 μm	medium	~10 μm
carrier mobility	very high (15 000 cm^2/V.s)	low (2–200 cm^2/V.s)	High (1100 cm^2/V.s)	low (300 cm^2/V.s)	high (1000–5000 cm^2/V.s)
quantity	low	large	medium	large	large
condition	on graphite surface	in solution	on SiC surface	in solution	on transition metal surface
references	[1,42,43]	[46,47]	[48,51]	[71,72]	[53–70]

[1] SLG (single layer graphene) [2] FLG (few layer graphene).

12.1.2 Comparison of Methods of Graphene Synthesis

Many methods for synthesizing graphene have been developed in the past decade. The most common ones include mechanical peeling of graphite [1,41–43], reduction of chemically synthesized graphene oxide (GO) [44–47], graphite intercalation, high temperature (~2000 K) sublimation of SiC single crystal [48–51], and the recently developed chemical vapor deposition (CVD) [52–70]. Among all these methods, the mechanical peeling method only leads to small areas covered with graphene. The ultrahigh vacuum annealing of single-crystal SiC may lead to better coverage but with a relatively small domain size and thus the concentration of linear defects must be high. Chemical synthetic routes through annealing of GO lack control over the number of graphene layers and normally introduce many point defects or holes into the graphene. A detailed comparison of these methods is shown in Table 12.1. Notice that only the CVD method has the potential to produce high quality graphene covering a large area with a controllable number of layers [53–70].

12.1.3 Graphene Chemical Vapor Deposition (CVD) Growth

12.1.3.1 The Status of Graphene CVD Growth

Historically, the CVD method was used to synthesize graphene more than 30 years ago [73]. Recently, it has aroused a great deal of research interest because of the boom in graphene material. An oversimplified scenario of graphene CVD growth is shown in Figure 12.2 [59] and the full process includes the following steps:

1. Carbon feedstock decomposes on the catalyst surface and releases C atoms or active C species to the catalyst surface at the elevated temperature (e.g., 1000–1500 K);

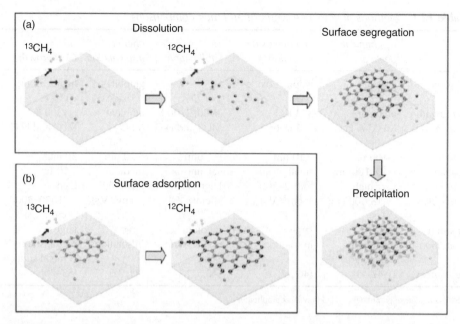

Figure 12.2 The illustration of the two growth mechanisms of graphene. (a) C atoms firstly dissolve into the catalyst and then precipitate onto the catalyst surface to form many-layer graphene upon cooling. (b) Decomposed C atoms stay on the catalyst surface and segregate into a single layer of graphene. The gray and black dots present different C isotopes used in experimental study. (Reprinted with permission from [59] © 2009 American Chemical Society)

2. The diffusive C species meet each other on the catalyst surface and form C clusters of various sizes;
3. Once a C cluster becomes matured or greater than the size of the 2D nucleus of graphene growth, the continuous graphene growth starts;
4. The matured graphene clusters grow larger and larger by continuing to adsorb decomposed C atoms around them and become an graphene island with a specific shape;
5. Islands formed on a catalyst surface grow larger and meet each other until the full surface is covered with graphene, which is normally multi-crystalline with grain boundaries appeared between single crystalline domains (islands).

One key issue of graphene CVD growth is the catalyst surface, on which all the reactions occur and the graphene is formed by the self-assembly of the decomposed C atoms. So the selection of proper catalyst is crucial for graphene synthesis. Experimentally, although some non-metal surfaces have been used as catalysts, the active transition metals are the primary choice. It has been proved that, many metals such as Cu, Ni, Au, Pt, Pd, Ru, Rh, Ir, and their alloys can be used as effective catalysts for graphene growth under different conditions [53–70,74–84]. A summary of the graphene growth on these metal surfaces is shown in Table 12.2. It can be clearly seen that the graphene growth behavior and the quality of the synthesized graphene are highly dependent on the catalyst used. Among

Table 12.2 Comparison of graphene CVD growth catalyzed by different transition metals

Metal	Temp.	C source	Number of layers	Carbon solubility	Epitaxial or not?	Maxim size (domain size)	Ref.
Ni	>800 K	hydrocarbon	SLG & FLG	high	yes	~5 mm	[74–76]
Co	>600 K	hydrocarbon	SLG & FLG	high	yes	~1–5 cm	[69,77]
Cu	~1300 K	hydrocarbon	mainly SLG	low	no	~80 cm (1 µm–1 mm)	[59–61,78]
Ru	~1400 K	CO or hydrocarbon	SLG & FLG	medium	yes	~1 mm (~100 µm)	[53,88–90]
Rh	>1100 K	hydrocarbon	SLG & FLG	medium	yes	~0.1 µm	[79, 80]
Ir	~1300 K	hydrocarbon	SLG & FLG	low	no	full coverage (~1 µm)	[54,63,81,82]
Pt	>1100 K	Hydrocarbon	SLG & FLG	low	yes	full coverage (~50 µm)	[55,65,83]
Pd	>1000 K	ethylene	SLG & FLG	medium	yes	~1–10 cm (~2 µm)	[84]

these metals, Cu and Ni are broadly used for the synthesis of large areas of graphene as a major step toward the commercial realization of graphene [56, 57, 59–61, 68]. It is amazing that high-quality single-layer graphene (SLG) over areas as large as 30 inches have been achieved on polycrystalline copper foils with the roll-to-roll technology [57].

Another important factor of graphene CVD growth is the used carbon feedstock. Although the solid polymer can also serve as the feedstock, [52] the most used ones are still various hydrocarbon gases, such as methane, ethylene, acetylene, benzene, and so on [53–70]. Although the feedstock is very important in graphene CVD growth, their roles are very similar as illustrated in a recent study. Tour and colleagues at Rice University developed an approach using six easily obtained raw carbon-containing materials (i.e., cookies, chocolate, grass, plastics, roaches, and dog feces) to grow graphene directly on the backside of a Cu foil [85]. This study clearly indicates that the main role of the feedstock is to provide C species to form graphene and the difference between them is very limited. But it should be noted that the activity of the feedstock is an important issue for graphene CVD growth control. For example, using active feedstock (e.g., C_2H_2) may lead to graphene formation at a lower temperature than that of using inactive feedstock (e.g., CH_4) [86].

Although graphene CVD growth is a new technology which was only used for 5–6 years, it was recognized as the most promising method for high quality, large area graphene synthesis. Compared to other methods of graphene synthesis (see Table 12.1), the CVD graphene synthesis has the following advantages:

1. It can be achieved at a relatively low temperature (i.e., ~1300 K or lower, which is considerably lower than the temperature required for SiC sublimation, that is, 1900–2300 K).
2. A single-layer or a few-layer (FLG) graphene of high quality is readily synthesizable owing to catalyst-assisted defect healing.

3. Very large area graphene can be easily synthesized (e.g., graphene in 100–1000 square inches).
4. Synthesized graphene can be easily transferrable onto other substrates for further processing.
5. There are numerous tunable experimental parameters, such as type of catalyst, pressure and type of feedstock and carrier gases, temperatures, and so on, are available. With these parameters, we can easily develop more than 100 thousands different parameter combinations (recipes). For example, we can use more than 20 different catalysts, more than 10 different feedstock gases, 10 different temperatures from 1000–1400 K, 10 different pressures from ultra-low limit to ambient pressure, and 10 different ratios between H_2 gas and hydrocarbon gas in experiments. So, there's still a lot of space to improve the quality of CVD synthesized graphene.

12.1.3.2 Phenomenological Mechanism

As discussed previously, to search for the optimum conditions for graphene CVD growth within such a large parameter space is practically not possible. So, only deep insight into the mechanism of growth can lead to the optimum synthesis of graphene. Up to now, the mechanisms of graphene CVD growth have drawn considerable attention. Experimentally, two distinct growth mechanisms have been proposed (Figure 12.2) [59]: (A) *Precipitated growth*: the decomposed C atoms dissolve into the catalyst first and then precipitate to the metal surface to aggregate into graphene during the subsequent cooling; (B) *Diffusive growth*: the decomposed C atoms remain on the metal surface and then incorporate into graphene directly. Mechanism (A) corresponds to those metals that interact strongly with C atoms and have the phases of metal carbide (e.g., Ni, Co, Mo); while growth mechanism (B) corresponds to those metals having weak interaction with C atoms (e.g., Cu, Au) and therefore, no stable metal-carbide phase. For mechanism (A), the continuous precipitation of C atoms from the bulk of the catalysts normally leads to uncontrollable multi-layer graphene growth, while mechanism (B) is known to be the best for the synthesis of monolayer graphene because the graphene formation on the catalyst surface blocks the interaction of feedstock gas with the catalyst, and thus the formation of the second and further layers is very difficult. Recently, by using very thin catalyst layer, SLG and FLG were also successfully synthesized with the framework of mechanism (A) [87].

12.1.3.3 Challenges in Graphene CVD Growth

Although great successes on graphene CVD growth have been achieved, many challenges still remain for the extensive application of graphene:

1. The size of single crystalline domain in a large area graphene is normally around 1–10 μm, which means high concentrated grain boundaries (GBs) exist in the graphene samples.
2. Since the properties of graphene have a strong dependence on the number of layers, therefore, controlling the number of graphene layers is another challenge in making full use of graphene's unique properties in application.
3. Graphene CVD growth can be achieved by using many different carbon feedstocks, among them CH_4 is mostly used and shows great advantages on single-layer graphene

growth. The differences between carbon feedstocks are not well addressed. As the feedstock decomposition occurs on the catalyst surface, the difference must be associated with the type of catalyst surface.

4. Selection and design of proper catalysts. As mentioned earlier, different catalysts, as well as the catalyst surface types, affect graphene growth behavior greatly. Pursuing one catalyst leading to a high quality, large area SLG is greatly desired.
5. Proper graphene growth conditions. For each combination of feedstock and catalyst, varying the temperature, the partial pressures of feedstock and hydrogen in the carrier gas will greatly influence the graphene growth. So, to determine the optimum condition for each combination of feedstock and catalyst is a big challenge because there are probably several hundred different combinations.
6. Graphene CVD growth is broadly named as the epitaxial process. Graphene normally interacts with the catalyst surface that has the weaker van der Waals (VDW) interaction. Theoretically, such a weak interaction should not be able to predetermine an orientation of the grown graphene. However, experimental observations show that there's normally one or a few preferred orientations for the graphene grown on each type of catalyst surface [53–55,90–92]. So, what is the mechanism of graphene orientation determination during growth is another big challenge.

To solve these challenges, doing unlimited experiments to test all possibilities is unrealistic. Only with deep insight into the mechanism of the graphene growth and proper understanding on the differences among potentially used catalysts, the activities of different feedstock, the role of hydrogen, and the required temperature for graphene growth, we can design the experiments in the most desired parameter space and achieve the success of high quality graphene growth by an affordable number of experimental trials.

12.2 The Initial Nucleation Stage of Graphene CVD Growth

Similar to the synthesis of carbon nanotubes, the CVD growth mechanism of graphene can be understood by a *Vapor-Liquid-Solid* (VLS) or *Vapor-Solid-Solid* (VSS) model [93–95]. In the growth process, the transition-metals (TMs) are generally used as the catalysts to accelerate the dissociation of vapor C feedstock gases and serve as the template surface for the graphene nucleation and growth. As mentioned earlier, a complete CVD growth process includes three relatively independent stages: the dissociation of C species and the formation of C atoms or small C clusters on TM surface; the nucleation of graphene; and the lateral expansion of graphene islands.

Among these three stages, the initial nucleation is a crucial step for the following reasons. First, the size of nucleus and nucleation barrier determine the incubation time (time required to form a stable graphene island on the catalyst surface). Secondly, the nucleation barrier, nucleus size, C concentration and diffusion of C atoms on the TM surface determine the density of graphene islands. Since the merging of two graphene islands would mostly lead to a grain boundary between them [34–39], it turns out that controlling the concentration of graphene islands on the TM surface during the nucleation stage is decisive for graphene quality control. In this section, we will introduce in detail the initial nucleation of graphene on TM surfaces.

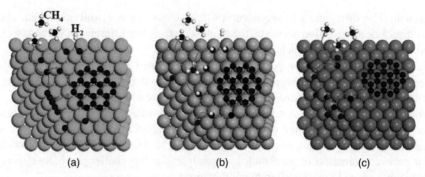

Figure 12.3 *Various C species appeared on different catalyst surfaces during graphene growth. (a) Single C atoms are the dominate C precursors for graphene growth on the metal surfaces who interact strongly with carbon, such as Ir(111) or Ru(0001) surfaces; (b) The dominating C species would be CH, CH2, CH3 for graphene growth on Cu(111) surface with the appearance of hydrogen; (c) For catalysts with a stable carbide phase, such as Ni, dissociated C atoms would be dissolved into the bulk at high temperatures and then precipitate to the surface for graphene nucleation and growth. The arrows represent the diffusion path of C species on metal surfaces*

12.2.1 C Precursors on Catalyst Surfaces

In the initial stage of graphene CVD growth, C feedstock molecules (e.g., CH_4, C_2H_2, C_6H_6) [48,60,61,96] dissociate on TM surfaces and then form C monomers or dimers due to TM-catalyzed dehydrogenation. The TM-C interaction is an important factor for determining the C species on TM surfaces during the initial graphene growth. On very active metal surfaces, such as Ir(111), Rh(111), and Ru(0001), the dissociated carbon atoms prefer to form C monomers (see Figure 12.3a), while carbon dimers are energetically more favorable on those less active metal surfaces, such as Cu(111) and Au(111) [97]. Due to the presence of hydrogen in the reaction environment, atomic carbon is thermodynamically less stable on the less active TM surfaces, such as Cu. Therefore, the C species for graphene growth on Cu surface should be dominated by CH_x (x = 1, 2, 3) species under the typical experimental conditions (see Figure 12.3b) [98]. For some transition metals who interact strongly with C atoms (e.g., Ni, Fe, Co, Mo ...) [99], they can form stable metal-carbide phases to store the decomposed C atoms (see Figure 12.3c). It's important to note that a single C atom tends to stay on the subsurface of Cu, which leads to a more stable formation and a higher diffusion barrier, while C clusters, such as the C dimer, prefer to stay on the metal surface, the nucleation of graphene always occurs on the surface [100].

12.2.2 The *sp* C Chain on Catalyst Surfaces

Once the concentration of C species (e.g., C monomers or dimers) on TM surfaces reaches a critical value, they will aggregate into C clusters and initiate the nucleation of graphene. On the route of nucleation, C clusters of various sizes are intermediates between C monomer and graphene nuclei. In a small C cluster C_N (N < 8), there is no enough C atoms to form a sp^2 carbon network containing at least two connected polygonal rings, so only two possible

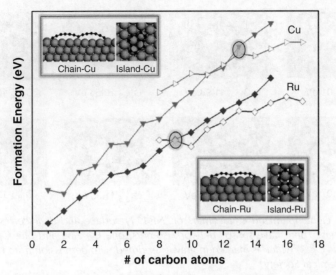

Figure 12.4 Formation energies of C chains and C sp^2 networks on Cu and Ru surfaces versus the number of carbon atoms. The solid and hollow triangles represent the energies of C chains and C sp^2 networks on Cu(111) surface, and solid and hollow rectangles represent the energies of C chains and C sp^2 networks on Ru(0001) surface, respectively. The circles denote the critical size of C clusters on metal surfaces

configurations, C chain and C ring, are stable [73,101–103]. In this size range, the C chain is thermodynamically more favorable than C ring.

The configuration of the C chains on a TM surface depends on the C-metal binding. On the Ni(111) surface [103], all atoms of a C chain tend to stick on the surface due to the strong C–Ni binding. Differently, as the C–Cu binding is relatively weak, the arc-shaped C chain (or C nanoarch) on the Cu surface is energetically more favorable instead [102]. Moreover, the C chains have a very low diffusion barrier and thus can diffuse quickly on the catalyst surface (less than 0.7 eV) [73]. Therefore, these C chains are potential C precursors to be attached directly onto the edge of a growing graphene. For example, it was proposed that the graphene islands grow predominantly by the attachment of five-atom clusters on Ir(111) and Ru(0001) surfaces [104–106]. The details will be discussed in the following section.

Once the size of the C cluster reaches a critical value, a structural transformation from C chain to sp^2 network occurs. Depending on the C-metal interactions, [107] the critical size of C cluster is different. As shown in Figure 12.4, the critical size for the Cu(111) surface, which interacts with C weakly; and the Ru(0001) surface, which represents a strong interaction with C, are 13 and 9, respectively.

12.2.3 The sp^2 Graphene Islands

Further increase of the C cluster size leads to a great number of possible isomers of sp^2 networks. Figure 12.5 shows eight C_{13} isomers on Ni(111) surface, including six sp^2 network isomers, one ring and one chain. It can be found that the C_{13-3} (a sp^2 network

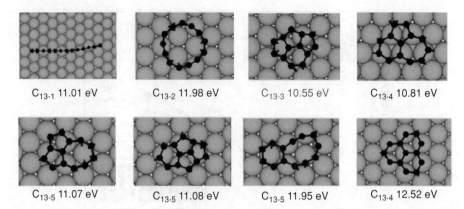

Figure 12.5 Eight optimized C_{13} isomers on Ni(111) surface and their corresponding formation energies. There have six sp^2 network isomers, one C_{13} ring and one C_{13} chain. The ground state is C_{13-3}, which is marked in gray. (Reprinted with permission from [103] © 2011 American Chemical Society)

configuration) is the most stable structure, while the chain and ring configurations possess relative higher formation energies [103]. As the cluster grows larger, the stability of the sp^2 network is further enhanced. Interestingly, the most stable sp^2 networks always contain a few pentagons and the full-hexagon isomers are energetically less favorable. The energy increment of a sp^2 C network relative to the pristine graphene originated primarily from its edge atoms, and thus a reduction in the number of these atoms is energetically preferred. Incorporating one or a few pentagons into a sp^2 structure alters its shape from flat to bowl-like, which normally results in a reduced circumference length or number of edge atoms and thus leads to a lower formation energy.

12.2.4 The Magic Sized sp^2 Carbon Clusters

During the initial nucleation of graphene, some experimental findings have shown that uniformed carbon clusters with diameters of about 1 nm were observed during the graphene growth on Ir(111), Rh(111), and Ru(0001) surfaces [91,108,109]. Although these clusters were suspected have a pure hexagonal structure, C_{24}, which contains seven six-membered rings (7-6MRs), recently they were theoretically proved to be a core-shell structured C_{21} cluster [107], which has three isolated pentagons with the C_{3V} symmetry. The high stability of C_{21} can be understood by the following reasons:

1. The closed core-shell structure. The ground structures of C_N on transition-metal surfaces can be classified into three groups (see Figure 12.6): The small clusters, such as C_{16}, C_{17}, and C_{18}, have a unclosed core-shell geometry; The medium sized clusters, such as C_{19}, C_{20}, C_{21}, and C_{24}, have a closed core-shell formation; The large clusters, such as C_{22}, C_{23}, C_{25}, and C_{26}, have a core-shelled structure with one or two additional rings. It has been generally accepted that the closed core-shell geometry are more stable than others.

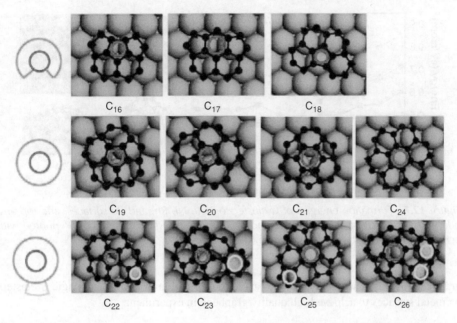

Figure 12.6 The ground state structures of C_{16}–C_{26} on TM surface. They can be classified into three groups: C_{16}–C_{18} have unclosed core-shell structures; C_{19}–C_{21} and C_{24} are closed core-shell structures and C_{22}, C_{23}, C_{25}, C_{26} own a core-shell geometry with one or two additional rings. (Reprinted with permission from [107] © 2012 American Chemical Society)

2. High symmetry. Among the four possible core-shell clusters, C_{19} and C_{20} own a pentagonal core while the core of C_{21} and C_{24} is a hexagon; of which, the C_{19} has the lowest C_s symmetry. In contrast, the free standing clusters, C_{20}, C_{21}, and C_{24}, have very high symmetries of C_{5v}, C_{3v}, and C_{6v}, respectively (see Figure 12.7). When binding to the Rh(111) or Ru(0001) surface, which has a C_{3v} symmetry, the symmetries of the supported C_{20} and C_{24} are reduced to C_S and C_3, respectively; but the C_{21} is able to maintain its C_{3v} symmetry. The high symmetry allows the C_{21} to achieve a strong binding between all its edge atoms and the metal surface and thus enhances its stability.
3. Strong cluster-metal binding. Owing to the incorporation of three pentagons, the C_{21} is highly curved, the other core-shell structures, such as C_{20} and C_{19} are slightly curved because of the pentagon core. In sharp contrast, the 7-6MRs C_{24} is completely flat. The energy tendency shows that a nanographene on a metal surface favors a dome-like shape to minimize its edge formation energy. Therefore the interaction between C_{21} and metal surface is significantly enhanced.

As there are many edge C atoms interacting strongly with the catalyst surface, the barrier of C_{21} cluster diffusion is quite high (~3 eV). At the nucleation stage, the fusion of C_{21} clusters on a metal surface with a high coverage results in high-density nuclei or graphene islands on the metal surfaces [91,108,109]. The latter coalescence of these islands during the growth that follows will inevitably lead to the grain boundary formation in the grown

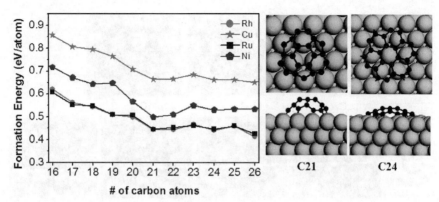

Figure 12.7 Formation energies of C clusters on Cu, Ni, Rh, and Ru surfaces, and top and side view of atomic structures of C_{21} and C_{24} clusters on Rh(111) surface. (Reprinted with permission from [107] © 2012 American Chemical Society)

graphene [34,92]. Thus, it is crucial to avoid the formation of uniformed magic C clusters on metal surfaces to achieve high quality graphene in experiment.

12.2.5 Nucleation of Graphene on Terrace versus Near Step

With the knowledge of C cluster evolution on the catalyst surface, we can consider the nucleation process of graphene there. We now consider the different nucleation behaviors of graphene on a flat transition-metal terrace, or near a step of the terrace on the Ni(111) surface. During the graphene CVD growth, further growth of a sp^2 C network leads to the formation of a graphene nucleus. To form such a graphene nucleus, the C cluster needs to be greater than the nucleation size, N^*, where its formation energy needs to overcome the maximum, which is called nucleation barrier, G^*. Based on the crystal nucleation theory, the nucleation barrier and nucleation size of the graphene nucleus can be viewed as the maximum point (N^*, G^*) of the curve of Gibbs free energy (G) versus cluster size (N). Noting the chemical potential difference between the crystal phase and feedstock as $\Delta\mu$, the Gibbs free energy $G(N)$ of the C cluster on TM surfaces is, [110]:

$$G(N) = E(N) - \Delta\mu \times N, \tag{12.1}$$

where $E(N)$ is the formation energy of C clusters on the TM surface. Following this, we can easily determine G^* and N^* as a function of $\Delta\mu$ for graphene nucleation on a transition metal terrace or near a step if the formation energy of C clusters is obtained.

The formation energy of a C_N cluster on transition metal surface or near a step edge can be calculated by the first-principles calculations based on the following definition,

$$E_N = E(C_N @M) - E(M) - N \times \varepsilon_G, \tag{12.2}$$

where $E(C_N@M)$ is the energy of a C_N cluster on a transition-metal substrate, $E(M)$ is the energy of the transition-metal substrate, and ε_G is the energy per carbon atom of graphene.

Following Eq. (12.2), Gao et al. [111] calculated the optimized atomic structures and formation energies of different sized C clusters on the Ni(111) terrace, or near a step,

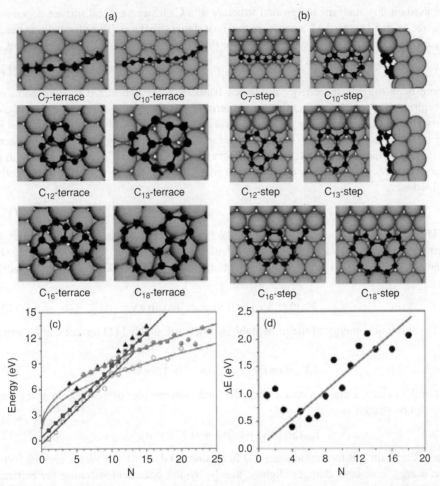

Figure 12.8 Ground state structures and energetics of the supported C clusters on a Ni(111) terrace and near a step edge. (a) The atomic structures of C_N on Ni(111) terrace, (b) the atomic structures of C_N on Ni(111) step, (c) the energies of the supported C_N on metal terrace and near a step as a function of the number of C atoms, (d) the energy difference between the optimized C_N on Ni(111) terrace and near a step edge, with the straight line providing a linear fit to the data to guide the eye. (Reprinted with permission from Ref. [111]. Copyright 2011 American Chemical Society)

respectively, with the density-functional theory (DFT) method (see Figure 12.8). Unlike the C clusters on Ni(111) terrace, because of the strong binding between C and the step on a catalyst surface, the most stable C clusters tend to have more C atoms attached to that step. This results in a crescent moon-shaped cluster ground structure, which is in sharp contrast to the circular shape of the C clusters on Ni(111) terrace [107,111]. The comparison of the formation energies of C clusters on the transition metal surface or near a metal step suggests that the metal step has the higher affiliation to C atoms and thus lowers the formation energies of C clusters.

Based on this analysis, the ground structure of a C cluster on metal surface depends on its size. At small sizes, the C chains are the most stable structures. Their formation energies E_{ch} increase linearly with cluster size N (see Figure 12.8c) and can be estimated using,

$$E_{ch} = b_1 \times N + a_1, \qquad (12.3)$$

where b_1 is roughly the energy difference between a sp^1 hybridized C atom and a sp^2 hybridized C in graphene, and the second term on the right-hand side of the formula, a_1, is the formation energy of the two chain ends that are passivated by the TM surfaces. For the larger C clusters, the sp^2 networks are the most energetically favorable. The energy increment of the sp^2 networks is mainly from the contribution of edge C atoms N_e in the cluster, and $N_e \sim N^{1/2}$. Thus, the formation energy E_{sp2} of the most stable sp^2 network is defined as:

$$E_{sp^2} = b_2 \times N^{1/2} + a_2 \qquad (12.4)$$

The parameters (a_1, a_2, b_1, b_2) in Eq. (12.3) and Eq. (12.4) can be fitted based on the formation energies of various C clusters calculated by the first-principles calculations. For example, on the Ni(111) terrace, the formation energy of the 1D C chains can be written as:

$$E_{ch}(\text{Terrace}) = 0.81 \times N + 0.40 \text{ eV}, \qquad (12.5)$$

The formation energy of the most stable sp^2 network on Ni(111) terrace can be written as:

$$E_{sp^2}(\text{Terrace}) = 2.4 N^{1/2} + 1.6 \text{ eV}. \qquad (12.6)$$

Similar to the C_N clusters on a terrace, the formation energies of the C chain near a metal step can be written as:

$$E_{ch}(\text{Step}) = -0.263 + 0.775 \times N \text{ eV}, \qquad (12.7)$$

where the chain end formation energy is further reduced to -0.13 eV/end, and the formation energy increment changes slightly, that is, by 3%, both demonstrating the enhanced chemical activity of the step edge. The formation energy of a sp^2 C network near a step edge can be written as:

$$E_{sp^2}(\text{Step}) = 1.992 \times N^{1/2} + 1.328 \text{ eV}. \qquad (12.8)$$

Using Eq. (12.5–12.8), the formation energies of any cluster on a Ni(111) terrace or near step edge can be obtained. We present the formation energy difference between the C clusters on a Ni(111) terrace and those near a metal step edge as a function of cluster size in Figure 12.8(d). Clearly, approaching a metal step always stabilizes the C cluster. The energy difference rises to 2 eV or higher at a size of $N > 12$. Such an energy difference is crucial in the graphene nucleation behavior displayed on a terrace or near a step edge.

Taking Eq. (12.5–12.8) into Eq. (12.1), the nucleation barrier G^* and nucleation size N^* on Ni(111) terrace or near a step can be determined. Figure 12.9(a) shows the Gibbs free energies of supported C_N clusters on Ni(111) terrace as a function of cluster size N. It can be clearly seen that the maximum point of the curve of G versus N (where the nucleation barrier, G^*, and nucleus, N^* are defined). Figure 12.9(b) presents nucleation barrier G^* and nucleus size N^* as a function of $\Delta\mu$ for both cases. We can find that the nucleus size, N^*

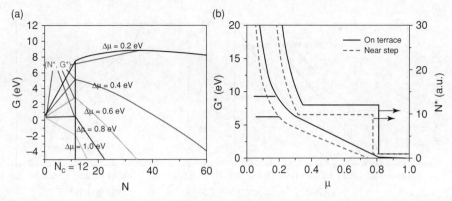

Figure 12.9 (a) The Gibbs free energies of C_N ground structures on Ni(111) terrace as a function of cluster size at $\Delta\mu = 0.2, 0.4, 0.6, 0.8, 1.0$ eV. (b) Nucleus size N^* and nucleation barrier G^* as a function of chemical potential difference between the C in graphene and C in feedstock, where the solid and dashed lines represent the nucleation on Ni(111) terrace and near a step edge, respectively. (Reprinted with permission from [111] © 2011 American Chemical Society)

exhibits stepwise behavior that stems from the ground structure transformation from the C chain to the sp^2 network. For the nucleation on the Ni(111) terrace (Ni(111) step edge), the nucleation barrier G^* and nucleation size N^* decrease with the increasing $\Delta\mu$ at $\Delta\mu < 5.77$ eV ($\Delta\mu < 4.47$ eV). In the $\Delta\mu$ range from 0.35 to 0.81 eV, the nucleation size N^* is maintained at $N^* = 12$ for the nucleation on Ni(111) terrace. Similarly, $N^* = 10$ in the range of $\Delta\mu$ from 0.32 to 0.78 eV for the nucleation on Ni(111) step edge. In the region of $\Delta\mu > 0.81$ eV for nucleation on a terrace or $\Delta\mu > 0.775$ eV for that near a step edge, the nucleation barrier goes to zero and the nucleus size drops to $N^* = 1$. This ultralow G^* and very small N^* imply that the nucleation of graphene on Ni(111) surface in this range of $\Delta\mu$ may occur from the deposited C clusters with any size. In this case, graphene nucleation or growth is dominated by the C deposition rate and C diffusion on the metal surface.

Based on the classical nucleation theory, the 2D nucleation rate is defined as:

$$J = \omega^* \Gamma^* N_1 \exp(-G^*/kT) = J_0 \exp(-G^*/kT), \tag{12.9}$$

where ω^* is the attachment rate of C atoms into a cluster of critical size, $\Gamma = [G^*/(4\pi kT^* N^{*2})]^{1/2}$ is the Zeldovich factor, and N_1 is the concentration of atoms/molecules [110]. To obtain the nucleation rate of graphene on a terrace or near a step edge, it requires an appropriate estimation of the pre-factor J_0. The attachment rate can be estimated as $\omega^* = N^*_{edge} p\, (\nu \exp(-E_b/kT))$, where N^*_{edge} is the number of attachment sites of the 2D nucleus. For a sp^2 network structure, $N^*_{edge} \sim (6 \times N^*)^{1/2} \sim 10$, and $\nu = 10^{13}$ s^{-1}. E_b is the barrier for attaching a C atom to the graphene. Considering the active sites on the bare graphene edge, E_b can be estimated as the barrier of a C atom diffused on the Ni surface, which is ~ 0.50 eV. The occupancy rate is $p \sim N_1$. Based on extensive experimental observations, the concentration of C monomers on a metal surface is roughly 0.01 monolayer graphene (MLG) [112]. Hence, N_1 is approximately equal to ~ 0.38 nm^2 and $p \sim 0.01$. To estimate the magnitude of the pre-factor, $J_0 = \omega \times \Gamma N_1$, assume that $N^* = 10$ and $kT \sim 0.1$ eV (which

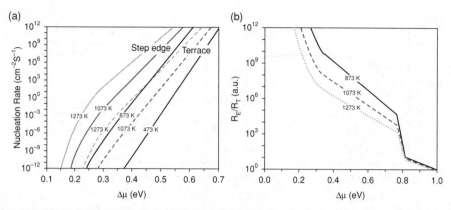

Figure 12.10 (a) Nucleation rates of graphene growth on a Ni(111) terrace, RT, or near a step edge, RE, as a function of $\Delta\mu$ at temperatures of 873, 1073, and 1273 K; (b) their ratios, RE/RT. (Reprinted with permission from [111] © 2011 American Chemical Society)

is consistent with the typical temperature of CVD graphene growth, i.e., 800–1000°C), we then have $J_0 \sim 4 * 10^{21}$ cm^{-2} s^{-1}. Thus, the nucleation rate can be rewritten as:

$$R_{\text{nul}} = J_0 \exp(-G^*/kT) \exp(-G^*/kT + 49) \, \text{cm}^{-2}\text{s}^{-1}. \quad (12.10)$$

Following Eq. (12.7), the nucleation rate of graphene on Ni(111) terrace or near a step edge can be plotted (see Figure 12.10a) as a function of $\Delta\mu$ at several typical experimental temperatures: 873 K, 1073 K, and 1273 K. It is clear that the nucleation rate is extremely sensitive to both temperature and $\Delta\mu$. A slight variation in temperature or $\Delta\mu$ may result in a dramatic change in the growth rate. For example, at 1073 K, altering the $\Delta\mu$ from 0.4 to 0.6 eV leads to a growth rate change on the terrace, R_T, of 12 orders of magnitude (from 10^{-4} to 10^8 cm^{-2} s^{-1}). Varying the temperature by ~200 K results in a change of six or more orders-of-magnitude in the R_T or R_E.

The differences in graphene nucleation on a terrace and near a step edge at these temperatures are shown in Figure 12.10(b). It can be found that the R_E/R_T ratio changes monotonically with either temperature or $\Delta\mu$. However, the difference vanishes at $\Delta\mu = 0.81$ eV. It is because any C monomer may lead to the graphene nucleation, namely, the nucleation size $N^* = 1$. In a typical growth region, for the range of $\Delta\mu$ from 0.30 to 0.75 eV at the temperature T = 1073 K, the ratio R_E/R_T varies from 10^4 to 10^8. For the nucleation on Ni(111) terrace, its great advantage is that the nucleation can start at any place of the metal surface while the nucleation near Ni(111) step must start from the step edge. Thus, the effective area of nucleation on terrace is far larger than that of step edge and the ratio of R_E/R_T is actually smaller than the present calculated values. In other words, the nucleation may occur on terrace with the appropriate $\Delta\mu$.

It is easy to understand that a very high nucleation rate will result in a large number of nuclei on a TM surface. The coalescence of these nuclei and growing graphene islands will result in many grain boundaries. Thus, a relatively low nucleation rate is preferred for high-quality graphene growth, and it can be achieved through the use of a lower temperature and a low $\Delta\mu$ (i.e., 0.3–0.5 eV). In addition, a high nucleation barrier notably diminishes the probability of graphene nucleation based on Eq. (12.9). Thus, for growing single crystal graphenes, the seeded growth, such as adding a small graphene patch onto the metal surface can help to avoid the nucleation stage of graphene growth, and multi-nucleation sites are consequently prohibited by the high nucleation barrier.

12.3 Continuous Growth of Graphene

After nucleation, nuclei are formed on the catalyst surface. In the following stage, feedstocks keep decomposing on the catalyst surface and the supplied C species diffuse to the edge of small nuclei to be adsorbed to the edge continuously. The addition leads to a continuous reduction of the system's free energy and the reverse process, the detachment of C from the growing island, is normally less important. In such a process, the growth of graphene is kinetically controlled because not all the possible configurations can be experienced, which is the key feature of the thermodynamics.

The fast growth rate is crucial for the synthesis of large area graphene islands to avoid linear defects, which is the key of growing graphene of high quality. Experimental investigations have reported many interesting observations on the graphene growth kinetics [105,112,113]. It has been found that the growth rate of graphene on Ru(0001) and Ir(111) surfaces is a nonlinear function of carbon concentration and thus graphene growth was considered as a process of adding C_5 clusters into the graphene edge [104, 105,112]. Graphene growth on a Cu catalyst surface was widely observed as a diffusion limited process, characterized by the branched or even fractal-shaped graphene islands on the Cu surface [57,59,114, 115]. In the meantime, these interesting experimental observations are still puzzles to us. To achieve a full understanding of these observations for the design of optimum experiments to grow high quality graphene in a large area, theoretical exploration is important.

To understand the mechanism of graphene's continuous growth, a precondition is to know the structure of the growing front of the graphene. In this section, starting from the stability of graphene edge and its passivation by metal surface, we will firstly explore the most stable graphene edge on a metal surface and then discuss the process of carbon atom being incorporated onto the graphene edge.

12.3.1 The Upright Standing Graphene Formation on Catalyst Surfaces

Free graphene edges are unstable in a vacuum due to the existence of dangling bonds at the edge, formation energies for free zigzag (ZZ) and armchair (AC) graphene edges are as high as 13.46 and 10.09 eV/nm, respectively [111]. Such unstable edges can be stabilized by attaching to a metal catalyst surface because the active surface metal atoms can saturate the dangling bonds [116]. Figure 12.11(a) gives the calculated relative formation energy of GNR on Ni(111) surface as a function of the tilted angle θ. It can be seen that, for both ZZ and AC edges, $\theta = 90$ degree corresponds the lowest formation energy of the potential energy surface (PES), demonstrating that GNRs prefer to stand vertically on the Ni(111) substrate surface [116]. Changing the tilt angle from the 30 to 90° will make the energy rise by $\sim$2 eV/nm, showing the robustness of the vertically standing GNR. Such a trend is due to the orientation dependence of the sp^2 hybridized σ bond. Each sp^2 hybridized edge

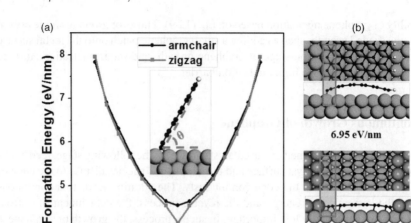
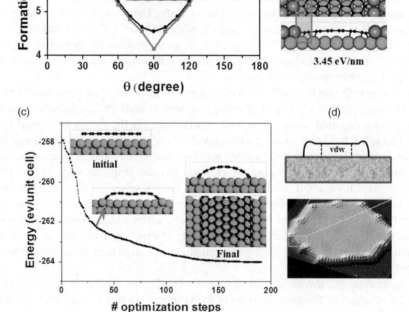

Figure 12.11 (a) The preference of an upright standing graphene nanoribbon (GNR) on an Ni(111) surface. (b) Optimized zigzag (ZZ) graphene edges on a Ni(111) terrace and near a metal step, along with the corresponding formation energies for each structure. (c) Formation of a curved GNR on Ni(111) surface and the total energy change during the structural optimization. (d) The scheme of graphene formation on metal substrate and the experimentally observed STM for graphene on Ir(111) surface. (Parts (a–c) Reprinted with permission from [111,116] © 2009 American Chemical Society; (d) Reprinted with permission from [117] © 2012 American Physical Society). See colour version in the colour plate section

C atom of a ZZ-GNR has a perpendicular dangling bond aimed towards the metal surface, and vertically attaching to the substrate results in the strongest binding.

The phenomena of graphene standing upright on a substrate can also be seen dynamically during the geometrical optimization process of free edged GNRs on Ni (111) surface, as shown in Figure 12.11(c). It should be noted that the van der Waals (vDW) interaction between the graphene wall and the metal substrate can't be neglected for a large sized graphene. In a large graphene island, the vDW interaction will force the central part of the graphene to be attached to the metal surface, leaving the edge part of the graphene curved down towards the substrate for a better binding. Such an expectation has been proved by high resolution STM images as shown in the bottom of Figure 12.11(d) [117]. The STM image in Figure 12.11(d) shows the atomic structure of a medium-sized graphene on Ir(111) surface, it can be seen that graphene near the edge bumps up more than those in the central area of the graphene.

The tendency of graphene to stand on metal substrate surfaces also contributes to the stability of the magic sized carbon cluster, C_{21}. Since a vertically upright graphene has the minimum edge formation energy, a small graphene island prefers to have a dome-like shape to minimize its edge formation energy [118]. As shown in Figure 12.7(b), the edge of C_{21} has a larger tilt angle than that of the 7-6MRs C_{24}. The formation energy difference resulted from the binding angle is ~ 2.0 eV/nm or ~ 0.5 eV/edge atom. Although the larger tilt angle also increases the curvature energy of the C_{21}, it can be compensated by the strong binding energy.

The upright standing formation of graphene on metal surfaces is also responsible for the preference of graphene nucleation near a metal step. As shown in Figure 12.11(b) the formation of a **ZZ** edge on Ni(111) terrace is about 6.95 eV/nm, while the formation energy of a **ZZ** edge at the Ni(111) step is notably reduced to 3.45 eV/nm [111]. This significant reduction in edge formation energy implies that graphene nucleation can be facilitated by a metal step edge, which is in agreement with many experimental observations [54,82,105].

12.3.2 Edge Reconstructions on Metal Surfaces

Previous investigations found that a free AC edge has no reconstruction due to the self-passivation by the triple C≡C bonds between neighboring outermost atoms, while a free-standing graphene **ZZ** edge tends to be reconstructed by turning every neighboring hexagon pair (6–6) on the edge into a pentagon-heptagon pair (5|7) [119–121]. In graphene CVD growth, since graphene edge on the TM surface is saturated by metal substrate, the edge configuration could be different to that in vacuum.

In general, two different structures can be reconstructed for each **ZZ** and **AC** edge, ZZ(57)/AC(677) and ZZ(ad)/AC(ad), as shown in Figure 12.12(a). Among them, the ZZ(57)/AC(677) can be achieved by rotating a C–C bond on the pristine ZZ/AC edge; and the AC(ad)/ZZ(ad) is obtained by adding one C atom to each unit cell of the pristine AC/ZZ edge. The AC(ad) edge on TM surfaces has two different potential configurations named as AC(ad)-I and AC(ad)-II according to the binding site difference of the added atom [119]. The formation energies of all these edges in vacuum and on three metal surfaces; Ni(111), Co(111), and Cu(111) are listed in Figure 12.12(b). Taking edges on Ni(111) surface as an example, for the three derivative configurations of **ZZ** edges, the **ZZ(ad)** has the highest formation energy and the **ZZ(57)** configuration, which is the most stable **ZZ** edge in vacuum, [122] has the second highest formation energy. In sharp contrast to that in vacuum, a pristine ZZ graphene edge on the Ni(111) surface has the lowest formation energy and thus there's no edge reconstruction on the ZZ edge. However, for three derivative **AC** edges on Ni(111) surface, the reconstructed **AC(ad)-I** edge is the most energetically favorable and the **AC(677)** edge is the most unfavorable one.

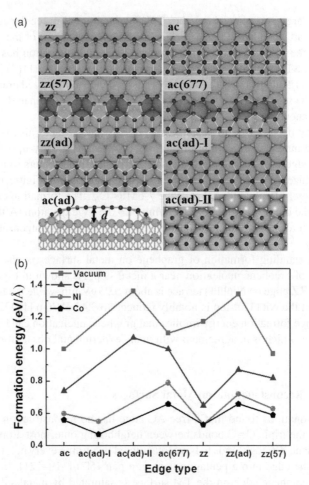

Figure 12.12 (a) Some typical supported graphene edges on TM surfaces. Top views of pristine ZZ and AC edges, reconstructed zz(57), reconstructed ac(677), zz(ad), two types of AC(ad): AC(ad)-I AC(ad)-II, and the side view of AC(ad)-I edge on Co(111) (the left-hand bottom one); (b) formation energies of various graphene edges in a vacuum and on TM surfaces. (Reprinted with permission from [119] © 2012 American Chemical Society)

The strength of the carbon-metal interaction has big effect on the graphene edge structure. On both Co(111) and Ni(111) surfaces, the average formation energy of graphene edge drops by about 50%, while the energy reduction on Cu is only 30%, which indicates that Co and Ni passivate the graphene edge more efficiently than Cu. More importantly, the energy drops are configuration dependent. For example, the **AC(ad)-I** edge prevails the pristine **AC** edge on Co and Ni surfaces, while the case is different for Cu(111) surface. On Cu(111) surface, the **AC(ad)-I** configuration is not stable and will be transformed into **AC(ad)-II** upon optimization, which is similar to that in vacuum. The additional C atom in **AC(ad)-II** binds to two edge C atoms of the armchair site and a pentagon is formed. However, on both Ni(111) and Co(111) surfaces, the C adatom only bonds to one edge C atom, presenting a very stable graphene edge configuration of **AC(ad)-I** in Figure 12.12. In short, different from the freestanding graphene edges in vacuum, the reconstructed edges with pentagon or heptagon are not stable on transition metal surfaces owing to passivation of the graphene edge by the metals.

12.3.3 Growth Rate of Graphene and Shape Determination

As a 2D crystal, the graphene growth has to follow the principles that govern the crystal growth. During the crystal growth, the crystal's shape is normally governed by the kinetics of atomic addition instead of the minimization of surface energies, which controls the grand structure of the sufficiently annealed crystal. The kinetic Wulff construction (KWC) was broadly used to construct the shape of a growing crystal. A simple understanding of the KWC is shown in Figure 12.13(a). Mimic the C_{6v} symmetry of the graphene, a dodecagon with two different types of edges represents a growing graphene island. It can be clearly seen that the edge that grows faster will gradually become shorter and shorter during the further growth and finally the edge that grows slower dominates the final shape of the growing island. Such an analysis clearly shows a big disadvantage of graphene growth compared to one-dimensional carbon nanotubes: that the growth rate must be slow.

Most experimental graphene islands are dominated by the ZZ edge, which implies that the ZZ edge is the edge type that grows slowest in the CVD environment. Such a hypothesis was studied theoretically by Shu et al. [123]. It was found that, on Cu(111) catalyst surface, the active catalyst atoms play a crucial role for the insertion of C atoms onto the graphene.

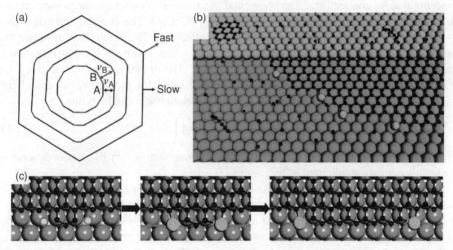

Figure 12.13 (a) The evolution of a growing graphene island. The initial graphene island has six fast growing edges (B) and six slow growing ones (A) of equal length. During the growth process, the slow growing edge of the island gradually dominates by circumference of the island. (b) The termination of armchair sites of graphene during growth. (c) The mode of graphene zigzag edge growth on Cu(111) surface. Reprinted with permission from [123] © 2012 American Chemical Society)

For each AC or AC-like site, the termination by a Cu adatom is energetically preferred and therefore the barrier of C addition onto the AC edge will be greatly reduced by the catalytic effort of the terminated Cu atom. In comparison, the ZZ edge tends not to be passivated and thus the barrier of the ZZ edge growth is significantly high. By classifying the graphene edge into AC-like and ZZ-like sites, the growth rate of an arbitrary graphene edge can be written as:

$$R(\theta) = 4(\sqrt{3}) \times R_{AC} \times \sin(\theta) + 2 \times R_{ZZ} \times \sin(30° - \theta) \qquad (12.11)$$

where θ is the angle of the orientation from the closest ZZ direction, R_{AC} and R_{ZZ} are the rate of C addition onto AC and ZZ sites, respectively. So, a conclusion which is in agreement with experimental observation that ZZ edge will dominate the circumference of a graphene islands can be easily drawn by simply applying the KWC analysis [123]. Very recently, Yakobson and co-workers considered the growth barriers on three different sites, AC, ZZ, and Kink sites, and have drawn another expression of graphene growth rate which is slightly different to Eq. (12.11) [124].

12.3.4 Nonlinear Growth of Graphene on Ru and Ir Surfaces

The carbon incorporation into the graphene edge is another important issue. Studies on this topic are very limited and there are only a few experimental and theoretical publications on this topic [104–106,112,123]. It has been found that graphene growth rate is proportional to the fifth of power of a carbon monomer's concentration. In this section, we will introduce the experimental measurement on graphene growth kinetics and the related theoretical explanation.

Loginova et al. [105] firstly reported the dependence of graphene growth rate on carbon concentration by conducting the experimental measurements under ultrahigh-vacuum conditions using the low-energy electron microscopy (LEEM). The concentration of carbon atoms was measured from the change of electron reflectivity. The graphene growth rate was measured as a function of supersaturation, $c - c^{eq}$, where c and c^{eq} are concentration of carbon atoms, and in equilibrium state, respectively. The graphene growth rate as a function of carbon supersaturation is shown in Figure 12.14(a). It clearly indicates a nonlinear relationship between the growth rate, v, and c. The relationship can be written as:

$$v = B\left[\left(\frac{c}{c^{eq}}\right)^n - 1\right] \qquad (12.12)$$

where $n = 4.8 \pm 0.5$, indicating that the growth of graphene is very like a consequence of the simultaneously attachment of five atoms.

Such nonlinear growth kinetics was originally explained by the nucleation theory. To initiate the graphene growth from a zigzag edge, as shown in the insert of Figure 12.13(a), at least three C atoms must be added to the edge to form new six-membered rings. But adding three carbon atoms produces an isolated ring, which may not provide a pathway to the attachment because the three C atoms are bonded only to two C atoms. In contrast, attaching five C atoms to the zigzag edge adds two adjacent six membered rings. If this configuration is the smallest stable "nucleus" for further island growth, then adding fewer C atoms would not lead to the further graphene growth. Once this stable nucleus forms, a new linear C chain can be quickly formed near the nucleus by adding just two C atoms. Thus, the edge will rapidly grow until reaching the next compact edge, where adding another five

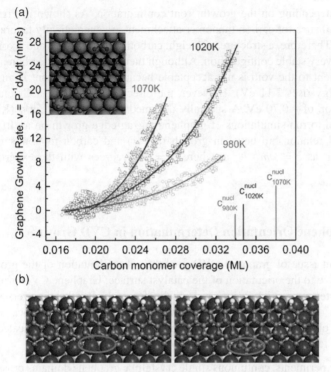

Figure 12.14 (a) Island growth rate as a function of C concentration. The solid lines fit Eq. (12.1) with three parameters: B, c_{eq} and n. At 980, 1020, and 1070 K, n = 4.9, 4.8, and 5.2, respectively. The vertical lines show the adatom concentration at island nucleation, c_{nucl}. The inset in the figure is the graphene nucleation on Ru(0001) surfaces; (b) graphene ribbon in which eight of the ten sites attached by carbon monomer attachment and other two site filled with no carbon (left) and a C_5 cluster (right) attached. (Reprinted with permission from [105] © 2008 IOP Publishing Ltd)

C adatoms is again necessary to nucleate a new "riser" along an island edge before further crystal growth can happen.

A more detailed C_5 addition mechanism is proposed by Li and coworkers based on their first principles calculations [106]. It is well-known that graphene grown on Ir(111) surface will form a moiré structure due to the lattice mismatch. The lattice mismatch between graphene and Ir(111) surface is ~11%, which means 10 Ir atoms correspond to 11 C atoms in a very repeatable unit cell. Such a mismatch makes the graphene edge has different adsorption sites on the substrate. Generally, there are three typical adsorption sites on Ir(111) surface: the top site, where a carbon atom is adsorbed on the top of a metal atom; the bridge site, where a carbon atom is adsorbed at the center of two neighboring metal atoms; and the hollow site, where carbon atom is adsorbed at the center of three metal atoms. Theoretical calculations found that the adsorption energy at a top site is 1.73 eV higher than the most stable hollow site and the adsorption on the bridge site is in between. Therefore, attachment of atomic carbon to a graphene zigzag edge can be energetically favorable or unfavorable depending on the growth front configuration. As shown in Figure 12.14(b), eight of the 10 sites for carbon monomer attachment are energetically favorable along the zigzag edge. Therefore, a structure with eight carbon atoms attached and a void at the three top sites is a very stable configuration. Although there may be many carbon atoms around, their attachment to the void is not acceptable because the attachment is an endothermic process (energy rises 2.11 eV). However, the direct attachment of a C_5 cluster leads to an energy drop of -0.70 eV. A specially designed kinetic Monte Carlo (KMC) method was also used to run simulations of graphene zigzag edge growth on an Ir(111) surface. The obtained relationship between growth rate (r) and carbon monomer concentration (m) is written as $r = am^b + c$, where $b = 5.17$ agrees with the experimental value very well.

12.4 Graphene Orientation Determination in CVD Growth

One important issue of graphene CVD growth is the orientation of the grown graphene layer in relative to the orientation of the catalyst surface. Graphene CVD growth has been broadly reported as an *epitaxial* process, which refers to the fact that the over-layer graphene normally has one or a few preferred orientations on the catalyst surface. However, the experimental observations of graphene orientation are much more complicated. The known experimental facts can be summarized as:

1. In many experiments, continuous single crystalline graphene domains crossing the grain borders of the polycrystalline catalyst surfaces were observed, usually with a change of the Moiré pattern. For example, single crystalline graphene domains have been observed crossing adjacent facets on Ni[125] or Cu[34] surfaces (Figure 12.15a).

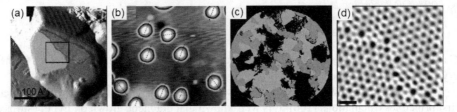

Figure 12.15 (a) Single crystalline graphene domain (light shading) formed on the (110) facet continues to grow over the adjacent (111) facet (darker shading) (Reprinted with permission from [125] © 2010 American Chemical Society); (b) Large scale STM topographs of $7C_6$ and coronene respectively (100 × 85 Å^2) with their identified molecular alignments with respect to Rh(111) indicated by dashed lines. (Reprinted with permission from [108] © 2011 American Chemical Society); (c) Spatial distribution of graphene domains determined from the DF LEEM images. The shaded areas show the graphene domains rotated by 0 and 30° with respect to the underlying Cu lattice, respectively. (Reprinted with permission from [126] © 2012 American Chemical Society); (d) Scanning electron microscope image of graphene transferred onto a TEM grid with over 90% coverage using novel, high-yield methods. (Reprinted with permission from [37] © 2012 Macmillan Publishers Limited)

2. High quality graphene with an average single crystalline domain size greater than that of the catalyst surface has been synthesized. For example, Li et al. have synthesized graphene with a domain size of up to a millimeter on a polycrystalline Cu foil [115].
3. During the initial nucleation stage, small graphene islands are observed well-aligned along a few specific directions on the catalyst surface [92,108] (Figure 12.15b).
4. Evidences of the orientation correlation between graphene and catalyst surface are frequently shown. For example, the abundance of low (∼7°) and high (∼30°) mismatching angles has been observed for graphene growth on Cu(111) and Cu(100) surfaces [61,126] and the graphene grown on Ir(111) surface tends to has a mis-orientation angle of ± 2.6° [127] (Figure 12.15c).
5. The statistics of the domain mismatching angles on both sides of a GB in graphene grown on a Cu surface shows a multi-peak shape, indicating an orientation correlation between adjacent graphene islands [37,57,128,129] (Figure 12.15d).

Among these observations, (1) and (2) undoubtedly indicate that the graphene CVD growth is not a strict epitaxial process in most cases. This can be easily understood by the weak vDW interaction between the graphene wall and the catalyst surfaces. On the other hand, (3–5) point to the opposite conclusion, that the orientation of the synthesized graphene is highly correlated with the crystalline orientation of the catalyst surface, which may be considered to be a consequence of epitaxial growth.

Theoretically, using Cu(111) substrates as example, Zhang et al. [130] first explored the orientation dependent graphene wall-catalyst (**GW–Cu**) and graphene edge-catalyst (**GE–Cu**) interactions. Their study showed that the graphene edge-catalyst interaction dominates the orientation of the growing graphene, and they named such graphene CVD growth behavior as *edge epitaxy* (**EE**).

As shown in Figure 12.16(b), $E_{GW\text{-}Cu}$ is in the range of 30–35 meV/C atom, of which the configuration of no rotation ($\theta = 0°$) corresponds to the strongest binding of 35 meV/C

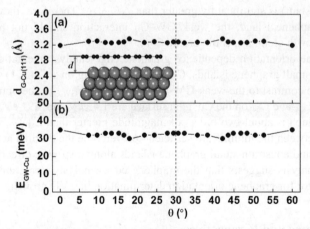

Figure 12.16 *The optimized distance of graphene from the top layer of the Cu(111) surface (a), and the binding energy of the graphene on the Cu(111) surface (b), as a function of rotation angle. (Reprinted with permission from [130] © 2012 American Chemical Society)*

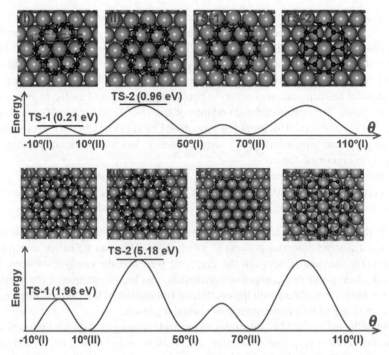

Figure 12.17 Energy profiles of C_{24} (a) and C_{54} (b) clusters rotating on the Cu(111) surfaces. (Reprinted with permission from [130] © 2012 American Chemical Society)

atom and the weakest binding, around 30 meV/C atom, appears at $\theta = 17.9°$. The amplitude of the fluctuation, $\Delta E_{\text{GW-Cu}}$, is only 5 meV/C atom. At a typical graphene CVD growth temperature, ($T \sim 1000°C$), the thermal activation energy $kT \sim 0.1$ eV (where k is the Boltzmann constant) is significantly greater than $\Delta E_{\text{GW-Cu}}$. Therefore, they conclude that for a small graphene island, the weak GW–Cu interaction should not be sufficient to constrain its growth along a specific orientation.

To explore the orientation dependence graphene edge-catalyst interactions, the energy profiles of two small graphene islands, C_{24} and C_{54}, rotating on a Cu(111) surface (Figure 12.17). In sharp contrast to the weak **GW-Cu** interaction ($E_{\text{GW-Cu}}$), the optimized **GE-Cu** interactions of C_{24} and C_{54} on the Cu(111) surface are 1.43 eV and 1.91 eV per edge atom, respectively, which is about two orders of magnitude greater than $E_{\text{GW-Cu}}$. Certainly such high barriers between the minima are sufficient to overcome the thermal activation energy ($kT \sim 0.1$ eV) and constrain small graphene islands along a specific direction. Therefore, the previous analysis suggests that the graphene edge-catalyst interaction, $E_{\text{GE-C}}$, is the dominating factor for graphene orientation determination in CVD growth.

12.5 Summary and Perspectives

Although the CVD growth of graphene has been extensively studied all over the world, the understanding of its mechanism is still very immature. That is mainly due to the fact that (1) theoretical efforts into this study are still limited; (2) the communication between theoreticians and experimentalists is not sufficient; (3) the graphene CVD growth is much more complicated than people's original expectations as it involves so many different catalysts types, so many different C feedstocks, so broad a range of pressure from 10^{-5} Torr to ambient, and so on. The study of graphene CVD growth mechanism is very interdisciplinary; knowledge of surface chemical reaction, surface diffusion, crystal surface reconstruction, theory of crystal growth, and low level physics are required. Because of these difficulties, we can't expect complete understanding of graphene CVD growth in a few years. It will probably be very like the mechanism of carbon nanotubes – most of us still say "it's a mystery" after 20 years of dedicated study. In the meantime, the goal of graphene growth – synthesis of high quality graphene is greatly desired and its realization is probably easier than the great challenge of carbon nanotube synthesis; chirality-selected growth.

So far, we have some understanding of the nucleation of graphene – the nucleation process, the barrier of nucleation and the role of defects on the catalyst surface. We also have some knowledge of the graphene expansion process that follows – the edge formation of graphene islands and the C adsorption onto the graphene edge, but our understanding is still very limited. In our experience, the most challenging and pressing issues regarding graphene CVD growth are:

1. The formation and the healing of defects in graphene CVD growth are crucial for synthesizing high quality graphene. How great the defect concentration in graphene depends on the activity of the catalyst, type of feedstock, the growth temperature, and other experimental parameters should be well understood for experimental design.
2. The differences between so many catalysts used in graphene synthesis should be carefully addressed. Many key processes, such as the capacity of decomposing the feedstock, the mobility of carbon diffusion on them, and the addition/etching of C atoms onto/from the graphene edge, are catalyst-dependent. Only with a thorough understanding of these issues, can we learn how to select proper catalysts for desired graphene synthesis.
3. The role of hydrogen in feedstock is another unknown issue for graphene study. This has been addressed in many experimental studies [131–133]. Experimental evidence shows that the hydrogen can change the morphology of grown graphene islands, the growth rate and the selective synthesis of SLG and FLG. But how these dramatic changes could happen via the addition of hydrogen into the carrier gas is unclear.
4. The evolution of a growing graphene island at different growth conditions is certainly related to the well-studied phenomenon, diffusion limited aggregation (DLA), and thus can be called *diffusion limited growth* (DLG). In the extreme regime of DLG, the graphene islands possess a fractal shape, which has been confirmed by many experimental observations, especially for those samples grown on Cu surface. Beyond this, it has been shown that the six-fold symmetry of graphene also plays a role in the growth, evidenced by the six-branched structure of the graphene islands. In another extreme regime, in attachment limited growth (ALG), the graphene islands normally present a regular hexagonal shape with smooth and straight edges. These edges are normally characterized as zigzag edges. Such an observation implies that the zigzag edge grows slower than others [123]. In the regime between DLG and ALG, the graphene islands show diverse shapes and are size dependent. The evolution of the shape of graphene

islands is associated with carbon feedstock dissociation, C diffusion, and C attachment and thus is experimentally controllable. Using these controls to accelerate the graphene growth and/or improve the graphene quality can only be achieved with a deep insight into the mechanisms.

Overall, there are great challenges and also great opportunities in the study of graphene CVD growth mechanisms. We hope the presently confused experimental facts can be gradually explained and the optimum design for graphene growth can be achieved by the studies of both theoreticians and experimentalists in the near future.

References

[1] K. S. Novoselov, A. K. Geim, S. V. Morozov, D. Jiang, Y. Zhang, S. V. Dubonos, et al., Electric field effect in atomically thin carbon films, *Science*, **306** (5696), 666–669 (2004).

[2] S. Ghosh, I. Calizo, D. Teweldebrhan, E. P. Pokatilov, D. L. Nika, A. A. Balandin, et al., Extremely high thermal conductivity of graphene: Prospects for thermal management applications in nanoelectronic circuits, *Applied Physics Letters*, **92** (15), 151911 (2008).

[3] J. H. Seol, I. Jo, A. L. Moore, L. Lindsay, Z. H. Aitken, M. T. Pettes, et al., Two-dimensional phonon transport in supported graphene, *Science*, **328** (5975), 213–216 (2010).

[4] A. A. Balandin, S. Ghosh, W. Z. Bao, I. Calizo, D. Teweldebrhan, F. Miao and C. N. Lau, Superior thermal conductivity of single-layer graphene, *Nano Letters*, **8** (3), 902–907 (2008).

[5] K. Saito, J. Nakamura and A. Natori, Ballistic thermal conductance of a graphene sheet, *Physical Review B*, **76** (11), 115409 (2007).

[6] N. M. R. Peres, J. M. B. L. dos Santos and T. Stauber, Phenomenological study of the electronic transport coefficients of graphene, *Physical Review B*, **76** (7), 073412 (2007).

[7] K. I. Bolotin, K. J. Sikes, Z. Jiang, M. Klima, G. Fudenberg, J. Hone, et al., Ultrahigh electron mobility in suspended graphene, *Solid State Communications*, **146** (9–10), 351–355 (2008).

[8] J. H. Chen, C. Jang, S. D. Xiao, M. Ishigami and M. S. Fuhrer, Intrinsic and extrinsic performance limits of graphene devices on SiO_2, *Nature Nanotechnology*, **3** (4), 206–209 (2008).

[9] A. K. Geim and K. S. Novoselov, The rise of graphene, *Nature Materials*, **6** (3), 183–191 (2007).

[10] C. Lee, X. D. Wei, J. W. Kysar and J. Hone, Measurement of the elastic properties and intrinsic strength of monolayer graphene, *Science*, **321** (5887), 385–388 (2008).

[11] F. Liu, P. M. Ming and J. Li, Ab initio calculation of ideal strength and phonon instability of graphene under tension, *Physical Review B*, **76** (6), 064120 (2007).

[12] H. Zhao, K. Min and N. R. Aluru, Size and chirality dependent elastic properties of graphene nanoribbons under uniaxial tension, *Nano Letters*, **9** (8), 3012–3015 (2009).

[13] R. Khare, S. L. Mielke, J. T. Paci, S. L. Zhang, R. Ballarini, G. C. Schatz and T. Belytschko, Coupled quantum mechanical/molecular mechanical modeling of the fracture of defective carbon nanotubes and graphene sheets, *Physical Review B*, **75** (7), 075412 (2007).
[14] K. Min and N. R. Aluru, Mechanical properties of graphene under shear deformation, *Applied Physics Letters*, **98** (1), 013113 (2011).
[15] H. Zhao and N. R. Aluru, Temperature and strain-rate dependent fracture strength of graphene, *Journal of Applied Physics*, **108** (6), 064321 (2010).
[16] C. N. R. Rao, A. K. Sood, K. S. Subrahmanyam and A. Govindaraj, Graphene: the new two-dimensional nanomaterial, *Angewandte Chemie-International Edition*, **48** (42), 7752–7777 (2009).
[17] A. H. Castro Neto, F. Guinea, N. M. R. Peres, K. S. Novoselov and A. K. Geim, The electronic properties of graphene, *Reviews of Modern Physics*, **81** (1), 109–162 (2009).
[18] M. J. Allen, V. C. Tung and R. B. Kaner, Honeycomb carbon: a review of graphene, *Chemical Reviews*, **110** (1), 132–145 (2010).
[19] F. Banhart, J. Kotakoski and A. V. Krasheninnikov, Structural defects in graphene, *ACS Nano*, **5** (1), 26–41 (2011).
[20] A. J. Stone and D. J. Wales, Theoretical studies of icosahedral C_{60} and some related species, *Chemical Physics Letters*, **128** (5–6), 501–503 (1986).
[21] D. A. J. Ma, A. Michaelides and E. Wang, Stone–Wales defects in graphene and other planar sp^2-bonded materials, *Physical Review B*, **80**, 033407 (2009).
[22] D. W. Boukhvalov and M. I. Katsnelson, Chemical functionalization of graphene with defects, *Nano Letters*, **8** (12), 4373–4379 (2008).
[23] J. C. Meyer, C. Kisielowski, R. Erni, M. D. Rossell, M. F. Crommie and A. Zettl, Direct imaging of lattice atoms and topological defects in graphene membranes, *Nano Letters*, **8** (11), 3582–3586 (2008).
[24] M. H. Gass, U. Bangert, A. L. Bleloch, P. Wang, R. R. Nair and A. K. Geim, Free-standing graphene at atomic resolution, *Nature Nanotechnology*, **3** (11), 676–681 (2008).
[25] M. M. Ugeda, I. Brihuega, F. Guinea and J. M. Gomez-Rodriguez, Missing atom as a source of carbon magnetism, *Physical Review Letters*, **104** (9), 096804 (2010).
[26] A. A. El-Barbary, R. H. Telling, C. P. Ewels, M. I. Heggie and P. R. Briddon, Structure and energetics of the vacancy in graphite, *Physical Review B*, **68** (14), 144107 (2003).
[27] C. Z. Wang, E. Yoon, N. M. Hwang, D. Y. Kim and K. M. Ho, D(i)ffusion, coalescence, and reconstruction of vacancy defects in graphene layers, *Physical Review Letters*, **95** (20), 205501 (2005).
[28] A. Hashimoto, K. Suenaga, A. Gloter, K. Urita and S. Iijima, Direct evidence for atomic defects in graphene layers, *Nature*, **430** (7002), 870–873 (2004).
[29] A. V. Krasheninnikov, P. O. Lehtinen, A. S. Foster and R. M. Nieminen, Bending the rules: Contrasting vacancy energetics and migration in graphite and carbon nanotubes, *Chemical Physics Letters*, **418** (1–3), 132–136 (2006).
[30] C. O. Girit, J. C. Meyer, R. Erni, M. D. Rossell, C. Kisielowski, L. Yang, et al., Graphene at the edge: stability and dynamics, *Science*, **323** (5922), 1705–1708 (2009).

[31] O. Lehtinen, J. Kotakoski, A. V. Krasheninnikov, A. Tolvanen, K. Nordlund and J. Keinonen, Effects of ion bombardment on a two-dimensional target: Atomistic simulations of graphene irradiation, *Physical Review B*, **81** (15), 153401 (2010).

[32] J. Kotakoski, A. V. Krasheninnikov, U. Kaiser and J. C. Meyer, From point defects in graphene to two-dimensional amorphous carbon, *Physical Review Letters*, **106** (10), 105505 (2011).

[33] M. M. Ugeda, I. Brihuega, F. Hiebel, P. Mallet, J. Y. Veuillen, J. M. Gomez-Rodriguez and F. Yndurain, Electronic and structural characterization of divacancies in irradiated graphene, *Physical Review B*, **85** (12), 121402 (2012).

[34] Q. K. Yu, L. A. Jauregui, W. Wu, R. Colby, J. F. Tian, Z. H. Su, et al., Control and characterization of individual grains and grain boundaries in graphene grown by chemical vapour deposition, *Nature Materials*, **10** (6), 443–449 (2011).

[35] O. V. Yazyev and S. G. Louie, Electronic transport in polycrystalline graphene, *Nature Materials*, **9** (10), 806–809 (2010).

[36] O. V. Yazyev and S. G. Louie, Topological defects in graphene: Dislocations and grain boundaries, *Physical Review B*, **81** (19), 195420 (2010).

[37] P. Y. Huang, C. S. Ruiz-Vargas, A. M. van der Zande, W. S. Whitney, M. P. Levendorf, J. W. Kevek, et al., Grains and grain boundaries in single-layer graphene atomic patchwork quilts, *Nature*, **469** (7330), 38–392 (2011).

[38] R. Grantab, V. B. Shenoy and R. S. Ruoff, Anomalous strength characteristics of tilt grain boundaries in graphene, *Science*, **330** (6006), 946–948 (2010).

[39] A. Mesaros, S. Papanikolaou, C. F. J. Flipse, D. Sadri and J. Zaanen, Electronic states of graphene grain boundaries, *Physical Review B*, **82** (20), 205119 (2010).

[40] J. Lahiri, Y. Lin, P. Bozkurt, I. I. Oleynik and M. Batzill, An extended defect in graphene as a metallic wire, *Nature Nanotechnology*, **5** (5), 326–329 (2010).

[41] A. C. Ferrari, J. C. Meyer, V. Scardaci, C. Casiraghi, M. Lazzeri, F. Mauri, et al., Raman spectrum of graphene and graphene layers, *Physical Review Letters*, **97** (18), 187401 (2006).

[42] K. S. Novoselov, A. K. Geim, S. V. Morozov, D. Jiang, M. I. Katsnelson, I. V. Grigorieva, et al., Two-dimensional gas of massless Dirac fermions in graphene, *Nature*, **438** (7065), 197–200 (2005).

[43] K. S. Novoselov, D. Jiang, F. Schedin, T. J. Booth, V. V. Khotkevich, S. V. Morozov and A. K. Geim, Two-dimensional atomic crystals, *Proceedings of the National Academy of Sciences of the United States of America*, **102** (30), 10451–10453 (2005).

[44] D. Yang, A. Velamakanni, G. Bozoklu, S. Park, M. Stoller, R. D. Piner, et al., Chemical analysis of graphene oxide films after heat and chemical treatments by X-ray photoelectron and micro-Raman spectroscopy, *Carbon*, **47** (1), 145–152 (2009).

[45] S. Stankovich, R. D. Piner, X. Q. Chen, N. Q. Wu, S. T. Nguyen and R. S. Ruoff, Stable aqueous dispersions of graphitic nanoplatelets via the reduction of exfoliated graphite oxide in the presence of poly(sodium 4-styrenesulfonate), *Journal of Materials Chemistry*, **16** (2), 155–158 (2006).

[46] G. Eda, G. Fanchini and M. Chhowalla, Large-area ultrathin films of reduced graphene oxide as a transparent and flexible electronic material, *Nature Nanotechnology*, **3** (5), 270–274 (2008).

[47] S. Stankovich, D. A. Dikin, R. D. Piner, K. A. Kohlhaas, A. Kleinhammes, Y. Jia, et al., Synthesis of graphene-based nanosheets via chemical reduction of exfoliated graphite oxide, *Carbon*, **45** (7), 1558–1565 (2007).

[48] C. Berger, Z. M. Song, X. B. Li, X. S. Wu, N. Brown, C. Naud, et al., Electronic confinement and coherence in patterned epitaxial graphene, *Science*, **312** (5777), 1191–1196 (2006).

[49] W. A. de Heer, C. Berger, X. S. Wu, P. N. First, E. H. Conrad, X. B. Li, et al., Epitaxial graphene, *Solid State Communications*, **143** (1–2), 92–100 (2007).

[50] T. Ohta, A. Bostwick, T. Seyller, K. Horn and E. Rotenberg, Controlling the electronic structure of bilayer graphene, *Science*, **313** (5789), 951–954 (2006).

[51] C. Berger, Z. M. Song, T. B. Li, X. B. Li, A. Y. Ogbazghi, R. Feng, et al., Ultrathin epitaxial graphite: 2D electron gas properties and a route toward graphene-based nanoelectronics, *Journal of Physical Chemistry B*, **108** (52), 19912–19916 (2004).

[52] Z. Z. Sun, Z. Yan, J. Yao, E. Beitler, Y. Zhu and J. M. Tour, Growth of graphene from solid carbon sources, *Nature*, **468** (7323), 549–552 (2010).

[53] P. W. Sutter, J. I. Flege and E. A. Sutter, Epitaxial graphene on ruthenium, *Nature Materials*, **7** (5), 406–411 (2008).

[54] A. T. N'Diaye, J. Coraux, T. N. Plasa, C. Busse and T. Michely, Structure of epitaxial graphene on Ir(111), *New Journal of Physics*, **10**, 043033 (2008).

[55] S. Y. Kwon, C. V. Ciobanu, V. Petrova, V. B. Shenoy, J. Bareno, V. Gambin, et al., Growth of semiconducting graphene on palladium, *Nano Letters*, **9** (12), 3985–3990 (2009).

[56] A. Reina, S. Thiele, X. T. Jia, S. Bhaviripudi, M. S. Dresselhaus, J. A. Schaefer and J. Kong, Growth of large-area single- and bi-layer graphene by controlled carbon precipitation on polycrystalline Ni surfaces, *Nano Research*, **2** (6), 509–516 (2009).

[57] X. S. Li, C. W. Magnuson, A. Venugopal, J. H. An, J. W. Suk, B. Y. Han, et al., Graphene films with large domain size by a two-step chemical vapor deposition process, *Nano Letters*, **10** (11), 4328–4334 (2010).

[58] X. S. Li, Y. W. Zhu, W. W. Cai, M. Borysiak, B. Y. Han, D. Chen, et al., Transfer of large-area graphene films for high-performance transparent conductive electrodes, *Nano Letters*, **9** (12), 4359–4363 (2009).

[59] X. S. Li, W. W. Cai, L. Colombo and R. S. Ruoff, Evolution of graphene growth on Ni and Cu by carbon isotope labeling, *Nano Letters*, **9** (12), 4268–4272 (2009).

[60] X. S. Li, W. W. Cai, J. H. An, S. Kim, J. Nah, D. X. Yang, et al., Large-area synthesis of high-quality and uniform graphene films on copper foils, *Science*, **324** (5932), 1312–1314 (2009).

[61] L. Gao, J. R. Guest and N. P. Guisinger, Epitaxial graphene on Cu(111), *Nano Letters*, **10** (9), 351–3516 (2010).

[62] J. M. Wofford, S. Nie, K. F. McCarty, N. C. Bartelt and O. D. Dubon, Graphene islands on Cu foils: the interplay between shape, orientation, and defects, *Nano Letters*, **10** (12), 4890–4896 (2010).

[63] J. Coraux, A. T. N'Diaye, C. Busse and T. Michely, Structural coherency of graphene on Ir(111), *Nano Letters*, **8** (2), 565–570 (2008).

[64] P. Sutter, J. T. Sadowski and E. Sutter, Graphene on Pt(111): Growth and substrate interaction, *Physical Review B*, **80** (24), 245411 (2009).

[65] E. Sutter, P. Albrecht and P. Sutter, Graphene growth on polycrystalline Ru thin films, *Applied Physics Letters*, **95** (13), 133109 (2009).

[66] Q. J. Wang and J. G. Che, Origins of distinctly different behaviors of Pd and Pt contacts on graphene, *Physical Review Letters*, **103** (6), 066802 (2009).

[67] A. B. Preobrajenski, M. L. Ng, A. S. Vinogradov and N. Martensson, Controlling graphene corrugation on lattice-mismatched substrates, *Physical Review B*, **78** (7), 073401 (2008).

[68] A. Reina, X. T. Jia, J. Ho, D. Nezich, H. B. Son, V. Bulovic, et al., Large area, few-layer graphene films on arbitrary substrates by chemical vapor deposition, *Nano Letters*, **9** (1), 30–35 (2009).

[69] D. Eom, D. Prezzi, K. T. Rim, H. Zhou, M. Lefenfeld, S. Xiao, et al., Structure and electronic properties of graphene nanoislands on Co(0001), *Nano Letters*, **9** (8), 2844–2848 (2009).

[70] E. Sutter, D. P. Acharya, J. T. Sadowski and P. Sutter, Scanning tunneling microscopy on epitaxial bilayer graphene on ruthenium (0001), *Applied Physics Letters*, **94** (13), 133101 (2009).

[71] J. Lu, J. X. Yang, J. Z. Wang, A. L. Lim, S. Wang and K. P. Loh, One-pot synthesis of fluorescent carbon nanoribbons, nanoparticles, and graphene by the exfoliation of graphite in ionic liquids, *ACS Nano*, **3** (8), 2367–2375 (2009).

[72] P. K. Ang, S. Wang, Q. L. Bao, J. T. L. Thong and K. P. Loh, High-throughput synthesis of graphene by intercalation – exfoliation of graphite oxide and study of ionic screening in graphene transistor, *ACS Nano*, **3** (11), 3587–3594 (2009).

[73] D. J. Cheng, G. Barcaro, J. C. Charlier, M. Hou and A. Fortunelli, Homogeneous nucleation of graphitic nanostructures from carbon chains on Ni(111), *Journal of Physical Chemistry C*, **115** (21), 10537–10543 (2011).

[74] C. Klink, I. Stensgaard, F. Besenbacher and E. Laegsgaard, An STM Study of carbon-induced structures on Ni(111) – evidence for a carbidic-phase clock reconstruction, *Surface Science*, **342** (1–3), 250–260 (1995).

[75] Y. S. Dedkov, A. M. Shikin, V. K. Adamchuk, S. L. Molodtsov, C. Laubschat, A. Bauer and G. Kaindl, Intercalation of copper underneath a monolayer of graphite on Ni(111), *Physical Review B*, **64** (3), 035405 (2001).

[76] A. M. Shikin, G. V. Prudnikova, V. K. Adamchuk, F. Moresco and K. H. Rieder, Surface intercalation of gold underneath a graphite monolayer on Ni(111) studied by angle-resolved photoemission and high-resolution electron-energy-loss spectroscopy, *Physical Review B*, **62** (19), 13202–13208 (2000).

[77] A. Varykhalov and O. Rader, Graphene grown on Co(0001) films and islands: Electronic structure and its precise magnetization dependence, *Physical Review B*, **80** (3), 035437 (2009).

[78] Y. Lee, S. Bae, H. Jang, S. Jang, S. E. Zhu, S. H. Sim, et al., Wafer-scale synthesis and transfer of graphene films, *Nano Letters*, **10** (2), 490–493 (2010).

[79] M. Sicot, S. Bouvron, O. Zander, U. Rudiger, Y. S. Dedkov and M. Fonin, Nucleation and growth of nickel nanoclusters on graphene Moire on Rh(111), *Applied Physics Letters*, **97** (7), 093115 (2010).

[80] B. Wang, M. Caffio, C. Bromley, H. Fruchtl and R. Schaub, Coupling epitaxy, chemical bonding, and work function at the local scale in transition metal-supported graphene, *ACS Nano*, **4** (10), 5773–5782 (2010).

[81] A. T. N'Diaye, S. Bleikamp, P. J. Feibelman and T. Michely, Two-dimensional Ir cluster lattice on a graphene moire on Ir(111), *Physical Review Letters*, **97** (21), 043033 (2006).
[82] J. Coraux, A. T. N'Diaye, M. Engler, C. Busse, D. Wall, N. Buckanie, et al., Growth of graphene on Ir(111), *New Journal of Physics*, **11** (023006) (2009).
[83] M. Gao, Y. Pan, L. Huang, H. Hu, L. Z. Zhang, H. M. Guo, et al., Epitaxial growth and structural property of graphene on Pt(111), *Applied Physics Letters*, **98** (3), 033101 (2011).
[84] T. A. Land, T. Michely, R. J. Behm, J. C. Hemminger and G. Comsa, STM investigation of single layer graphite structures produced on Pt(111) by hydrocarbon decomposition, *Surface Science*, **264** (3), 261–270 (1992).
[85] G. D. Ruan, Z. Z. Sun, Z. W. Peng and J. M. Tour, Growth of graphene from food, insects, and waste, *ACS Nano*, **5** (9), 7601–7607 (2011).
[86] Y. G. Yao, Z. Li, Z. Y. Lin, K. S. Moon, J. Agar and C. P. Wong, Controlled growth of multilayer, few-layer, and single-layer graphene on metal substrates, *Journal of Physical Chemistry C*, **115** (13), 5232–5238 (2011).
[87] T. Iwasaki, H. J. Park, M. Konuma, D. S. Lee, J. H. Smet and U. Starke, Long-range ordered single-crystal graphene on high-quality heteroepitaxial Ni thin films grown on MgO(111), *Nano Letters*, **11** (1), 79–84 (2011).
[88] S. Marchini, S. Gunther and J. Wintterlin, Scanning tunneling microscopy of graphene on Ru(0001), *Physical Review B*, **76** (7), 075429 (2007).
[89] E. Starodub, S. Maier, I. Stass, N. C. Bartelt, P. J. Feibelman, M. Salmeron and K. F. McCarty, Graphene growth by metal etching on Ru(0001), *Physical Review B*, **80** (23), 235422 (2009).
[90] Y. Pan, H. G. Zhang, D. X. Shi, J. T. Sun, S. X. Du, F. Liu and H. J. Gao, Highly ordered, millimeter-scale, continuous, single-crystalline graphene monolayer formed on Ru (0001), *Advanced Materials*, **21** (27), 2739–2739 (2009).
[91] Y. Cui, Q. A. Fu, H. Zhang and X. H. Bao, Formation of identical-size graphene nanoclusters on Ru(0001), *Chemical Communications*, **47** (5), 1470–1472 (2011).
[92] B. I. Yakobson and F. Ding, Observational geology of graphene, at the nanoscale, *ACS Nano*, **5** (3), 1569–1574 (2011).
[93] F. Ding, K. Bolton and A. Rosen, Nucleation and growth of single-walled carbon nanotubes: A molecular dynamics study, *Journal of Physical Chemistry B*, **108** (45), 17369–17377 (2004).
[94] A. Gorbunov, O. Jost, W. Pompe and A. Graff, Solid-liquid-solid growth mechanism of single-wall carbon nanotubes, *Carbon*, **40** (1), 113–118 (2002).
[95] G. R. Desiraju and J. Hulliger, Current opinion in solid state and materials science molecular crystals and materials, *Current Opinion in Solid State & Materials Science*, **5** (2–3), 105–106 (2001).
[96] Z. C. Li, P. Wu, C. X. Wang, X. D. Fan, W. H. Zhang, X. F. Zhai, et al., Low-temperature growth of graphene by chemical vapor deposition using solid and liquid carbon sources, *ACS Nano*, **5** (4), 3385–3390 (2011).
[97] H. Chen, W. G. Zhu and Z. Y. Zhang, Contrasting behavior of carbon nucleation in the initial stages of graphene epitaxial growth on stepped metal surfaces, *Physical Review Letters*, **104** (18), 186101 (2010).

[98] W. H. Zhang, P. Wu, Z. Y. Li and J. L. Yang, First-principles thermodynamics of graphene growth on Cu surfaces, *Journal of Physical Chemistry C*, **115** (36), 17782–17787 (2011).

[99] J. Lahiri, T. Miller, L. Adamska, I. I. Oleynik and M. Batzill, Graphene growth on Ni(111) by transformation of a surface carbide, *Nano Letters*, **11** (2), 518–522 (2011).

[100] S. Riikonen, A. V. Krasheninnikov, L. Halonen and R. M. Nieminen, The role of stable and mobile carbon adspecies in copper-promoted graphene growth, *Journal of Physical Chemistry C*, **116** (9), 5802–5809 (2012).

[101] P. Wu, W. H. Zhang, Z. Y. Li, J. L. Yang and J. G. Hou, Communication: Coalescence of carbon atoms on Cu (111) surface: Emergence of a stable bridging-metal structure motif, *Journal of Chemical Physics*, **133** (7), 071101 (2010).

[102] R. G. Van Wesep, H. Chen, W. G. Zhu and Z. Y. Zhang, Communication: Stable carbon nanoarches in the initial stages of epitaxial growth of graphene on Cu(111), *Journal of Chemical Physics*, **134** (17), 171105 (2011).

[103] J. F. Gao, Q. H. Yuan, H. Hu, J. J. Zhao and F. Ding, Formation of carbon clusters in the initial stage of chemical vapor deposition graphene growth on Ni(111) surface, *Journal of Physical Chemistry C*, **115** (36), 17695–17703 (2011).

[104] A. Zangwill and D. D. Vvedensky, Novel growth mechanism of epitaxial graphene on metals, *Nano Letters*, **11** (5), 2092–2095 (2011).

[105] E. Loginova, N. C. Bartelt, P. J. Feibelman and K. F. McCarty, Evidence for graphene growth by C cluster attachment, *New Journal of Physics*, **10**, 093026 (2008).

[106] P. Wu, H. J. Jiang, W. H. Zhang, Z. Y. Li, Z. H. Hou and J. L. Yang, Lattice mismatch induced nonlinear growth of graphene, *Journal of the American Chemical Society*, **134** (13), 6045–6051 (2012).

[107] Q. H. Yuan, J. F. Gao, H. B. Shu, J. J. Zhao, X. S. Chen and F. Ding, Magic carbon clusters in the chemical vapor deposition growth of graphene, *Journal of the American Chemical Society*, **134** (6), 2970–2975 (2012).

[108] B. Wang, X. F. Ma, M. Caffio, R. Schaub and W. X. Li, Size-selective carbon nanoclusters as precursors to the growth of epitaxial graphene, *Nano Letters*, **11** (2), 424–430 (2011).

[109] P. S. E. Y. J. Lu, C. K. Gan, P. Wu and K. P. Loh, Transforming C60 molecules into graphene quantum dots, *Nature Nanotechnology*, **6**, 247–252 (2011).

[110] I. V. Markov, *Crystal Growth for Beginners: Fundamentals of Nucleation, Crystal Growth, and Epitaxy*, 2nd edn, World Scientific Publishing Co. Pte. Ltd: Singapore (2003).

[111] J. F. Gao, J. Yip, J. J. Zhao, B. I. Yakobson and F. Ding, Graphene nucleation on transition metal surface: structure transformation and role of the metal step edge, *Journal of the American Chemical Society*, **134** (22), 9534–9534 (2012).

[112] E. Loginova, N. C. Bartelt, P. J. Feibelman and K. F. McCarty, Factors influencing graphene growth on metal surfaces, *New Journal of Physics*, **11**, 063046 (2009).

[113] K. F. McCarty, P. J. Feibelman, E. Loginova and N. C. Bartelt, Kinetics and thermodynamics of carbon segregation and graphene growth on Ru(0001), *Carbon*, **47** (7), 1806–1813 (2009).

[114] S. Nie, J. M. Wofford, N. C. Bartelt, O. D. Dubon and K. F. McCarty, Origin of the mosaicity in graphene grown on Cu(111), *Physical Review B*, **84** (15), 155425 (2011).

[115] X. S. Li, C. W. Magnuson, A. Venugopal, R. M. Tromp, J. B. Hannon, E. M. Vogel, et al., Large-area graphene single crystals grown by low-pressure chemical vapor deposition of methane on copper, *Journal of the American Chemical Society*, **133** (9), 2816–2819 (2011).

[116] Q. H. Yuan, H. Hu, J. F. Gao, F. Ding, Z. F. Liu and B. I. Yakobson, Upright standing graphene formation on substrates, *Journal of the American Chemical Society*, **133** (40), 16072–16079 (2011).

[117] D. Subramaniam, F. Libisch, Y. Li, C. Pauly, V. Geringer, R. Reiter, et al., Wavefunction mapping of graphene quantum dots with soft confinement, *Physical Review Letters*, **108** (4), 046801 (2012).

[118] P. Lacovig, M. Pozzo, D. Alfè, P. Vilmercati, A. Baraldi and S. Lizzit, Growth of dome-shaped carbon nanoislands on ir(111): the intermediate between carbidic clusters and quasi-free-standing graphene, *Physical Review Letters*, **103** (16), 166101 (2009).

[119] J. F. Gao, J. J. Zhao and F. Ding, Transition metal surface passivation induced graphene edge reconstruction, *Journal of the American Chemical Society*, **134** (14), 6204–6209 (2012).

[120] J. M. H. Kroes, M. A. Akhukov, J. H. Los, N. Pineau and A. Fasolino, Mechanism and free-energy barrier of the type-57 reconstruction of the zigzag edge of graphene, *Physical Review B*, **83** (16), 165411 (2011).

[121] T. Wassmann, A. P. Seitsonen, A. M. Saitta, M. Lazzeri and F. Mauri, Structure, stability, edge states, and aromaticity of graphene ribbons, *Physical Review Letters*, **101** (9), 096402 (2008).

[122] P. Koskinen, S. Malola and H. Häkkinen, Self-passivating edge reconstructions of graphene, *Physical Review Letters*, **101** (11), 115502 (2008).

[123] H. B. Shu, X. S. Chen, X. M. Tao and F. Ding, Edge Structural stability and kinetics of graphene chemical vapor deposition growth, *ACS Nano*, **6** (4), 3243–3250 (2012).

[124] Y. L. V. I. Artyukhov, and B. I. Yakobson, Equilibrium at the edge and atomistic mechanisms of graphene growth, *Proceedings of the National Academy of Sciences of the United States of America*, doi: 10.1073/pnas.1207519109 (2012).

[125] Y. Murata, V. Petrova, B. B. Kappes, A. Ebnonnasir, I. Petrov, Y. H. Xie, et al., Moire superstructures of graphene on faceted nickel islands, *ACS Nano*, **4** (11), 6509–6514 (2010).

[126] Y. Ogawa, B. S. Hu, C. M. Orofeo, M. Tsuji, K. Ikeda, S. Mizuno, et al., Domain structure and boundary in single-layer graphene grown on Cu(111) and Cu(100) films, *Journal of Physical Chemistry Letters*, **3** (2), 219–226 (2012).

[127] A. T. N'Diaye, J. Coraux, T. N. Plasa, C. Busse and T. Michely, Structure of epitaxial graphene on Ir(111), *New Journal of Physics*, **10**, 43033 (2008).

[128] K. Kim, Z. Lee, W. Regan, C. Kisielowski, M. F. Crommie and A. Zettl, Grain boundary mapping in polycrystalline graphene, *ACS Nano*, **5** (3), 2142–2146 (2011).

[129] J. H. An, E. Voelkl, J. W. Suk, X. S. Li, C. W. Magnuson, L. F. Fu, et al., Domain (grain) boundaries and evidence of "twinlike" structures in chemically vapor deposited grown graphene, *ACS Nano*, **5** (4), 2433–2439 (2011).

[130] X. Y. Zhang, Z. W. Xu, H. Li, H. Xin and F. Ding, How the orientation of graphene is determined during chemical vapor deposition growth, *Journal of Physical Chemistry Letters*, **3** (19), 2822–2827 (2012).

[131] I. Vlassiouk, M. Regmi, P. F. Fulvio, S. Dai, P. Datskos, G. Eres and S. Smirnov, Role of hydrogen in chemical vapor deposition growth of large single-crystal graphene, *ACS Nano*, **5** (7), 6069–6076 (2011).

[132] T. R. Wu, G. Q. Ding, H. L. Shen, H. M. Wang, L. Sun, D. Jiang, et al., Triggering the continuous growth of graphene toward millimeter-sized grains, *Advanced Functional Materials*, DOI: 10.1002/adfm.201201577 (2012).

[133] L. X. Liu, H. L. Zhou, R. Cheng, W. J. Yu, Y. Liu, Y. P. Chen, et al., High-yield chemical vapor deposition growth of high-quality large-area AB-stacked bilayer graphene, *ACS Nano*, **6** (9), 8241–8249 (2012).

13

From Graphene to Graphene Oxide and Back

Xingfa Gao,[a] Yuliang Zhao,[a] and Zhongfang Chen[b]

[a]Key Laboratory for Biomedical Effects of Nanomaterials and Nanosafety, Institute of High Energy Physics, Chinese Academy of Sciences, China
[b]Department of Chemistry, Institute for Functional Nanomaterials, University of Puerto Rico, USA

13.1 Introduction

Despite its short history of realization [1], graphene has attracted much attention because of its intriguing physical properties, as well as its potential in diverse applications [2–4]. It remains a challenge to produce high-quality graphene on a large scale. Several laboratory methods have been proposed, each with its own advantages and shortcomings [5–10]. Among them, the chemical oxidation of graphite followed by the reduction treatment yields chemically converted graphene. This synthetic method has intriguing features: it is cheap and easily scalable to high-volume production. Particularly, being a solution-based approach, it is well-suited for chemical modification and subsequent processing. Consequently, the reduction of graphene oxide is widely used for device applications such as preparation of graphene-based composites, thin films, and paper-like materials.

Figure 13.1 summarizes the chemical procedures for the syntheses of chemically converted graphene from graphite. Three steps are mainly involved: oxidation to transfer graphite to graphite oxide, exfoliation to form graphene oxide (single-layered graphite oxide), and reduction to form chemically converted graphene [11–13]. Four kinds of oxygen groups are known to exist in graphene oxide: epoxide (-O-), hydroxyl (-OH), carbonyl (-C=O) and carboxyl (-COOH) [14–17]. Epoxide and hydroxyl, located on the basal plane

---

Graphene Chemistry: Theoretical Perspectives, First Edition. Edited by De-en Jiang and Zhongfang Chen.
© 2013 John Wiley & Sons, Ltd. Published 2013 by John Wiley & Sons, Ltd.

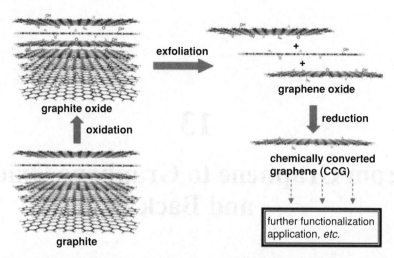

Figure 13.1 *Schematic illustration for the chemical processes from graphite to chemically converted graphene by oxidation, exfoliation and reduction procedures*

of graphene oxide, are the major components; carbonyl and carboxyl, distributed at the edges of graphene oxide, are minor. It has been reported that the presence of five-membered-ring and six-membered-ring lactols at the edges of graphene oxide are most likely responsible for the minor oxygen-containing impurities of graphene oxide [18]. After reduction treatment, the oxygen atoms in graphene oxide are greatly reduced. However, there still exists a large amount of residual oxygen functionality, structural disorder and defects inherited from the pristine graphene oxide [19, 20]. To improve production of graphene-based functional materials via chemical reduction, it is therefore of great importance to know the underlying mechanisms for graphene oxidation, the structures of graphene oxides and the mechanisms for the reduction of graphene oxide. Great effort has been made in elucidating the structures of graphene oxides using computational approaches [21–27]. Here, we will review theoretical studies on the topic of graphene chemistry. Especially, we will focus on theoretical insights into the oxidation mechanisms of graphene to form graphene oxide and the reduction mechanisms of graphene oxide back to chemically converted graphene.

13.2 From Graphene to Graphene Oxide

13.2.1 Modeling Using Cluster Models

13.2.1.1 Oxidative Etching of Armchair Edges

Using the structure of Figure 13.2(a) as the model, Sendt and Haynes studied the chemical reactions of O_2 on the dehydrogenated armchair-edge of graphene [28]. Calculations at the B3LYP/6-31G(d) level of theory showed that the bond length of C_2-C_3 is 1.24 Å, equal to that of the triple bond of benzyne. Therefore, the dehydrogenation converts the edge to triple bonds (Figure 13.2a), satisfying the octet rule. Figure 13.2(b) presents the reaction energy profile for O_2 chemisorption on the C_2-C_3. It suggests that the formation of quinone

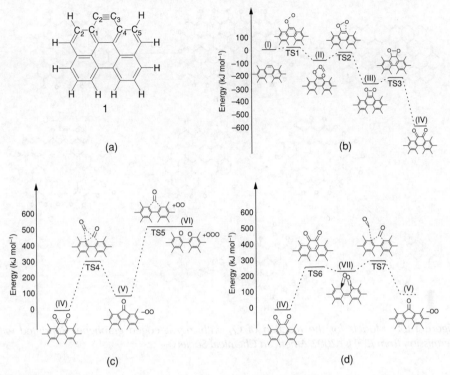

Figure 13.2 Potential energy surface for O_2 chemisorption on C_2 and C_3, calculated at the level of B3LYP/6-31G(d). Reprinted with permission from [28] © 2005 Elsevier

structure IV is thermodynamically favorable, with a small energy barrier (4.3 kcal/mol) and a large exothermicity (−137.6 kcal/mol). Structures II and III exist on the transformation path, but both intermediates are unlikely to have a significant lifetime because of the small reaction energy barriers of TS2 and TS3 (Figure 13.2b). The reaction occurs mainly at a triplet state from I to II, and a singlet from III to IV. The surface crossing from triplet to singlet takes place near the geometry of TS2, which the triplet and singlet species have the same energy.

Quinone structure IV is possible to undergo two successive CO desorptions to give the graphene diradical VI. As shown in Figure 13.2(c), the formation of the five-membered ring ketone (V) from IV (i.e., the first CO desorption) is 20.2 kcal/mol endothermic with an energy barrier of 70.5 kcal/mol. However, the second CO desorption is approximately 103.6 kcal/mol endothermic with a significant high barrier and a negligible reverse barrier. Therefore, the second CO desorption occurs only at high temperatures. The ketone structure (V) can also be formed via a stepwise CO desorption as shown in Figure 13.2(d).

13.2.1.2 Oxidative Etching of Zigzag Edges

Zhu et al. proposed the oxidation and gasification mechanism for graphene fragments with bare zigzag edges (Figure 13.3) [29]. Owning to the large exothermicity for O_2

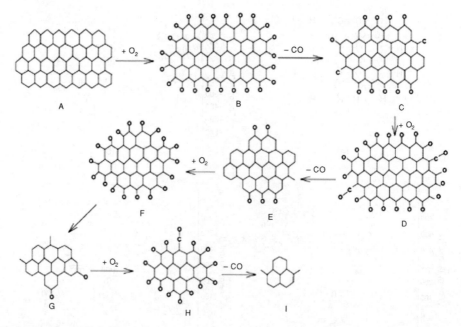

Figure 13.3 Models for the reaction of O_2 with zigzag-edged graphene. Reprinted with permission from [29] © 2002 American Chemical Society

chemisorption on the bare edges (A), nearly all the edge sites can chemisorb O atoms (B). The *o*-quinone C—C bonds of B, which are calculated to have lowed bond strengths provide rational starting points for gasification. The CO desorption occurring on B reduces the carbon surface and results in new active sites on C. Therefore, more O atoms will be captured to form new *o*-quinone units in D. As the reaction proceeds from D to E to F to G to H to I, the graphene flake shrinks until it is completely consumed.

This *o*-quinone-oriented O_2-gasification model for graphene flakes is applicable in conditions where the O_2 partial pressures (P_{O2}) are medium. Figure 13.4 shows the unified model for graphene flakes at different P_{O2}s. At a low P_{O2}, *o*-quinone units, which are more subject to CO desorption, will not form. Instead, O_2 dissociatively chemisorbs on A, giving B with semiquinone units (Figure 13.4). The release of CO from these semiquinone structures requires more energy than that from *o*-quinone owing to the larger bond energy of semiquinone (316 kJ/mol). At medium P_{O2}, the reaction takes place via the mechanism of Figure 13.3, forming *o*-quinone units as gasification centers. If the P_{O2} is further increased, the formation of epoxy groups on the basal plane of graphene flake will occur, which add more active sites and accelerate the gasification process.

13.2.1.3 Linear Oxidative Unzipping

To rationalize the experimentally observed fault lines and cracks of on graphene oxide, Li *et al.* have proposed the linear oxidation mechanism [30]. The application of this mechanism to a coronene molecule is shown in Figure 13.5. The addition of a single O

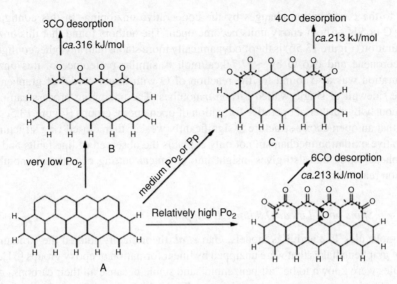

Figure 13.4 The unified O_2 gasification mechanism proposed by Zhu et al. Reprinted with permission from [29] © 2002 American Chemical Society

atom to coronene forms an epoxy group (Figure 13.5a). In this process, the total energy of the system is lowered by ~55.3 kcal/mol. Simultaneously, the epoxy C–C bond is stretched from 1.42 Å (the pristine carbon lattice) to 1.58 Å. It was found that the addition of the second O atom takes place more preferably at the opposite site of the epoxy group (Figure 13.5b). Importantly, when double epoxy structures are formed on the opposite sites of a coronene hexagonal ring, the formation energy of one epoxy is increased by the other,

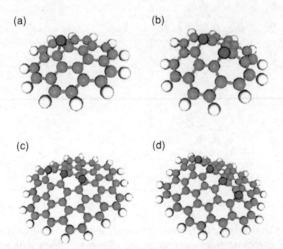

Figure 13.5 Coronene molecules attached with (a) one, (b) two, (c) three, and (d) four epoxy groups, respectively

owing to the confirmation changes by the cooperative unzipping. In this configuration, the C–C bonds of both epoxy units become open. The authors found that this unzipping configuration (Figure 13.5b) is thermodynamically more stable than any other configuration for a coronene and two O by ~27.7 kcal/mol. A similar preference of this unzipping configuration was also found for the reaction of O with larger model of graphene flake and the sidewall of single-walled carbon nanotubes. The unzipping process can proceed continuously by further cooperative formation of open epoxy groups (Figure 13.5c and d). Given that an open epoxy structure is significantly weaker than closed-ring structures, the cooperative oxidation mechanism not only explains the observed of line faults and cracks of graphene oxide but also gives insight into chemical cutting of carbon nanotubes via oxidation reactions.

13.2.1.4 Spins upon Linear Oxidative Unzipping

Using molecules **1**, **2** and **3** as models, Gao *et al.* theoretically studied the structures and spins of graphene flakes that were unzipped by linear formation of epoxy groups [31]. These molecules were known to be "all-benzenoid" and stable because all their carbons could be put in isolated sextets (i.e., a circle as shown in Figure 13.6), according to Clar's sextet rule. The linear oxidation of **1** along the dash line led to five products, which had different spins at ground states. Among them, the triplet one had the lowest energy. The linear oxidations of **2** and **3** led to the quintet and triplet as the lowest-energy isomers, respectively. The following general conclusion was drawn. When an all-benzenoid graphene fragment is linearly unzipped into oxygen-joined fragments, the oxidized benzenoid rings (aromatic sextets) selectively adopt the low-spin ($\Delta S = 0$) or high-spin conformation ($\Delta S = 1$) to yield the thermally most stable isomer. The selection of the conformation depends simply

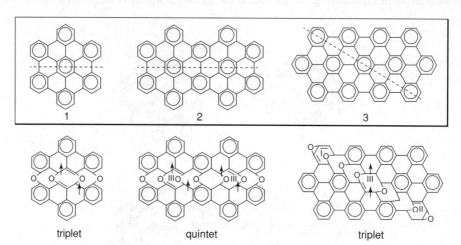

Figure 13.6 *All-benzenoid graphene fragments and the ground-state oxides. The dashed lines (oxidizing lines) of graphene fragments show the directions along which they are oxidized. Reprinted with permission from [31] © 2009 American Chemical Society*

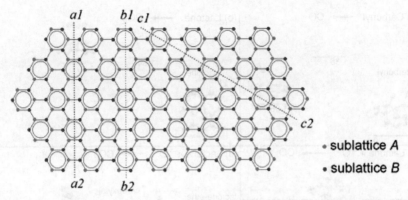

Figure 13.7 *Oxidation-unzipping of graphene fragments along dashed lines a1−a2, b1−a2 and c1−c2. See colour version in the colour plate section. Reprinted with permission from [31] © 2009 American Chemical Society*

on the position of the aromatic sextets: the inner ones prefer the high-spin conformation, whereas the peripheral ones prefer the low-spin conformation. Therefore, the resulting most stable isomer has a total spin whose value equals the number of inner aromatic sextets (n_i) along the oxidizing line. The graphene fragments contained in this isomer have a ferromagnetic spin coupling. Due to the tautomerization between the high-spin and low-spin conformations, there also exist other possible isomers with higher energies and with spins at ground state ranging from 0 to ($n_i - 1$). For example, in Figure 13.7, linear oxidations along *a1-a2*, *b1-b2*, and *c1-c2* will lead to ground-state oxides with $S = 2$, 1, and 2, respectively, and these oxides will isomerize to give at least ($n_i - 1$) higher-energy isomers with $S = 0$ through ($n_i - 1$), respectively.

13.3 Modeling Using PBC Models

13.3.1 Oxidative Creation of Vacancy Defects

Sun *et al.* theoretically studied the mechanisms for atomic oxygens creating vacancy defects on the graphene basal plane [32]. Among various configurations of oxygen distribution, the carbonyl−ether (i.e., 2A), lactone−ether (i.e., 3A and 3A') and ether−lactone−ether structures (i.e., 4A and 4A') were found to have the lowest energies for 2 O, 3 O and 4 O series, respectively (Figure 13.8). Interestingly, the latter two kinds of structures can undergo further gasification encountering only small activation energy barriers: 0.97 eV for 3A to dissociate to CO, 0.61 eV for 3A' to dissociate to CO_2, 0.50 eV for 4A to dissociate to CO, and 0.28 eV for 4A' to dissociate to CO_2. Therefore, lactone−ether and ether−lactone−ether structures were identified to be the main precursors for oxidative gasification of the surface of defect-free graphene. In contrast, the desorption of CO/CO_2 from the carbonyl−ether (2A) and lactone (2I) structures has activation energies larger than 1.8 eV. Compared with lactone−ether and ether−lactone−ether, carbonyl−ether and lactone are less oxygen-abundant. The excessive atomic oxygen thus plays an important

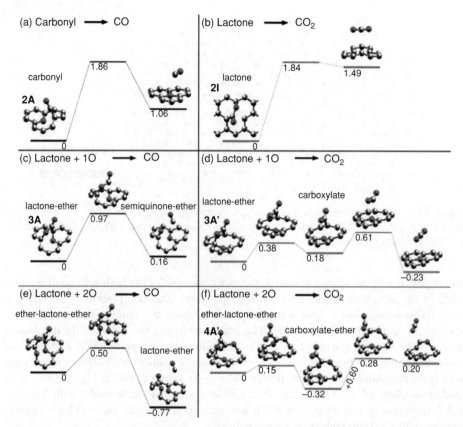

Figure 13.8 Minimum-energy reaction paths for CO (left panel) and CO_2 (right panels) formation at graphene basal plane starting from the lowest-energy carbonyl (a), lactone (b), lactone–ether (c,d), and ether–lactone–ether (e,f) precursors. Reprinted with permission from [32] © 2011 American Chemical Society

role in lowering the activation energies for the oxidative gasification. Larciprete et al. [46] proposed that the ether–lactone precursor can be formed through the transformation from an epoxy–ether structure with the assistance of an additional epoxy.

13.3.2 Oxidative Etching of Vacancy Defects

The previously mentioned gasification models concerning the reactions of O_2 with graphene edges give CO as the main products. To account for the experimental observation of complete combustion under atmospheric oxygen pressure, which yields CO_2 rather than CO, Carlsson et al. proposed a two-step oxidation mechanism, which starts from the vacancy defects in the graphene basal plane [33]. Figure 13.9 shows the results. As reported by other researchers [34, 35], the dissociation adsorption of O_2 on the basal plane of defect-free graphene forms epoxy groups. This process is highly endothermic and thermodynamically disfavored (see route G of Figure 13.9a), in agreement with the experimentally observed

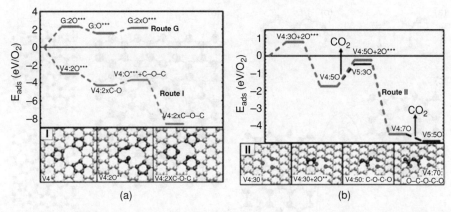

Figure 13.9 Calculated reaction energy profiles for O_2 and graphene. (a) The adsorption energy for dissociative O_2 adsorption on the defect-free graphene (route G) and at a bare four-atom (V4) vacancy (Route I). (b) The adsorption energy for dissociative O_2 adsorption at an oxygen-saturated V4:3O vacancy (Route II). Reprinted with permission from [33] © 2009 American Physical Society

inertness of graphene basal plane towards O_2 at low temperatures [36]. However, if vacancies exist on graphene, these defective sites provide very reactive sites for oxidation. Route I of Figure 13.9(a) shows the reaction energy profile for O_2 with a four-atom carbon vacancy (V4) of graphene. To minimize the dangling bonds, V4 rearranges to form three pentagons as the metastable structure [37]. This structure strongly reacts with molecular O_2 to form either (C–O–C) or carbonyl groups, releasing a large amount of heat. The addition of O can proceed until complete saturation. The reaction of O_2 with the bare vacancies of graphene is the first step of the mechanism of Carlsson et al. [33]. The oxygen-saturated sites are less reactive than bare vacancies but still more reactive than defect-free graphene. Above the critical temperature, additional oxygens attach these oxygen-saturated structures, which is the second step of the Carlsson mechanism (Figure 13.9b). These reactions are also exothermic and form oxygen-rich structures like lactone (C–O–C=O) and the anhydride (O=C–O–C=O). The lactone either releases CO_2 or form anhydride by a further adsorption of O_2, whereas anhydride rapidly decomposes to CO_2. The desorption of CO_2 creates new sites for O_2 dissociation/addition, driving the etching reaction further.

13.3.3 Linear Oxidative Unzipping

Sun et al. found that the stability order ether-pair (Figure 13.10a) > epoxy–pair (Figure 13.10b), which was calculated using coronene as the graphene model, was reversed when the calculations were done using extended graphene models [38]. The different stability order was ascribed to the different types of graphene models used for the calculations: strain energy associated with the reactions can be released relatively easily in cluster models while the strain cannot be released as easily when using PBC models. This implies that the linear oxidative unzipping mechanism proposed in Ref. [30] is valid only for small graphene flakes. Starting from the epoxy-pair, Sun et al. proposed the nine-membered ring ether–trimer (NET) structure (see the circled structure in Figure 13.10c). This NET

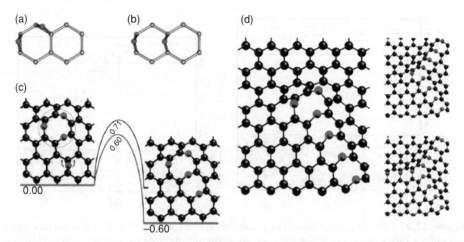

Figure 13.10 (a) Ether–pair. (b) Epoxy pair. (c) Oxygen diffusion to the nine-member ring ether–trimer (NET). (d) Schematic representation of further nucleation and growth of oxygen-driven unzipping of graphene. Reprinted with permission from [38] © 2012 American Chemical Society

structure was found to be thermodynamically more stable than the linear ether–trimer. It can serve as the center of linear oxidative unzipping of large graphene sheets. Figure 13.10(c) shows the diffusion of an oxygen atom to the NET. This oxygen diffusion initiates the unzipping, encountering an activation energy barrier of only 0.6 eV. Figure 13.10(d) shows the structures of the unzipped graphene along the three oxygen edges of the NET, respectively.

13.3.4 Linear Oxidative Cutting

It was found that the mechanical strength of graphene is weakened by ∼16% by the introduction of epoxy frauds during oxidation [39]. This suggests that the mechanical strength of graphene is not critically affected by the formation of linear frauds of ethers. The probable reason is that the graphene sheets still remain linked by the oxygen atoms after the rupture of the C—C bonds. Thus the cooperative mechanism for the formation of linear frauds cannot eventually account for the breakup of graphene oxides. To study how graphene can be cut during oxidation, Li *et al.* studied the subsequent oxidation process of graphene oxide after the formation of linear ethers (Figure 13.11a), and proposed a new "tear-from-the-edge" mechanism [40].

Through DFT calculations, Li *et al* found that the formation of ether structures on graphene basal plane can further activate the carbons where the ethers are attached. For example, the energy of the epoxy-pair structure of Figure 13.11(b) was calculated to be lower than that of the structure with the second epoxy group added far from the epoxy chain by 2.71 eV. The formation of an epoxy pair neighboring an already existing epoxy pair was calculated to be 0.78 eV lower in energy than that forming an isolated epoxy group. This suggests the creation of new epoxy pair is still energetically favored even with the existing epoxy pairs. For a short epoxy chain, the formation of an epoxy pair by adding a new O

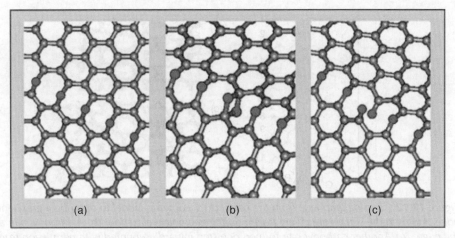

Figure 13.11 Oxygen chains on graphene. (a) A graphene sheet with an epoxy chain. (b) An epoxy pair or (c) a carbonyl pair is formed in the epoxy chain. Reprinted with permission from [40] © 2009 American Chemical Society

to an existing epoxy was comparable in energy (within 0.1 eV) to the extension of the ether chain by adding a new O on fresh carbon–carbon bond neighboring existing ether. Further calculations found that a carbonyl pair has an even lower energy than an epoxy pair by 0.48 eV and that an epoxy pair can transform to a carbonyl pair with activation energy of 0.76 eV. If there is a neighboring epoxy pair, this activation energy is lowered to 0.45 eV, which reflects a substantial reaction rate for the transition from epoxy pair to carbonyl pair. For two neighboring epoxy pairs, their simultaneous transformation to carbonyl pairs has a relatively high activation barrier energy (1.07 eV). However, after the complete of first transition, the conversion of the second epoxy to carbonyl encounters only a small barrier of 0.26 eV. Therefore, the cooperative formation of linear ethers by Li *et al.* [30] and the subsequent formation of epoxy pairs and carbonyl pairs by Li *et al.* [40] have given a more thorough, atomistic-scaled insight into the fracture of graphene caused by oxidation.

The possibility of direct formation of carbonyl pairs by breaking the C–C bonds at graphene edges inspired Li *et al.* to propose a new oxidation-cutting progress: the "tear-from-the-edge" progress [40]. Using the slab model in Figure 13.12(a), the authored computationally studied their proposed progress. In the model, the leftmost carbons were saturated by hydrogens and were frozen during the calculations, whereas the remaining carbons were allowed to relax. The rightmost carbons were saturated by hydroxyl groups. It was found that the formation of carbonyl pair at positions A6 and A7 (Figure 13.12b) is more favorable in energy (1.08) than that at positions B7 and B8. However, the formation of another carbonyl pair one step inward (Figure 13.12c) is energetically less favorable than carbonyl pairs formed at new edge carbons. Therefore, the "tear-from-the-edge" progress was unlikely to proceed and fully cut the graphene sheet. The unsuccessfulness of the "tear-from-the-edge" mechanism in turn indicated epoxy pairs to be the critical intermediate species responsible for the experimentally observed oxidation-cutting phenomena.

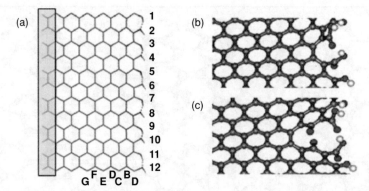

Figure 13.12 *Tearing graphene from the edge. (a) Geometric model to describe a graphene edge. Atoms in the square area were frozen during geometry relaxation. (b) A carbonyl pair at the edge. (c) Another carbonyl pair formed one step inward. Reprinted with permission from [40] © 2009 American Chemical Society*

13.4 From Graphene Oxide back to Graphene

13.4.1 Modeling Using Cluster Models

13.4.1.1 Cluster Models for Graphene Oxide

Gao *et al.* used ethene, benzene and polycyclic aromatic hydrocarbons (PAHs) attached with oxygen functionalities as the models of graphene oxide to study their reduction reactions by hydrazine (N_2H_4) and heat, which are among the most efficient graphene oxide reduction method established experimentally so far [41]. The ethene, benzene and PAH models are parallel with the experimental observation that oxygen functionalities form islands and lines on the basal plane of graphene oxide, dividing the graphene oxide sheet into small in-plane aromatic domains. Figure 13.13 shows the models (i.e., **1** through **5**) used in the study. These models have increasing sizes, and thus can be used to study the properties of the reduction reactions with respective to the sizes of aromatic domains in graphene oxide. PAHs **3**, **4** and **5** contain two different types of C—C bonds: bonds located at the edges and bonds in the interior aromatic domains. Therefore, **3**, **4** and **5** can also be used to investigate the edge effects on the reduction reactions. Then, **1** through **5** were attached with the four oxygen-containing groups known to be abundant in graphene oxide, that is, epoxy (—O—), hydroxyl (—OH), carbonyl (—CHO) and carboxyl (—COOH), serving as the cluster models for graphene oxide.

13.4.1.2 Hydrazine De-Epoxidation

Attaching epoxy groups to the interior C—C bonds of **3**, **4** and **5** yields **3a**, **4a** and **5a**, while attaching epoxy groups to the edge C—C bonds of **1**, **2**, **4** and **5** yields **1b**, **2b**, **4b** and **5b**. Density functional theory calculations at the M05-2X/6-31G(d) level of theory found that these epoxy groups can be removed by N_2H_4 in the gas phase via three different mechanisms: Routes 1, 2 and 3 (Figure 13.14).

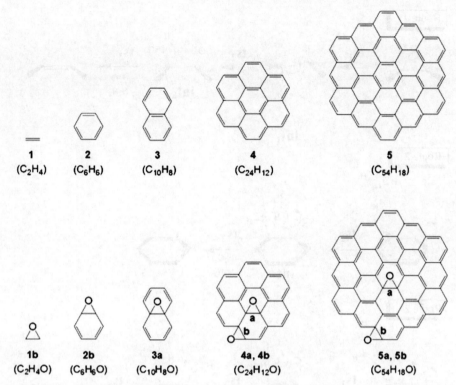

Figure 13.13 Diagrams of ethene (**1**), benzene (**2**), PAHs (**3**, **4**, **5**) and their epoxides (**1b**, **2b**, **3a**, **4a**, **4b**, **5a** and **5b**). Reprinted with permission from [41] © 2010 American Chemical Society

The reduction of **1b** by N_2H_4 proceeds exclusively via Route 1. Route 1 consists of three steps. First, N_2H_4 attacks the epoxy carbon from the backside of the oxygen, leading to the opening of the epoxy ring and the formation of the first intermediate **int$_1$**. **int$_1$** has an inter-molecular charge separation, which undergoes the second step, that is, the intermolecular H transfer from N_2H_4 to oxygen to form **int$_2$**. The last step of Route 1 involves another intermolecular H transfer from $-N_2H_3$ to $-OH$ group, which gives water (H_2O), cis-diazene (N_2H_2) and ethene (**1**) as the de-epoxidation products. Table 13.1 shows the total energies and Gibbs free energies (25°C) with respect to reactants for the transition states, intermediates and products involved in the reaction **1b** + N_2H_4 via Route 1. **int$_2$** is located at a deep well of the reaction energy profile: the total energy of **int$_2$** is lower than those of **ts$_2$** and **ts$_3$** by 72.2 and 83.7 kcal/mol, respectively; the Gibbs free energy of **int$_2$** is lower than those of **ts$_2$** and **ts$_3$** by similar gaps, 72.6 and 83.6 kcal/mol, respectively. This suggests the high kinetically stability of **int$_2$**. Furthermore, **int$_2$** has lower total energies and Gibbs free energies than the reactant or the de-epoxidation product. **int$_2$** is thus the most favorable product for the reaction of **1b** + N_2H_4. These results suggest that hydrazine cannot reduce **1b**, which instead introduces hydrazine alcohols into the graphene moiety as new impurities.

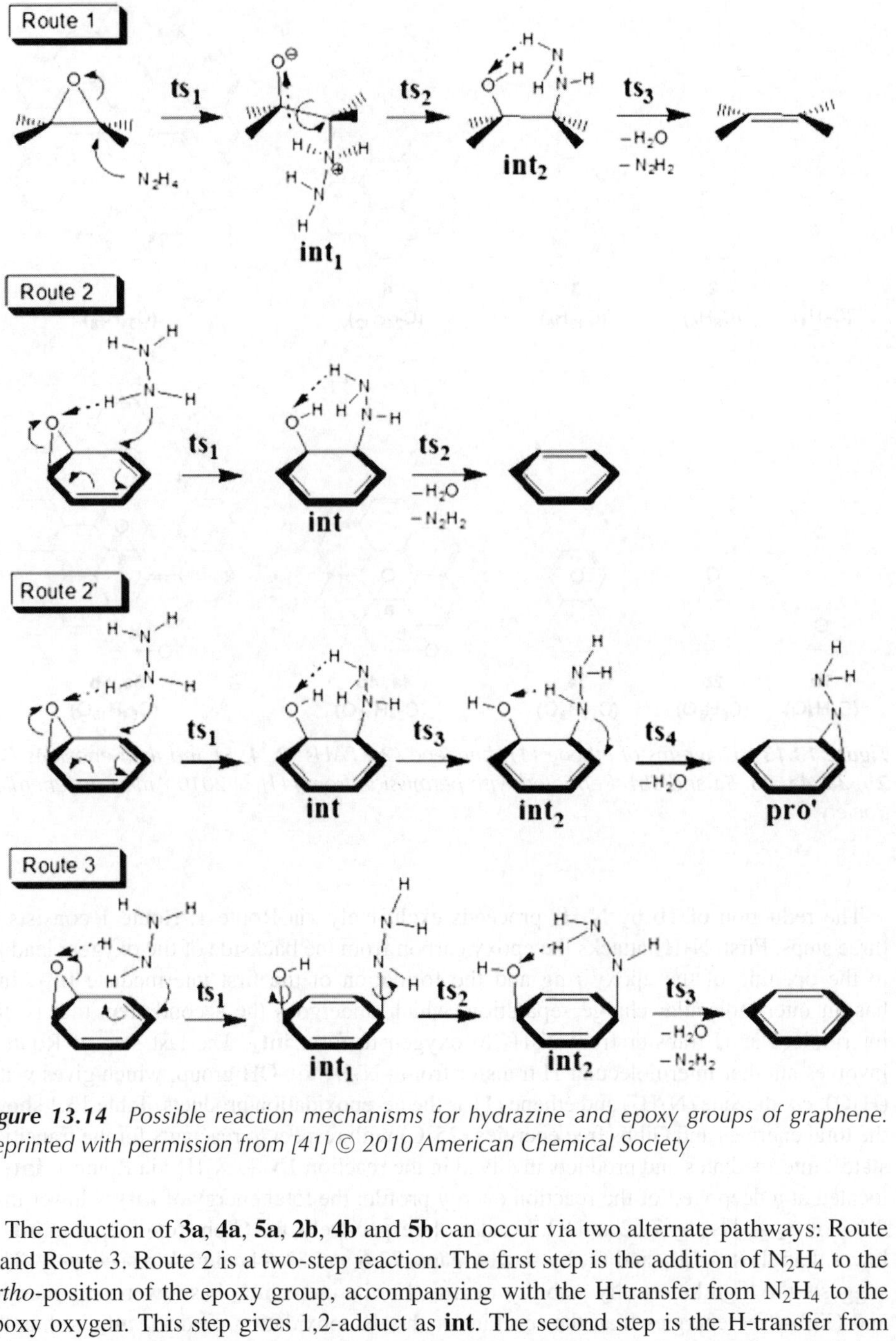

Figure 13.14 Possible reaction mechanisms for hydrazine and epoxy groups of graphene. Reprinted with permission from [41] © 2010 American Chemical Society

The reduction of **3a**, **4a**, **5a**, **2b**, **4b** and **5b** can occur via two alternate pathways: Route 2 and Route 3. Route 2 is a two-step reaction. The first step is the addition of N_2H_4 to the *ortho*-position of the epoxy group, accompanying with the H-transfer from N_2H_4 to the epoxy oxygen. This step gives 1,2-adduct as **int**. The second step is the H-transfer from –N_2H_3 to –OH, leading to the formation and leaving of H_2O and *cis*-N_2H_2, and giving the de-epoxidation product. Route 3 is a three-step reaction similar to Route 1: (1) N_2H_4 addition accompanying with the first H-transfer, (2) adjusting the relative position of –OH

Table 13.1 Reaction enthalpies (kcal/mol) and Gibbs free energies (25°C, in parentheses, kcal/mol) for stationary points involved in hydrazine de-epoxidations of **3a**, **4a**, **1b**, **2b** and **4b** in gas phase. M05-2X/6-31G(d) level of theory.[a] Reprinted with permission from [41] © 2010 American Chemical Society

	reactant	ts$_1$	int	ts$_2$	product
			De-epoxidation via Route 1		
1b + N$_2$H$_4$	0.0 (0.0)	36.0 (47.6)	33.1 (44.7)	35.0 (47.4)	9.7 (−0.4)
	reactant	ts$_1$	int	ts$_2$	product
			De-epoxidation via Route 2		
3a + N$_2$H$_4$	0.0 (0.0)	29.1 (42.4)	−18.1 (−5.8)	33.1 (45.3)	−33.6 (−44.1)
4a + N$_2$H$_4$	0.0 (0.0)	27.1 (41.2)	−8.5 (4.9)	9.6 (22.4)	−46.1 (−57.0)
5a + N$_2$H$_4$	0.0 (0.0)	21.2 (34.9)	−10.0 (3.5)	13.2 (25.5)	−41.5 (−53.2)
2b + N$_2$H$_4$	0.0 (0.0)	20.0 (33.6)	−28.8 (−16.7)	32.2 (44.1)	−22.1 (−33.1)
4b + N$_2$H$_4$	0.0 (0.0)	35.4 (47.8)	−34.4 (−21.6)	43.9 (58.9)	−9.5 (−20.6)
5b + N$_2$H$_4$	0.0 (0.0)	35.3 (48.8)	−43.3 (−29.8)	29.5 (42.3)	−12.1 (−22.6)
	reactant	ts$_1$	int	ts$_2$	int$_2$	ts$_3$	product
				De-epoxidation via Route 3			
3a + N$_2$H$_4$	0.0 (0.0)	35.8 (48.3)	−15.2 (−2.5)	−8.0 (4.9)	−17.6 (−5.6)	17.1 (29.6)	−33.6 (−44.1)
4a + N$_2$H$_4$	0.0 (0.0)	35.6 (49.3)	2.2 (15.5)	10.7 (24.3)	8.0 (20.8)	11.0 (23.8)	−46.1 (−57.0)
5a + N$_2$H$_4$	0.0 (0.0)	24.5 (38.0)	−15.7 (−2.3)	−6.9 (6.8)	−8.8 (3.9)	4.0 (15.7)	−41.5 (−53.2)
2b + N$_2$H$_4$	0.0 (0.0)	25.9 (38.5)	−24.3 (−11.9)	−20.2 (−7.7)	−25.8 (−13.9)	17.8 (30.1)	−22.1 (−33.1)
4b + N$_2$H$_4$	0.0 (0.0)	49.8 (62.5)	0.3 (13.3)	9.5 (22.5)	3.3 (15.9)	21.0 (33.9)	−9.5 (−20.6)
5b + N$_2$H$_4$	0.0 (0.0)	55.9 (69.2)	3.4 (17.1)	12.9 (26.8)	6.7 (19.9)	20.9 (33.7)	−12.1 (−22.6)

[a] For definitions of **3a**, **4a**, **1b**, **2b**, and **4b**, see Figure 13.13; for definitions of Routes 1, 2 and 3, see Figure 13.14.

and $-N_2H_3$ groups through a σ-bond rotation, and (3) the second H-transfer leading to the de-oxygenation.

Table 13.1 shows the energies of stationary points for hydrazine reduction of **3a**, **4a**, **5a**, **2b**, **4b** and **5b** via Route 2 and Route 3. The second and first steps serve as the rate-determining steps for hydrazine de-epoxidation of **3a** via Route 2 and Route 3, respectively. These two steps have Gibbs free-energy barrier ($G^{\neq}_{25°C}$) of 51.1 and 48.3 kcal/mol, respectively. Therefore, Route 3 is kinetically more favorable than Route 2 for the hydrazine and **3a**. As for hydrazine de-epoxidation of **4a** (**5a**), the first step is rate determining, irrespective of via Route 2 or Route 3. The de-epoxidation of **4a** (**5a**) via Route 2 has $G^{\neq}_{25°C}$ of 41.2 (34.9) kcal/mol. This barrier is lower than the corresponding barrier in Route 3, which is 49.3 (38.0) kcal/mol. Therefore, Route 2 is kinetically more favorable than Route 3 in the de-epoxidation of **4a** (**5a**) with hydrazine. These results suggest that Route 3 dominates hydrazine reductions when the epoxies being attacked are located at the interior of a small aromatic domain, but Route 2 is expected to prevail if those interior epoxies belong to a large aromatic domain (see Figure 13.15).

Hydrazine reductions of **3a**, **4a**, and **5a** are exothermic. The enthalpy changes (ΔH) for **3a**, **4a**, and **5a** are respectively -33.6, -46.1, and -41.5 kcal/mol. The corresponding changes in the Gibbs free energy at room temperature ($\Delta G_{25°C}$) – given respectively as -4.1, -57.0, and -53.2 kcal/mol – are even more negative. Meanwhile, the intermediates

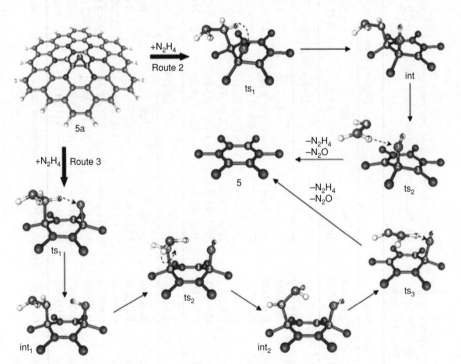

Figure 13.15 Local atomic structures for stationary points involved in hydrazine de-epoxidations of **5a** via Route 2 and Route 3. Reprinted with permission from [41] © 2010 American Chemical Society

involved in these reductions, the hydrazine alcohols, are thermally much less stable than the corresponding reduction products. The hydrazine de-epoxidations of **3a**, **4a**, and **5a** are therefore thermodynamically spontaneous processes. The reaction energy barriers ($G^{\neq}{}_{25°C}$) for these processes are respectively 48.3, 41.2, and 34.9 kcal/mol. Consequently, epoxy groups at the interior of an aromatic domain of graphene oxide will be removed by hydrazine at reaction temperatures of 100–150 °C. The order of energy barriers – **3a** > **4a** > **5a** – suggests that interior epoxy groups belonging to larger aromatic domains are kinetically more facile to be reduced than those belonging to smaller aromatic domains.

Similar analysis of energetic data suggests that the reactions of **2b**, **4b** and **5b** with hydrazine occur preferentially via Routes 3, 2 and 2, respectively. Similar to the case of **1b**, the de-epoxidations of **4b** and **5b** involve stable hydrazino alcohol intermediates. These intermediates have Gibbs free energies of formation of -21.6, and -29.8 kcal/mol at room temperature, respectively, which are even lower than those for the corresponding reduction products (i.e., -20.6, and -22.6 kcal/mol). Therefore, the hydrazine de-epoxidation for **4b**, and **5b** will stop with the formation of a hydrazino alcohol. The reaction energy barriers for the de-epoxidations of **2b**, **4b** and **5b** are 44.0, 62.5, and 69.2 kcal/mol, respectively, showing an increasing behavior concomitant with increased size of aromatic domain. Consequently, edge epoxy groups belonging to larger aromatic domains are kinetically more difficult to reduce by N_2H_4. This is opposite to reduction of interior epoxy groups.

13.4.1.3 Thermal De-Hydroxylation

The calculations did not find reaction pathways corresponding to N_2H_4 de-hydroxylation. Hydrazine plays a minor role, if any, in the de-hydroxylation of graphene oxide. Instead, two reduction routes without hydrazine were found: Routes 4 and 5 (Figure 13.16). In Route 4, the hydroxyl reduction takes place in two steps and yields a monoxide (CO) and a graphene sheet with a vacancy. By contrast, in Route 5, the hydroxyl group directly leaves

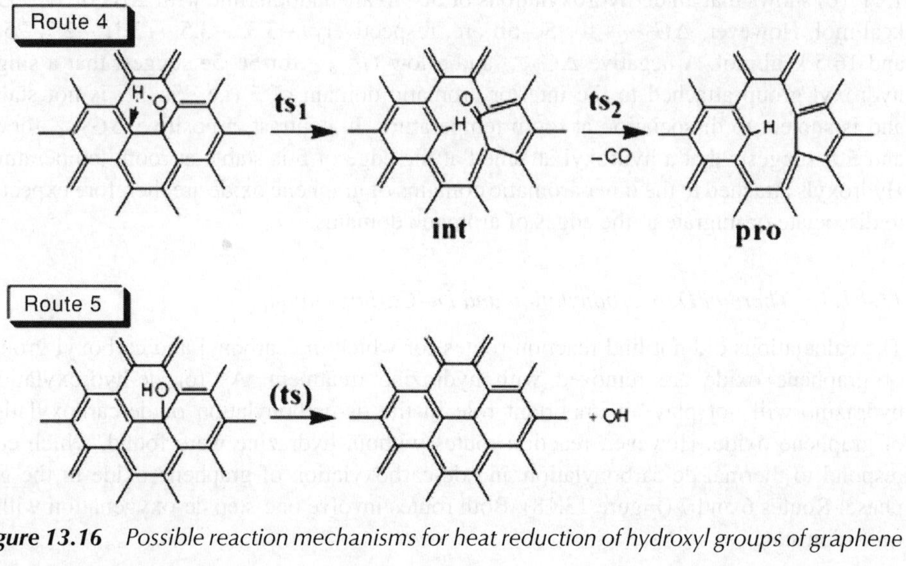

Figure 13.16 Possible reaction mechanisms for heat reduction of hydroxyl groups of graphene

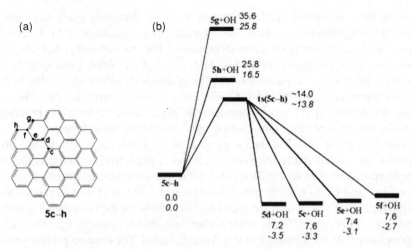

Figure 13.17 (a) **5c–h**, PAH with a hydroxyl group attached at different positions. (b) Reaction energy profiles for detaching of the hydroxyl groups of **5c–h**. Values in normal and italic fonts respectively show relative enthalpies (kcal/mol) and relative Gibbs free energies (kcal/mol) at room temperature (298 K). Reprinted with permission from [41] © 2010 American Chemical Society

the graphene sheet, producing an OH radical and a graphene radical. Route 4 has a large energy barrier and is less possible to dominate the de-hydroxylation process.

To study the edge effect on thermal de-hydroxylation, hydroxyl groups were attached to six different positions – c through h – of **5**, giving structures **5c** through to **5h** (Figure 13.17a). The de-hydroxylations of interior hydroxyls, that is, **5c** through **5f**, via Route 5 have a transition state with $G^{\neq}_{25\,°C}$ of approximately 13.8 kcal/mol. The de-hydroxylations of edge hydroxyls, that is, **5g** and **5h**, via Route 5 involve no transition states. Figure 13.17(b) shows that all de-hydroxylations of **5c–5h** are endothermic with ΔHs of 7.2–35.6 kcal/mol. However, $\Delta G_{25\,°C}$s for **5c–5h** are, respectively, −3.3, −3.5, −3.1, −2.7, 25.8, and 16.5 kcal/mol. A negative $\Delta G_{25\,°C}$ and a low $G^{\neq}_{25\,°C}$ for **5c–5e** suggest that a single hydroxyl group attached to the interior aromatic domain of **5** (i.e., **5c–5f**) is not stable and is subject to dissociation at room temperature. In contrast, a positive $\Delta G_{25\,°C}$ for **5g** and **5 h** suggests that a hydroxyl attached at the edge of **5** is stable at room temperature. Hydroxyls attached to the inner aromatic domains of graphene oxide are therefore expected to dissociate or migrate to the edges of aromatic domains.

13.4.1.4 Thermal De-Carbonylation and De-Carboxylation

The calculations did not find reaction routes for which the carbonyl and carboxyl groups of graphene oxide are removed with hydrazine treatment. As for de-hydroxylation, hydrazine will not play an important role in the de-carbonylation or de-carboxylation of graphene oxide. However, reaction routes without hydrazine were found, which correspond to thermal de-carbonylation and de-carboxylation of graphene oxide in the gas phase: Routes 6 and 7 (Figure 13.18). Both routes involve one-step de-oxygenation with a transition state.

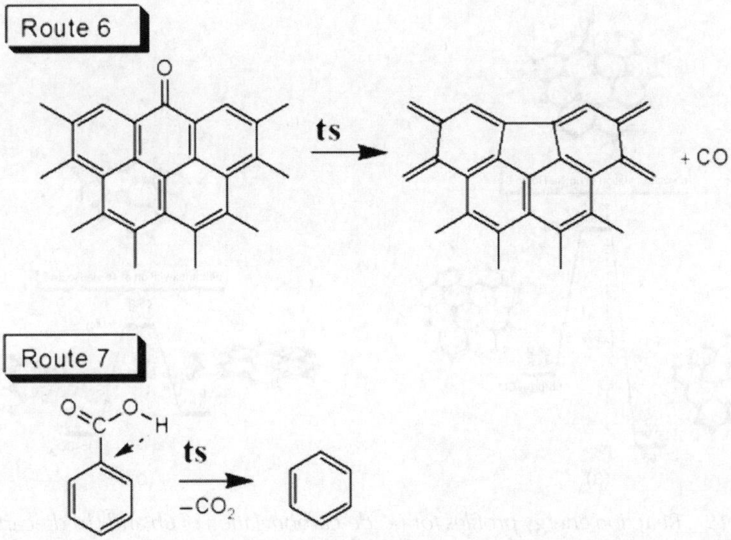

Figure 13.18 Possible reactions for de-oxygenation of carbonyl and carboxyl groups of graphene. Reprinted with permission from [41] © 2010 American Chemical Society

Figure 13.19(a) presents the reaction energy profile for the de-carbonylation of **6b** via Route 6. The thermal de-carbonylation in gas phase is endothermic ($\Delta H = 56.2$ kcal/mol). The corresponding $\Delta G_{25°C}$ and $G^{\neq}_{25°C}$ are 44.5 and 133.0 kcal/mol, respectively, meaning that the de-carbonylation of **6b** via Route 6 hardly occurs at room temperature. Figure 13.19(b) also shows the reaction energy profile for the de-carboxylation of **4c** via Route 7. Actually, $\Delta G_{25°C} = -14.3$ kcal/mol and $G^{\neq}_{25°C} = 65.7$ kcal/mol for thermal de-carboxylation. This de-carboxylation is expected to occur slowly at room temperature. These results suggest that the carbonyl groups of graphene oxide will not be removed, but the carboxyl groups are expected to be reduced slowly at temperatures of 100–150°C.

13.4.1.5 Temperature Effect on De-Epoxidation and De-Hydroxylation

As described previously, epoxy (hydroxyl) groups located at the interior of an aromatic domain of graphene oxide can be eliminated using hydrazine (thermally annealing) with relative ease, but those located at the edges of graphene oxide are difficult to remove. Therefore, epoxies and hydroxyls at the edges of aromatic domain deserve special attention. Figure 13.20(a) presents the reaction energy profile for the hydrazine de-epoxidation of **5b** via Route 2 for various temperatures. When the temperature increases from -175 to $1125°C$, ΔG decreases from -10 to -60 kcal/mol. The value of $G_1^{\neq}$ for the first step of reaction increases from 40 to 100 kcal/mol (Figure 13.20a). That increase in $G_1^{\neq}$ results from the fact that the first step in the reaction relies upon the capture of a hydrazine molecule on the moiety **5b** ($\Delta S < 0$). An increased temperature reduces the probability for such capturing; consequently, it will not improve the efficiency of hydrazine de-epoxidation.

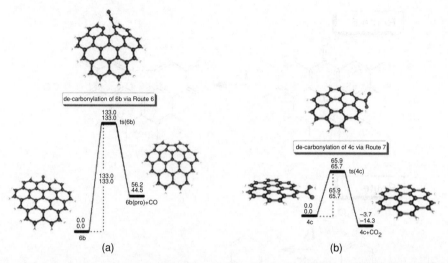

Figure 13.19 Reaction energy profiles for (a) de-carbonylation of **6b** and (b) de-carboxylation of **4c** in a vacuum. The calculation used the M05-2X/6-31G(d) level of theory. Values in roman and italic print respectively, depict relative enthalpies (H$_{rel}$, in kcal/mol) and relative Gibbs free energies (G$_{rel}$, in kcal/mol) at room temperature (25°C). Schematic structures of stationary points are also shown. Reprinted with permission from [41] © 2010 American Chemical Society

Figure 13.20(b) shows the reaction energy profile for the thermal de-hydroxylation of **5 h** via *Route 5* at various temperatures. When the temperature increases, the ΔG decreases from positive to negative (Figure 13.20b). The critical temperature (T_c), defined as the point at $\Delta G = 0$, is 650°C, showing that raising the temperature will make the de-hydroxylation of graphene oxide spontaneous. When the temperature is greater than 650°C, even the hydroxyls attached to the edges of graphene oxide can be readily removed.

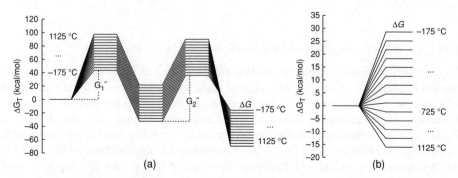

Figure 13.20 Relative Gibbs free energies (G$_{rel}$s) for stationary points involved in (a) hydrazine de-epoxidation of **5b** and (b) thermal de-hydroxylation of **5 h** at 14 different temperatures: $-175.0°C$, $-75.0°C$, $25.0°C$ ... $1125.0°C$. Reprinted with permission from [41] © 2010 American Chemical Society

13.4.1.6 Residual Groups of Graphene Oxide Reduced by Hydrazine and Heat

The N-containing compounds might be hydrazino alcohols formed by the reactions of hydrazine with (1) epoxy groups that are located at the edge of graphene oxide via Route 2, and (2) epoxy groups that are sticking out of graphene oxide via Route 1 (Figure 13.14). Epoxy groups of the second type are rare in graphene oxide [17]. Therefore, the hydrazino alcohols formed by the hydrazine reaction of the first type are expected to be responsible for the N-containing compounds observed experimentally. In contrast to this, Stankovich et al. recently proposed that the N-containing compounds might be amino-aziridines formed by the reaction of hydrazine with epoxies [42]. Route 2' is the pathway for the formation of amino-aziridines through the reactions of hydrazine and epoxies. As shown in Figure 13.14, Route 2 and Route 2' are identical in the first H-transfer step but that they subsequently deviate from one another. The values of $G_{25°C}^{\neq}$ and $\Delta G_{25°C}$ for **4a** and hydrazine via Route 2' are 33.4 and −4.1 kcal/mol, respectively; the corresponding values via Route 2 are, respectively, 17.5 and −57.0 kcal/mol (Table 13.1). Therefore, the formation of amino-aziridine via Route 2' is much less competitive than the de-epoxidation via Route 2 for the reaction of **4a** with hydrazine. For the reaction of **1b** (**4b**) with hydrazine, $\Delta G_{25°C}$ for the formation of amino-aziridine is only −8.8 (−4.5) kcal/mol. $\Delta G_{25°C}$ for the formation of hydrazine alcohol is −24.1 (−20.7) kcal/mol. The formation of hydrazino alcohol is therefore much more likely than that of amino-aziridine in the reactions of **1b** and **4b** with hydrazine, which suggests that the N-containing species observed in CMGs[42–44] are likely to be hydrazino alcohols instead of amino-aziridines.

Table 13.2 shows the response of oxygen functionalities of graphene oxide to hydrazine treatment, heat treatment, and their combination. To summarize, the residual groups of CMG prepared through the hydrazine reduction of graphene oxide at room temperature are **B'**, **C'**, and hydrazino alcohols converted from **A'** and probably **D**. The residual groups of CMG prepared through the thermal annealing of graphene oxide at temperatures of 700–1200°C are **A**, **A'** and **C**. In fact, **A'** and **C** are only residual groups of CMGs obtained from the hydrazine reduction of graphene oxide followed by thermal annealing at temperatures of 700–1200°C.

Table 13.2 Status of oxygen-containing groups upon treatment of hydrazine and heat. Reprinted with permission from [41] © 2010 American Chemical Society

groups[a]	hydrazine[b]	heat[c]	hydrazine[b] and heat[c]
A	removed	not removed	removed
A'	converted to hydrazino alcohol	not removed	not removed
B	removed	removed	removed
B'	not removed	removed	removed
C	not removed	not removed	not removed
D	partly removed	removed	removed

[a]See Figure 13.21 for the definition of each oxygen functionality. [b]Hydrazine treatment at room temperature. [c]Thermal annealing at temperatures of 700–1200°C.

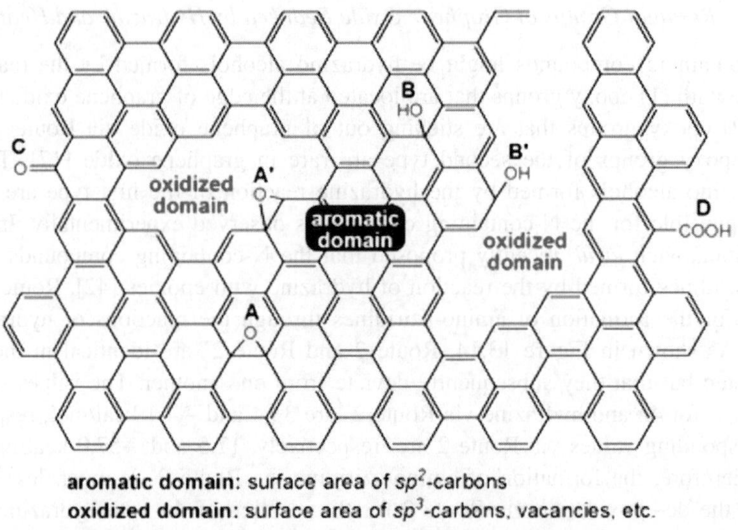

Figure 13.21 *Schematic illustration of oxygen-containing groups in graphene oxide: **A**, epoxy located at the interior of an aromatic domain of graphene oxide; **A′**, epoxy located at the edge of an aromatic domain; **B**, hydroxyl located at the interior of an aromatic domain; **B′**, hydroxyl at the edge of an aromatic domain; **C**, carboxyl at the edge of an aromatic domain; and **D**, carboxyl at the edge of an aromatic domain. Reprinted with permission from [41] © 2010 American Chemical Society*

13.4.2 Modeling Using Periodic Boundary Conditions

13.4.2.1 Hydrazine De-Epoxidation

Kim et al. used a graphene slab consisting of 32 carbon atoms with periodic boundary conditions (PBC) to model the graphene surface (Figure 13.22a) [45]. One oxygen atom was added to the graphene slab, forming the unit cell of graphene epoxy, as shown in Figure 13.22(b). Using the local density approximation (LDA) of density functional theory method, Kim et al. investigated the de-oxygenation of this graphene epoxy model by hydrazine. Besides Route 2 of Figure 13.14, Kim et al. found Route 8 for this process. The first step of Route 8 is N_2H_4 transferring a hydrogen atom to epoxy, leading to the ring opening of epoxy. However, the resultant $-N_2H_3$ group does not form a covalent bond with the graphene substrate. Instead, it stays around the $-OH$ group due to the attractive interaction of the hydroxyl hydrogen with the nitrogen atom of N_2H_3. The energy barrier of this step was calculated at only 10.2 kcal/mol. *Ab initio* molecular dynamic (AIMD) simulation showed that Route 8 is similar to Route 2 and Route 3 in the second step (Figure 13.14): the transfer of hydrogen from N_2H_3 to OH leads to the formation and leaving of H_2O. According to AIMD simulation, the second step has an exothermicity of 37.2 kcal/mol. This step occurs in less than 1 ps, and the energy barrier is thus insignificant. Kim et al. also located the reaction mechanism similar to that of Route 2 for the hydrazine reduction of graphene epoxy using the PBC model. The energy barriers of the first and third steps were 10.6 and 6.7 kcal/mol, respectively.

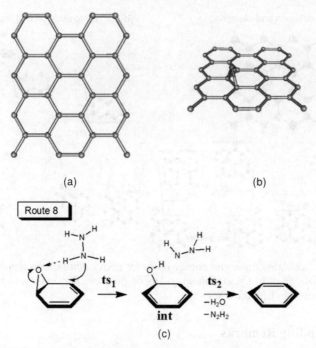

Figure 13.22 (a) A graphene slab consisting of 32 carbon atoms. (b) A graphene slab with an epoxy group. (c) A possible hydrazine reduction mechanism. Reprinted with permission from [41] © 2010 American Chemical Society

These results obtained using the LDA/PBC-model imply that epoxy of graphene can be readily reduced by hydrazine. The PBC model represents an epoxy situated in the interior of a large aromatic domain. Thus, the conclusion of the PBC model is, in general, consistent with that obtained using cluster models, which suggests that epoxy located in the interior of graphene's aromatic domain can be easily reduced.

13.4.2.2 Thermal De-Epoxidation

Larciprete et al. investigated the thermal reduction of epoxy groups using the PBC model [46]. Figure 13.23 shows the reaction energy profiles for the migration and desorption of two epoxy groups on graphene surface. As shown in Figure 13.23, the migration of epoxy to form a neighboring epoxy pair is an exothermic process, with a change in energy of −11.3 kcal/mol. The energy barrier of each step along the migration path is less than 17 kcal/mol. Hence, the migration process is facile even at room temperature. A neighboring epoxy pair can further undergo a two-step reaction to recombine to form an oxygen molecule, which causes the removal of the two epoxy oxygens from the graphene surface (Figure 13.23b). The first step is the rate-limiting step of the recombination, having an energy barrier of 26.1 kcal/mol. This migration and recombination pathway has received some support from experiments.

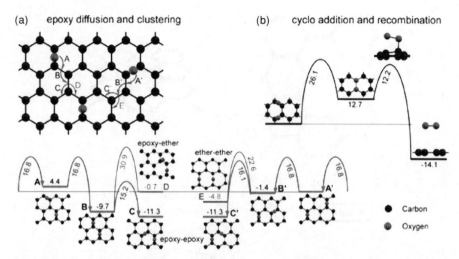

Figure 13.23 Calculated minimum energy paths for epoxy diffusion on graphene. (a) Epoxy diffusion and clustering. (b) Cycloaddition and recombination. Reprinted with permission from [46] © 2011 American Chemical Society

13.5 Concluding Remarks

Graphene oxide and chemically converted graphene are amorphous materials, which contain miscellaneous chemical groups and substructures. To a large extent, their chemical components and accurate structures still remain unclear. Therefore, the macroscopically observed chemistry of these materials is actually an overlap of diverse chemical processes. On the other hand, high-level applications of these materials require accurately addressing the structures and the associated physical and chemical properties. Computation thus serves as an indispensable approach to exploring the atomistic mechanisms of graphene chemistry, which in turn can offer guidance for subsequent experiments. In the past few years, great progress has been made in understanding the oxidation mechanisms of graphene, as well as the mechanisms for the reduction of graphene oxide back to graphene. However, a big gap still exists between the experimental observations and theoretical predictions. Persistent efforts in both experiment and theory towards the solutions are thus desired.

Acknowledgement

This work was supported in China by MOST 973 program of China (2012CB934001 and 2012CB932504), and CAS Hundreds Elite Program (Y1515530U1), and in US by Department of Defense (Grant W911NF-12-1-0083) and NSF (Grant EPS-1010094).

References

[1] K. S. Novoselov, A. K. Geim, S. V. Morozov, D. Jiang, Y. Zhang, S. V. Dubonos, et al., Electric field effect in atomically thin carbon films, *Science* **306**, 666 (2004).

[2] A. K. Geim and K. S. Novoselov, The rise of graphene, *Nat. Mater.* **6**, 183 (2007).
[3] D. A. Dikin, S. Stankovich, E. J. Zimney, R. D. Piner, G. H. B. Dommett, G. Evmenenko, et al., Preparation and characterization of graphene oxide paper, *Nature* **448**, 457 (2007).
[4] K. P. Loh, Q. Bao, P. K. Ang and J. Yang, The chemistry of graphene, *Chem. Soc. Rev.* **20**, 2277 (2010).
[5] M. J. Allen, V. C. Tung and R. B. Kaner, Honeycomb carbon: a review of graphene, *Chem. Rev.*, 132 (2010).
[6] M. Choucair, P. Thordarson and J. A. Stride, Gram-scale production of graphene based on solvothermal synthesis and sonication, *Nat. Nanotechnol.* **4**, 30 (2009).
[7] K. Müllen and J. P. Rabe, Nanographenes as active components of single-molecule electronics and how a scanning tunneling microscope puts them to work, *Acc. Chem. Res.* **41**, 511 (2008).
[8] S. Park and R. S. Ruoff, Chemical methods for the production of graphenes, *Nat. Nanotechnol.* **4**, 217 (2009).
[9] A. N. Obraztsov, Chemical vapour deposition: Making graphene on a large scale, *Nat. Nanotechnol.* **4**, 212 (2009).
[10] L. Yan, Y. B. Zheng, F. Zhao, S. Li, X. Gao, B. Xu, et al., Chemistry and physics of a single atomic layer: strategies and challenges for functionalization of graphene and graphene-based materials, *Chem. Soc. Rev.* **41**, 97 (2012).
[11] X. Gao, D.-e. Jiang, Y. Zhao, S. Nagase, S. B. Zhang and Z. Chen, Theoretical insights into the structures of graphene oxide and its chemical conversions between graphene, *J. Comput. Theor. Nanosci.* **8**, 2406 (2011).
[12] N. Leconte, J. I. Moser, P. Ordejón, H. Tao, A. I. Lherbier, A. Bachtold, et al., Damaging graphene with ozone treatment: a chemically tunable metal-insulator transition, *ACS Nano* **4**, 4033 (2010).
[13] M. Z. Hossain, J. E. Johns, K. H. Bevan, H. J. Karmel, Y. T. Liang, S. Yoshimoto, et al., Chemically homogeneous and thermally reversible oxidation of epitaxial graphene, *Nat. Chem.* **4**, 305 (2012).
[14] D. R. Dreyer, S. Park, C. W. Bielawski and R. S. Ruoff, The chemistry of graphene oxide, *Chem. Soc. Rev.* **39**, 228 (2010).
[15] A. Lerf, H. He, M. Forster and J. Klinowski, Structure of graphite oxide revisited, *J. Phys. Chem. B* **102**, 4477 (1998).
[16] T. Szabó, O. Berkesi, P. Forgó, K. Josepovits, Y. Sanakis, D. Petridis and I. Dékány, Evolution of surface functional groups in a series of progressively oxidized graphite oxides, *Chem. Mater.* **18**, 2740 (2006).
[17] W. Cai, R. D. Piner, F. J. Stadermann, S. Park, M. A. Shaibat, Y. Ishii, et al., Synthesis and solid-state NMR structural characterization of ^{13}C-labeled graphite oxide, *Science* **321**, 1815 (2008).
[18] W. Gao, L. B. Alemany, L. Ci and P. M. Ajayan, New insights into the structure and reduction of graphite oxide, *Nat. Chem.* **1**, 403 (2009).
[19] G. Eda, G. Fanchini and M. Chhowalla, Large-area ultrathin films of reduced graphene oxide as a transparent and flexible electronic material, *Nat. Nanotechnol.* **3**, 270 (2008).

[20] J. I. Paredes, S. Villar-Rodil, P. Solís-Fernández, A. Martínez-Alonso and J. M. D. Tascón, Atomic force and scanning tunneling microscopy imaging of graphene nanosheets derived from graphite oxide, *Langmuir* **25**, 5957 (2009).

[21] D. W. Boukhvalov and M. I. Katsnelson, Modeling of graphite oxide, *J. Am. Chem. Soc.* **130**, 10697 (2008).

[22] J.-A. Yan, L. Xian and M. Y. Chou, Structural and electronic properties of oxidized graphene, *Phys. Rev. Lett.* **103**, 086802 (2009).

[23] H. J. Xiang, S.-H. Wei and X. G. Gong, Structural motifs in oxidized graphene: a genetic algorithm study based on density functional theory, *Phys. Rev. B* **82**, 035416 (2010).

[24] L. Wang, Y. Y. Sun, K. Lee, D. West, Z. F. Chen, J. J. Zhao and S. B. Zhang, Stability of graphene oxide phases from first-principles calculations, *Phys. Rev. B* **82**, 161406 (2010).

[25] L. Liua, L. Wang, J. Gao, J. Zhaoa, X. Gao and Z. Chen, Amorphous structural models for graphene oxides, *Carbon* **50**, 1690 (2012).

[26] N. Lu, D. Yin, Z. Li and J. Yang, Structure of graphene oxide: thermodynamics versus kinetics, *J. Phys. Chem. C* **115**, 11991 (2011).

[27] D. K. Samarakoon and X.-Q. Wang, Twist-boat conformation in graphene oxides, *Nanoscale* **3**, 192 (2011).

[28] K. Sendt and B. S. Haynes, Density functional study of the chemisorption of O_2 on the armchair surface of graphite, *Proc. Combust. Inst.* **30**, 2141 (2004).

[29] Z. H. Zhu, J. Finnerty, G. Q. Lu and R. T. Yang, A comparative study of carbon gasification with O_2 and CO_2 by density functional theory calculations, *Energy Fuels* **16**, 1359 (2002).

[30] J.-L. Li, K. N. Kudin, M. J. McAllister, R. K. Prud'homme, I. A. Aksay and R. Car, Oxygen-driven unzipping of graphitic materials, *Phys. Rev. Lett.* **96**, 176101 (2006).

[31] X. Gao, L. Wang, Y. Ohtsuka, D.-e. Jiang, Y. Zhao, S. Nagase and Z. Chen, Oxidation unzipping of stable nanographenes into joint spin-rich fragments, *J. Am. Chem. Soc.* **131**, 9663 (2009).

[32] T. Sun, S. Fabris and S. Baroni, Surface precursors and reaction mechanisms for the thermal reduction of graphene basal surfaces oxidized by atomic oxygen, *J. Phys. Chem. C* **115**, 4730 (2011).

[33] J. M. Carlsson, F. Hanke, S. Linic and M. Scheffler, Two-Step mechanism for low-temperature oxidation of vacancies in graphene, *Phys. Rev. Lett.* **102**, 166104 (2009).

[34] S.-P. Chan, G. Chen, X. G. Gong and Z.-F. Liu, Oxidation of carbon nanotubes by singlet O_2, *Phys. Rev. Lett.* **90**, 086403 (2003).

[35] D. Lamoen and B. N. J. Persson, Adsorption of potassium and oxygen on graphite: a theoretical study, *J. Chem. Phys.* **108**, 3332 (1998).

[36] F. Stevens, L. A. Kolodny and T. P. Beebe, Kinetics of graphite oxidation: monolayer and multilayer etch pits in HOPG studied by STM, *J. Phys. Chem. B* **102**, 10799 (1998).

[37] J. M. Carlsson and M. Scheffler, Structural, electronic, and chemical properties of nanoporous carbon, *Phys. Rev. Lett.* **96**, 046806 (2006).

[38] T. Sun and S. Fabris, Mechanisms for oxidative unzipping and cutting of graphene, *Nano Lett.* **12**, 17 (2012).

[39] J. T. Paci, T. Belytschko and G. C. Schatz, Computational studies of the structure, behavior upon heating, and mechanical properties of graphite oxide, *J. Phys. Chem. C* **111**, 18099 (2007).

[40] Z. Li, W. Zhang, Y. Luo, J. Yang and J. G. Hou, How graphene is cut upon oxidation?, *J. Am. Chem. Soc.* **131**, 6320 (2009).

[41] X. Gao, J. Jang and S. Nagase, Hydrazine and Thermal reduction of graphene oxide: reaction mechanisms, product structures, and reaction design, *J. Phys. Chem. C* **114**, 832 (2010).

[42] S. Stankovich, D. A. Dikin, R. D. Piner, K. A. Kohlhaas, A. Kleinhammes, Y. Jia, et al., Synthesis of graphene-based nanosheets via chemical reduction of exfoliated graphite oxide, *Carbon* **45**, 1558 (2007).

[43] V. C. Tung, M. J. Allen, Y. Yang and R. B. Kaner, High-throughput solution processing of large-scale graphene, *Nat. Nanotechnol.* **4**, 25 (2009).

[44] D. Yang, A. Velamakanni, G. Bozoklu, S. Park, M. Stoller, R. D. Piner, et al., Chemical analysis of graphene oxide films after heat and chemical treatments by X-ray photoelectron and micro-Raman spectroscopy, *Carbon* **47**, 145 (2009).

[45] M. C. Kim, G. S. Hwang and R. S. Ruoff, Epoxide reduction with hydrazine on graphene: a first principles study, *J. Chem. Phys.* **131**, 064704 (2009).

[46] R. Larciprete, S. Fabris, T. Sun, P. Lacovig, A. Baraldi and S. Lizzit, Dual path mechanism in the thermal reduction of graphene oxide, *J. Am. Chem. Soc.* **133**, 17315 (2011).

14

Electronic Transport in Graphitic Carbon Nanoribbons

Eduardo Costa Girão,[a] Liangbo Liang,[b] Jonathan Owens,[b] Eduardo Cruz-Silva,[b,c] Bobby G. Sumpter,[d] and Vincent Meunier[b]

[a]*Departamento de Física, Universidade Federal do Piauí, Brazil*
[b]*Department of Physics, Applied Physics, and Astronomy, Rensselaer Polytechnic Institute, USA*
[c]*Department of Polymer Science and Engineering, University of Massachusetts, USA*
[d]*Center of Nanophase Materials Sciences, Oak Ridge National Laboratory, USA*

14.1 Introduction

Known carbon nanostructures span all four spatial dimensionalities and exhibit properties intimately related to different expressions of quantum spatial confinement. The discovery of fullerenes (0D) [1], carbon nanotubes (1D) [2], and the more recent experimental isolation of graphene sheets (2D) [3] have complemented the allotrope family tree long occupied by the centuries-old 3D forms of carbon (graphite and diamond). Discoveries made over the past 30 years have unveiled a number of emerging phenomena and paved the way to the possibility of devising a spectrum of diverse carbon nanostructures where elementary low-dimensional building blocks are assembled into systems with ever-increasing complexity. For instance, the most recent isolation of graphene from graphite has triggered an unprecedented material science research activity in the study of a wide variety of graphene derivatives, such as graphitic nanoribbons (GNRs).

At the same time as new materials are discovered and characterized, the confluence of scalable mathematical algorithms, novel electronic structure methods, advanced computing tools and exascale computational performance has enabled the development of predictive

---

Graphene Chemistry: Theoretical Perspectives, First Edition. Edited by De-en Jiang and Zhongfang Chen.
© 2013 John Wiley & Sons, Ltd. Published 2013 by John Wiley & Sons, Ltd.

tools for the discovery of novel phenomena and principles for the design of new molecular electronics with breakthrough properties. In that respect, theoretical methods have now evolved to a point where the properties of materials can be successfully predicted based essentially on their atomic structure and with limited or no experimental input. As a consequence, computational science capabilities can substantially accelerate the search for novel materials and devices, especially when integrated with experimental effort. Enabled by these computational capabilities, exploration of how atomic scale structure and quantum mechanical effects impact electron transport and other electronic processes within nanostructures and across interfaces, provides critical information for developing the next generation of materials and electronic devices. In this chapter we review recent progress in describing electron transport in graphitic nanoribbons (GNRs) from a theoretical and computational perspective and how research results published in the literature provide unique physicochemical insight that is applicable to the development of novel materials and devices.

The rest of this chapter is organized as follows: we review a general theoretical framework for electronic structure and transport calculations in Section 14.2, we introduce the general properties of GNRs in Section 14.3, provide a short overview of synthesis and processing approaches in Section 14.4, and then show how transport properties can be tailored for targeted applications in Section 14.5. We conclude this chapter with an overview of the current understanding of thermoelectric properties of GNRs in Section 14.6.

14.2 Theoretical Background

14.2.1 Electronic Structure

14.2.1.1 Density Functional Theory

Density functional theory (DFT) usually treats systems on the basis of the ground state electron density. While this approach reduces the N-body problem from the determination of a many-body wave-function depending on 3N variables to a simpler three-variable density, approximations must be made in the expression of exchange and correlation, for which a number of local, semi-local, and non-local expressions exist. This allows the practical treatment of systems made up of hundreds of carbon atoms. As such, DFT has been widely used to study carbon nanostructures, including GNRs where spin-polarized DFT is typically needed to investigate edge effects. Unfortunately DFT is a theory which, in practical terms, allows one to deal with the system's ground state, so that it cannot strictly be used to compute electronic band gaps without using post-DFT treatments such as quasiparticle GW. Reviews on density functional theory (DFT) can be found in a number of places, including the recent review by two of the present authors [4]. Thus, we will redirect the interested readers to the literature on DFT and elaborate on semi-empirical methods that make it possible to study larger systems (typically thousands of atoms) in a more systematic way. However, DFT remains the method of choice to establish new phenomena and to benchmark results obtained using those semi-empirical approaches.

14.2.1.2 Semi-Empirical Methods

Tight-binding (TB) calculations are known to provide a good description of the electronic properties of carbon based materials [5, 6], especially for geometries with moderate to low curvature. TB can be made very accurate (e.g., compared to DFT) when considering high-order neighbor interactions [7,8] and enables simulation of very large systems [9]. For sp^2 carbon systems, it is usual to restrict the basis set to an orthogonal basis involving only the π-orbitals. This works particularly well in flat systems where the remaining orbitals are involved quasi-exclusively in forming the strong σ-bonds and are essentially decoupled from the π-orbitals. It follows that the major features of the electronic states close to the Fermi level have a strong π-character [10]. This π-orbital description implicitly considers that carbon atoms on the edges are saturated with hydrogens. The tight-binding hopping integrals are usually fit to experimental data or *ab initio* calculations [7,8]. For instance, the third-neighbor parametrization proposed by Gunlycke and White with $\gamma_1 = 3.2$ eV, $\gamma_2 = 0$ eV and $\gamma_3 = 0.3$ eV for the first-, second-, and third-nearest neighbor hopping integrals, respectively, was recently shown to result in an excellent agreement with DFT calculations for a varied set of systems [8, 11]. Finally, the different carbon coordination number at the edges is satisfactorily accounted for by including a $\Delta\gamma_1 = 0.2$ eV correction to the γ_1 parameter for the frontier atoms [8]. The various hopping integrals are schematically defined in Figure 14.1.

While spin-orbit and hyperfine couplings are weak in light elements such as carbon [12–14], it is also known that non-spin-polarized zigzag edges in graphene present an unstable high charge density which can be resolved by electronic charge polarization [15]. The proper description of the physics of edge states is fundamental to the successful development of nanodevices based on carbon nanoribbons as they can play a dominant role not only in infinite crystalline systems [15–17], but also in small finite-sized structures [18–20]. Such border effects persist in long one-dimensional systems even for lengths longer than 70 nm [18]. It follows that a correct description of zigzag edges in graphene must include spin interactions.

The addition of a Hubbard-like term to the Hamiltonian has been successful in describing the most relevant physical aspects of magnetic states in a number of graphitic carbon nanostructures, including finite nanoislands [21], where the results are shown to be consistent with Lieb's theorem [22]. Fundamental aspects of defect-induced magnetism in graphene and GNRs [23] as well as the magnetic states of zigzag edged GNRs [14, 15, 24] are also

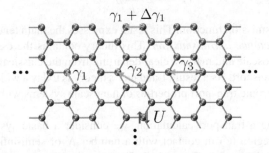

Figure 14.1 Schematic representation of the first-, second-, and third-nearest neighbor hopping integrals (γ_1, γ_2 and γ_3, respectively) on an armchair edged graphitic structure. The different chemical environment at the edges is accounted for by a modified first neighbor interaction for these frontier atoms

captured by such a simple approach. The Hubbard Hamiltonian $\hat{H}_U$ is written in terms of the number operators $\hat{n}_i^\uparrow$ and $\hat{n}_i^\downarrow$ for the spin-orbitals from atom i as:

$$\hat{H}_U = U \sum_i \hat{n}_i^\uparrow \hat{n}_i^\downarrow, \qquad (14.1)$$

where the single parameter U corresponds to the on-site repulsion. Due to the complexity of solving this equation, a mean-field interaction is usually employed and this Hamiltonian is written as:

$$\hat{H}_U = U \sum_i \left(\langle \hat{n}_i^\uparrow \rangle \hat{n}_i^\downarrow + \hat{n}_i^\uparrow \langle \hat{n}_i^\downarrow \rangle \right) \qquad (14.2)$$

where the densities $\langle \hat{n}_i^\uparrow \rangle$ and $\langle \hat{n}_i^\downarrow \rangle$ are determined self-consistently. A value of $U = 0.92\gamma_1$, as determined from DFT calculations, is reported in Ref. [11]. The resulting tight-binding+U (TBU) model is then used to predict properties of GNRs. Results show a remarkable and systematic agreement between DFT and TBU for states close to the Fermi energy.

14.2.2 Electronic Transport at the Nanoscale

When describing electronic transport at the nanoscale, we take advantage of the fact that the device size is typically smaller than the phase coherence length [25, 26] and quantum interference effects become prominent, demanding a full departure from the macroscopic description of Ohm's law. In a perfect crystalline system like a pristine GNR, for instance, electrons transmit through the system without suffering collisions, thereby marking a clear manifestation of ballistic transport. When we relax the defect free assumption (by considering doping and other structural defects like vacancies or reconstructions), electrons traveling through the system experience scattering events, thereby increasing the probability that some electrons will not be transmitted. Hence, a complete description of the transport at the nanoscale has to consider the transmission and reflection of the electronic waves. These aspects are captured by the Landauer formalism, which results in the following equation for the conductance G between two terminals (labeled by n and m) in the low bias and zero temperature limits [27, 28]:

$$G^{n \to m} = \frac{2e^2}{h} T^{n \to m}. \qquad (14.3)$$

where T is the transmission function. This result expresses the main tenet of the formalism, namely that *conductance is transmission*. This theory contains the essence of electronic transport at the nanoscale and marks a clear departure from the classical top-down description of electronic transport. As a result, Landauer's theory is today a widely used formalism for computing electronic transport properties of nanoscale systems with the aid of Green's function theory.

When performing a transport calculation, we consider a basic system made up of a central scattering region ($\mathcal{C}$) in contact with a number N of semi-infinite terminals ($\mathcal{L}_l$, $l = 1, 2, \ldots, N$) as depicted in Figure 14.2. Each terminal is composed by the repetition of a characteristic unit along a specific (periodic) direction.

The terminals do not need to be identical and we assume that each of the terminal's layers is indistinguishable from one in the corresponding bulk system. In addition, it is

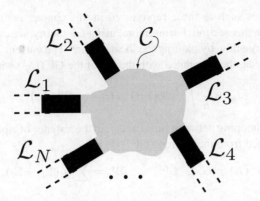

Figure 14.2 Basic system for electronic transport calculations. A central scattering region C coupled to the semi-infinite terminals $\mathcal{L}_1, \mathcal{L}_2, \mathcal{L}_3, \mathcal{L}_4, \ldots, \mathcal{L}_N$

common to adopt the principal layer (PL) concept to the electrodes, so that the terminal's PL is defined in such a way as to interact only with the first-neighboring layers. This allows the problem to be written in a convenient and numerically tractable way. In practice, Green's functions (GFs) provide a systematic way to determine the quantum transmission for the Landauer formula. Even though the Hamiltonian for the extended open system is infinite and non-periodic, we can still treat the conductor's GF (G_C) in terms of a modified Hamiltonian representing the central conductor by:

$$G_C^{r,a} = \left((E \pm i\eta)I_C - H_C - \sum_l \Sigma_l \right)^{-1} \qquad (14.4)$$

with:

$$\Sigma_l = h_{Cl} g_l h_{lC}, \quad g_l = (\epsilon I_l - H_l)^{-1}, \qquad (14.5)$$

where $\eta \to 0^+$ is a small complex number added to (subtracted from) the energy to avoid singularities in the calculation of the retarded (advanced) GF. Here, H_C (H_l) [h_{Cl} and h_{lC}] is the Hamiltonian describing C ($\mathcal{L}_l$) [the interaction between C and $\mathcal{L}_l$] and the so-called self-energy matrices (Σ_i) can be interpreted as effective potentials which describe the effects of the semi-infinite terminals on C. In other words, the sum of the H_C Hamiltonian with the self-energies represents a finite central conductor satisfying the boundary conditions corresponding to the extended system [27,28]. The Hamiltonian matrices are obtained from a properly truncated DFT or TB calculation performed in a localized basis. Even though the Σ_l terms depend on infinite matrices when they are expressed in a localized-basis, we can write them in terms of finite matrices describing the interface between C and the first PL of $\mathcal{L}_l$ and determine them by recursive methods [29,30]. Finally, the quantum transmission is obtained by [27,28]:

$$T^{n \to m} = Tr(\Gamma_n G_C^r \Gamma_m G_C^a), \qquad (14.6)$$

with:

$$\Gamma_l = i(\Sigma_l - \Sigma_l^\dagger). \qquad (14.7)$$

Conductance curves such as those reproduced in this chapter are computed using the equations provided in this section. Furthermore, using GF theory, we can attain insight into the profile of the charge flow by plotting the local current. The current I_{ij} between any two atomic sites, i and j, can be calculated with the aid of the GF ($G^<$) with [31, 32]:

$$I_{ij} = \frac{e}{h} \int_{-\infty}^{\infty} dE \left(t_{ij} G_{ji}^<(E) - t_{ji} G_{ij}^<(E) \right), \tag{14.8}$$

where $t_{ij} = t_{ji}^*$ is the hopping between sites i and j. In the absence of electronic correlations, $G^<$ is directly obtained from the retarded GF (G^r) by:

$$G^<(E) = G^r \Sigma^< G^{r,\dagger}; \quad \Sigma^< = \sum_l f_l (\Sigma_l^\dagger - \Sigma_l), \tag{14.9}$$

where the sum runs over all the terminals, and f_l is the corresponding Fermi distribution used to fill the terminal states according to the chemical potential μ_l.

14.3 From Graphene to Ribbons

14.3.1 Graphene

Graphite is an abundant and well-known material. Yet, its isolation into individual *graphene* sheets was only accomplished in 2004 [3]. This result marked the starting point of intense research on both experimental and theoretical investigations, making graphene a focal point in materials science. Graphene's structure is composed of a honeycomb lattice, where carbon atoms are arranged on a bidimensional hexagonal motif. Its electronic structure can be accurately described within a tight-binding model. Setting $E_F = 0$ and considering only first-neighbors interactions, we write:

$$E(\mathbf{k}) = \pm \gamma \sqrt{3 + 2 \cos \mathbf{k} \cdot \mathbf{a}_1 + 2 \cos \mathbf{k} \cdot \mathbf{a}_2 + 2 \cos \mathbf{k} \cdot (\mathbf{a}_1 - \mathbf{a}_2)}. \tag{14.10}$$

where $\mathbf{a}_1$ and $\mathbf{a}_2$ are graphene lattice vectors and $\mathbf{k}$ is a quasi-momentum chosen in the first Brillouin zone. The corresponding dispersion relation is plotted in Figure 14.3 over the graphene's hexagonal Brillouin Zone (BZ) using different visualization methods.

A notable characteristic of the $E - \mathbf{k}$ relation is the presence of a conical structure at the K and K' points (vertices of the BZ where the valence and conduction bands meet).

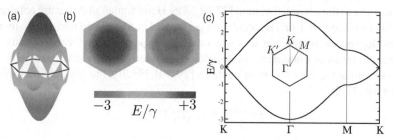

Figure 14.3 *Electronic band structure for graphene over the Brillouin zone in a 3D (a) and 2D (b) representations and along the high symmetry lines (c). The Fermi level is at $E = 0$*

Due to this local linear relation for low-energy levels, the electrons behave as massless Dirac fermions leading to the onset of Klein tunneling (where an electron enters a potential barrier with unity transmission probability) [33]. This effect was predicted by theory for a graphene $p - n$ junction [34] and further confirmed by experiments [35].

14.3.2 Graphene Nanoribbons

Graphene is regarded as a promising candidate for replacing silicon technology as silicon approaches its miniaturization and performance limit [36]. One of the reasons for such interest in graphene is its high electronic mobility and low contact resistance [37, 38]. Even though graphene itself has a rich physics, both chemical and physical modifications can be used to further tune its properties [39]. For example, ideal infinite graphene is not a semi-conducting system and the absence of an energy gap is a fundamental impediment for some applications in nanoelectronics. To remedy this issue, it is possible to modify graphene's structure to induce an opening of the energy bands around the Fermi energy. One widely studied approach is to induce quantum confinement along one in-plane direction, thus creating structures called graphene nanoribbons, which are the topic of this chapter. GNRs present electronic properties strongly dependent on their width and edge structure. For the most symmetric cases of armchair (n-AGNR, Figure 14.4 a) and zigzag (n-ZGNR, Figure 14.4 b) edged ribbons, their width is trivially obtained from the number n of $C-C$ dimer lines or zigzag strips, respectively.

Theory predicts that AGNRs present a semi-conducting character with a band gap Δ_n strongly dependent on the number n of $C-C$ lines along its width. While the gap of an n-AGNR with $n \to \infty$ tends to zero (so as to recover the graphene result), the Δ_n versus n curve has three different branches such that $\Delta_{3i+1} \geq \Delta_{3i} \geq \Delta_{3i+2}$ [17]. In Figure 14.5 we show the calculated band gaps for various AGNR width as a function of n using the tight-binding method described in Section 14.2.

Compared to AGNRs, ZGNRs present a richer set of properties. While spin polarization is absent in AGNRs, ZGNRs possess ferromagnetically polarized edges with two

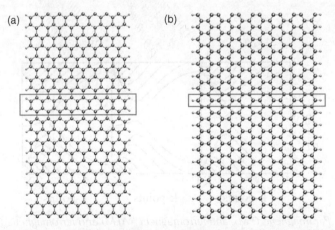

Figure 14.4 Basic structures for AGNRs (a) and ZGNRs (b). The boxes indicate the GNRs unit cells. The periodic direction is along the vertical axis

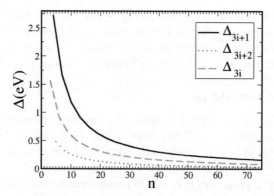

Figure 14.5 *Electronic band gap Δ for an AGNR as a function of the number n of C–C lines. The three families correspond to $n = 3i + j$ with $j = 0, 1, 2$*

possibilities for edge-to-edge polarization. These two possibilities correspond to parallel (ferromagnetic–FM) and anti-parallel (anti-ferromagnetic–AFM) alignments, the latter being the overall ground state. The spin polarization for these two states together with the non-polarized paramagnetic (PM) spin distribution are illustrated in Figure 14.6 with their corresponding band structures calculated using the TBU method.

One observes that the PM state has two two-fold degenerate bands around the Fermi energy which meet and become a flat four-fold degenerate band that extends along one third of the BZ and whose energy value approaches $E_F = 0$ as the ribbon width increases

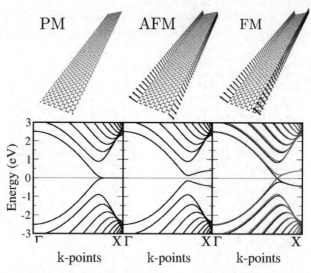

Figure 14.6 *Paramagnetic (PM), anti-ferromagnetic (AFM) and ferromagnetic (FM) states in a ZGNR and their corresponding band structures (the middle line is the Fermi energy, and black/grey lines correspond to spin up/down levels*

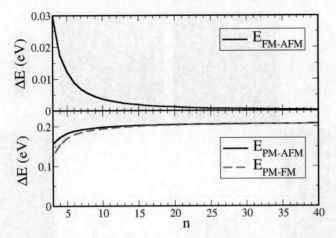

Figure 14.7 *Energy difference between any pair of different magnetic states in a ZGNR with n = 3, ..., 40. The calculation was performed using the TB band-structure energy. The difference in energy indicates that the magnetic ordering is easily destroyed at room temperature, especially for wide enough ZGNRs*

[14]. These states are strongly localized along the edges (therefore the four-fold degeneracy: two due to the spin and two due to the two symmetric edges), producing a high concentration of low energy electrons [15]. Such edge states are predicted to be responsible for a paramagnetic behavior of ZGNRs at low temperatures [40], while a diamagnetic behavior is expected at high-temperature. This high density indeed produces an instability (paramagnetic instability) which gives rise to the two lower energy magnetic states. One observes that the spin up and down polarizations along the opposite edges are located on different graphene sublattices for the AFM case which turns out to be the ground state [15]. In the FM case, edge atoms belonging to both sub-lattices present the same spin orientation and this ends up raising slightly the FM energy compared with AFM. While AFM is lower in energy, a remarkably interesting fact about these AFM and FM states is the small energy difference between them. In the upper panel of Figure 14.7 we show the TBU band-energy difference between the AFM and FM states for one ZGNR cell as a function of n. It is clear that this difference vanishes as the ribbon width increases, indicating a lowering in the edge-to-edge interaction as they are farther apart. A possible switching property based on this low energy difference is an interesting aspect that motivates the concept of a ZGNR based magnetic sensor [41]. These states are considerably more stable than the PM states [15]. As shown in the lower panel of Figure 14.7, this paramagnetic instability approaches its upper limit as the the ribbon's width is increased.

The existence of these rich magnetic properties opens up a number of exciting possibilities for the use of finite pieces of graphene in nanoelectronics and spintronics. For instance, it has been shown that ZGNRs present a half-metallicity behavior (where the electronic structure has a metallic character for spin-up levels and is semi-conducting for spin-down levels, or vice- versa) which can be tuned using a gate voltage [16]. The importance of the details of the conductance and valence states in ZGNRs in their various spin-distributions is further demonstrated by a plot of the local current distribution using the formalism embodied by

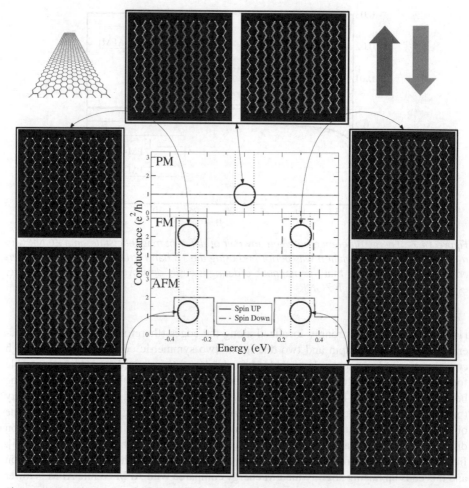

Figure 14.8 Quantum conductance as a function of energy (center) for the PM, FM and AFM states in a 12-ZGNR (full blue curves correspond to spin-up and red dashed lines to spin-down). For each state we present local current plots corresponding to $\mu_1 = -0.35$ eV and $\mu_2 = -0.25$ eV and to $\mu_1 = +0.25$ eV and $\mu_2 = +0.35$ eV for both the AFM and FM states and $\mu_{1/2} = +/-0.05$ eV for the PM case (chemical potential windows marked by vertical black dotted lines), both to spin-up (blue) and -down (red). See colour version in the colour plate section

Eq. 14.8. Figure 14.8 is an overview of the spin-polarized conductance at various chemical potentials for the PM, FM, and AFM states. Here, the structures are connected to perfect electrodes made of the same materials and it follows that these representations are similar to the spatial distributions of Bloch states. For example, in the ground state (AFM), a remarkable 1D localization of the states along the edge can be seen with spin-up and spin-down channels localized at opposite sides the structure. Note that experimental evidence indicates that edges are usually made up of a mixture of armchair and zigzag edges, thereby forming what's typically referred to as *chiral* edges (see, e.g., Refs. [42, 43] and references therein).

14.4 Graphene Nanoribbon Synthesis and Processing

A large number of systems studied theoretically are often deemed unrealistic due to their narrow sizes and/or to their highly specific atomic structure. While graphene's isolation is a recent experimental achievement, the synthesis of narrow nanoribbons has also been significantly improved over the past few years. However, these accomplishments need to present additional quality for technological applications: their clean production in bulk quantity needs to be practical. In addition, large-scale deployment of GNR-based electronics calls for the possibility of packing billions of such devices onto a centimeter-wide chip wafer. Graphene pieces with well-defined size and edge structure are essential for such applications [44]. This demands a high level of control over the synthesis and manipulation of nanostructures in order to stitch them together in specific and highly ordered conformations. To address this pressing issue, a series of top-down and bottom-up approaches have been developed to obtain GNRs and other graphene like systems [37]. The former route is mainly concentrated on etching pre-synthesized graphene and lithography methods [45–47] as well as carbon nanotube unzipping [48–50]. However, in many cases, the existence of stochastic defect distributions linked to the presence of these topological irregularities in graphene prevents faithful reproducibility in the fabrication process [37, 46].

At the same time, as it is difficult to create GNRs with defect free edges, it has been shown that the intriguing electronic and magnetic properties of GNRs are strongly related to clean edges. Edge defects can effectively suppress some of the most important properties of GNRs for applications [51]. Unfortunately, most GNR synthesis approaches do not allow full control over clean edge formation. Moving toward the ultimate goal of atomically sharp edges, bottom-up approaches are successful alternatives for large-scale production of GNRs. They are typically based on chemical vapor decomposition (CVD) techniques and on the assembling of complex structures by fusing small molecular building blocks [37, 43, 52–55]. The general strategy is to heat a gaseous solution of molecular precursors over a given substrate in order to promote growth of more complex structures. These approaches take advantage of surface assisted reactions which reduce the degrees of freedom of the molecular building blocks and can result in narrow highly crystalline systems. During this synthesis process, a number of aspects have to be carefully taken into account as they can prevent the success of the synthesis protocol to obtain the desired product. The interplay between diffusion energies and surface coupling is one of these factors [37, 53, 55–57]. In addition, recent scanning tunneling microscopy images have indicated the presence of non-hexagonal rings, such as pentagons, at the edge of CVD-grown GNRs [43], while joule-heating post-synthesis approaches were shown capable of cleaning up rough structures into perfectly crystalline edged-systems [58].

The choice of the molecular precursors is a critical determining feature in such pre-programmed processes for obtaining graphitic ribbons. The experiment conducted by

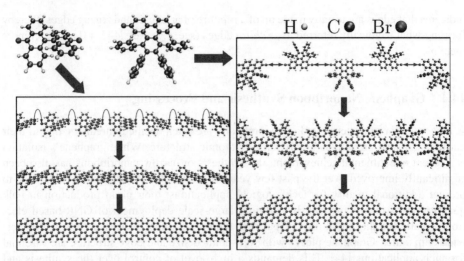

Figure 14.9 *Illustration of the bottom-up approach to obtain graphene nanoribbons with clean armchair edges and wiggle-like structures based on specific choices of the precursor monomer [54]. In the first thermally activated stage, the halogens are removed from the molecules and diffusion of the radicals over the substrate surface allows the formation of a linear polymer chain. In the following stage, a dehydrogenation process takes place, resulting in the assembly of narrow and crystalline ribbons*

Cai et al., [54] is a remarkable example of this aspect. While conducting a surface-assisted pre-programmed reaction (see Figure 14.9), they showed that 10,10'-dibromo-9,9'-bianthryl molecules used as precursor yield straight and narrow crystalline carbon nanoribbons with an armchair edge geometry. In addition, starting with a 6,11-dibromo-1,2,3,4-tetraphenyltriphenylene molecule precursor, the final result is a more complex chevron-like (or wiggle-like) ribbon [54], as illustrated in Figure 14.9. This approach is also capable of forming other complex structures including three-terminal GNR junctions.

14.5 Tailoring GNR's Electronic Properties

Carbon based materials are sought as candidates for integrated nanoelectronics to overcome the shortcomings of silicon and to maintain the current development trends that have followed Moore's law for the past 45 years. Graphene nanoribbons are among the most promising materials for this quest, due to their unique structure-dependent electronic properties [40, 59, 60], as reviewed in Section 14.3. GNRs can be either semi-conducting or metallic, depending on their edge geometry and magnetic order [15, 17]. However, the energy gap is strongly dependent on the nanoribbon width [17]. For experimentally feasible nanoribbon widths, semiconductive GNRs have energy band-gaps on the order of a few hundredths to a few tenths of eV, which are comparable to thermal energies and are impractical for integrated nanoelectronic applications. Several methods have been explored to increase the energy gap and modulate the electronic properties of GNRs, ranging from defect-based to chemical-based methods.

14.5.1 Defect-Based Modifications of the Electronic Properties

14.5.1.1 Non-Hexagonal Rings

Terrones and Mackay proposed that pentagon and heptagon (5–7) pairs could be introduced in planar graphitic structures without modifying the long-range planarity of the structure, as opposed to the curvature changes introduced by only pentagons or heptagons [61]. 5–7 pairs and other higher order rings, such as 5–8–5 groups, could be found as grain boundaries between graphitic domain with different orientations, and similar structures were later observed experimentally by Simonis and coworkers [62]. Koskinen et al. proposed that un-passivated zigzag edges in graphene nanoribbons could undergo a bond rotation that would change the hexagons at the zigzag edge into a sequence of pentagons and heptagons, transforming the zigzag edge into an armchair one [63]. Later, the same group reported the observation of such edges [64] from analysis of high resolution transmission electron microscope (HRTEM) images obtained by Girit et al. [65]. In a similar study, Dubois et al. later investigated the effects of pentagon and heptagon rings at the edges of armchair graphene nanoribbons, showing that defects break the aromaticity of the edge atoms and severely affect the conductance of the system [66].

Defects play an important role in the science of sp^2 carbon materials and offer an additional degree of freedom to tune their properties. Extended lines of defects (ELDs – Figure 14.13b) and grain boundaries (GBs) are natural defects that can be found in some synthesized graphene samples and they can assume a highly crystalline organization. Both extended line defects [67] and grain boundaries [68] in graphene are predicted theoretically to have interesting electronic and transport properties stemming from the interface between systems of varying properties. These structures do not necessarily represent a problem for the goal for producing high crystalline structures, but instead open a set of new possibilities to modify and tune the properties of graphene and their GNRs to suit new applications [62, 67–71]. Botello-Mendez et al. used a chain of 5–7 pairs in order to create a seamless interface between an armchair and a zigzag nanoribbon, creating hybrid graphene nanoribbons that display new properties emerging from the presence of both zigzag and armchair edges [72]. When stitched along the periodic direction, these hybrid graphene nanoribbons display half-metallicity (Figure 14.10), as the zigzag edge presents a spin polarized edge state, while the armchair edge is non-magnetic, resulting in a spin polarized conductor. On the other hand, if stitched transverse to the periodic direction, these hybrid graphene nanoribbons have both zigzag and armchair domains, and the electronic transport across these junctions is driven by tunneling phenomena.

An even more profound example of the impact of defects in carbon nanostructures are the Haeckelite lattices (Figure 14.13b). In these systems, an extended distribution of Stone–Wales defects is introduced over the graphene sheet so as to produce a perfectly threefold coordinated flat structures without any hexagons [73]. Even though experimental realization of these structures is still lacking, they present a set of new properties that can be further expanded, in particular in their tubular [73] and nanoribbon versions. The existence of such exciting properties motivates future experimental studies aimed at their synthesis.

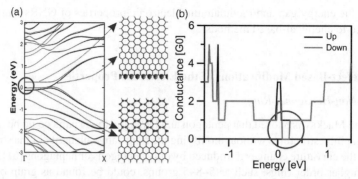

Figure 14.10 (a) Band structure of a hybrid graphene nanoribbon with both zigzag and armchair edges. The arrows point to the wave-function plot of the states around the Fermi energy, displaying the edge state associated to the zigzag edge. (b) Electronic transport plot for this nanoribbon, revealing that the nanoribbon can be used as a spin filter on an energy window just above the Fermi energy

14.5.1.2 Edge and Bulk Disorder

In order to provide accurate estimates of the stability of GNR-based electronic devices, an accurate account of the defects and their effects on transport properties must be taken, as these are inherent to any large scale production system. Several groups have studied the effects of edge and bulk disorder on the transport properties of graphene nanoribbons [74–79]. Specifically, Areshkin et al. used a recursive model to remove edge atoms to create vacancies at the edges that span several layers of atoms [74]. They also found that zigzag edged GNRs are more resistant to edge degradation than their armchair counterparts. While the zigzag edge is able to withstand large (50%) edge defect concentrations up to four edge layers deep, in the armchair case only a 10% erosion of the outer edge layer led to suppression of electronic transport.

Furthermore, two similar works used a GNR model with a random edge disorder for widths similar to experimentally available devices [75, 78, 80]. They found that even moderate edge roughness is enough to cause localized scattering centers leading to Anderson localization and creating an electron transport gap, in accordance with the experimental results [80].

Finally, atomic vacancies have also been proposed to modify local properties [81], resulting in an enhancement of the system's reactivity and opening a set of new possibilities for the physics and chemistry of these structures.

14.5.2 Electronic Properties of Chemically Doped Graphene Nanoribbons

14.5.2.1 Substitutional Doping of Graphene Nanoribbons

Substitutional doping by non-carbon atoms has been widely studied as a method for tailoring the electronic properties of other carbon-based materials such as fullerenes and carbon nanotubes. The rationale behind this interest is that ions of atoms with similar size to carbon (boron and nitrogen, in particular) could be easily inserted into the graphitic hexagonal lattice. In addition, it was earlier found that boron and nitrogen could respectively introduce acceptor or donor states in carbon nanotubes [82]. Substitutional doping of graphene has

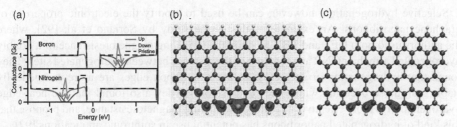

Figure 14.11 (a) Electronic transport of boron and nitrogen doped graphene nanoribbons. The arrows indicate the donor-like (boron) and acceptor-like (nitrogen) states. Local density of states (LDOS) plots for the nitrogen (b) and boron (c) localized states at the energies indicated by the arrows on the plot (a)

been experimentally achieved by different methods, such as electrothermal reactions [83], hydrocarbon pyrolysis in presence of ammonia [84], and arc discharge [85], among others.

A number of theoretical studies of the effect of substitutional doping on graphene nanoribbons electronic transport have been carried out by several groups. One of the earliest studies of such effects were carried by Martins and coworkers who focused on doped narrow graphene nanoribbons, finding that introduction of doping atoms that interact with the carbon π states creates a spin anisotropy in the electronic transport around the energy corresponding to the localized state near the dopant [86, 87]. It was proposed that substitutional doping could be used to create graphene-based spin polarized conductors. Additionally, other work studied the effects of boron and nitrogen doping on wider graphene nanoribbons [88, 89], and Cruz-Silva *et al.* examined the ramifications of substitutional doping with phosphorus on GNRs [90]. A surprising effect of doping found in these cases is that for zigzag GNRs, both nitrogen and boron could induce donor or acceptor states depending on their position within the nanoribbon. As observed in Figure 14.11, the energies of the localized states created by the doping atoms change as their position in the nanoribbon is closer to the edges, until finally crossing over the Fermi energy and moving to the other side of the energy spectrum. This effect can be explained by the increased exchange interaction due to the localization of the dopant-induced states and the zigzag edge states [90].

14.5.2.2 Chemical Functionalization of Graphene Nanoribbons

A different pathway for modifying the electronic and transport properties of graphene nanoribbons is through the use of chemical functionalization. As mentioned before, modifications of the carbon π network, by doping or chemical interactions, result in changes of the electronic structure near the Fermi energy. Such changes allow tailoring of the electronic transport properties at low voltages [87]. There are a variety of techniques to enable functionalization of graphene such as hydrogenation [91, 92], oxidation [93–95], and chemical attachment of metallic atoms [96].

Fully hydrogenated graphene (graphane) was first theoretically proposed by Sofo et al. [97], and experimentally achieved by Elias et al. [91]. On an sp^2 carbon atom, hydrogen can only bond to the free π_z orbital, forcing a change to an sp^3 hybridization on the host atom and forming a σ bond. Since full hydrogenation of graphene results in the rehybridization of all carbon atoms, graphane can be thought of as a single diamond layer, and is a wide gap semiconductor (as is diamond).

Selective hydrogenation, however, can be used to modify the electronic properties of graphene nanoribbons. An example of this is the study by Soriano et al. [92], where the authors used both DFT and TBU to study the tunneling magnetoresistance of a doubly hydrogenated armchair GNR as a function of the distance between hydrogenated sites. They found that hydrogenation of sites close to the nanoribbon edges are more energetically stable than those occuring within the central atoms. They also found that the near-edge hydrogenation is more effective at creating a tunneling magnetoresistance, and propose that this kind of hydrogenated nanoribbons has potential use in spintronic applications [92].

Oxygen, as well as other functional groups that attach to sp^2-based carbon nanostructures, can modify the $\pi-$ orbital network. It was shown early on that by attaching oxygen, hydroxyl (OH), or imine (NH) groups at the edges of a zigzag GNR, it could be possible to indirectly close the energy gap in the antiferromagnetic ground state of zigzag GNRs [93]. Using localized Wannier functions, Cantele et al. showed that the states induced by oxygen at the Fermi energy are due to lone pair states [95]. Another approach to study the effects of oxidation is based on a decimation method to perform quantum transport calculations on real-length graphene nanoribbons (up to 600 nm) [94]. They found that the elastic mean free path for these nanoribbons strongly decays with increasing number of hydrogen and hydroxyl functional groups attached to the nanoribbon. Other attempts to modify the electronic structure of GNRs are based on connecting third row transition metal atoms to manipulate the electronic transport properties of GNRs [96, 98].

14.5.3 GNR Assemblies

14.5.3.1 Nanowiggles

It has been pointed out that systems such as the segmented structure obtained using a surface-assisted bottom-up approach [54] can be viewed as an example of a more general set of structures called graphitic nanowiggles (GNWs) [11], since they present, in general, a wiggly edge conformation. Such GNWs can be viewed as a successive repetition of GNR sectors which can be either armchair (A) or zigzag (Z) edged (resulting in the four achiral GNW classes illustrated in Figure 14.12). Their electronic structure [11, 99] and thermoelectric transport properties [100, 101] have been studied using an array of computational approaches. Regarding their electronic structure, GNWs with at least one

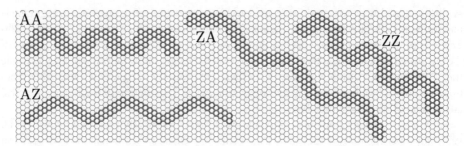

Figure 14.12 *The four classes of achiral GNWs: AA, AZ, ZA and ZZ, depending on when the parallel and oblique sectors, respectively, present either an armchair (A) or zigzag (Z) edge. The figure represents three unit cells from each case*

zigzag sector present a set of multiple magnetic states with a rather larger diversity compared to the corresponding set from ZGNRs. Further, depending on how individual sectors are arranged together, they can present a large, quasi-continuous, range of band-gaps [11].

14.5.3.2 Antidots and Junctions

In addition to GNWs, graphene and GNRs have been assembled into several new structures. Porous systems like graphene antidot lattices, for instance, have been shown to allow a controlled manipulation of graphene's electronic properties [102] as well as on carbon nanoribbons [103, 104]. Other proposals exploit the interplay between armchair and zigzag edges in more complex ribbon geometries to demonstrate spin-filter devices and geometry-dependent controlling approaches for the localization of magnetic edge-states [105–110]. Junctions composed of A- and ZGNRs [111] and 1D GNR-superlattices [108, 109] (Figure 14.13a) are also new structures suitable for embedding in new electronic nanodevices.

14.5.3.3 GNR Rings

While GNR-made nanorings are expected to be more stable than carbon nanotube-based tori, these systems impose boundary conditions on the electronic states, resulting in fascinating transport properties which include energy selectivity rules for the electronic

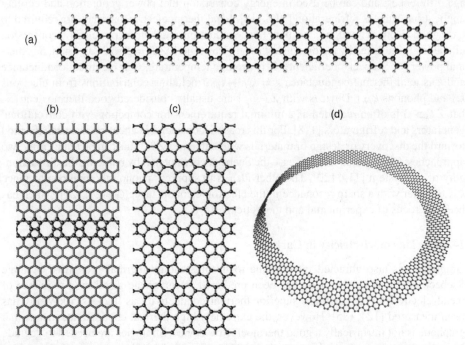

Figure 14.13 Atomic models for the assembling of graphene and GNRs into new structures: (a) 1D GNR-based superlattice; (b) extended line of defects; (c) Haeckelite lattice; (d) Möbius-like GNR nanoring

paths on the structure [112]. In a separate theoretical prediction, Möbius rings (Figure 14.13d) have been predicted to present specific UV/vis spectra signatures [113] and characteristic charge localization due to curvature [114].

14.5.3.4 GNR Stacking

The stacking of multiple GNRs is predicted to have a major influence on the electronic structure and magnetic states of these multi-layer systems due to an interplay between intralayer and interlayer coupling [115]. On the other hand, the behavior of the electronic current as a function of temperature and device length is now well understood in terms of *ab initio* calculations [116]. Note also that the stacking of two GNRs bound by van der Waals forces can be used as a rheostat with cross-GNR conductance being finely tuned by the relative angle between the two GNRS [111]. In this case, multi-terminal transport formalism must be employed (Eq. 14.6) to evaluate electronic transmission between any two sides of one layer and the corresponding ones of the second layer.

14.6 Thermoelectric Properties of Graphene-Based Materials

14.6.1 Thermoelectricity

The thermoelectric effect is the direct conversion between temperature gradients and voltage differences, and can be used in energy conversion like power generation and cooling applications such as refrigeration [117, 118]. Ideal thermoelectric materials are required to have good electric conduction yet poor heat conduction and their thermoelectric conversion efficiency is indicated by the dimensionless figure of merit $ZT = S^2 GT/k$, where S is thermal power (or Seebeck coefficient), T is average temperature, G is electrical conductance and k is total thermal conductance $k = k_{el} + k_{ph}$ (including contributions from electrons k_{el} and phonons k_{ph}). Devices with $ZT > 1$ are usually considered good thermoelectrics, but $ZT > 3$ is often regarded as a minimal requirement for competing with conventional generators and refrigerators [118]. For this reason, considerable attention has been devoted toward the discovery or design of materials with enhanced ZT. This can be realized by two approaches. One method is to increase the Seebeck coefficient S by reducing the dimensionality of the system [119, 120]. The other idea is to suppress phonon thermal conductance k_{ph} together with a sharp resonance in the electronic conductance G [121, 122], which has been the focus of experimental and theoretical efforts.

14.6.2 Thermoelectricity in Carbon

Graphene has also attracted interest due to its outstanding thermal properties. A huge Seebeck coefficient (30 mV/K) has been predicted for graphene gated by a sequence of metal electrodes [123] while a superior thermal conductivity as high as 5 kW/mK has been measured [124, 125]. However, the extremely high thermal conductance means that graphene is not intrinsically a good thermoelectric device. Strategies are needed to degrade its thermal conductance while preserving or enhancing the high electronic conductance and thermopower characteristics. Several enabling routes have already been examined, such as the use of isotopes [126], strain [127], edge passivation [128, 129], random hydrogen vacancies [130], atomic vacancies and defects [120, 131, 132], molecular junctions [121, 133], multi-juctions GNRs and superlattices [100, 101, 122, 132], and edge disorder [134].

It was discovered that AGNRs show lower thermal conductance than ZGNRs of similar width, suggesting that AGNRs are better candidates to be optimized for thermoelectric performance [100, 129]. However, peak ZT values for pristine AGNRs at room temperature are far less than 1. Theoretical calculations by Chen et al. [100] indicate that peak ZT value for the narrowest AGNR is less than 0.4 and it generally decreases with a width increase. Atomic vacancies and Stone–Wales defects have been introduced to suppress phonon thermal conductance. A 0.23% double vacancy or Stone–Wales concentration leads to a 80% reduction in thermal conductance of GNRs from pristine ones [131]. However, corresponding ZT values at room temperature are not significantly enhanced (less than 0.5) due to the concomitant reduction of the electrical conductance [132] reported that the reduction of phonon thermal conductance by random vacancies can be fully compensated by the concomitant reduction of electrical conductance, which results in a reduction of ZT compared to perfect AGNRs. However, a periodic distribution of vacancies inside the ribbon may be seen as a reduction of the effective width, which tends to reduce the thermal conductance while enhancing the thermopower, leading to higher ZT. Additionally, multi-junctions GNRs have been proposed to degrade phonon thermal conductance [100, 101, 122]. A specific patterning of a mixed GNR obtained by alternating armchair and zigzag sections of different widths has been shown to provide high thermoelectric performance with a ZT factor reaching unity at room temperature (Figure 14.14) [122]. Besides the reduction of phonon conductance, the resonant tunneling of electrons between these sections retains high electron conductance and Seebeck coefficient. Finally, the introduction of edge disorder has been predicted to dramatically reduce phonon thermal transport while only weakly decreasing electronic conduction [134], resulting in ZT higher than 1 at room temperature in the diffusive limit (Figure 14.15).

Although theoretical calculations indicate that GNRs can be modified to have good thermoelectric performance, those modifications are experimentally challenging. The newly synthesized wiggle-type GNRs, so called graphitic nanowiggles (see Section 14.5.3 and

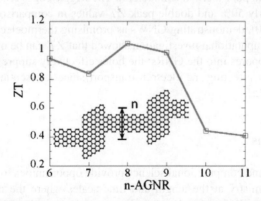

Figure 14.14 Maximum ZT value as a function of the number of dimers n in armchair sections in the multijuction structure [122]. Reprinted with permission from [122] © 2011 the American Physical Society

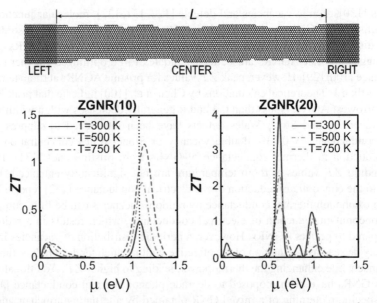

Figure 14.15 *Thermoelectric properties in edge-disordered GNR. The central region with edge disorder has length L and is connected to semi-infinite ZGNRs free of disorder. Thermoelectric figure of merit, ZT, versus chemical potential μ at different temperatures T for ZGNR10 and ZGNR20 having lengths L = 0.25 nm and 4 μm, respectively [134]. Reprinted with permission from [134] © 2010 the American Physical Society*

Figure 14.12), have been fabricated with an atomically precise bottom-up approach using surface-assisted coupling of molecular precursors into linear polyphenylenes. Their intrinsic electronic and magnetic properties have been demonstrated for a number of arrangements and edge symmetries [11]. The attractiveness of these structure is their high-order (making them good electronic conductors) with regularly repeated junctions between different domain (making them poor thermal conductors). The wiggle edges reduce phonon thermal conductance by nearly 50% and double peak ZT values in comparison to straight GNRs of similar width [100], demonstrating GNWs as promising thermoelectric materials. Note also that a recent computational investigation showed that ZT can be optimized up a value of 5 by drilling nanopores into the GNRs: the holes effectively suppress phonon transport without significantly affecting the electron transport since it essentially flows along the edges of ZGNRs [135].

14.7 Conclusions

Modern theoretical and computational science provide opportunities to explore fundamental physics and chemistry at the length and time scales where the materials properties and behavior are controlled and is thereby a key to perform predictive design. When integrated with experimental studies at the same time and length scales, validated discoveries are possible, enabling potential breakthroughs in areas such as nanoelectronics. In this chapter we have reviewed recent research in the area of electronic transport of GNRs and GNR assemblies. Electron transport in GNRs is a rapidly moving field and the present review is not an attempt to be exhaustive. Instead, we have presented an overview of the major research directions that can provide a viable pathway towards the fine-tuning of the properties of GNR systems for future applications in nano electronics optoelectronics and spintronics.

Acknowledgements

ECG acknowledges support from Brazilian agencies CAPES (process 0327-10-7) and CNPq (process 140887/2008-3). ECS was supported in part by PHASE, an EFRC funded by the US-DOE under Award Number DE-SC0001087. VM and BGS acknowledge support from the Center for Nanophase Materials Sciences (CNMS), sponsored at ORNL by the Scientific User Facilities Division, Office of Basic Energy Sciences, U.S. DOE. VM is also supported by New York State under NYSTAR contract C080117. Some of the original computations used resources of the Oak Ridge Leadership Computing Facility and the NCCS.

References

[1] Kroto HW, Heath JR, Obrien SC, Curl RF, Smalley RE. C-60 – Buckminsterfullerene. *Nature*. 1985;318(6042):162–163.
[2] Iijima S. Helical microtubules of graphitic carbon. *Nature*. 1991;354(6348):56–58.
[3] Novoselov K, Geim A, Morozov S, Jiang D, Zhang Y, Dubonos S, et al. Electric field effect in atomically thin carbon films. *Science*. 2004;306(5696):666–669.
[4] Sumpter BG, Meunier V. Can computational approaches aid in untangling the inherent complexity of practical organic photovoltaic systems? *Journal of Polymer Science Part B: Polymer Physics*. 2012.
[5] Milnera M, Kürti J, Hulman M, Kuzmany H. periodic resonance excitation and intertube interaction from quasicontinuous distributed helicities in single-wall carbon nanotubes. *Physical Review Letters*. 2000;84(6):1324–1327.
[6] Saito R, Dresselhaus G, Dresselhaus MS. Trigonal warping effect of carbon nanotubes. *Physical Review B*. 2000;61(4):2981–2990.
[7] Reich S, Maultzsch J, Thomsen C, Ordejón P. Tight-binding description of graphene. *Physical Review B*. 2002;66:035412.
[8] Gunlycke D, White CT. Tight-binding energy dispersions of armchair-edge graphene nanostrips. *Physical Review B*. 2008;77:115116.
[9] Girão EC, Souza Filho AG, Meunier V. Electronic transport properties of carbon nanotoroids. *Nanotechnology*. 2011;22(7):075701.
[10] Saito R, Dresselhaus G, Dresselhaus MS. *Physical Properties of Carbon Nanotubes*. London: Imperial College Press; 1998.
[11] Girão EC, Liang L, Cruz-Silva E, Souza Filho AG, Meunier V. Emergence of atypical properties in assembled graphene nanoribbons. *Physical Review Letters*. 2011;107:135501.

[12] Yazyev OV. Hyperfine interactions in graphene and related carbon nanostructures. *Nano Letters*. 2008;8(4):1011–1015.

[13] Fischer J, Trauzettel B, Loss D. Hyperfine interaction and electron-spin decoherence in graphene and carbon nanotube quantum dots. *Physical Review B*. 2009;80:155401.

[14] Yazyev OV. Emergence of magnetism in graphene materials and nanostructures. *Reports on Progress in Physics*. 2010;73(5):056501.

[15] Pisani L, Chan JA, Montanari B, Harrison NM. Electronic structure and magnetic properties of graphitic ribbons. *Physical Review B*. 2007;75(6):064418.

[16] Son YW, Cohen ML, Louie SG. Half-metallic graphene nanoribbons. *Nature*. 2006;444(7117):347–349.

[17] Son YW, Cohen ML, Louie SG. Energy gaps in graphene nanoribbons. *Physical Review Letters*. 2006;97(21):216803.

[18] Hod O, Peralta JE, Scuseria GE. Edge effects in finite elongated graphene nanoribbons. *Physical Review B*. 2007;76:233401.

[19] Hod O, Barone V, Scuseria GE. Half-metallic graphene nanodots: A comprehensive first-principles theoretical study. *Physical Review B*. 2008;77:035411.

[20] Jiang D, Dai S. Circumacenes versus periacenes: HOMO-LUMO gap and transition from nonmagnetic to magnetic ground state with size. *Chemical Physics Letters*. 2008;466(1–3):7275.

[21] Fernández-Rossier J, Palacios JJ. Magnetism in graphene nanoislands. *Physical Review Letters*. 2007;99:177204.

[22] Lieb EH. Two theorems on the Hubbard model. *Physical Review Letters*. 1989;62:1201–1204.

[23] Palacios JJ, Fernández-Rossier J, Brey L. Vacancy-induced magnetism in graphene and graphene ribbons. *Physical Review B*. 2008;77:195428.

[24] Fernández-Rossier J. Prediction of hidden multiferroic order in graphene zigzag ribbons. *Physical Review B*. 2008;77:075430.

[25] Ferrier M, Angers L, Rowe ACH, Guéron S, Bouchiat H, Texier C, et al. Direct measurement of the phase-coherence length in a GaAs/GaAlAs square network. *Physical Review Letters*. 2004;93:246804.

[26] Miao F, Wijeratne S, Zhang Y, Coskun UC, Bao W, Lau CN. Phase-coherent transport in graphene quantum billiards. *Science*. 2007;317(5844):1530–1533.

[27] Di Ventra M. *Electrical Transport in Nanoscale Systems*. Cambridge University Press; 2008.

[28] Datta S. *Quantum Transport: Atom to Transistor*. Cambridge University Press; 2005.

[29] Lopez-Sancho MP, Lopez-Sancho JM, Rubio J. Highly convergent schemes for the calculation of bulk and surface Green-Functions. *Journal of Physics F: Metal Physics*. 1985;15(4):851–858.

[30] Sancho MPL, Sancho JML, Rubio J. Quick iterative scheme for the calculation of transfer-matrices – application to Mo(100). *Journal of Physics F - Metal Physics*. 1984;14(5):1205–1215.

[31] Caroli C, Combesco R, Nozieres P, Saintjam D. Direct calculation of tunneling current. *Journal of Physics Part C Solid State Physics*. 1971;4(8):916.

[32] Kazymyrenko K, Waintal X. Knitting algorithm for calculating Green functions in quantum systems. *Physical Review B*. 2008;77(11):115119.

[33] Das Sarma S, Adam S, Hwang EH, Rossi E. Electronic transport in two-dimensional graphene. *Reviews of Modern Physics.* 2011;83(2):407–470.
[34] Cheianov VV, Fal'ko VI. Selective transmission of Dirac electrons and ballistic magnetoresistance of np junctions in graphene. *Physical Review Letters.* 2006 Jul;74:041403.
[35] Huard B, Sulpizio JA, Stander N, Todd K, Yang B, Goldhaber-Gordon D. Transport measurements across a tunable potential barrier in graphene. *Physical Review Letters.* 2007;98(23):236803.
[36] Schwierz F. Graphene transistors. *Nature Nanotechnology.* 2010;5(7):487–496.
[37] Palma CA, Samori P. Blueprinting macromolecular electronics. *Nature Chemistry.* 2011;3(6):431–436.
[38] Xia F, Perebeinos V, Lin Ym, Wu Y, Avouris P. The origins and limits of metal-graphene junction resistance. *Nature Nanotechnology.* 2011;6(3):179–184.
[39] Novoselov K. Mind the gap. *Nature Materials.* 2007;6(10):720–721.
[40] Wakabayashi K, Fujita M, Ajiki H, Sigrist M. Electronic and magnetic properties of nanographite ribbons. *Physical Review B.* 1999;59(12):8271–8282.
[41] Munoz-Rojas F, Fernandez-Rossier J, Palacios JJ. Giant Magnetoresistance in Ultra-small Graphene Based Devices. *Physical Review Letters.* 2009;102(13):136810.
[42] Tao C, Jiao L, Yazyev OV, Chen YC, Feng J, Zhang X, et al. Spatially resolving edge states of chiral graphene nanoribbons. *Nature Physics.* 2011 AUG;7(8):616–620.
[43] Pan M, Girão EC, Jia X, Bhaviripudi S, Li Q, Kong J, et al. Topographic and spectroscopic characterization of electronic edge states in CVD grown graphene nanoribbons. *Nano Letters.* 2012;12(4):1928.
[44] Barone V, Hod O, Scuseria GE. Electronic structure and stability of semiconducting graphene nanoribbons. *Nano Letters.* 2006;6(12):2748–2754.
[45] Wang X, Dai H. Etching and narrowing of graphene from the edges. *Nature Chemistry.* 2010;2(8):661–665.
[46] Yang R, Zhang L, Wang Y, Shi Z, Shi D, Gao H, et al. An anisotropic etching effect in the graphene basal plane. *Advanced Materials.* 2010;22(36):4014–4019.
[47] Xie L, Jiao L, Dai H. Selective etching of graphene edges by hydrogen plasma. *Journal of the American Chemical Society.* 2010;132(42):14751–14753.
[48] Kosynkin DV, Higginbotham AL, Sinitskii A, Lomeda JR, Dimiev A, Price BK, et al. Longitudinal unzipping of carbon nanotubes to form graphene nanoribbons. *Nature.* 2009;458(7240):872.
[49] Jiao L, Zhang L, Wang X, Diankov G, Dai H. Narrow graphene nanoribbons from carbon nanotubes. *Nature.* 2009;458(7240):877–880.
[50] Morelos-Gomez A, Vega-Diaz SM, Gonzalez VJ, Tristan Lopez F, Cruz-Silva R, Fujisawa K, et al. Clean nanotube unzipping by abrupt thermal expansion of molecular nitrogen: graphene nanoribbons with atomically smooth edges. *ACS Nano.* 2012;6(3):2261–2272.
[51] Jia X, Campos-Delgado J, Terrones M, Meunier V, Dresselhaus MS. Graphene edges: a review of their fabrication and characterization. *Nanoscale.* 2011;3(1):86–95.
[52] Campos-Delgado J, Romo-Herrera JM, Jia X, Cullen DA, Muramatsu H, Kim YA, et al. Bulk production of a new form of sp2 carbon: crystalline graphene nanoribbons. *Nano Letters.* 2008;8(9):2773–2778.

[53] Grill L, Dyer M, Lafferentz L, Persson M, Peters MV, Hecht S. Nano-architectures by covalent assembly of molecular building blocks. *Nature Nanotechnology.* 2007;2(11):687–691.

[54] Cai J, Ruffieux P, Jaafar R, Bieri M, Braun T, Blankenburg S, et al. Atomically precise bottom-up fabrication of graphene nanoribbons. *Nature.* 2010;466(7305):470–473.

[55] Bieri M, Nguyen MT, Groening O, Cai J, Treier M, Ait-Mansour K, et al. Two-dimensional polymer formation on surfaces: insight into the roles of precursor mobility and reactivity. *Journal of the American Chemical Society.* 2010;132(46):16669–16676.

[56] Lu J, Yeo PSE, Gan CK, Wu P, Loh KP. Transforming C_{60} molecules into graphene quantum dots. *Nature Nanotechnology.* 2011;6(4):247–252.

[57] Lafferentz L, Ample F, Yu H, Hecht S, Joachim C, Grill L. Conductance of a single conjugated polymer as a continuous function of its length. *Science.* 2009;323(5918):1193–1197.

[58] Jia X, Hofmann M, Meunier V, Sumpter BG, Campos-Delgado J, Romo-Herrera JM, et al. Controlled formation of sharp zigzag and armchair edges in graphitic nanoribbons. *Science.* 2009;323(5922):1701–1705.

[59] Castro Neto AH, Guinea F, Peres NMR, Novoselov KS, Geim AK. The electronic properties of graphene. *Rev Mod Phys.* 2009 Jan;81:109–162.

[60] Nakada K, Fujita M, Dresselhaus G, Dresselhaus MS. Edge state in graphene ribbons: Nanometer size effect and edge shape dependence. *Phys Rev B.* 1996 Dec;54:17954–17961.

[61] Terrones H, Mackay AL. The geometry of hypothetical curved graphite structures. *Carbon.* 1992;30(8):1251–1260.

[62] Simonis P, Goffaux C, Thiry P, Biro L, Lambin P, Meunier V. STM study of a grain boundary in graphite. Surface Science. 2002;511(1-3):319–322.

[63] Koskinen P, Malola S, Häkkinen H. Self-passivating edge reconstructions of graphene. *Phys Rev Lett.* 2008 Sep;101:115502.

[64] Koskinen P, Malola S, Häkkinen H. Evidence for graphene edges beyond zigzag and armchair. *Phys Rev B.* 2009 Aug;80:073401.

[65] Girit ÇÖ, Meyer JC, Erni R, Rossell MD, Kisielowski C, Yang L, et al. Graphene at the edge: stability and dynamics. *Science.* 2009;323(5922):1705–1708.

[66] Dubois SM, Lopez-Bezanilla A, Cresti A, Triozon F, Biel B, Charlier J, et al. Quantum transport in graphene nanoribbons: effects of edge reconstruction and chemical reactivity. *ACS Nano.* 2010;4(4):1971–1976.

[67] Botello-Mendez AR, Declerck X, Terrones M, Terrones H, Charlier JC. One-dimensional extended lines of divacancy defects in graphene. *Nanoscale.* 2011;3(7):2868–2872.

[68] Yazyev OV, Louie SG. Electronic transport in polycrystalline graphene. *Nature Materials.* 2010;9(10):806.

[69] Lahiri J, Lin Y, Bozkurt P, Oleynik II, Batzill M. An extended defect in graphene as a metallic wire. *Nature Nanotechnology.* 2010;5(5):326–329.

[70] Lin X, Ni J. Half-metallicity in graphene nanoribbons with topological line defects. *Physical Review B.* 2011;84(7):075461.

[71] Ajayan PM, Yakobson BI. Graphene: Pushing the boundaries. *Nature Materials.* 2011;10(6):415–417.

[72] Botello-Mendez AR, Cruz-Silva E, Lopez-Urias F, Sumpter BG, Meunier V, Terrones M, et al. Spin polarized conductance in hybrid graphene nanoribbons using 5-7 defects. *ACS Nano.* 2009;3(11):3606–3612.
[73] Terrones H, Terrones M, Hernandez E, Grobert N, Charlier J, Ajayan P. New metallic allotropes of planar and tubular carbon. *Physical Review Letters.* 2000;84(8):1716–1719.
[74] Areshkin DA, Gunlycke D, White CT. Ballistic transport in graphene nanostrips in the presence of disorder: importance of edge effects. *Nano Letters.* 2007;7(1):204–210.
[75] Evaldsson M, Zozoulenko IV, Xu HY, Heinzel T. Edge-disorder-induced Anderson localization and conduction gap in graphene nanoribbons. *Physical Review B.* 2008;78(16):161407.
[76] Li TC, Lu SP. Quantum conductance of graphene nanoribbons with edge defects. *Physical Review B.* 2008;77(8):085408.
[77] Lherbier A, Biel B, Niquet YM, Roche S. Transport length scales in disordered graphene-based materials: Strong localization regimes and dimensionality effects. *Physical Review Letters.* 2008;100(3):036803.
[78] Mucciolo ER, Neto AHC, Lewenkopf CH. Conductance quantization and transport gaps in disordered graphene nanoribbons. *Physical Review B.* 2009;79(7):075407.
[79] Cresti A, Roche S. Range and correlation effects in edge disordered graphene nanoribbons. *New Journal of Physics.* 2009;11:095004.
[80] Han MY, Ozyilmaz B, Zhang YB, Kim P. Energy band-gap engineering of graphene nanoribbons. *Physical Review Letters.* 2007;98(20):206805.
[81] Amorim RG, Fazzio A, Antonelli A, Novaes FD, da Silva AJR. Divacancies in graphene and carbon nanotubes. *Nano Letters.* 2007;7(8):2459–2462.
[82] Choi HJ, Ihm J, Louie SG, Cohen ML. Defects, quasibound states, and quantum conductance in metallic carbon nanotubes. *Physical Review Letters.* 2000;84(13):2917–2920.
[83] Wang X, Li X, Zhang L, Yoon Y, Weber PK, Wang H, et al. N-doping of graphene through electrothermal reactions with ammonia. *Science.* 2009;324(5928):768–771.
[84] Qu L, Liu Y, Baek JB, Dai L. Nitrogen-doped graphene as efficient metal-free electrocatalyst for oxygen reduction in fuel cells. *ACS Nano.* 2010;4(3):1321–1326.
[85] Subrahmanyam KS, Panchakarla LS, Govindaraj A, Rao CNR. Simple method of preparing graphene flakes by an arc-discharge method. *The Journal of Physical Chemistry C.* 2009;113(11):4257–4259.
[86] Martins TB, Miwa RH, da Silva AJR, Fazzio A. Electronic and transport properties of boron-doped graphene nanoribbons. *Physical Review Letters.* 2007;98(19):196803.
[87] Martins TB, da Silva AJR, Miwa RH, Fazzio A. σ- and π-Defects at graphene nanoribbon edges: building spin filters. *Nano Letters.* 2008;8(8):2293–2298.
[88] Biel B, Blase X, Triozon F, Roche S. Anomalous doping effects on charge transport in graphene nanoribbons. *Phys Rev Lett.* 2009 Mar;102:096803.
[89] Zheng XH, Rungger I, Zeng Z, Sanvito S. Effects induced by single and multiple dopants on the transport propertiesin zigzag-edged graphene nanoribbons. *Phys Rev B.* 2009 Dec;80:235426.

[90] Cruz-Silva E, Barnett ZM, Sumpter BG, Meunier V. Structural, magnetic, and transport properties of substitutionally doped graphene nanoribbons from first principles. *Physical Review B*. 2011;83:155445.

[91] Elias DC, Nair RR, Mohiuddin TMG, Morozov SV, Blake P, Halsall MP, et al. Control of graphene's properties by reversible hydrogenation: evidence for graphane. *Science*. 2009;323(5914):610–613.

[92] Soriano D, Munoz-Rojas F, Fernandez-Rossier J, Palacios JJ. Hydrogenated graphene nanoribbons for spintronics. *Physical Review B*. 2010;81(16):165409.

[93] Gunlycke D, Li J, Mintmire JW, White CT. Altering low-bias transport in zigzag-edge graphene nanostrips with edge chemistry. *Applied Physics Letters*. 2007;91(11):112108.

[94] Lopez-Bezanilla A, Triozon F, Roche S. Chemical functionalization effects on armchair graphene nanoribbon transport. *Nano Letters*. 2009;9(7):2537–2541.

[95] Cantele G, Lee YS, Ninno D, Marzari N. Spin channels in functionalized graphene nanoribbons. *Nano Letters*. 2009;9(10):3425–3429.

[96] Gorjizadeh N, Farajian AA, Esfarjani K, Kawazoe Y. Spin and band-gap engineering in doped graphene nanoribbons. *Physical Review B*. 2008;78(15):155427.

[97] Sofo JO, Chaudhari AS, Barber GD. Graphane: A two-dimensional hydrocarbon. *Phys Rev B*. 2007 Apr;75:153401.

[98] Rigo VA, Martins TB, da Silva AJR, Fazzio A, Miwa RH. Electronic, structural, and transport properties of Ni-doped graphene nanoribbons. *Physical Review B*. 2009;79(7):075435.

[99] Costa Girão E, Liang L, Cruz-Silva E, Souza Filho AG, Meunier V. Structural and electronic properties of carbon nanowiggles. *Phys Rev B*, in press. 2012;.

[100] Chen Y, Jayasekera T, Calzolari A, Kim KW, Nardelli MB. Thermoelectric properties of graphene nanoribbons, junctions and superlattices. *Journal of Physics-Condensed Matter*. 2010;22(37).

[101] Huang W, Wang JS, Liang G. Theoretical study on thermoelectric properties of kinked graphene nanoribbons. *Physical Review B*. 2011;84(4):045410.

[102] Pedersen TG, Flindt C, Pedersen J, Mortensen NA, Jauho AP, Pedersen K. Graphene antidot lattices: Designed defects and spin qubits. *Physical Review Letters*. 2008;100(13):136804.

[103] Hatanaka M. Band structures of porous graphenes. *Chemical Physics Letters*. 2010;488(4-6):187–192.

[104] Baskin A, Kral P. Electronic structures of porous nanocarbons. *Scientific Reports*. 2011;1:36.

[105] Saffarzadeh A, Farghadan R. A spin-filter device based on armchair graphene nanoribbons. *Applied Physics Letters*. 2011;98(2):023106.

[106] Wang ZF, Shi QW, Li Q, Wang X, Hou JG, Zheng H, et al. Z-shaped graphene nanoribbon quantum dot device. *Applied Physics Letters*. 2007;91(5):053109.

[107] Hancock Y, Saloriutta K, Uppstu A, Harju A, Puska MJ. Spin-dependence in asymmetric, v-shaped-notched graphene nanoribbons. *Journal of Low Temperature Physics*. 2008;153(5-6):393–398.

[108] Sevinçli H, Topsakal M, Ciraci S. Superlattice structures of graphene-based armchair nanoribbons. *Physical Review B*. 2008;78(24):245402.

[109] Topsakal M, Sevincli H, Ciraci S. Spin confinement in the superlattices of graphene ribbons. *Applied Physics Letters.* 2008;92(17):173118.
[110] Ma Z, Sheng W. A spin-valve device based on dumbbell-shaped graphene nanoislands. *Applied Physics Letters.* 2011;99(8):083101.
[111] Botello-Mendez AR, Cruz-Silva E, Romo-Herrera JM, Lopez-Urias F, Terrones M, Sumpter BG, et al. Quantum transport in graphene nanonetworks. *Nano Letters.* 2011;11(8):3058–3064.
[112] Girão EC, Souza Filho AG, Meunier V. Electronic transmission selectivity in multi-terminal graphitic nanorings. *Applied Physics Letters.* 2011;98(11):112111.
[113] Caetano EWS, Freire VN, dos Santos SG, Albuquerque EL, Galvão DS, Sato F. Defects in graphene-based twisted nanoribbons: structural, electronic, and optical properties. *Langmuir.* 2009;25(8):4751–4759.
[114] Korte AP, van der Heijden GHM. Curvature-induced electron localization in developable Möbius-like nanostructures. *Journal of Physics: Condensed Matter.* 2009;21(49):495301.
[115] Kharche N, Zhou Y, O'Brien KP, Kar S, Nayak SK. Effect of layer stacking on the electronic structure of graphene nanoribbons. *ACS Nano.* 2011;5(8):6096–6101.
[116] Padilha JE, Lima MP, da Silva AJR, Fazzio A. Bilayer graphene dual-gate nanodevice: An *ab initio* simulation. *Physical Review B.* 2011;84:113412.
[117] Bell LE. Cooling, heating, generating power, and recovering waste heat with thermoelectric systems. *Science.* 2008;321(5895):1457–1461.
[118] Vining CB. An inconvenient truth about thermoelectrics. *Nature Materials.* 2009;8(2):83–85.
[119] Dresselhaus MS, Chen G, Tang MY, Yang R, Lee H, Wang D, et al. New directions for low-dimensional thermoelectric materials. *Advanced Materials.* 2007;19(8):1043–1053.
[120] Gunst T, Markussen T, Jauho AP, Brandbyge M. Thermoelectric properties of finite graphene antidot lattices. *Physical Review B.* 2011;84(15):155449.
[121] Saha KK, Markussen T, Thygesen KS, Nikolić BK. Multiterminal single-molecule–graphene-nanoribbon junctions with the thermoelectric figure of merit optimized via evanescent mode transport and gate voltage. *Physical Review B.* 2011;84(4):041412.
[122] Mazzamuto F, Nguyen VH, Apertet Y, Caër C, Chassat C, Saint-Martin J, et al. Enhanced thermoelectric properties in graphene nanoribbons by resonant tunneling of electrons. *Physical Review B.* 2011;83(23):235426.
[123] Dragoman D, Dragoman M. Giant thermoelectric effect in graphene. *Appl Phys Lett.* 2007;91:203116.
[124] Balandin AA, Ghosh S, Bao W, Calizo I, Teweldebrhan D, Miao F, et al. Superior thermal conductivity of single-layer graphene. *Nano Letters.* 2008;8(3):902–907.
[125] Seol JH, Jo I, Moore AL, Lindsay L, Aitken ZH, Pettes MT, et al. Two-dimensional phonon transport in supported graphene. *Science.* 2010;328(5975):213–216.
[126] Mingo N, Esfarjani K, Broido D, Stewart D. Cluster scattering effects on phonon conduction in graphene. *Physical Review B.* 2010;81(4):045408.
[127] Wei N, Xu L, Wang HQ, Zheng JC. Strain engineering of thermal conductivity in graphene sheets and nanoribbons: a demonstration of magic flexibility. *Nanotechnology.* 2011;22:105705.

[128] Hu J, Schiffli S, Vallabhaneni A, Ruan X, Chen YP. Tuning the thermal conductivity of graphene nanoribbons by edge passivation and isotope engineering: A molecular dynamics study. *Applied Physics Letters.* 2010;97:133107.

[129] Tan ZW, Wang JS, Gan CK. first-principles study of heat transport properties of graphene nanoribbons. *Nano Letters.* 2011;11(1):214–219.

[130] Ni X, Liang G, Wang JS, Li B. Disorder enhances thermoelectric figure of merit in armchair graphane nanoribbons. *Applied Physics Letters.* 2009;95:192114.

[131] Haskins JB, Kinaci A, Sevik C, Sevincli H, Cuniberti G, Cagin T. Control of thermal and electronic transport in defect-engineered graphene nanoribbons. *ACS Nano.* 2011;5:3779–3787.

[132] Mazzamuto F, Saint-Martin J, Nguyen V, Chassat C, Dollfus P. Thermoelectric performance of disordered and nanostructured graphene ribbons using Green's function method. *Journal of Computational Electronics.* 2012;11:67–77. 10.1007/s10825-012-0392-0.

[133] Nikolic BK, Saha KK, Markussen T, Thygesen KS. First-principles quantum transport modeling of thermoelectricity in single-molecule nanojunctions with graphene nanoribbon electrodes. *Journal of Computational Electronics.* 2012 Mar;11(1):78–92.

[134] Sevinçli H, Cuniberti G. Enhanced thermoelectric figure of merit in edge-disordered zigzag graphene nanoribbons. *Phys Rev B.* 2010 Mar;81:113401.

[135] Chang PH, Nikolic BK. Edge currents and nanopore arrays in zigzag and chiral graphene nanoribbons as a route toward high-ZT thermoelectrics. *Phys Rev B.* 2012 Jul(4); 16;86.

15
Graphene-Based Materials as Nanocatalysts

Fengyu Li[a] and Zhongfang Chen[b]

[a]Department of Physics and Department of Chemistry, Institute for Functional Nanomaterials, University of Puerto Rico, USA

[b]Department of Chemistry, Institute for Functional Nanomaterials, University of Puerto Rico, USA

15.1 Introduction

Nanocatalysis is one of the most exciting subfields in the emerging area of nanoscience. Graphene has brought many breakthroughs, and nanocatalysis is no exception. Some graphene derivatives can be good catalysts themselves, for example, graphene oxide (GO) is a mild and efficient catalyst for the generation of aldehydes or ketones from various alcohols, alkenes, and alkynes [1], and it is highly possible that further exploitation of surface modifications and edge defects can expand GO's catalytic activity to other reactions [2]. Graphene or reduced graphite oxide (rGO), and GO can serve as catalyst supports, which often exhibit synergistic and/or extra contributions with the supported active components to catalytic reactions.

In this chapter, we review recent efforts in developing graphene-based nanomaterials as versatile and efficient catalysts, with emphasis on the theoretical studies on their catalytic performance for electrocatalysis, photocatalysis, and CO oxidation.

15.2 Electrocatalysts

A fuel cell is an electrochemical device that oxidizes fuel (e.g., hydrogen, methanol, etc.) at the anode and reduces oxygen from air at the cathode to produce electricity [3,4]. Since fuel cells have very low or even zero emissions of harmful greenhouse gases (e.g., CO_2, NO_x, SO_x, etc.), they show promise for green energy supply.

Graphene Chemistry: Theoretical Perspectives, First Edition. Edited by De-en Jiang and Zhongfang Chen.
© 2013 John Wiley & Sons, Ltd. Published 2013 by John Wiley & Sons, Ltd.

The electrocatalysts which catalyze oxygen reduction on the cathode and fuel oxidation reactions on the anode are key components in a fuel cell, and searching for low-cost, high-performance electrocatalysts is an urgent task for fuel cells development. So far, carbon nanomaterial supported noble metal (particularly Pt) nanoparticles (NPs) are still the best choice as advanced nanoelectrocatalysts for fuel cells.

The discovery of graphene, the unique two-dimensional (2D) carbon nanomaterial, offers unprecedented opportunities for electrocatalysis. Among others, graphene can support the dispersal of metal NPs well, and promises to enhance the electrocatalytic activity of NPs in fuel cells; new catalytic active sites can be introduced, for example, by substituting some carbon atoms in the graphene skeleton by heteroatoms.

Experimentally impressive progress has been achieved in developing graphene-based nanomaterials for electrocatalysis. Pt NPs have been deposited on graphene using several advanced techniques, such as using ethylene glycol as both reductive and dispersing agent [5], and using $NaBH_4$ [6] or H_2 [7] as reductants; bimetallic NPs such as Pt/Ru [8], Pt–Co [9], and Pt–Pd [10] have also been attached to graphene. All these NP-graphene hybrids exhibit higher electrocatalytic activity towards methanol oxidation reaction (MOR). Moreover, it has been found that nitrogen-doped graphene (N-graphene) itself presents high electrocatalytic activity and remarkably good tolerance for crossover effects in oxygen reduction reactions (ORR) [11], and a novel class of monolithic Fe_3O_4 NPs supported on three-dimensional (3D) N-doped graphene aerogels have better durability than the commercial Pt/C catalyst for the ORR in alkaline media [12]. Additionally, other non-metal elements (B, S, P, I, and Si, etc.) used to dope/co-dope graphene have been demonstrated good ORR catalytic activity [13–18], or are expected to be as effective as the ORR catalyst [19], graphene oxides with oxygen-containing groups on the zigzag edges (rather than on the armchair edges or surface sites) have been revealed to be active for ORR [20].

In the following sections, we will focus on theoretical investigations on designing N-graphene and graphyne-based materials for oxygen reduction reaction.

15.2.1 N-Graphene

With large-scale production, the cost of N-graphene is expected to decrease tremendously. Thus, N-graphene may replace platinum in the long run. To facilitate this process, a fundamental understanding of N-graphene and its catalytic activity is necessary, to which density functional theory (DFT) studies have contributed greatly in last few years.

Through studying the ORR behavior of the cluster models of N-graphene, $C_{45}NH_{20}$ and $C_{45}NH_{18}$ (Figure 15.1) in acidic environments, Zhang et al. [21] found that the spin density

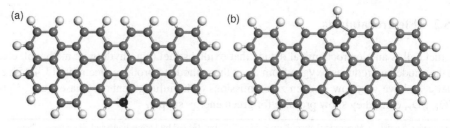

Figure 15.1 Cluster models of N-graphene $C_{45}NH_{20}$ (a) and $C_{45}NH_{18}$ (b) in Ref. [21]

Table 15.1 Electrochemical O_2 reductions in acidic and alkaline aqueous solution

	four-electron	two-electron
Acidic aqueous solution	$O_2 + 4H^+ + 4e^- \rightarrow H_2O$	$O_2 + 2H^+ + 2e^- \rightarrow H_2O_2$
		$H_2O_2 + 2H^+ + 2e^- \rightarrow 2H_2O$
Alkaline aqueous solution	$O_2 + H_2O + 4e^- \rightarrow 4OH^-$	$O_2 + H_2O + 2e^- \rightarrow HO_2^- + OH^-$
		$HO_2^- + H_2O + 2e^- \rightarrow 3OH^-$

and charge density of atoms are the major determining factors for catalytic active sites, and the active sites, which are close to the N dopant, have either high positive spin density or high positive atomic charge density. Further studies by the same group revealed that the higher N-doping concentration (including 2–4 N atoms in the cluster models of the same size) reduces the number of catalytic active sites per N atom, and Stone–Wales defects can strongly promote ORR on N-graphene [22].

Generally, O_2 could be reduced following two pathways (Table 15.1). One is a direct four-electron (4e) pathway in which O_2 is reduced to water in an acidic environment or OH^- in an alkaline environment. The other is a two-electron (2e) pathway in which O_2 is partly reduced to H_2O_2 in an acidic environment or OOH^- in an alkaline environment. Experimentally, some nitrogen-treated carbon nanotube electrodes have been found to catalyze the 4e electroreduction of O_2 to water in acid; [23] Wu et al.'s recent work indicates ORR undergoes two sequential 2e processes with HO_2^- as the intermediate at the nitrogen-doped graphene electrode in the alkaline environment [24]. Theoretically, both 4e and 2e mechanisms have been considered.

In an acidic environment, the proton provided by the acidic media facilitates the transformation of O_2 to the OOH^+ intermediate, which is further protonated into H_2O_2 or H_2O via a 2e or 4e pathway.

Sidik et al.'s joint experimental and theoretical studies [25] showed that oxygen electroreduction to H_2O_2 is predominantly a 2e process in an acid electrolyte, and computational studies based on a cluster model (Figure 15.2) suggest that the active sites are the carbon radical sites formed adjacent to the substitutional N in a graphite sheet. In view of this, they

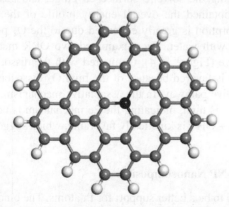

Figure 15.2 N-substituted graphite model $C_{41}NH_{16}$ in Ref. [25]

suggested that the substitutional N atoms far from the graphene edges will be more active than those at the periphery.

Later, Kurak et al. [26] proposed a 2e ORR pathway based on a slab model possessing two neighbor N atoms on the zigzag edge (Figure 15.3a), and attributed the lack of an apparent pathway for the direct 4e reduction to the following reasons: (1) the possible existence of some other catalytic sites involving substituent N, (2) the peroxide pathway may be followed by O_2 and H_2O generation when peroxy intermediates disproportionate, (3) the contribution of possible transition metal impurities to direct 4e reduction. Another probable reason not pointed out by the authors is that the adopted slab model may be incorrect since it is not energetically favorable: two nitrogen atoms are very unlikely to locate at two neighboring zigzag sites, as recognized previously [27] and supported by our test calculations (Figure 15.3).

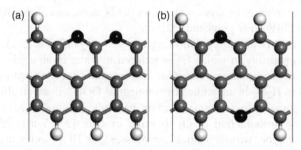

Figure 15.3 $C_{16}H_4N_2$ translational cell in Ref. [26] (a) and the energetically more preferred cell (b) (by 1.16 eV at the level of PBE/TNP)

Zhang et al. [21] proposed two different 4e ORR pathways on N-graphene in an acidic environment. In the more favored pathway, the adsorbed OOH and proton translate into two OH molecules, followed by further reduction with a proton to produce H_2O. In the less favored route, the adsorbed OOH is converted into O and H_2O, and then the resulting O is first protonated to OH, which further reacts with the environmental H to generate H_2O.

The reaction mechanisms, both 4e and 2e, in the alkaline aqueous environment were also examined. By taking the solvent, surface coverage, and adsorbates into consideration, Yu et al. [28] obtained the overall energy profile of the ORR pathway on N-graphene. The O_2 adsorption is greatly enhanced due to the O_2 polarization induced by the hydrogen bonding with water. They examined two ORR mechanisms, namely dissociative and associative (Figure 15.4). Compared with the dissociative mechanism, the associative mechanism is favored because of the high O_2 dissociation barrier ($O_{2(ads)} \rightarrow 2O_{(ads)}$) in the dissociative pathway. In the associative mechanism, the desorption barrier of $OOH_{(ads)}$ into OOH^- is high, indicating the 2e mechanism is not favored; while reaction $OOH_{(ads)} \rightarrow O_{(ads)} + OH^-$ is kinetically favorable, which suggests the presence of the 4e ORR pathway.

15.2.2 N-Graphene-NP Nanocomposites

N-graphene is expected to be a better support for Pt atoms. The binding energy between Pt and carbon atoms in N-graphene can be doubled, as Groves et al. [29] found by examining

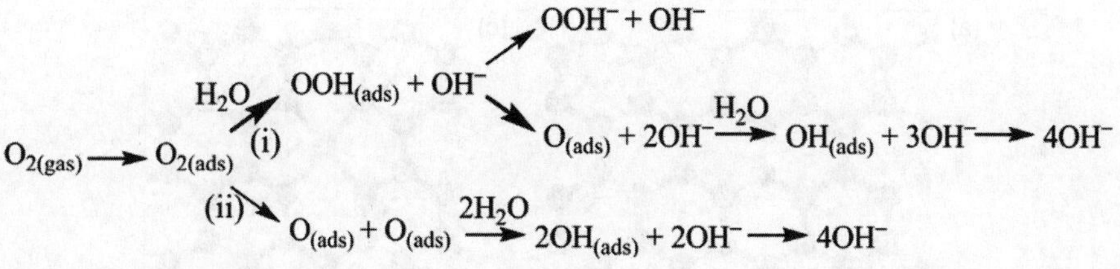

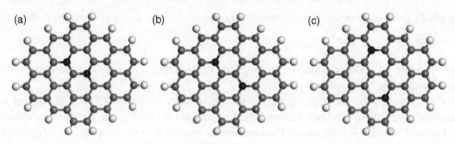

Figure 15.4 Scheme of the ORR reaction on N-graphene in an alkaline aqueous environment where (i) is the associative mechanism and (ii) is the dissociative mechanism in Ref. [28]. Bold arrows are used to indicate the preferred route

the three cluster models in Figure 15.5. The stronger binding of Pt to N-graphene is not only helpful to prevent Pt nanoclusters from migrating and forming larger particles, but also indicates improved catalytic durability of Pt.

Figure 15.5 Structural models of two nitrogen-doped graphene with the two N separation of 2.50 Å (a), 3.79 Å (b) and 5.16 Å (c) [29]

15.2.3 Non-Pt Metal on the Porphyrin-Like Subunits in Graphene

Tough platinum has been the best choice for high activity ORR but the crucial disadvantages are also evident: Pt is expensive and has low CO tolerance, with a large overpotential loss. All these factors hinder its applications in commercial fuel cells. Thus, developing non-Pt catalysts with high efficiency is crucial. A recent breakthrough is that the Fe-porphyrin-like carbon nanotubes with highly dense Fe-N_4 moieties developed by Lee *et al.* [30] exhibit extreme structural stability and excellent ORR catalytic activity. A similar idea was used to design materials with non-Pt metals embedded into the porphyrin-like subunits in graphene.

Kattel *et al.* [31] examined the ORR activity of graphitic Ni–N_x ($x = 2, 4$) and Ni–N_2 edge defect motifs in an Ni–N_x/C electrocatalyst (Figure 15.6). Their computations showed that O_2 and peroxide both chemisorb to the Ni–N_2 edge site but not to graphitic Ni–N_2 or Ni–N_4 sites, indicating that ORR in an Ni–N_x/C electrocatalyst occurs predominantly on edge sites via a successive $2 \times 2e^-$ process in both alkaline and acidic media. They predicted that the Ni–N_x/C-based electrocatalysts perform better in an alkaline medium as compared to an acidic medium. Moreover, the presence of magnetism in the Ni–N_2 edge site improves catalytic activity, suggesting that the magnetic state of the catalytic sites

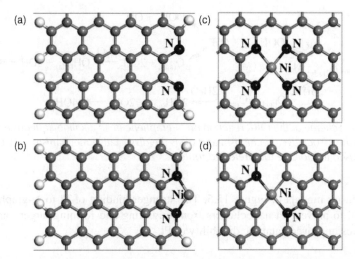

Figure 15.6 N-edge defect, Ni–N$_2$ edge defect and Ni–N$_x$ (x = 2, 4) defect motifs in Ref. [31]

may provide an additional degree of freedom in designing efficient non-platinum ORR electrocatalysts.

Very recently, the same group examined the stability of Co–N$_x$/C (x = 2, 4) and its catalytic mechanism for ORR in both alkaline and acidic media (the models are similar to Figure 15.6c,d) [32]. Their DFT computations predict that the graphitic Co–N$_4$ defect motifs are energetically more favorable than the Co–N$_2$ defect motifs, thus are dominant in Co–N$_x$/C electrocatalysts. Both Co–N$_4$ and Co–N$_2$ defects are active for the reduction of O$_2$ to peroxide. However, the interaction between peroxide and the Co–N$_4$ defect is not as strong as that on the Co–N$_2$ motif, thus they proposed a 2 × 2e$^-$ dual site ORR mechanism for a Co–N$_4$ defect, and a 2 × 2 e$^-$ single site ORR mechanism for a Co–N$_2$ defect.

We believe that these proposed systems can be realized experimentally without big difficulty, considering the recent experimental finding that the gold clusters can nucleate and grow on the nitrogen-induced defective graphene [33].

15.2.4 Graphyne

Graphyne is a 2D periodic carbon allotrope with a one-atom-thick sheet of carbon built from triple- and double-bonded units of two sp- and sp^2-hybridized carbon atoms.

Recent DFT studies suggested that graphyne itself is a promising metal-free electrocatalyst for fuel cells in an acidic environment [34]. It is predicted that the distribution of the charge density at each carbon atom on the graphyne plane is not uniform, and that the positively charged carbon atoms help the adsorption of O$_2$ and OOH$^+$, thus facilitating ORRs. In the acidic media, O$_2$ is chemisorbed preferably on graphyne in the form of OOH$^+$, the sequential H$^+$ adsorption leads to the formation of OH and H$_2$O. The ORR process proceeds with the formation of an O–C bond between oxygen and graphyne, the breakage of the O–O bond and the formation of an O–H bond, which is accompanied by a spontaneous electron transfer along a 4e pathway (Figure 15.7).

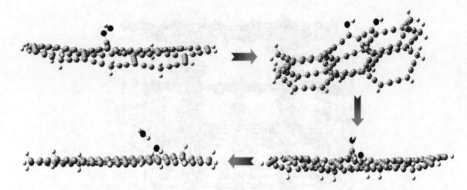

Figure 15.7 Illustration of ORR proceeding on graphyne in a 4e pathway. Reprinted with permission from [34] © 2012 American Chemical Society

15.3 Photocatalysts

Graphene-based semiconductors show great potential as photocatalysts [35–38], since they can offer desirable efficiency for separating electron-hole pairs, thus resulting in improved photocatalytic activities. So far, such photocatalysts are used for degrading pollutants (dyes, bacteria, and volatile organic pollutants), water splitting, CO_2 photoreduction, as well as photocatalytic "selective" redox reactions [39–41], among others.

In contrast to the extensive experimental studies, there have been only a few theoretical investigations into the photocatalytic performance of graphene-based semiconductor nanocomposites, partly due to the big challenges in simulating photocatalytic processes. Currently, theoretical efforts mainly focus on the electronic and optical properties, which serve as the initial but important step toward a more in-depth understanding of the photocatalytic activity of graphene-based nanocomposites.

15.3.1 TiO$_2$-Graphene Nanocomposite

Graphene–TiO_2 composites display excellent photovoltaic performance, which may be contributed to the charge transfer between graphene and the TiO_2 interface. To understand the interface interactions between TiO_2 and graphene, Du and co-workers [42] performed DFT calculations on a complex of a graphene monolayer and a rutile $TiO_2(110)$ surface (Figure 15.8). Due to the large difference of work functions between graphene and titania, the graphene/titania interface forms a charge-transfer complex, which implies efficient hole doping in titania-supported graphene. Most interestingly, it is expected that the valence electrons are able to directly excite from graphene into the titania conduction band under visible light irradiation, and thus well-separated electron-hole pairs can be produced. The wavelength-dependent photocurrent generation of the graphene/titania photoanode they measured experimentally verifies the visible light-response of graphene/TiO_2, and thus proves the role of graphene as the sensitizer to TiO_2, as proposed by DFT studies. Such nanocomposites may be useful for other photovoltaic and photocatalytic applications.

On the other hand, charge separation competes with energy losses (electron–phonon relaxation), and the latter can accelerate the electron-hole pair recombination inside the

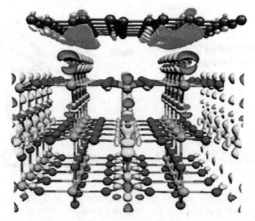

Figure 15.8 Three-dimensional charge density difference for the graphene/titania nanocomposite. Reprinted with permission from [42] © 2012 American Chemical Society

metallic graphene. Bearing this in mind, Prezhdo and co-workers [43] investigated the mechanisms, time scales of electron transfer and energy relaxation processes of graphene–TiO_2 complex. By means of nonadiabatic molecular dynamics and time-dependent DFT computations, they found that the photoinduced electron injection from graphene into TiO_2 occurs ultrafast due to the strong electronic donor-acceptor coupling between graphene and TiO_2: the absorbed photons stir up electrons from the graphene ground state located within the TiO_2 energy gap to the excited states in the TiO_2 conduction band (Figure 15.9), as a result, the electrons move from graphene into TiO_2, and simultaneously transfer energy in the vibrational modes of graphene into the TiO_2 vibrational modes. Moreover, the electron injection is several times faster than electron–phonon energy relaxation, which explains well the experimentally reported high direct light-to-current conversion efficiencies in the graphene–TiO_2 solar cells.

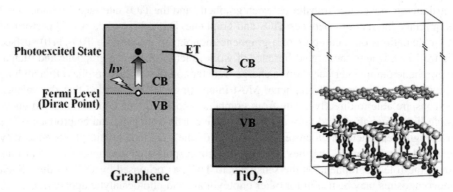

Figure 15.9 Schematic of the photoinduced electron injection process. Reprinted with permission from [43] © 2012 American Chemical Society

15.3.2 Graphitic Carbon Nitrides (g-C_3N_4)

Wang *et al.* recently found that graphitic carbon nitrides (g-C_3N_4) semiconductors (with the band gap of 2.7 eV) can produce hydrogen from water under visible light irradiation in the presence of a sacrificial donor [44]. However, its wide band gap and low conductivity hinders the application in photocatalysis. Accordingly, theoretical efforts have been devoted to improving its photocatalytic performance by either doping or integrating g-C_3N_4 into hybrids.

As one may expect based on the semiconducting character of crystalline g-C_3N_4, doping is an effective approach to modify its energy gap, thus tunes its visible light absorption. In this regard, Ma *et al.* [45] examined the substitutional and interstitial sites for nonmetal (sulfur and phosphorus) dopants, and suggested that doping nonmetal impurities is able to modify the photoelectrochemical properties of the crystalline g-C_3N_4, since such doping can reduce the energy gap. Their results identified that P-interstitial doping and S-substitutional doping are energetically preferred, and the former shows a prominent potential due to the appearance of a new channel for the migration of photogenerated carriers (Figure 15.10).

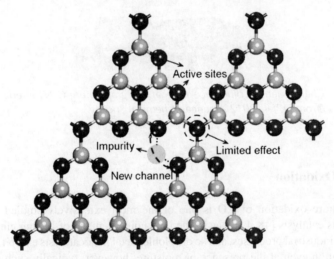

Figure 15.10 Scheme of nonmetal doping in the crystal g-C_3N_4. Reprinted with permission from [45] © 2012 American Chemical Society

Pairing up g-C_3N_4 monolayers to bilayers is another approach to tuning optical properties. By means of DFT computations, Wu *et al.* [46] found that a g-C_3N_4 bilayer has much better visible light adsorption than a single layer. Due to the interlayer coupling, the optical adsorption threshold of bilayer significantly shifts downward by 0.8 eV, and can be further engineered by the external electric field.

Integrating optically active g-C_3N_4 into electronically active graphene may bring unexpected photocatalytic performance. Inspired by this idea, Du *et al.* [47] investigated the interface between g-C_3N_4 and graphene using dispersion-corrected DFT computations. The electronic properties of both components were altered due to the significant charge transfer from graphene to the g-C_3N_4, particularly it forms a well-defined electron–hole puddle on graphene (Figure 15.11). Compared to a pure g-C_3N_4 monolayer, the hybrid graphene/g-C_3N_4 complex exhibits an enhanced optical absorption in the visible region, thus serves as a promising novel candidate for photovoltaic and photocatalytic applications.

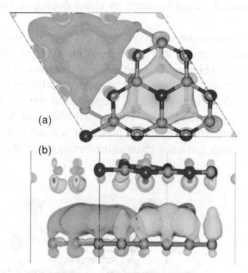

Figure 15.11 *Charge density difference plots at the graphene/g-C_3N_4 interface. Reprinted with permission from [47] © 2012 American Chemical Society*

15.4 CO Oxidation

Low-temperature oxidation of CO is one of the most extensively studied reactions of heterogeneous catalysis [48, 49], due to the impending demand of lowing emissions from automobiles, industrial processes, and so on. Noble metal nanocatalysts exhibit high activity for CO oxidation even at the presence of moisture; however, typically such catalysts are only efficient at high temperatures (>100°C) [50]. A shining exception is gold nanoparticles supported on transition-metal oxides, which have been reported to be highly active at low temperatures both experimentally [51–56] and theoretically [57–60].

Though the noble metal based catalysts are efficient for CO oxidation, the cost and limited naturally available amount impede their general use in large-scale production. To overcome these problems, we need to search for low-cost nonnoble metal catalysts, and suitable substrates that can enhance the stabilities and improve the catalytic properties of nanocatalysts. Graphene and graphene oxide turn out to be good candidates for substrates; for example, recent experiments demonstrated that the Pt clusters deposited on graphene have high activity for CO oxidation at room temperature [7, 61, 62].

However, perfect graphene is not the ideal substrate for the metal atom/cluster to develop stable catalysts, because the interaction between the perfect graphene and metal is weak, and the diffusion barrier for the absorbed metal atom/cluster is quite small, which lead to the metal clustering problem [63,64]. A simple approach to avoid the clustering problem is to increase the binding energy between graphene and metal by taking advantage of the defective sites on graphene, that is, to form metal-embedded graphene. A single-carbon-vacancy defect has usually been adopted in theoretical studies: in such models, the metal atom/cluster is anchored on the top of the vacancy site. The carbon vacancies or dangling bonds of carbon atoms can improve the stability of supported metal atoms/clusters on graphene. Other approaches to enhance the stability of metal-graphene complexes include using graphene oxide instead of graphene as the substrate, exerting mechanical strain and an electric field. In this section, we will discuss the recent theoretical efforts (by means of DFT computations) in developing graphene-based catalysts for low-temperature CO oxidation.

15.4.1 Metal-Embedded Graphene

Since gold clusters have high activity for CO oxidation, gold was the first metal to be embedded into graphene theoretically. Feng's group [65] predicted that Au-embedded graphene has a high activity for CO oxidation. The first step of CO oxidation is most likely to proceed with the Langmuir–Hinshelwood (LH) reaction ($CO + O_2 \rightarrow OOCO \rightarrow CO_2 + O$) with a low activation energy (0.31 eV), while the second step is predicted to proceed via the Eley–Rideal (ER) reaction ($CO + O \rightarrow CO_2$) with an even lower barrier (0.18 eV). The authors rationalized that the high activity may be attributed to the electronic resonance among electronic states of CO, O_2, and the Au atom, particularly, among the d states of the Au atom and the antibonding $2\pi^*$ states of CO and O_2.

Similarly, embedding Pt atoms into defective graphene also enhances the binding strength between Pt atom and the graphene substrate, and improves the capability of CO tolerance. Pt-embedded graphene shows extremely high catalytic activity for CO oxidation (both LH and ER reactions with modest energy barriers less than 0.6 eV) [66].

The high catalytic performance also holds true for small Au_8 and Pt_4 clusters anchored on the single-carbon-vacancy defect in graphene (Figure 15.12): compared to those clusters supported on defect-free graphene, these embedded clusters enhance O_2 adsorption and improve the catalytic activity for CO oxidation [67].

Some efforts have also been devoted to replacing noble metals by inexpensive earth-abundant metals on graphene support for low-temperature CO oxidation. For instance, Li et al. [68] explored the catalytic performance of graphene embedded with Fe for CO oxidation, and found that such low-cost Fe-embedded graphene shows good catalytic activity for the CO oxidation via the more favorable ER mechanism with a two-step route (see Figure 15.13). Similarly, Song et al. theoretically designed the Cu-embedded graphene, which shows comparable catalytic behavior [69].

Very recently, Wannakao investigated CO oxidation with N_2O as an oxidizing agent over an Fe-embedded graphene catalyst [70]. During the oxidation process, the adsorbed N_2O is dissociated into O and N_2 with the activation energy of 8.0 kcal/mol, then the CO molecule reacts with the chemisorbed O on the Fe atom and generates CO_2 with an even smaller activation energy barrier (4.0 kcal/mol). Note that the oxidation by N_2O is more favorable than that by O_2 on the Fe- and Cu-embedded graphene catalysts ($E_a \approx 12-13$ kcal/mol) [68,69].

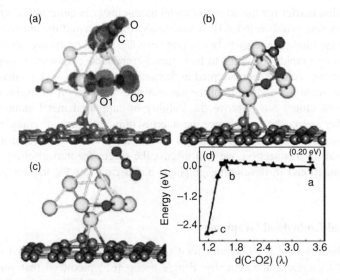

Figure 15.12 LH type of CO oxidation catalyzed by the 3D isomer of Au$_8$ on the defective graphene. Reprinted with permission from [67] © 2009 The American Institute of Physics

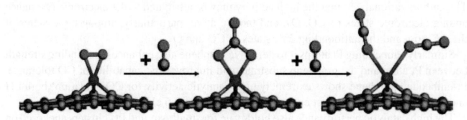

Figure 15.13 Scheme of CO oxidation using the Fe-graphene as the catalyst. Reprinted with permission from [68] © 2010 American Chemical Society

15.4.2 Metal-Graphene Oxide

As discussed previously, the metal-embedded graphene exhibits good performance for CO oxidation, and in principle can be prepared, for example, by trapping the metal atoms in the defective graphenes [63], however, to synthesize such materials is not an easy process at all.

In this regard, Li *et al.* [71] examined the catalytic performance of the more accessible Fe-anchored graphene oxide system (Fe–GO) (Figure 15.14). Their computations showed that Fe atoms bind strongly to the oxygen sites with the binding energy as high as 7.09 eV, which excludes metal clustering. Adsorption of O_2 on Fe-GO is energetically preferred over that of CO by 0.55 eV, and the reaction of CO with O_2 is followed by the ER mechanism. The low reaction barriers indicate the CO oxidation process can occur at a relatively low temperature.

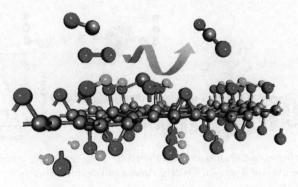

Figure 15.14 Scheme of CO oxidation on an Fe-anchored GO in Ref. [71]. See colour version in the colour plate section. Reprinted with permission from [71] © 2012 American Chemical Society

Meanwhile, Yang *et al.* investigated the catalytic properties of Au_8 clusters supported on GO (Au_8/GO) [72], in which the chosen GO model has only epoxy groups. The rather low reaction barrier (<0.25 eV) of CO oxidation catalyzed by the 3D Au_8 cluster supported on GO substrate for both LH and ER mechanisms, suggests that CO oxidation can occur at room temperature, which demonstrates the importance of the GO substrate (see Figure 15.15).

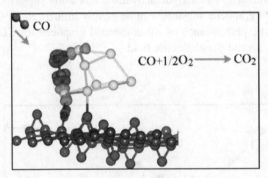

Figure 15.15 Scheme of Au_8/GO model for CO oxidation. Reprinted with permission from [72] © 2012 American Chemical Society

15.4.3 Metal-Graphene under Mechanical Strain

Mechanical strain can alter the electron distribution of the pristine graphene, and is expected to increase the binding energy between the metal atom/cluster and defect-free graphene. Following this line, Zhou *et al.* [73] found that a considerable tensile strain (around 5%) applied to graphene can significantly stabilize the system containing Au_{16} and Au_8 clusters on the graphene sheet, and greatly reduce the reaction barrier of CO oxidation catalyzed by those nanoclusters from around 3.0 eV (without strain) to less than 0.2 eV (Figure 15.16).

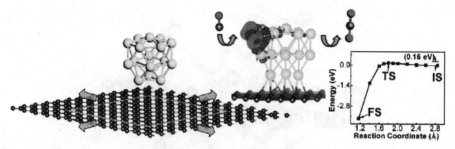

Figure 15.16 Scheme of CO oxidation catalyzed by Au_{16}–graphene under a tensile strain. Reprinted with permission from [73] © 2010 American Chemical Society

15.4.4 Metal-Embedded Graphene under an External Electric Field

The catalytic performance of the Au-embedded graphene can be further improved. Recently, Zhang *et al.* studied the effect of an external electric field on the interaction between O_2 and Au-embedded graphene [74]. Their simulations showed that the O–O distance (d_{O-O}) and the adsorption energy (E_{ad}) of O_2 are dramatically changed with electric field: E_{ad} is enhanced and d_{O-O} is extended under a negative electric field, whereas E_{ad} is decreased and d_{O-O} is shortened under a positive electric field (Figure 15.17). Because d_{O-O} of the adsorbed O_2 correlates with its catalytic activation, this study suggests that O_2 adsorption onto a Au-embedded graphene substrate can be greatly influenced by the external electric field, and the catalytic performance of Au-embedded graphene for CO oxidation may be effectively tuned by an additional electric field.

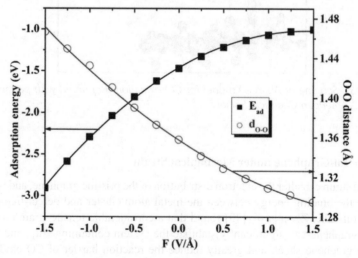

Figure 15.17 E_{ad} and d_{O-O} shifts under different electric fields. Reprinted with permission from [74] © 2012 American Chemical Society

15.4.5 Porphyrin-Like Fe/N/C Nanomaterials

Considering the CO poisoning for most noble-metal catalysts, and inspired by the recent findings that Fe/N/C electrocatalysts can avoid CO poisoning [75], Zhang et al. [76] investigated the CO tolerance and the CO oxidation on Fe/N/C active sites in nanocarbon materials (Figure 15.18), including Fe-N4 and Fe-N3 porphyrin-like carbon nanotubes (T-FeN4 and T-FeN3), Fe-N4 porphyrin-like graphene (G-FeN4), and Fe-N2 nanoribbons (R-FeN2). They found that CO adsorption and oxidation exhibit structural selectivity: CO adsorption is energetically more favorable than O_2, and CO oxidization to CO_2 is kinetically less favorable on T-FeN4 and G-FeN4, suggesting that these two configurations may be poisoned by CO adsorption; in stark contrast, O_2 prefers to adsorb on T-FeN3 and R-FeN2 energetically, and CO can be easily oxidized by O_2. Thus, they identified that T-FeN3 and R-FeN2 are not only CO tolerant, but also show a good catalytic performance for CO oxidation.

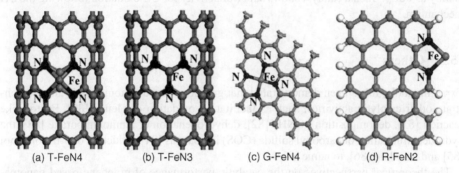

Figure 15.18 Structures of T-FeN4, T-FeN3, G-FeN4, and R-FeN2. Reprinted with permission from [76] © 2012 American Chemical Society

15.4.6 Si-Embedded Graphene

Nonmetal embedded graphenes may also be effective catalysts for CO oxidation. A recent example is the metal-free Si-embedded graphene for CO oxidation in the presence of O_2 or N_2O [77]. When using O_2 as the oxidizing agent, the first CO_2 is produced via the LH mechanism with a barrier of 0.48 eV; while when N_2O serves as the oxidizing agent, the first step, N_2O dissociation to generate N_2 and the adsorbed O on Si, has a much lower (only 0.02 eV) activation energy. The second step is the same for both cases, that is, CO reacts with the O on Si via the ER mechanism with a barrier of 0.57 eV.

15.4.7 Experimental Aspects

In spite of extensive theoretical efforts, few experimental studies for the catalytic performance of metal or metal oxide supported on graphene for CO oxidation have been reported, and in those experimental studies there are, graphene was obtained by reducing graphene oxide.

Yoo et al. [7] prepared small Pt clusters on graphene sheets (GNS) from a mixture of a platinum precursor [Pt(NO$_2$)$_2$·(NH$_3$)$_2$] and GNS powder and demonstrated that such complexes exhibit high catalytic activity for CO oxidation, and the GNS display a remarkable modulation to the catalytic performances of Pt clusters. Furthermore, Zhang et al. [78] integrated Pt, Ni, and Pt–Ni nanoparticles with graphene by an impregnation method, and tested their catalytic performance for CO oxidation. Their results showed that pure graphene is inactive, graphene supported nickel shows poor activity, while graphene-supported platinum exhibits good activity, and a graphene-supported Pt-Ni alloy is more active owing to the formation of a Pt–Ni alloy.

Very recently, Li et al. [79] prepared a graphene supported palladium (Pd) catalyst using the conventional impregnation and hydrogen reduction method. Their DFT computations and the catalyst characterization using Raman and X-ray photoelectron spectroscopy confirmed that the oxygen containing groups (in the reduced graphene oxide) play an important role in stabilizing Pd clusters on graphene, and the first layer of the metal particle presents mainly as PdO$_x$. Such a catalyst shows superior activity for CO oxidation following the LH mechanism.

15.5 Others

Experimentally, it has been demonstrated that graphene-based nanomaterials also demonstrate other catalytic activities, such as facilitating oxidation of olefins [80], Fenton-like reaction [81], decomposition of NH$_3$ [82], dehydrogenation of ammonia borane [83], the hydrodesulfurization of carbonyl sulfide (COS) in coal gas [84], reduction of 4-nitrophenol [85] and triiodide [86], to name a few.

The theoretical perspectives on the catalytic performance of graphene-based nanomaterials are surely beyond ORR, photocatalysis, and CO oxidation as we summarized in Sections 15.2–15.4, but these other aspects have been rarely explored. Below we introduce two examples: propene epoxidation and nitromethane combustion.

15.5.1 Propene Epoxidation

Pulido et al. [87] investigated the mechanism of propene epoxidation by H$_2$/H$_2$O/O$_2$ mixtures catalyzed by Au-embedded graphene. They found that the formation of gold hydroperoxide H–Au–OOH intermediates from H$_2$ and O$_2$ is barrierless, and the subsequent reaction with propene yields propene oxide (PO) and H$_2$O thermodynamically and kinetically favorable. The Au-embedded graphene not only has high catalytic activity, but also has high selectivity towards the desired PO due to the suppression of the main competing routes, namely propene hydrogenation and O$_2$ dissociation (Figure 15.19).

15.5.2 Nitromethane Combustion

Liu et al. [88] performed *ab initio* molecular dynamics simulations on nitromethane (NM) combustion on functionalized graphene sheets (FGSs) with oxygen-containing groups (hydroxyls, ethers, and carbonyls) and carbon divacancies substituted by two oxygens, and showed these FGSs to be effective catalysts for NM combustion (Figure 15.20). FGSs

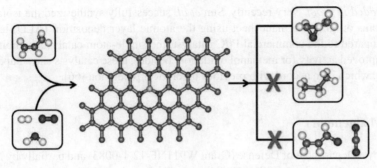

Figure 15.19 Scheme of propene epoxidation by $H_2/H_2O/O_2$ mixtures over the Au-embedded graphene. Reprinted with permission from [87] © 2012 American Chemical Society

initiate the NM decomposition via the exchange of protons or oxygens between the oxygen-containing functional groups and NM, and strongly enhance the rate of successive reactions, ultimately forming H_2O, CO_2, and N_2.

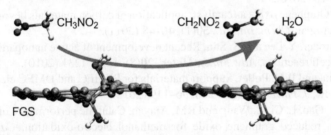

Figure 15.20 Scheme of FGS in the decomposition reaction of nitromethane. Reprinted with permission from [88] © 2012 American Chemical Society

15.6 Conclusion

In this chapter, we summarized the recent progress in developing graphene-based nanomaterials as efficient and versatile catalysts. The graphene, reduced graphene oxide, graphene oxides, and their derivatives exhibit synergistic and/or extra contributions to electrocatalytic, photocatalytic, and other catalytic reactions. Overall, graphene-based nanomaterials are versatile and scalable, and are adaptable to a wide variety of catalytic processes.

However, we are also well aware of the great challenges and complexity in designing high-performance graphene-based nanocatalysts. Ideally, adopted models should consider various inevitable factors in realistic experiments, such as impurities, defects, temperature, and pressure; the dopant content, doping sites, functional groups, the size/morphology/composition of nanoparticles, as well as the interface/interaction in graphene-based nanomaterials which also need to be investigated in depth. In these regards, further efforts are required to understand the mechanisms and improve catalytic performance of graphene-based materials for various important chemical reactions.

Note added in proof. Very recently, Sun *et al.* successfully synthesized the isolated single Pt atoms on graphene nanosheet using the atomic layer deposition (ALD) technique [89]. Compared to the commercial Pt/C catalyst, the single-atom catalysts exhibit significantly improved activity for methanol oxidation; besides, these catalysts show superior CO tolerance, which is in accord with previous theoretical prediction [66].

Acknowledgements

Support by a Department of Defense (Grant W911NF-12-1-0083) and partially by National Science Foundation (Grant EPS-1010094) is gratefully acknowledged.

References

[1] D.R. Dreyer, H.P. Jia and C.W. Bielawski, Graphene oxide: a convenient carbocatalyst for facilitating oxidation and hydration reactions, *Angew. Chem. Int. Ed.*, **49**(38), 6813–6816 (2010).

[2] J. Pyun, Graphene oxide as catalyst: application of carbon materials beyond nanotechnology, *Angew. Chem. Int. Ed.*, **50**(1), 46–48 (2011).

[3] V. Mazumder, Y. Lee and S. Sun, Recent development of active nanoparticle catalysts for fuel cell reactions, *Adv. Funct. Mater.*, **20**(8), 1224–1231 (2010).

[4] S. Sharma and B.G. Pollet, Support materials for PEMFC and DMFC electrocatalysts – A review, *J. Power Sources*, **208**, 96–119 (2012).

[5] Y. Li, W. Gao, L. Ci, C. Wang and P.M. Ajayan, Catalytic performance of Pt nanoparticles on reduced graphene oxide for methanol electro-oxidation, *Carbon*, **48**(4), 1124–1130 (2010).

[6] B. Seger and P.V. Kamat, trocatalytically active graphene-platinum nanocomposites. Role of 2-D carbon support in PEM fuel cells, *J. Phys. Chem. C*, **113**(19), 7990–7995 (2009).

[7] E. Yoo, T. Okata, T. Akita, M. Kohyama, J. Nakamura and I. Honma, Enhanced electrocatalytic activity of Pt subnanoclusters on graphene nanosheet surface, *Nano Lett.*, **9**(6), 2255–2259 (2009).

[8] S. Bong, Y.-R. Kim, I. Kim, S. Woo, S. Uhm, J. Lee and H. Kim, Graphene supported electrocatalysts for methanol oxidation, *Electrochem. Commun.*, **12**(1), 129–131 (2010).

[9] J. Shen, B. Yan, M. Shi, H. Ma, N. Li and M. Ye, Fast and facile preparation of reduced graphene oxide supported Pt-Co electrocatalyst for methanol oxidation, *Mater. Res. Bulletin*, **47**(6), 1486–1493 (2012).

[10] S. Guo, S. Dong and E. Wang, Three-dimensional Pt-on-Pd bimetallic nanodendrites supported on graphene nanosheet: facile synthesis and used as an advanced nanoelectrocatalyst for methanol oxidation, *ACS Nano*, **4**(1), 547–555 (2010).

[11] L.T. Qu, Y. Liu, J.B. Baek and L.M. Dai, Nitrogen-doped graphene as efficient metal-free electrocatalyst for oxygen reduction in fuel cells, *ACS Nano*, **4**(3), 1321–1326 (2010).

[12] Z. Wu, S. Yang, Y. Sun, K. Parvez, X. Feng and K. Müllen, 3D nitrogen-doped graphene aerogel-supported Fe_3O_4 nanoparticles as efficient electrocatalysts for the oxygen reduction reaction, *J. Am. Chem. Soc.*, **134**(22), 9082–9085 (2012).

[13] Z.-H. Sheng, H.-L. Gao, W.-J. Bao, F.-B. Wang and X.-H. Xia, Synthesis of boron doped graphene for oxygen reduction reaction in fuel cells, *J. Mater. Chem.*, **22**, 390–400 (2012).

[14] S. Wang, L. Zhang, Z. Xia, A. Roy, D.W. Chang, J.-B. Baek and L. Dai, BCN graphene as efficient metal-free electrocatalyst for the oxygen reduction reaction, *Angew. Chem. Int. Ed.*, **51**(17), 1–5 (2012).

[15] Z. Yang, Z. Yao, G. Li, G. Fang, H. Nie, Z. Liu, *et al.*, Sulfur-doped graphene as an efficient metal-free cathode catalyst for oxygen reduction, *ACS Nano*, **6**(1), 205–211 (2012).

[16] J. Liang, Y. Jiao, M. Jaroniec and S.Z. Qiao, Sulfur and nitrogen dual-doped mesoporous graphene electrocatalyst for oxygen reduction with synergistically enhanced performance, *Angew. Chem. Int. Ed.*, **51**(46), 11496–11500 (2012).

[17] Z. Yao, H. Nie, Z. Yang, X. Zhou, Z. Liu and S. Huang, Catalyst-free synthesis of iodine-doped graphene via a facile thermal annealing process and its use for electrocatalytic oxygen reduction in an alkaline medium, *Chem. Commun.*, **48**(7), 1027–1029 (2012).

[18] M. Kaukonen, A.V. Krasheninnikov, E. Kauppinen and R.M. Nieminen, Doped graphene as a material for oxygen reduction reaction in hydrogen fuel cells: A computational study, *ACS Catal.*, **3**, 159–165 (2013).

[19] Y. Chen, X. Yang, Y. Liu, J. Zhao, Q. Cai and X. Wang, Can Si-doped graphene activate or dissociate O_2 molecule? *J. Mol. Graphics Model.*, **39**, 126–132 (2013).

[20] D. Deng, L. Yu, X. Pan, S. Wang, X. Chen, P. Hu, *et al.*, Size effect of graphene on electrocatalytic activation of oxygen, *Chem. Commun.*, **47**, 10016–10018 (2011).

[21] L.P. Zhang and Z.H. Xia, Mechanisms of oxygen reduction reaction on nitrogen-doped graphene for fuel cells. *J. Phys. Chem. C*, **115**(22), 11170–11176 (2011).

[22] L. Zhang, J. Niu, L. Dai and Z. Xia, Effect of microstructure of nitrogen-doped graphene on oxygen reduction activity in fuel cells, *Langmuir*, **28**(19), 7542–7550 (2012).

[23] D. Yu, Q. Zhang and L.M. Dai, Highly efficient metal-free growth of nitrogen-doped single-walled carbon nanotubes on plasma-etched substrates for oxygen reduction, *J. Am. Chem. Soc.*, **132**(43), 15127–15129 (2010).

[24] J. Wu, D. Zhang, Y. Wang and B. Hou, Electrocatalytic activity of nitrogen-doped graphene synthesized via a one-pot hydrothermal process towards oxygen reduction reaction, *J. Power Sources*, **227**, 185–190 (2013).

[25] R.A. Sidik, A.B. Anderson, N.P. Subramanian, S.P. Kumaraguru, and B.N. Popov, O_2 reduction on graphite and nitrogen-doped graphite: Experiment and theory, *J. Phys. Chem. B*, **110**(4), 1787–1793 (2006).

[26] K.A. Kurak and A.B. Anderson, Nitrogen-treated graphite and oxygen electroreduction on pyridinic edge sites, *J. Phys. Chem. C*, **113**(16), 6730–6734 (2009).

[27] S.F. Huang, K. Terakura, T. Ozaki, T. Ikeda, M. Boero, M. Oshima, *et al.*, First-principles calculation of the electronic properties of graphene clusters doped with nitrogen and boron: analysis of catalytic activity for the oxygen reduction reaction, *Phys. Rev. B*, **80**(23), 235410 (2009).

[28] L. Yu, X. Pan, X. Cao, P. Hu and X. Bao, Oxygen reduction reaction mechanism on nitrogen-doped graphene: a density functional theory study, *J. Catal.*, **282**(1), 183–190 (2011).

[29] M.N. Groves, A.S.W. Chan, C. Malardier-Jugroot and M. Jugroot, Improving platinum catalyst binding energy to graphene through nitrogen doping, *Chem. Phys. Lett.*, **481**(4–6), 214–219 (2009).

[30] D.H. Lee, W.J. Lee, W.J. Lee, S.O. Kim, and Y.-H. Kim, Theory, synthesis, and oxygen reduction catalysis of Fe-porphyrin-like carbon nanotube, *Phys. Rev. Lett.*, **106**(17), 175502–4 (2011).

[31] S. Kattel, P. Atanassov and B. Kiefer, Density functional theory study of Ni–N_x/C electrocatalyst for oxygen reduction in alkaline and acidic media, *J. Phys. Chem. C*, **116**(33), 17378–17383 (2012).

[32] S. Kattel, P. Atanassov and B. Kiefer, Catalytic activity of Co–N_x/C electrocatalysts for oxygen reduction reaction: a density functional theory study, *Phys. Chem. Chem. Phys.*, **15**, 148–153 (2013).

[33] H.Y. Koo, H.-J. Lee, Y.-Y. Noh, E.-S. Lee, Y.-H. Kim and W.S. Choi, Gold nanoparticle-doped graphene nanosheets: Sub-nanosized gold clusters nucleate and grow at the nitrogen-induced defects on graphene surfaces, *J. Mater. Chem.*, **22**, 7130–7135 (2012).

[34] P. Wu, P. Du, H. Zhang and C. Cai, Graphyne as a promising metal-free electrocatalyst for oxygen reduction reactions in acidic fuel cells: a DFT study, *J. Phys. Chem. C*, **116**(38), 20472–20479 (2012).

[35] P.V. Kamat, Graphene-based nanoassemblies for energy conversion, J. Phys. Chem. Lett., **2**(3):242–251 (2011).

[36] X. Huang, X. Qi, F. Boey and H. Zhang Graphene-based composites, *Chem. Soc. Rev.*, **41**, 666–686 (2012).

[37] N. Zhang, Y. Zhang and Y.J. Xu, Recent progress on graphene-based photocatalysts: current status and future perspectives, *Nanoscale*, **4**, 5792–5813 (2012).

[38] Q. Xiang, J. Yu and M. Jaroniec, Graphene-based semiconductor photocatalysts. *Chem. Soc. Rev.*, **41**, 782–796 (2012).

[39] Y. Zhang, Z.R. Tang, X. Fu and Y.J. Xu, Engineering the unique 2D mat of graphene to achieve graphene-TiO_2 nanocomposite for photocatalytic selective transformation: what advantage does graphene have over its forebear carbon nanotube? *ACS Nano*, **5**(9), 7426–7435 (2011).

[40] Y. Zhang, N. Zhang, Z.R. Tang and Y.J. Xu, Improving the photocatalytic performance of graphene of graphene-TiO_2 nanocomposites via a combined strategy of decreasing defects of graphene and increasing interfacial contact, *Phys. Chem. Chem. Phys.*, **14**, 9167–9175 (2012).

[41] Y. Zhang, N. Zhang, Z.R. Tang and Y.J. Xu, Graphene transforms wide band gap ZnS to a visible light photocatalyst. The new role of graphene as a macromolecular photosensitizer, *ACS Nano*, **6**(11), 9777–9789 (2012).

[42] A. Du, Y.H. Ng, N.J. Bell, Z. Zhu, R. Amal and S.C. Smith, Hybrid graphene/titania nanocomposite: Interface charge transfer, hole doping, and sensitization for visible light response, *J. Phys. Chem. Lett.*, **2**(8), 894–899 (2011).

[43] R. Long, N.J. English and O.V. Prezhdo, Photo-induced charge separation across the graphene–TiO_2 interface is faster than energy losses: a time-domain *ab initio* analysis, *J. Am. Chem. Soc.*, **134**(34), 14238–14248 (2012).

[44] X.C. Wang, K. Maeda, A. Thomas, K. Takanabe, G. Xin, K. Domen and M. Antonietti, A metal-free polymeric photocatalyst for hydrogen production from water under visible light, *Nat. Mater.*, **8**(1), 76–82 (2009).

[45] X. Ma, Y. Lv, J. Xu, Y. Liu, R. Zhang and Y. Zhu, A strategy of enhancing the photoactivity of g-C_3N_4 via doping of nonmetal elements: a first-principles study, *J. Phys. Chem. C*, **116**(44), 23485–23493 (2012).

[46] F. Wu, Y. Liu, G. Yu, D. Shen, Y. Wang and E. Kan, Visible-light-absorption in graphitic C_3N_4 bilayer: enhanced by interlayer coupling, *J. Phys. Chem. Lett.*, **3**(22), 3330–3334 (2012).

[47] A. Du, S. Sanvito, Z. Li, D. Wang, Y. Jiao, T. Liao, Q. Sun, *et al.*, Hybrid graphene and graphitic carbon nitride nanocomposite: Gap opening, electron–hole puddle, interfacial charge transfer, and enhanced visible light response, *J. Am. Chem. Soc.*, **134**(9), 4393–4397 (2012).

[48] M. Shelef and R.W. McCabe, Twenty-five years after introduction of automotive catalysts: what next? *Catal. Today*, **62**(1), 35–50 (2000).

[49] M.V. Twigg, Progress and future challenges in controlling automotive exhaust gas emissions, *Appl. Catal. B: Environ.*, **70**(1–4), 2–15 (2007).

[50] S.H. Oh and G.B. Hoflund, Low-temperature catalytic carbon monoxide oxidation over hydrous and anhydrous palladium oxide powders, *J. Catal.*, **245**(1):35–44 (2007).

[51] M. Haruta, N. Yamada, T. Kobayashi and S. Iijima, Gold catalysts prepared by coprecipitation for low-temperature oxidation of hydrogen and of carbon monoxide, *J. Catal.*, **115**(2), 301–309 (1989).

[52] M. Haruta, S. Tsubota, T. Kobayashi, H. Kageyama, M.J. Genet and B. Delmon, Low-temperature oxidation of CO over gold supported on TiO_2, α-Fe_2O_3, and Co_3O_4, *J. Catal.*, **144**(1), 175–192 (1993).

[53] M. Haruta, Size- and support-dependency in the catalysis of gold, *Catal. Today*, **36**(1), 153–166 (1997).

[54] F. Cosandey and T.E. Madey, Growth, morphology, interfacial effects and catalytic properties of Au on TiO_2, *Surf. Rev. Lett.*, **8**(1–2), 73–93 (2001).

[55] M. Valden, X. Lai and D.W. Goodman, Onset of catalytic activity of gold clusters on titania with the appearance of nonmetallic properties, *Science*, **281**(5383), 1647–1650 (1998).

[56] M. Daté, M. Okumura, S. Tsubota and M. Haruta, Role of moisture in the catalytic activity of supported gold nanoparticles, *Angew. Chem. Int. Ed.*, **43**(16), 2129–2132 (2004).

[57] Y. Cao, N. Shao, Y. Pei and X.C Zeng, Icosahedral crown gold nanocluster $Au_{43}Cu_{12}$ with high catalytic activity, *Nano Lett.*, **10**(3), 1055–1062 (2010).

[58] O. Lopez-Acevedo, K.A. Kacprzak, J. Akola and H. Häkkinen, Quantum size effects in ambient CO oxidation catalysed by ligand-protected gold clusters, *Nature Chem.*, **2**, 329–334 (2010).

[59] Z. Liu, X. Gong, J. Kohanoff, C. Sanchez and P. Hu, Catalytic role of metal oxides in gold-based catalysts: a first principles study of CO oxidation on TiO_2 supported Au, *Phys. Rev. Lett.*, **91**(26), 266102 (2003).

[60] D. Widmann, Y. Liu, F. Schuth and R.J. Behm, Support effects in the Au-catalyzed CO Oxidation–Correlation between activity, oxygen storage capacity, and support reducibility, *J. Catal.*, **276**(2), 292–305 (2010).

[61] B. Seger and P.V. Kamat, Electrocatalytically active graphene-platinum nanocomposites. Role of 2-D carbon support in PEM fuel cells, *J. Phys. Chem. C*, **113**(19), 7990–7995 (2009).

[62] K. Tanaka, M. Shou and Y. Yuan, Low temperature PROX reaction of CO catalyzed by dual functional catalysis of the Pt supported on CNT, CNF, graphite, and amorphous-C with Ni–MgO, Fe, and Fe-Al_2O_3: oxidation of CO via HCOO intermediate, *J. Phys. Chem. C*, **114**(40), 16917–16923 (2010).

[63] J.A. Rodríguez-Manzo, O. Cretu and F. Banhart, Trapping of metal atoms in vacancies of carbon nanotubes and graphene, *ACS Nano*, **4**(6), 3422–3428 (2010).

[64] R. Zan, U. Bangert, Q. Ramasse and K.S. Novoselov, Interaction of metals with suspended graphene observed by transmission electron microscopy, *J. Phys. Chem. Lett.*, **3**(7), 953–958 (2012).

[65] Y.H. Lu, M. Zhou, C. Zhang and Y.P. Feng, Metal-embedded graphene: a possible catalyst with high activity, *J. Phys. Chem. C*, **113**(47), 20156–20160 (2009).

[66] Y. Tang, Z. Yang and X. Dai, A theoretical simulation on the catalytic oxidation of CO on Pt/graphene, *Phys. Chem. Chem. Phys.*, **14**, 16566–16572 (2012).

[67] M. Zhou, A.H. Zhang, Z.X. Dai, C. Zhang and Y.P. Feng, Greatly enhanced adsorption and catalytic activity of Au and Pt clusters on defective graphene, *J. Chem. Phys.*, **132**(19), 194704 (2010).

[68] Y. Li, Z. Zhou, G. Yu, W. Chen and Z. Chen, CO catalytic oxidation on iron-embedded graphene: computational quest for low-cost nanocatalysts, *J. Phys. Chem. C*, **114**(14), 6250–6254 (2010).

[69] E.H. Song, Z. Wen and Q. Jiang, CO catalytic oxidation on copper-embedded graphene, *J. Phys. Chem. C*, **115**(9), 3678–3683 (2011).

[70] S. Wannakao, T. Nongnual, P. Khongpracha, T. Maihom and J. Limtrakul, Reaction mechanisms for CO catalytic oxidation by N_2O on Fe-embedded graphene, *J. Phys. Chem. C*, **116**(32), 16992–16998 (2012).

[71] F. Li, J. Zhao and Z. Chen, Fe-anchored graphene oxide: a low-cost and easily accessible catalyst for low-temperature CO oxidation, *J. Phys. Chem. C*, **116**(3), 2507–2514 (2012).

[72] M. Yang, M. Zhou, A. Zhang and C. Zhang, Graphene oxide: an ideal support for gold nanocatalysts, *J. Phys. Chem. C*, **116**(42), 22336–22340 (2012).

[73] M. Zhou, A. Zhang, Z. Dai, Y. Feng and C. Zhang, Strain-enhanced stabilization and catalytic activity of metal nanoclusters on graphene, *J. Phys. Chem. C*, **114**(39), 16541–16546 (2010).

[74] T. Zhang, Q. Xue, M. Shan, Z. Jiao, X. Zhou, C. Ling and Z. Yan, Adsorption and catalytic activation of O_2 molecule on the surface of Au-doped graphene under an external electric field, *J. Phys. Chem. C*, **116**(37), 19918–19924 (2012).

[75] L. Birry, J.H. Zagal and J.-P. Dodelet, Does CO poison Fe-based catalysts for ORR? *Electrochem. Commu.*, **12**(5), 628–631 (2010).

[76] P. Zhang, X.F. Chen, J.S. Lian and Q. Jiang, Structural selectivity of CO oxidation on Fe/N/C catalysts, *J. Phys. Chem. C*, **116**(33), 17572–17579 (2012).

[77] J. Zhao, Y. Chen and H. Fu, Si-embedded graphene: an efficient and metal-free catalyst for CO oxidation by N_2O or O_2, *Theor. Chem. Acc.*, **131**, 1242 (2012).

[78] C. Zhang, W. Lv, Q. Yang and Y. Liu, Graphene supported nanoparticles of Pt–Ni for CO oxidation, *Appl. Surf. Sci.*, **258**(20), 7795–7800 (2012).

[79] Y. Li, Y. Yu, J. Wang, J. Song, Q. Li, M. Dong and C. Liu, CO oxidation over graphene supported palladium catalyst, *Appl. Catal. B: Environ.*, **125**, 189–196 (2012).

[80] H.P. Jia, D.R. Dreyer and C.W. Bielawski, C–H oxidation using graphite oxide, *Tetrahedron*, **67**(24), 4431–4434 (2011).

[81] Y. Zhao, W. Chen, C. Yuan, Z. Zhu and L. Yan, Hydrogenated graphene as metal-free catalyst for Fenton-like reaction, *Chin. J. Chem. Phys.*, **25**(3), 335–338 (2012).

[82] X Liu, C. Menga and Y. Han, Unique reactivity of Fe nanoparticles-defective graphene composites toward NH_x ($x = 0, 1, 2, 3$) adsorption: a first-principles study, *Phys. Chem. Chem. Phys.*, **14**, 15036–15045 (2012).

[83] (a) Ö. Metina, E. Kayhanb, S. Özkara and J.J. Schneider, Palladium nanoparticles supported on chemically derived graphene: an efficient and reusable catalyst for the dehydrogenation of ammonia borane, *Int. J. Hydrogen Energy*, **37**(10), 8161–8169 (2012). (b) B. Kılıç, S. Şencanlı and Ö. Metin, Hydrolytic dehydrogenation of ammonia borane catalyzed by reduced graphene oxide supported monodisperse palladium nanoparticles: high activity and detailed reaction kinetics, *J. Mol. Catal. A: Chem.*, **361–362**, 104–110 (2012).

[84] W. Xu, X. Wang, Q. Zhou, B. Meng, J. Zhao, J. Qiu and Y. Gogotsi, Low-temperature plasma-assisted preparation of graphene supported palladium nanoparticles with high hydrodesulfurization activity, *J. Mater. Chem.*, **22**, 14363–14368 (2012).

[85] X. Li, X. Wang, S. Song, D. Liu and H. Zhang, Selectively deposited noble metal nanoparticles on Fe_3O_4/graphene composites: stable, recyclable, and magnetically separable catalysts, *Chem. Eur. J.*, **18**(24), 7601–7607 (2012).

[86] C.J. Liu, S.Y. Tai, S.W. Chou, Y.C. Yu, K.D. Chang, S. Wang, *et al.*, Facile synthesis of MoS_2/graphene nanocomposite with high catalytic activity toward triiodide reduction in dye-sensitized solar cells, *J. Mater. Chem.*, **22**, 21057–21064 (2012).

[87] A. Pulido, M. Boronat and A. Corma, Propene epoxidation with $H_2/H_2O/O_2$ mixtures over gold atoms supported on defective graphene: a theoretical study, *J. Phys. Chem. C*, **116**(36), 19355–19362 (2012).

[88] L.M. Liu, R. Car, A. Selloni, D.M. Dabbs, I.A. Aksay and R.A. Yetter, Enhanced thermal decomposition of nitromethane on functionalized graphene sheets: *ab initio* molecular dynamics simulations, *J. Am. Chem. Soc.*, **134**(46), 19011–19016 (2012).

[89] S. Sun, G. Zhang, N. Gauquelin, N. Chen, J. Zhou, S. Yang, W. Chen, X. Meng, D. Geng, M.N. Banis, R. Li, S. Ye, S. Knights, G.A. Botton, T.-K. Sham and X. Sun, Single-atom catalysis using Pt/graphene achieved through atomic layer deposition, *Scientific Reports*, **3**, 1775–1779 (2013).

16
Hydrogen Storage in Graphene

Yafei Li and Zhongfang Chen
Department of Chemistry, Institute for Functional Nanomaterials,
University of Puerto Rico, USA

16.1 Introduction

Globally, using petroleum and other fossil fuels has been causing serious air pollution and enormous releases of greenhouse gases. The petroleum reserve is limited, the dependence on foreign oil is a big threat to national security, and the rising cost of petroleum is deteriorating everyone's quality of life. Therefore, searching for an abundant, renewable and clean alternative energy resource is urgent and critical.

Among various alternative energy resources, hydrogen is very promising for its apparent advantages over fossil fuels. For example, the staggering energy content of H_2 (142 MJ/kg) is about three orders higher than that of petroleum, and more importantly its combustion product is harmless water vapor. In recent years, hydrogen has been attracting many interests from the scientific community, governments and enterprises [1, 2]. However, the development of hydrogen economy requires an economic, safe, lightweight and high-capacity storage medium [3–7]. The US Department of Energy (DOE) has set a series of targets for hydrogen storage materials, including 6.0 wt% and 9.0 wt% gravimetric density and 45 g l^{-1} and 81 g l^{-1} volumetric capacity in 2010 and 2015 respectively [8], and reversible hydrogen uptake and release at ambient temperature and under moderate pressure.

To achieve high hydrogen capacity, hydrogen storage media should only be considered among systems composed of lighter elements. Complex hydrides such as alanates, amides/imides and borohydrides can release high amount of hydrogen that meets the DOE target, but there are still crucial problems to solve, for instance, the hydrogen release or dehydrogenation temperature is still high and the kinetics is slow [2]. Metal–organic frameworks (MOFs) are proposed for hydrogen storage [9, 10], however, the binding between

---

Graphene Chemistry: Theoretical Perspectives, First Edition. Edited by De-en Jiang and Zhongfang Chen.
© 2013 John Wiley & Sons, Ltd. Published 2013 by John Wiley & Sons, Ltd.

H_2 and MOFs is usually very weak (the highest observed dihydrogen adsorption enthalpy is −13.5 kJ/mol) [11], and the storage capacity in available MOFs is too low. Carbon nanotubes (CNTs) were once considered a promising hydrogen storage material because of the possibility of reversibility, fast kinetics, and high capacity (large surface areas), however, it has been found that the hydrogen storage capacity in CNTs greatly diminishes at room temperature and ambient pressure [12].

To achieve the reversible hydrogen uptake and release at ambient conditions, the ideal H_2 adsorption energy should be in the range of 0.2–0.4 eV/H_2 [13, 14], which is intermediate between physisorption and chemisorption. Metal-decorated nanostructures, a kind of hydrogen sorbent, were proposed to satisfy these requirements based on density functional theory (DFT) computations. Among others, Zhao et al. [15–18] showed that Sc-coated B-doped fullerenes $C_{48}B_{12}[ScH]_{12}$ can store up to 8.77 wt% H_2 with the binding energy of ∼0.3 eV/H_2, while Yildirim et al. [19] found that up to 8 wt% of hydrogen can be stored in Ti-coated CNTs; Durgun et al. [20] demonstrated that a single ethylene molecule can form a stable complex with two transition metals (TM) such as Ti, the resulting TM−ethylene complex can adsorb up to 10 hydrogen molecules, reaching to gravimetric storage capacity of 14 wt%; Lee et al. [21] also reported that Ti-decorated cis-polyacetylene has reversibly usable gravimetric and volumetric hydrogen storage capacity of 7.6 wt% and 63 kg/m^3, respectively, near ambient conditions.

In these pioneering studies, transition metal (TM) atoms were assumed to be homogeneously distributed on the substrate. However, it is very difficult, if not impossible, to experimentally realize these predicted uniformly coated homogeneous monolayers, since TM atoms tend to form clusters on the surface of carbon nanostructures [22–24], and consequently the hydrogen storage capacity drops dramatically. To avoid the perplexing clustering problem, Shevlin et al. [25] proposed to firmly emplace the TM atoms in a carbon matrix by defecting the support, while Sun et al. [26] proposed to utilize Li atoms to coat C_{60} uniformly, taking advantage of the larger binding energy between Li and C_{60} than the cohesive energy of lithium bulk metal; however, the rather weak H_2 adsorption energy is a concern for Li and other alkali metal [27, 28] decorated nanostructures. To conquer this problem, Meng et al. [29] proposed metal-diboride (MB_2) nanotubes as a potential hydrogen storage medium, where the metal atoms are a natural part of the tubular structures, thus avoiding metal clustering. For example, the TiB_2 nanotubes show a hydrogen storage capacity of 5.5 wt% with binding energies of 0.2–0.6 eV. In recent years, calcium (Ca) [30, 33] has been suggested to be an ideal decorating element for various nanostructures. It is because that the cohesive energy of Ca is rather low (∼1.8 eV), and the Ca atom can adsorb multiple H_2 molecules with a binding energy exceeding 0.2 eV. For example, Yoon et al. [30] found that the notorious clustering can be prevented in Ca coated C_{60} system, and $Ca_{32}C_{60}$ has a hydrogen uptake of >8.4 wt%.

Graphene [34, 35], a single layer of carbon atoms tightly packed in a honeycomb sublattice, has been considered as a promising candidate for hydrogen storage material ever since its experimental realization due to the following reasons. First, graphene consists of light element carbon, which is in favor of achieving high gravimetric density. Second, graphene has an ultrahigh surface area (2630 m^2/g), which is rather favorable for hydrogen storage applications. Third, graphene is stable and robust [36] thus can be utilized in some harsh conditions and transported for long distances. Fourth, graphene is mechanically flexible

[37] and has very high conductivity [38], which would allow the integration of a hydrogen-storage module into graphene based devices. In particular, graphene can be now produced on a large and cost-effective scale by either top down (such as exfoliation from bulk) or bottom up (atom by atom growth) techniques [39], which significantly decreases the cost.

Not surprisingly, there has been enormous interest in developing graphene-based materials for hydrogen storage from both the experimental and theoretical communities [40,41]. In this chapter, we will review the recent progress in this exciting field, hydrogen storage in both molecular and atomic forms will be discussed.

16.2 Hydrogen Storage in Molecule Form

16.2.1 Hydrogen Storage in Graphene Sheets

H_2 is a non-polar molecule and the interaction between H_2 and the pristine graphene is mainly the instantaneous dipole–dipole induced forces (called London forces), which are derived from the overlapping the σ orbital of H_2 and the π states of graphene. Many theoretical studies [42–44] have demonstrated that the spatial distribution of H_2 molecule on graphene surface is quite delocalized and H_2 actually exhibits free lateral movement. Patchkovskii et al. [45] revealed a slightly attractive (-1.2 kJ/mol) H_2–graphene free reaction energy at room temperature. This tiny free reaction energy corresponds to an equilibrium constant of $K_{eq} = \exp(-\Delta F/RT) \approx 1.6$, indicating that the abundance of H_2 can only be increased by 60% on graphene at room temperature according to the equilibrium equation. Therefore, the surface of graphite is not suitable for hydrogen storage. This theoretical finding explains well the rather small gravimetric density of hydrogen in single-layered graphene as found in many experiments. For example, Ma et al. [46] evaluated the hydrogen storage behavior of the pristine single-layer graphene in a powder form, however, only 0.4 wt% and 0.2 wt% hydrogen uptake was obtained at 77 K under 100 kPa and room temperature under 6 MPa, respectively. By reducing a colloidal suspension of exfoliated graphite oxide, Srinivas et al. [47] successfully synthesized graphene-like nanosheets, which gave hydrogen adsorption capacities of about 1.2 wt% and 0.1 wt% at 77 K and 298 K, respectively.

Therefore, to increase the hydrogen storage capacity of graphene, the first urgent task is to enhance the interaction between H_2 molecules and graphene. In this regard, Patchkovskii et al. [45] proposed that H_2 molecules can be stored between graphene layers. They systematically investigated the physisorption capabilities of a graphene bilayer system by ab initio computations, and revealed that the interlayer distance plays a key role on determining the hydrogen storage capacity: when two layers of graphene are separated by 6 Å, one monolayer of H_2 molecules can be accommodated in between, yielding a 3–4 wt% hydrogen storage capacity at 10 MPa, and the H_2 adsorption energy is also optimal (10 kJ/mol). This storage capacity is encouraging but still below the DOE's target. Remarkably, Patchkovskii et al. [45] further demonstrated that it is possible to store two layers of hydrogen molecules between graphene layers if the interlayer distance is 8 Å (Figure 16.1), which would lead to a hydrogen storage capacity of 5.0–6.5 wt% under technologically acceptable conditions.

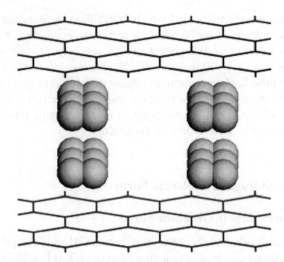

Figure 16.1 *Schematic of storing two layers of hydrogen molecules between spaced graphene sheets*

Inspired by the exciting prediction of Patchkovskii *et al.* [45], many experimental efforts have been devoted to increasing the interlayer distance of graphene layers towards a better hydrogen storage capacity. Having graphite oxide silylated with alkyltrichlorosialne and then with methyltrichlorosilane, Matsuo *et al.* [48] successfully synthesized silsesquioxane bridged graphene layers. The interlayer distance of the pillared graphene layers locates in the range of 1.34–1.6 nm. The hydrogen storage capacity of pillared graphene layers at ambient temperature can reach 0.6 wt% and show a relatively high total hydrogen storage of 14 g l^{-1} at 20 MPa. Especially, the H_2 adsorption energy on pillared graphene layers (8–11 kJ/mol) is much higher than that on the surface of single-layered graphene, indicating that sandwiching H_2 molecules between graphene layers can truly enhance the adsorption energy. However, since the interlayer distance is much larger than the optimal value, the hydrogen uptake is still very low.

Recently, Jin *et al.* [49] realized the facile organic functionalization and cross-linking of thermally exfoliated graphene (TEG) layers by using an *in situ* diazonium reaction in super acids (Figure 16.2). The functionalized groups (aryl, phenyl) can serve as interlayer spacers between TEG layers. The interlayer distance of the organic cross-linked TEG layers is about 0.96 nm, which is significantly higher than that of original TEG layers (0.76 nm). As expected, the functionalized TEG layers show enhanced hydrogen uptake (1.9 wt%) at 77K and 2 bar when compared to the original TEG material and normal carbonaceous adsorbents.

16.2.2 Hydrogen Storage in Metal Decorated Graphene

Though accommodating H_2 molecules within layered graphene can enhance the adsorption energy and lead to reasonable hydrogen uptake, the present hydrogen storage capacity is still far below the goals of DOE at practical conditions. In light of the weak H_2-graphene

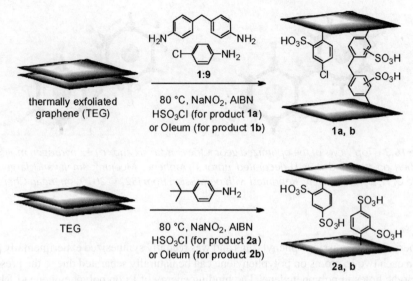

Figure 16.2 *Schematic of functionalization and cross-linking of teg sheets in chlorosulfonic acid and oleum. Reprinted with permission from [49] © 2011 American Chemistry Society*

interaction, decorating graphene with metals, including Li, Ca and transition metals, has been proposed by many studies towards enhancing the hydrogen binding energies.

16.2.2.1 Lithium Decorated Graphene

Li as decorating metal for graphene was first studied by Ataca *et al.* [50]. By using first-principles computations they predicted that Li atoms can be adsorbed on both sides of graphene to form a uniform and stable configuration. With the low ionization energy of 2s electrons, Li can easily donate its 2s electrons to graphene, which converts graphene from semi-metallic to metallic. Especially, Li decorated graphene can serve as a promising medium for hydrogen storage. Compared with pristine graphene, the adsorption energy of H_2 on Li-doped graphene is significantly enhanced: the hydrogen adsorption on H_2 is due to the polarization mechanism, since the charge transfer from Li to graphene leaves the Li atom in a cationic state. Li atoms can form a denser coverage on graphene forming the 2×2 pattern, and one Li atom can adsorb up to four H_2 molecules with an average adsorption energy of ~0.20 eV. Consequently, a gravimetric density of 12.8 wt%, much higher than the goal of DOE, could be expected. This finding was further supported by the reexamination by Zhou *et al.* [51], whose results indicate that hydrogen storage capacity can be further increased to 16 wt% by adjusting the coverage of Li atoms on graphene to the $\sqrt{3} \times \sqrt{3}$ pattern at both sides.

However, at a high coverage, the binding energy of Li will be significantly weakened by the strong electrostatic interaction between Li cations, which makes the whole system thermodynamically unstable, and consequently lowers the adsorption energy of H_2. To overcome this problem, Du *et al.* [52] theoretically designed a system with Li atoms adsorbed on the two dimensional polyphenylene, which structurally resembles graphene

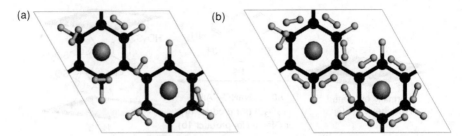

Figure 16.3 *Top views of the optimized geometries for (a) six and (b) 12 hydrogen molecules adsorbed on two and four Li-decorated porous graphene. Medium, small, and large balls denote C, H, Li, respectively. Reprinted with permission from [52] © 2010 American Chemistry Society*

with periodically "missing" phenyl rings and was recently synthesized experimentally [53]. Herein each two Li atoms on polyphenylene can be naturally separated due to the presence of periodic holes in polyphenylene. The binding energy of Li on polyphenylene (1.81 eV) is much larger that on a (2 × 2) graphene (0.86 eV) at the same theoretical level [50]. The adsorption energy for the first hydrogen molecule (0.27 eV) is much larger than that in Li-doped graphene (0.09 eV). The maximum number of absorbed Li atoms in one unit cell of polyphenylene is two and four for one and both sides, respectively (Figure 16.3). Each Li atom on polyphenylene can adsorb three H_2 molecules with an average binding energy of 0.24 eV, yielding up to a 12 wt% hydrogen storage capacity.

16.2.2.2 Calcium Decorated Graphene

Due to the low cohesive energy and low ionization energy of 4s electrons, Ca was considered to be an ideal decorating element to make graphene serve as hydrogen storage medium. The underlying mechanism for Ca-decorated graphene as hydrogen storage material is as same as Li; however, Ca deposited on graphene can be polarized more pronouncedly. Therefore, Ca would have more storage interaction with H_2 molecules.

Acata *et al.* [54] theoretically demonstrated that Ca atoms can be bound to both sides of graphene to reach 25% coverage, each adsorbed Ca can absorb up to four hydrogen molecules, with an average adsorption energy of 0.29 eV. At the maximum coverage (25%), this system can yield a hydrogen storage capacity of 8.4 wt%. However, on pristine graphene, the binding energy of Ca is much lower than the cohesive energy of Ca, and the barrier for Ca diffusion is rather low (~0.1 eV), thus Ca decorated graphene is not a stable system.

To enhance the binding energy between Ca and graphene, Lee *et al.* [55] proposed to use graphene nanoribbons (GNRs) instead of the graphene monolayer since the edge sites of GNRs are much more reactive than the inner sites. Their computations revealed that the clustering of Ca atoms can be safely hindered on the zigzag edges or B-doped armchair edges of GNRs. One Ca atom can bind up to six H_2 molecules with an average adsorption energy of ~0.2 eV, and a Ca-decorated zigzag graphene nanoribbon (ZGNR) can reach the gravimetric capacity of ~5 wt% hydrogen (Figure 16.4).

Figure 16.4 *The optimized atomic geometries of maximum number of adsorbed H_2 molecules for a Ca-decorated ZGNR. Medium, small, and large balls denote C, H, Ca, respectively. Reprinted with permission from [55] © 2010 American Chemistry Society*

16.2.2.3 Transition Metal Decorated Graphene

Transition metals were also considered as decorating elements for hydrogen storage [56–58], which takes advantage of the Kubas interaction. For example, DFT simulations [56,57] demonstrated that a Ti atom on graphene can bind up to four H_2 molecules. With Ti adsorbed on both sides of graphene, this complex can store hydrogen molecules up to 7.8 wt%.

However, transition metals usually have large cohesive energies and prefer to clustering, which is detrimental to the hydrogen storage capacity. Many approaches to enhance the binding energy between transition metals and graphene substrates have been explored. For instance, Zhou *et al.*'s DFT computations [59] demonstrated that an applied strain can not only stabilize the supported transition metal atoms but also prevent them from clustering. In particular, a tensile strain of 10% in graphene increases the adsorption energy of Ti atom by around 70% and the gravimetric density of hydrogen storage up to 9.5 wt%, with an average adsorption energy of 0.2 eV/H_2. Wang *et al.* [60] suggested graphene oxide as a substrate, in which transition metals such as Ti can strongly bind to epoxies and hydroxyl groups to prevent clustering. The estimated theoretical gravimetric and volumetric densities for Ti-decorated graphene oxide are 4.9 wt% and 64 g l^{-1}, respectively (Figure 16.5). Recently, Wu *et al.* [61] demonstrated that the bonding energy between Sc and the edge carbon atoms of GNRs is significantly greater than the cohesive energy of bulk Sc, thus hindering the clustering of Sc. The Sc-terminated GNRs can bind multiple H_2 molecules with the average H_2 adsorption energy ranging from 0.17 to 0.23 eV. The predicted highest hydrogen storage capacity of Sc-terminated GNRs is >9 wt%.

16.2.3 Hydrogen Storage in Graphene Networks

Many theoretical efforts also have been devoted to designing new graphene-based materials. Readers are encouraged to read Chapter 8 in which Guo *et al.* review the latest progress

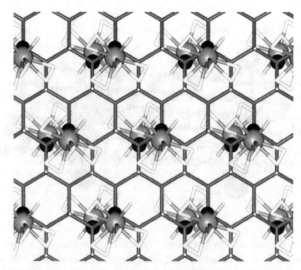

Figure 16.5 Optimized structure of Ti-decorated graphene oxide with fully loaded H_2 molecules. Gray and black balls denote Ti and O, respectively, Gray and white sticks denote C and H, respectively. Reprinted with permission from [60] © 2009 American Chemistry Society

on graphene-based architecture and assemblies. The hydrogen storage capability has been explored for several kinds of graphene networks.

16.2.3.1 Covalently Bonded Graphene

Park *et al.* [62] proposed a new class of carbon-based three-dimensional (3D) solid structures, covalently bonded graphenes (CBGs) which consist of sp^3 hybridized carbon atoms and graphene fragments, and could lead to high hydrogen capacity (Figure 16.6). By means of DFT and MP2 computations, they demonstrated that the interaction between H_2 and CBGs is much stronger than that on an isolated graphene with an increase of at least 20%

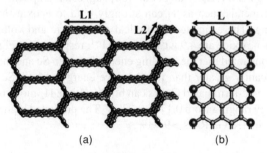

Figure 16.6 (a) Geometric structure of CBG (L1, L2). L1 and L2 denote the unit of the sidewall length. (b) Width of the side plane, as denoted by L, is given as a half integer starting from 1. For this particular size, L is 3. Gray circles indicate the four coordinated carbon atoms. Reprinted with permission from [62] © 2007 American Chemistry Society

in binding energy. CBGs could have a hydrogen uptake up to ~8 wt%, depending on the length of L1 and L2. Very recently, Martínez-Mesa et al. [63] reported the similar results on carbon foams. However, the H_2 adsorption energy on CBGs or carbon foams (~2.3 kcal/mol) is still far away from the requirement for ambient-temperature applications.

To increase H_2 adsorption energies, Park et al. [62] suggested to introduce transition metals into the graphene networks. Especially, CBGs with appropriate pore sizes are capable of holding transition metals without clustering, which are very promising candidate for hydrogen storage at ambient conditions. For instance, Ti-decorated CBGs can give a hydrogen storage capacity of 4 wt% or higher depending on their pore size, with the average adsorption energy exceeding 0.2 eV.

In 2003, Viculis et al. [64] synthesized a new carbon material, namely carbon nanoscroll (CNS), which shows a spiral form and can be obtained by twisting a graphite sheet. The structural properties of CNS are quite similar to multi-walled carbon nanotubes (MWC-NTs); what different is that the interlayer distance of CNS is variable while that of MWCNTs is nearly constant (~3.6 Å). Therefore, CNS was expected to be a promising material for hydrogen storage by the time it was first synthesized.

In 2007, Mpourmpakis et al. [65] investigated the nature of the H_2 interaction with a CNS model and the hydrogen storage capacity of CNS by means of first-principles quantum chemical calculations combined with classic Monte Carlo simulations. It was found that pure carbon nanoscrolls cannot accumulate hydrogen because the interlayer distance is too small. However, increasing the interlayer distance to approximately 7 Å combined with lithium doping (Figure 16.7) can lead to a 3 wt% hydrogen uptake at ambient temperature and pressure. This encouraging theoretical outcome remains to be verified experimentally.

Theoretically, Dimitrakakis et al. [66] designed a new 3D network nanostructure, namely pillared graphene, which consists of parallel graphene layers at a variable distance connected by vertically aligned CNTs (Figure 16.8). In this case, the interlayer distance can be controlled by modulating the length of CNTs. By using a multiscale theoretical approach, they showed that the interaction between pillared graphene and H_2 molecule is very weak, even on the junction sites of this novel material. However, if pillared graphene is doped with lithium, the H_2 binding energy can be significantly enhanced, and the doped pillared

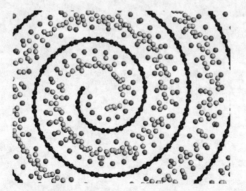

Figure 16.7 Hydrogen adsorption on lithium doped CNS structures of 7 Å interlayer distance with maximum Li ion to carbon atom ratio, at room temperature and 10 MPa pressure. Reprinted with permission from [65] © 2007 American Chemistry Society

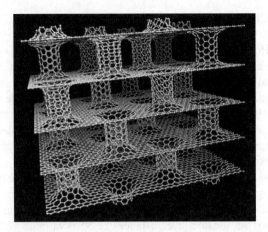

Figure 16.8 *Geometric structure of pillared graphene. Reprinted with permission from [66]* © *2008 American Chemistry Society*

graphene can store 41 g H_2/l under ambient conditions, almost reaching the DOE volumetric requirement for mobile applications.

Similarly, Tylianakis *et al.* [67] designed Li–doped pillared graphene oxides which can satisfy the DOE's gravimetric and volumetric H_2 uptake targets at low pressures and at 77 K. This kind of materials can be synthesized by substituting the OH groups of oxidized graphitic materials with alkoxide O–Li groups.

By means of joint theoretical and experimental efforts, Cheng *et al.* [68] successfully achieved a novel class of fluoride anions (F^-) containing graphite intercalation compounds (GICs), which exhibit significantly higher H_2 adsorption energies than commonly available porous carbon-based materials (Figure 16.9). The underlying mechanism of the unusually

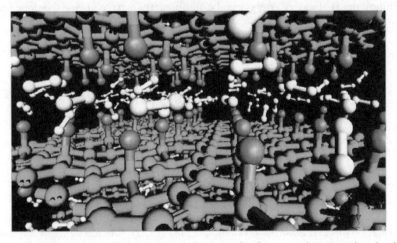

Figure 16.9 *Fully optimized structures of the partially fluorinated GIC with adsorbed H_2 molecules. Different shaded balls represent C, H, and F atoms, respectively. Reprinted with permission from [68]* © *2009 American Chemistry Society*

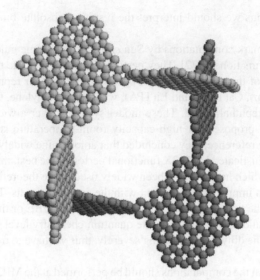

Figure 16.10 Geometric structures of open carbon frameworks. Reprinted with permission from [69] © 2012 American Chemistry Society

strong H_2 binding energy is that the semi-ionic C—F bonds in GICs can donate electrons to the σ^* orbital of H_2. GICs can give ~4 wt% hydrogen uptake, however, the H_2 adsorption energies diminish at very high H_2 densities. To conquer this problem, Cheng et al. [68] suggested doping the graphitic host with appropriate species, such as nitrogen, to promote high levels of charge transfer from graphene to F^- anions.

Quite recently, Kuchta et al. [69] proposed a type of open carbon frameworks (OCFs) with exceptional hydrogen storage capacity. Different from the traditional model of parallel graphene sheets hosting physisorbed H_2 layers in slit-shaped pores of variable width, OCFs are featured with locally planar units (unbounded or bounded fragments of graphene sheets) and variable ratios of in-plane to edge atoms (Figure 16.10). By means of grand canonical Monte Carlo simulations with appropriately chosen adsorbent–adsorbate interaction potentials, they carefully investigated the adsorption of H_2 molecules on OCFs. Since OCFs have ultrahigh surface area, their hydrogen storage capacity is higher than in slit-shaped pores. For OCFs with a ratio of in-plane to edge sites ≈ 1 and surface areas 3800–6500 m^2/g, the maximum hydrogen storage capacity of 7.5–8.5 wt% can be obtained at 77 K. If OCFs could be synthesized, they would provide hydrogen uptake at the level required for mobile application.

16.2.4 Notes to Computational Methods

Accurately calculating the dihydrogen binding energies is crucial to the modeling of hydrogen sorbents. However, we are still in the lack of adequate first-principles computational methods, nonbonding interactions are not well treated by classical density functional theory (DFT) [70]. Typically, the local density approximation (LDA) overestimates while the generalized gradient approximation (GGA) underestimates the dihydrogen binding energies [71]. Note that the theoretical studies summarized above mostly employed either LDA or GGA methods, thus we should interpret the resulting absolute binding energy values with caution.

The recent benchmark computations by Sun *et al.* provide some guidelines in choosing appropriate density functionals [72]. They performed highly accurate calculations at MP2 and CCSD(T) level of theory for the dihydrogen binding on four representative systems, namely Ti(Et), Sc(Cp), Ca(TPA), and Li(TPA), where Et = ethylene, Cp = cyclopentadienyl, and TPA = terephthalic acid. These model systems cover a wide range of sorbent materials previously proposed for high-capacity room-temperature storage. Using these high level data as the reference, they concluded that among nine widely used density functionals, the most sophisticated M05–2X functional performs the best, importantly, PBE and PW91 functionals, which have already been widely used in the theoretical modeling of H_2 sorbents, also give an impressive agreement with the accurate results. These critical evaluations indicate that the previous predictions using either the PBE or the PW91 functional are expected to be valid even at the accurate quantum chemistry level. On the other hand, LDA overestimates the dihydrogen binding severely, thus we have to treat the LDA results with great care.

We recommend that the computations should be performed at the MP2 or even high levels (such as coupled cluster method) whenever possible, for example, for small cluster models, zero-point energy (ZPE) correction and basis set superposition error (BSSE) should also be considered when targeting at accurate H_2 static binding energies. The small model computations will gain us deep chemical insights, which will guide us to design infinite systems. When the system is beyond the capability of MP2 method, the density functionals, which have shown good performance in treating noncovalent interactions (chosen by benchmark computation on the specific system), should be employed [73]. New generation density functionals, such as M05-2X [74], M08-HX [75], B2PLYP-D [76], and B97-D [77], which show outstanding performance in treating non-covalent interactions, will greatly facilitate our theoretical design and screening graphene-based hydrogen storage materials.

16.3 Hydrogen Storage in Atomic Form

16.3.1 Graphane

Chemical storage of hydrogen in graphene was first proposed by Sofo *et al.* [78] in 2007. They demonstrated theoretically that by adding a hydrogen atom to each carbon atom of graphene, the two-dimensional (2D) infinite hydrocarbon with the formula of C_nH_n, graphane, is thermodynamically stable and can form experimentally (Figure 16.11). Theoretically, graphane has a very high volumetric and gravimetric hydrogen density. The volumetric hydrogen capacity of 0.12 kg H_2/l is higher than the DOE target of 0.081 kg H_2/l for the year 2015. Additionally, the gravimetric capacity of 7.7 wt% H is higher than the 6.0 wt% DOE target for 2010. In 2009, Elias *et al.* [79] successfully synthesized graphane by exposing graphene to hydrogen plasma discharge. The experimentally characterized structural and electronic properties of graphane are the same as those predicted theoretically.

Although graphene can react exothermically with hydrogen plasma to produce graphane, this method is not economic in terms of hydrogen storage since it is not easy to obtain

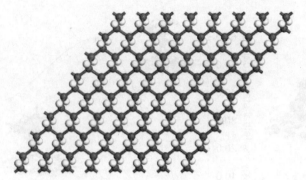

Figure 16.11 Geometric structure of graphane in the chair conformation. The carbon atoms are shown in gray and the hydrogen atoms in white

hydrogen plasma. Direct dissociation of H_2 molecule on graphene is also unrealistic due to the rather high binding energy of H_2 (~4.8 eV). Thus, the chemical storage of hydrogen in graphene has to be directed through some different paths.

Experimentally, Ryu *et al* [80]. demonstrated that hydrogenation of graphenes can also be realized by dissociating hydrogen silsesquioxane on graphene. Especially, chemisorbed hydrogen atoms can be detached by thermal annealing at 100~200°C, suggesting the possibility of using few-layer graphene for chemical storage of hydrogen. Recently, Subrahmanyam *et al.* [81] obtained hydrogenated graphene containing up to 5 wt% of hydrogen through Birch reduction of graphene oxide with lithium in liquid ammonia. The formation of sp^3 C−H bonds can be vigorously certified by spectroscopic studies. The attached hydrogen atoms can readily release on heating to 500°C or on irradiation with UV or laser radiation.

16.3.2 Chemical Storage of Hydrogen by Spillover

Recently, the chemical storage of hydrogen in graphene by spillover mechanism has been studied extensively. In hydrogen spillover [82, 83], H atoms are created on a metal surface, in which H_2 molecules can be more easily dissociated, and subsequently migrate to a receptor material to be stored.

By means of DFT computations, Chen *et al.* [84] studied the dissociation of molecular hydrogen on the Pt_6 cluster. By approaching the fully saturated Pt_6 cluster toward the graphene sheet, they demonstrated that the migration of H atoms from Pt cluster catalyst to graphene is facile at ambient conditions with a small energy barrier, and that the H atoms can be either physisorbed or chemisorbed on graphene surface. However, the process is slightly endothermic, and especially, the diffusion of chemisorbed H atoms on graphene is rather difficult since it requires breaking strong C−H bond. The theoretical studies performed by Froudakis *et al.* [85] are closer to realistic experimental environment, in which Pt_4 cluster is supported on the graphene substrate. They found that a very large energy barriers (~60 kcal/mol) has to be overcome first for H atoms to migrate from the Pt_4 cluster to the graphene surface (Figure 16.12). In agreement with Chen *et al.* [84], they also found that the chemisorbed H atoms cannot diffuse in this state due to the high energy barrier (31.2 kcal/mol), instead, the H atoms can enter the physisorption state first and then diffuse freely

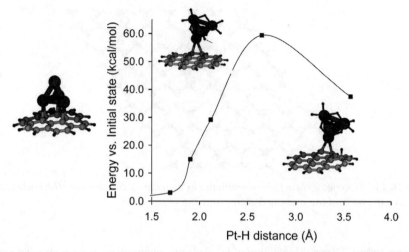

Figure 16.12 Left: optimized structure of Pt_4 cluster on coronene. Right: Energy scan for the migration of a H atom from the fully saturated Pt_4 cluster to the coronene surface. Medium, small, and large balls denote C, H, Pt, respectively. Reprinted with permission from [85] © 2009 American Chemical Society

on the surface of graphene. When two physisorbed H atoms are located on adjacent rings, they spontaneously recombine to form a H_2 molecule, indicating that it is impossible to have high coverage with physisorbed H atoms.

In light of the previous discussions, we can conclude that the main problems for hydrogen storage using spillover mechanism are: (1) it is very difficult for H atoms to transfer from saturated metal catalyst to pristine graphene, mostly due to the weaker interaction between atomic H and graphene than that between H and metal catalyst; (2) large barrier for the diffusion of chemisorbed H atoms on graphene. To conquer these two problems, many approaches have been proposed theoretically.

Singh et al. [86] pointed out that it is energetically unfavorable for the spillover to occur on pristine graphene surface, however, if graphene is partially hydrogenated, the reactivity at the interface areas, especially at the zigzag interface, is much higher than pristine graphene, as a result the interaction between atomic H and graphene can be notably enhanced (Figure 16.13). The diffusion barrier for hydrogen from the Pd cluster to the hydrogenated graphene is rather low, suggesting that the spillover can easily occur below room temperature. Especially, the process does not require full saturation of the cluster because thermodynamically spillover becomes favorable even before a cluster saturates.

The interaction between atomic H and graphene can also be enhanced via doping with foreign atoms. Wu et al. [87] demonstrated that B-doping can substantially enhance the adsorption strength for both H atoms and the metal cluster on graphene sheet. The firmly bound catalytic metal on B-doped graphene can effectively dissociate H_2 molecules into H atoms, and the resulting H atom can easily migrate from the bridge site of the H-saturated metal to the supporting graphene sheet. They also showed that BC_3 sheet has low activation barriers for both H migration and diffusion processes.

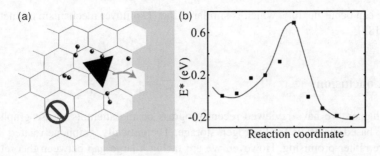

Figure 16.13 (a) Fully relaxed Pd cluster saturated with H, next to the hydrogenated phase (gray area). (b) Plot of energies of intermediate images for the barrier calculation versus the reaction coordinate. Black lines, balls, and triangles denote C, H, Pd, respectively. Reprinted with permission from [86] © 2009 American Chemistry Society

Psofogiannakis et al.'s theoretical studies [88] suggested that some O-containing groups on graphene can also significantly facilitate the spillover. With the help of a chemisorbed oxygen atom in the epoxide form, a hydrogen atom can easily migrate from Pt_4 cluster to graphene and eventually produce a hydroxyl group (Figure 16.14a). The DFT computations indicated that the migration event is exothermic by 0.67 eV, and only a 0.4 eV energy barrier needs to be crossed for the migration to take place. The energy barrier for the migration of a H atom from the hydroxyl group to an adjacent O (epoxide) atom on graphene is as low as 0.33 eV (Figure 16.14b), indicating that the diffusion process of H atoms on

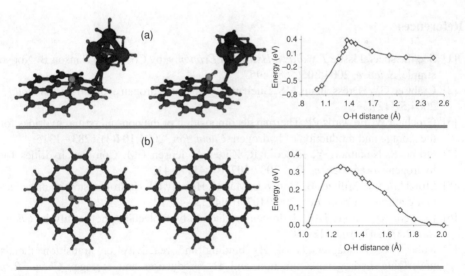

Figure 16.14 Optimized structures of the initial and final states and DFT-calculated reaction coordinate scans for the migration of H (a) from a Pt_4 cluster to epoxide O on a model graphite surface; (b) from the OH group to an epoxide O, both adsorbed on graphite (H diffusion) Reprinted with permission from [88] © 2009 American Chemistry Society

graphene can occur flexibly, which is similar to the H spillover mechanism on metal oxide surfaces [89].

16.4 Conclusion

In this chapter, we have reviewed recent progress on the attempts to use graphene and graphene-based materials for hydrogen storage. Theoretically, graphene-related nanomaterials are rather promising. However, we are facing a large gap between theoretical and experimental studies. For molecular hydrogen storage, it is now urgent to figure out methods to engineer the interlayer distance of graphene sheets more feasibly and efficiently in order to enhance the interaction between graphene and H_2 molecules. While for atomic hydrogen storage, the main challenges lie on the large barrier for hydrogen atoms to transition from metal clusters to graphene and the difficulties for H atoms to diffuse on graphene. Overall, hydrogen storage in graphene-related nanomaterials is still a very young but promising field, and tremendous efforts are desired to achieve the final goal of hydrogen economy.

Acknowledgements

Support by Department of Defense (Grant W911NF-12-1-0083) and National Science Foundation (Grant EPS-1010094) is gratefully acknowledged.

References

[1] See the special issue *Towards a Hydrogen Economy*, by Coontz R, Hanson B. Not so simple. *Science*. 2004;**305**(5686):957.
[2] Crabtree GW, Dresselhaus MS, Buchanan MV. The hydrogen economy. *Phys. Today*. 2004;**57**(12):39.
[3] Grochala W, Edwards PP. Thermal decomposition of the non-interstitial hydrides for the storage and production of hydrogen. *Chem. Rev.* 2004;**104**(3):1283–1315.
[4] Orimo S, Nakamori Y, Eliseo J R, Züttel A, Jensen CM. Complex hydrides for hydrogen storage. *Chem. Rev.* 2007;**107**(10):4111–4132.
[5] Struzhkin VV, Militzer B, Mao WL, Mao H, Hemley RJ. Hydrogen storage in molecular clathrates. *Chem. Rev.* 2007;**107**(10):4133–4151.
[6] Esswein AJ, Nocera DG. Hydrogen production by molecular photocatalysis. *Chem. Rev.* 2007;**107**(10):4022–4047.
[7] Kubas GJ. Fundamentals of H_2 binding and reactivity on transition metals underlying hydrogenase function and H_2 production and storage. *Chem. Rev.* 2007;**107**(10):4152–4205.
[8] Satyapal S, Petrovic J, Read C, Thomas G, Ordaz G. The U.S. Department of Energy's National Hydrogen Storage Project: Progress towards meeting hydrogen-powered vehicle requirements. *Catalysis Today*. 2007;**120**(3):246–256.

[9] Rosi NL, Eckert J, Eddaoudi M, Vodak DT, Kim J, O'Keefe M and Yaghi OM. Hydrogen storage in microporous metal-organic frameworks. *Science*. 2003;**300**(5622):1127–1129.

[10] Rowsell JLC, Yaghi OM. Strategies for hydrogen storage in metal–organic frameworks. *Angew. Chem. Int. Ed*. 2005;**44**(30):4670–4679.

[11] Vitillo JG, Regli L, Chavan S, Ricchiardi G, Spoto G, Dietzel PDC, et al. Role of exposed metal sites in hydrogen storage in MOFs. *J. Am. Chem. Soc*. 2008;**130**(26): 8386–8396.

[12] For the debate see Dagani R. *Chem. Eng. News*. 2008; **80**(1):25.

[13] Lochan R, Head-Gordon M. Computational studies of molecular hydrogen binding affinities: the role of dispersion forces, electrostatics, and orbital interactions. *Phys. Chem. Chem. Phys*. 2006;**8**:1357–1370.

[14] Zhao Y, Zhang SB, Kim YH, and Heben MJ, "Theory of hydrogen storage in nanoscale materials" in *The Oxford Handbook of Nanoscience and Technology*, Oxford University Press, 2010; Vol. **III**. 699–735.

[15] Zhao YF, Kim YH, Dillon AC, Heben MJ, Zhang SB. Hydrogen storage in novel organometallic buckyballs. *Phys. Rev. Lett*. 2005;**94**(15):155504.

[16] Zhao YF, Lusk, MT, Dillon AC, Heben MJ, Zhang SB. Boron-based organometallic nanostructures: hydrogen storage properties and structure stability. *Nano Lett*. 2008;**8**(1):157–161.

[17] Zhao YF, Heben MJ, Dillon AC, Simpson, LJ, Blackburn, JL, Dorn, HC, Zhang SB. Nontrivial tuning of the hydrogen-binding energy to fullerenes with endohedral metal dopants. *J. Phys. Chem. C*. 2007;**111**(35):13275–13279.

[18] Zhao YF, Dillon AC, Kim YH, Heben MJ, Zhang SB. Self-catalyzed hydrogenation and dihydrogen adsorption on titanium carbide nanoparticles. *Chem. Phys. Lett*. 2006;**425**(4):273–277.

[19] Yildirim T, Ciraci S. Titanium-decorated carbon nanotubes as a potential high-capacity hydrogen storage medium. *Phys. Rev. Lett*. 2005;**94**(17):175501.

[20] Durgun E, Ciraci S, Zhou W, Yildirim T. Transition-metal-ethylene complexes as high-capacity hydrogen-storage media. *Phys. Rev. Lett*. 2006;**97**(22):226102.

[21] Lee H, Choi WI, Ihm J. Combinatorial search for optimal hydrogen-storage nanomaterials based on polymers. *Phys. Rev. Lett*. 2006;**97**(5):056104.

[22] Sun Q, Wang Q, Jena P, Kawazoe Y. Clustering of Ti on a C_{60} Surface and its effect on hydrogen storage. *J. Am. Chem. Soc*. 2005;**127**(42):14582–14583.

[23] Krasnov PO, Ding F, Sing AK, Yakobson BI. Clustering of Sc on SWNT and reduction of hydrogen uptake: ab-initio all-electron calculations. *J. Phys. Chem. C*. 2007;**111**(49):17977–17980.

[24] Li S, Jena P. Comment on Combinatorial search for optimal hydrogen-storage nanomaterials based on polymers. *Phys. Rev. Lett*. 2006;**97**(20):209601.

[25] Shevlin SA, Guo ZX. High-capacity room-temperature hydrogen storage in carbon nanotubes via defect-modulated titanium doping. *J. Phys. Chem. C*. 2008;**112**(44):17456–17464.

[26] Sun Q, Jena P, Wang Q, Marquez M. First-principles study of hydrogen storage on $Li_{12}C_{60}$. *J. Am. Chem. Soc*. 2006;**128**(30):9741–9745.

[27] Chandrakumar KRS, Ghosh SK. Alkali-metal-induced enhancement of hydrogen adsorption in C60 fullerene: An ab initio study. *Nano Lett*. 2008;**8**(1):13–19.

[28] Zhao YF. Kim YH, Simpson LJ, Dillon AC, Wei SH, Heben MJ. Opening space for H_2 storage: Cointercalation of graphite with lithium and small organic molecules. *Phys. Rev. B*. 2009;**78**(14):144102.

[29] Meng S, Kaxiras E, Zhang ZY. Metal–diboride nanotubes as high-capacity hydrogen storage media. *Nano Lett.* 2007;**7**(3):663–667.

[30] Yoon M, Yang SY, Hicke C, Wang E, Geohegan D, Zhang ZY. Calcium as the superior coating metal in functionalization of carbon fullerenes for high-capacity hydrogen storage. *Phys. Rev. Lett.* 2008;**100**(20):206806.

[31] Wang Q, Sun Q, Jena P, Kawazoe Y. Theoretical study of hydrogen storage in Ca-coated fullerenes. *J. Chem. Theory Comput.* 2009; **5**(2):374–379.

[32] Yang XB, Zhang RQ, Ni J. Stable calcium adsorbates on carbon nanostructures: applications for high-capacity hydrogen storage. *Phys. Rev. B*. 2009;**79**(7):075431.

[33] Li M, Li YF, Zhou Z, Shen PW, Chen ZF. Ca-coated boron fullerenes and nanotubes as superior hydrogen storage materials. *Nano Lett.* 2009;**9**(5):1944–1948.

[34] Novoselov KS, Geim AK, Morozov SV, Jiang D, Zhang Y, Dubonos SV, et al. Electric field effect in atomically thin carbon films. *Science*. 2004;**306**(5696):666–669.

[35] Novoselov KS, Jiang D, Schedin F, Booth TJ, Khotkevich VV, Morozov SV, Geim AK. Two-dimensional atomic crystals. *Proc. Natl. Acad. Sci. U.S.A.* 2005;**102**(30):10451–10453.

[36] Balandin AA, Ghosh S, Bao W, Calizo I, Teweldebrhan D, Miao F, Lau CN. Superior thermal conductivity of single-layer graphene. *Nano. Lett.* 2008;**8**(2):902–907.

[37] Lee CG, Wei XD, Kysar JW, Hone J. Measurement of the elastic properties and intrinsic strength of monolayer graphene. *Science* 2008;**321**(5887):385–388.

[38] Morozov SV, Novoselov KS, Katsnelson MI, Schedin F, Elias D, Jaszczak JA, Geim AK. Giant intrinsic carrier mobilities in graphene and its bilayer. *Phys.Rev. Lett.* 2008;**100**(1):016602.

[39] Sega M. Selling graphene by the ton. *Nat. Nanotechnol.* 2009;**4**(1):612–614.

[40] Lu Y, Feng YP. Adsorptions of hydrogen on graphene and other forms of carbon structures: first principle calculations. *Nanoscale.* 2011;**3**(6):2444–2453.

[41] Pumera M. Graphene-based nanomaterials for energy storage. *Energy Environ. Sci.* 2011;**4**:668–674.

[42] Darkrim F, Vermesse J, Malbrunot P, Levesque D. Monte Carlo simulations of nitrogen and hydrogen physisorption at high pressures and room temperature. comparison with experiments. *J. Chem. Phys.* 1999;**110**(8):4020–4027

[43] Deng WQ, Xu X, Goddard WA. New alkali doped pillared carbon materials designed to achieve practical reversible hydrogen storage for transportation. *Phys. Rev. Lett.* 2004;**92**(16):1666103.

[44] Wang QY, Johnson JK. Hydrogen adsorption on graphite and in carbon slit pores from path integral simulations. *Mol. Phys.* 1998;**95**(2):299–309.

[45] Patchkovskii S, Tse JS, Yurchenko SN, Zhechkov L, Heine T, Seifert G. Graphene nanostructures as tunable storage media for molecular hydrogen. *Proc. Natl. Acad. Sci. U.S.A.* 2005;**102**(30):10439–10444.

[46] Ma LP, Wu ZS, Li J, Wu ED, Ren WC, Cheng HM. Hydrogen adsorption behavior of graphene above critical temperature. *Int. J. Hydrogen Energy.* 2009;**34**(5):2329–2332.

[47] Srinivas G, Zhu Y, Piner R, Skipper N, Ellerby M, Ruoff R. Synthesis of graphene-like nanosheets and their hydrogen adsorption capacity. *Carbon.* 2010;**48**(3):630–635.

[48] Matsuo Y, Ueda S, Konishi K, Marco-Lozar JP, Lozano-Castelló D, Cazorla-Amorós D. Pillared carbons consisting of silsesquioxane bridged graphene layers for hydrogen storage materials. *Int. J. Hydrogen Energy.* 2012;**37**(14):10702–10707.
[49] Jin Z, Lu W, O'Neill KJ, Parilla PA, Simpson LJ, Kittrell C, Tour JM. Nano-engineered spacing in graphene sheets for hydrogen storage. *Chem. Mater.* 2011; **23**(4):923–925.
[50] Ataca C, Aktürk E, Ciraci S, Ustunel H. High-capacity hydrogen storage by metallized graphene. *Appl. Phys. Lett.* 2008; **94**(4):043123.
[51] Zhou WW, Zhou JJ, Shen J, Ouyang C, Shi S. First-principles study of high-capacity hydrogen storage on graphene with Li atoms. *J. Phys. Chem. Soli.* 2012;**73**(2):245–251.
[52] Du AJ, Zhu ZH, Smith SC. Multifunctional porous graphene for nanoelectronics and hydrogen storage: new properties revealed by first principle calculations. *J. Am. Chem. Soc.* 2010;**132**(9):2876–2877.
[53] Bieri M, Treier M, Cai J, At-Mansour K, Ruffieux P, Gröning P, *et al.* Porous graphenes: Two-dimensional polymer synthesis with atomic precision. *Chem. Commun.* 2009;**45**:6919–6921.
[54] Ataca C, Aktürk E, Ciraci S. Hydrogen storage of calcium atoms adsorbed on graphene: first-principles plane wave calculations. *Phys. Rev. B.* 2009;**79**(4):041406(R).
[55] Lee H, Ihm J, Cohen ML, Louie SG. Calcium-decorated graphene-based nanostructures for hydrogen storage. *Nano Lett.* 2010;**10**(3):793–798.
[56] Durgun E, Ciraci S, Yildirim T. Functionalization of carbon-based nanostructures with light transition-metal atoms for hydrogen storage. *Phys. Rev. B.* 2008;**77**(8):085405.
[57] Liu Y, Ren L, He Y, Cheng HP. Titanium-decorated graphene for high-capacity hydrogen storage studied by density functional simulations. *J. Phys.: Condens. Matter.* 2010; **22**(44):445301.
[58] Kim G, Jhi SH, Park N, Louie SG, Cohen ML. Optimization of metal dispersion in doped graphitic materials for hydrogen storage. *Phys. Rev. B.* 2008;**77**(8):085408.
[59] Zhou M, Lu Y, Zhang C, Feng YP. Strain effects on hydrogen storage capability of metal-decorated graphene: a first-principles study. *Appl. Phys. Lett.* 2010;**97**(10):103109.
[60] Wang L, Lee K, Sun YY, Lucking M, Chen Z, Zhao JJ, Zhang SB. Graphene oxide as an ideal substrate for hydrogen storage. *ACS Nano.* 2009;**3**(10):2995–3000.
[61] Wu M, Gao Y, Zhang Z, Zeng XC. Edge-decorated graphene nanoribbons by scandium as hydrogen storage media. *Nanoscale.* 2012;**4**(3):915–920.
[62] Park N, Hong S, Kim G, Jhi SH. Computational study of hydrogen storage characteristics of covalent-bonded graphenes. *J. Am. Chem. Soc.* 2007;**129**(29):8999–9003.
[63] Martínez-Mesa A, Zhechkov L, Yurcheko SN, Heine T, Seifert G, Rubayo-Soneira J. Hydrogen physisorption on carbon foams upon inclusion of many-body and quantum delocalization effects. *J. Phys. Chem. C.* 2012;**116**(36):19543–19553.
[64] Viculis LM, Mack JJ, Kaner RB. A chemical route to carbon nanoscrolls. *Science.* 2003;**299**(5611), 1361.
[65] Mpourmpakis G, Tylianakis E, Froudakis GE. Carbon nanoscrolls: a promising material for hydrogen storage. *Nano Lett.* 2007;**7**(7):1893–1897.
[66] Dimitrakakis GK, Tylianakis E, Froudakis GE. Pillared graphene: a new 3-D network nanostructure for enhanced hydrogen storage. *Nano Lett.* 2008;**8**(10):3166–3170.

[67] Tylianakis E, Psofogiannakis GM, Froudakis GE. Li-doped pillared graphene oxide: a graphene-based nanostructured material for hydrogen storage. *J. Phys. Chem. Lett.* 2010;**1**(16):2459–2464.

[68] Cheng H, Sha X, Chen L, Cooper AC, Foo ML, Lau GC, et al. An enhanced hydrogen adsorption enthalpy for fluoride intercalated graphite compounds. *J. Am. Chem. Soc.* 2009;**131**(49):17732–17733.

[69] Kuchta B, Firlej L, Mohammadhosseini A, Boulet P, Beckner M, Romanos J, Pfeifer P. Hypothetical high-surface-area carbons with exceptional hydrogen storage capacities: open carbon frameworks, *J. Am. Chem. Soc.* 2012;**134**(36):15130–15137.

[70] Lochan RC, Head-Gordon M. Computational studies of molecular hydrogen binding affinities: the role of dispersion forces, electrostatics, and orbital interactions. *Phys. Chem. Chem. Phys.* 2006;**8**:1357–1370.

[71] Kim YH, Zhao Y, Williamson A, Heben MJ, Zhang SB. Nondissociative adsorption of H_2 molecules in light-element-doped fullerenes. *Phys. Rev. Lett.* 2006;**96**(1): 16102.

[72] Sun YY, Lee K, Wang L, Kim YH, Chen W, Chen Z, Zhang SB. Accuracy of density functional theory methods for weakly bonded systems: the case of dihydrogen binding on metal centers, *Phys. Rev. B.* 2010;**82**:073401.

[73] Zhang CG, Zhang R, Wang ZX, Zhou Z, Zhang SB, Chen Z. Ti-Substituted boranes as hydrogen storage materials: a computational quest for ideal combination of stable electronic structure and optimal hydrogen uptake, *Chem. Eur. J.* 2009;**15**(24):5910–5919.

[74] Zhao Y, Truhlar D. Density functionals with broad applicability in chemistry. *Acc. Chem. Res.* 2008;**41**(22):157–167.

[75] Zhao, Y, Truhlar, DG. Exploring the limit of accuracy of the global hybrid meta density functional for main-group thermochemistry, kinetics, and noncovalent interactions. *J. Chem. Theory Comput.* 2008;**4**(11):1849–1868.

[76] Schwabe T, Grimme S. Theoretical thermodynamics for large molecules: walking the thin line between accuracy and computational cost. *Acc. Chem. Res.* 2008;**41**(4):569–579.

[77] Grimme S, Antony J, Ehrlich S, Krieg HA. Consistent and accurate ab initio parametrization of density functional dispersion correction (DFT-D) for the 94 elements H−Pu. *J. Chem. Phys.* 2010;**132**(15):154104.

[78] Sofo JO, Chaudhari AS, Barber GD. Graphane: A two-dimensional hydrocarbon. *Phys. Rev. B.* 2007;**75**(15):153401.

[79] Elias DC, Nair RR, Mohiuddin TMG, Morozov SV, Blake P, Halsall MP, et al. Control of graphene's properties by reversible hydrogenation: evidence for graphane. *Science.* 2009;**323**(5914):610–613.

[80] Ryu SM, Han MY, Maultzsch J, Heinz TF, Kim P, Steigerwald ML, Brus LE. Reversible basal plane hydrogenation of graphene. *Nano Lett.* 2008;**8**(12): 4597–4602.

[81] Subrahmanyam KS, Kumar P, Maitra U, Govindaraj A, Hembram KPSS, Waghmareb UV, Rao CNR. Chemical storage of hydrogen in few-layer graphene. *Proc. Natl. Acad. Sci. U.S.A.* 2011;**108**(7):2674–2677.

[82] Conner WC, Falconer JL. Spillover in heterogeneous catalysis. *Chem. Rev.* 1995;**95**(3):759–788.

[83] Prins R. Hydrogen Spillover. Facts and fiction. *Chem. Rev.* 2012; **112**(5): 2714–2738.
[84] Chen L, Cooper AC, Pez GP, Cheng H. Mechanistic study on hydrogen spillover onto graphitic carbon materials. *J. Phys. Chem. C* 2007;**111**(51):18995–19000.
[85] Psofogiannakis GM, Froudakis GE. DFT study of the hydrogen spillover mechanism on Pt-Doped graphite. *J. Phys. Chem. C* 2009;**113**(33):14908–14915.
[86] Singh AK, Ribas MA, Yakobson B. H-spillover through the catalyst saturation: an ab initio thermodynamics study. *ACS Nano*. 2009;**3**(7):1657–1662.
[87] Wu HY, Fan X, Kuo JL, Deng WQ. DFT Study of hydrogen storage by spillover on graphene with boron substitution. *J. Phys. Chem. C*. 2011;**115**(18):9241–9249.
[88] Psofogiannakis GM, Froudakis GE. DFT study of hydrogen storage by spillover on graphite with oxygen surface groups. *J. Am. Chem. Soc.* 2009;**131**(30):15133–15135.
[89] Chen L, Cooper AC, Pez GP, Chen HS. On the mechanisms of hydrogen spillover in MoO_3. *J. Phys. Chem. C*. 2008;**112**(6):1755–1758.

17

Linking Theory to Reactivity and Properties of Nanographenes

Qun Ye, Zhe Sun, Chunyan Chi, and Jishan Wu
Department of Chemistry, National University of Singapore, Singapore

17.1 Introduction

Graphene is a nanomaterial with simple atomic structure but intriguing and largely unexplored chemistry and physics. The chemistry and physics of graphene have become hot topics in both theoretical and experimental science ever since its first isolation in 2004 [1]. A large variety of applications for this fantastic material has been proposed in anticipation of future technological revolutions. The electronic and magnetic properties of graphene are of particular interest to theoretical investigations due to their tight relationship with the property of π electrons. The properties of the π electron of the infinite graphene network can be partially realized by investigations on finite and definitive fragments. Such "hot" fragments, including acenes [2] and graphene nanoribbons [3] have been investigated theoretically to probe the potential role the edge structure plays in the bulk property of graphene. In this aspect, synthetic organic chemistry has made a contribution in providing experimental analysis of graphene nanofragments with definitive structures and well-defined edge structures.

Synthesis of graphene nanofragments, or polycyclic hydrocarbons, can be dated back to the nineteenth century and until now, a large pool of polycyclic hydrocarbons with various shapes, sizes, and edge structures have been synthesized and their synthesis and properties are well summarized [4, 7b]. The information collected for these model compounds has provided us with valuable guidelines to understand the properties of graphene and their origins. Acenes, for example, are molecules which have been intensively investigated both

---

Graphene Chemistry: Theoretical Perspectives, First Edition. Edited by De-en Jiang and Zhongfang Chen.
© 2013 John Wiley & Sons, Ltd. Published 2013 by John Wiley & Sons, Ltd.

experimentally and theoretically in order to understand the properties of the zigzag edges [2]. So far, the acenes up to nonacene have been synthesized in the lab with convincing evidence to support their existence [5]. While at the same time, calculation results have pointed out that as a singlet biradical state would be present in higher order acenes and in polyacenes, a polyradical state might be possible [2]. All these theoretical investigations as well as the EPR signal found in nonacene samples, provide insights in the presence of unpaired electrons in the graphene framework.

The edge structures have been found to provide many intriguing properties to the graphene nanofragments [6]. The theoretical considerations on edge effects are well illustrated in the rest of the chapters of this book. In this chapter, major attention focuses on the chemical properties of edge structures and how the edge structures are correlated to the chemical and physical properties of the molecules. The chemistry of both the zigzag edges and armchair edges present in different graphene nanofragments are introduced in this section. The structure-property relationship is also discussed in order to understand the chemistry and properties of these graphene nanofragments.

17.2 Nanographenes with Only Armchair Edges

Hexa-*peri*-hexabenzocoronene (HBC) based molecules are typical nanographene fragments with a periphery of only armchair edges. The HBC molecule has a D_{6h} symmetry similar to a benzene ring and can be regarded as a "superbenzene". According to Clar's aromatic sextet rule, seven benzenoid aromatic sextet rings can be drawn for HBC and no additional isolated double bond exists in this structure. This feature results in high stability compared to other linear and *meta*-annulated aromatics such as acenes and phenes. The detailed synthetic strategies and procedures for this novel core have been well summarized in the literature [7a,7b] and will not be covered here. The unsubstituted HBC core has poor solubility which hampers its electronic applications. Recently, processing methods have been developed to exfoliate unsubstituted HBC molecules to form dispersion in organic solvents, similar to that of graphene monolayer [7c]. A commonly used method to solubilize HBC is to attach solubilizing functional groups (alkyl, alkoxyl, alkyl phenyl, etc.) onto the periphery of the HBCs as shown in Figure 17.1. Discotic liquid crystal materials can be prepared based on such a structural design and molecules from **1** to **3** all possess liquid crystalline phase with suitable chains attached.

The single crystal of the HBC core was obtained by slow cooling of its solution in molten pyrene [8]. The crystal information is shown in Figure 17.2. Bond distances agree with the local molecular symmetry and reflect closely distances expected from Clar's aromatic sextet rule. It is noted that, for HBC, the peripheral C–C bonds are slightly shorter than those in the interior region. At the armchair edge of HBC, there is far less of a diene nature compared to normal dienes or the bay region of perylene.

The oxidation and reduction of the HBC core are also worth mentioning. Generation of the HBC radical cation is achieved by oxidation of the substituted HBC with triphenyl amine radical cation salt or antimony pentachloride ($SbCl_5$) [9]. The red HBC$^{•+}$ radical cation is highly persistent at room temperature and has not shown any decomposition during a 24h period. The radical anions of HBC were generated by reduction with potassium in THF solution (see Scheme 17.1) [10]. Anions of HBC from monoanion (monoradical),

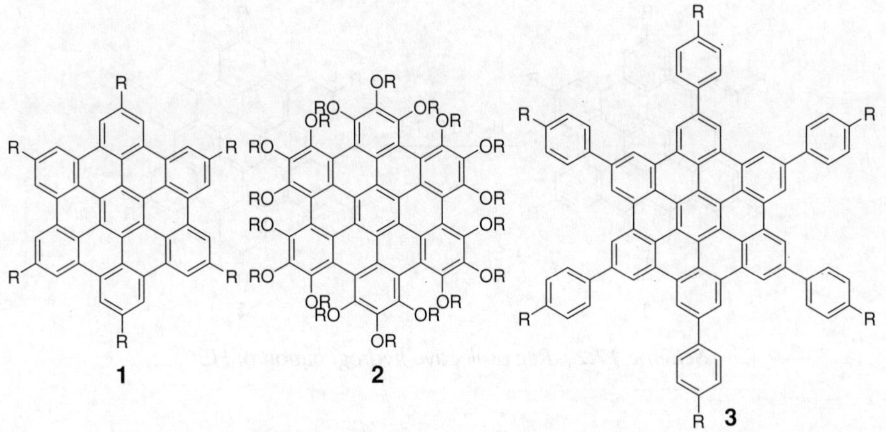

Linking Theory to Reactivity and Properties of Nanographenes   395

Figure 17.1 Soluble derivatives of HBC

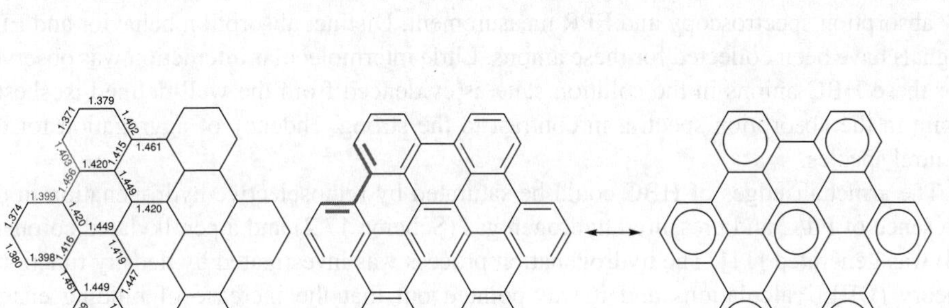

Figure 17.2 Bond length information of HBC crystal

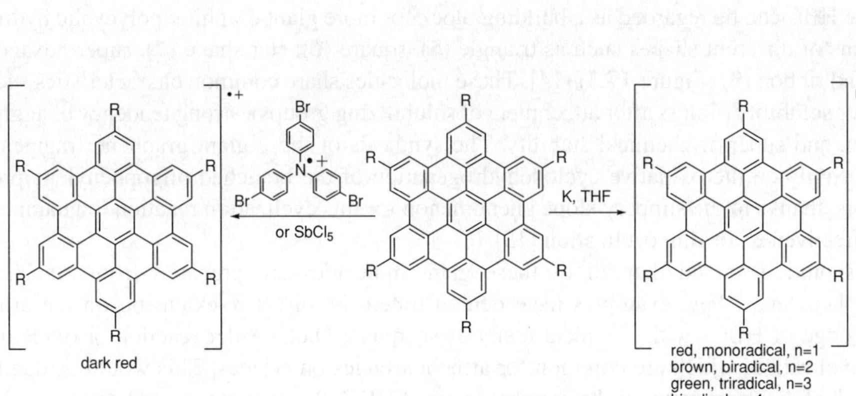

red, monoradical, n=1
brown, biradical, n=2
green, triradical, n=3
biradical, n=4

dark red

Scheme 17.1 Generation of radical cation and anion of HBC

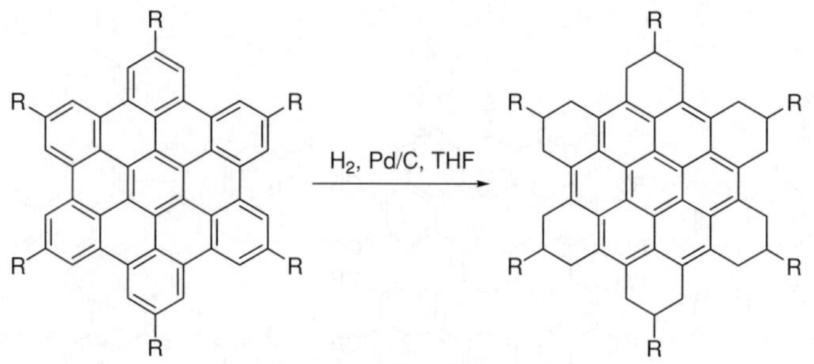

Scheme 17.2 Regioselective hydrogenation of HBC

dianion (biradical), trianion (triradical) up to tetraanion (biradical) have been characterized by absorption spectroscopy and EPR measurement. Distinct absorption behavior and EPR signals have been collected for these anions. Little intermolecular interaction was observed for these HBC anions in the solution state as evidenced from the well-defined isosbestic point in the absorption spectra, in contrast to the strong tendency of aggregation for the neutral species.

The armchair edges of HBC could be saturated by regioselective hydrogenation in the presence of Pd/C and pressured hydrogen gas (Scheme 17.2) and a peralkylated coronene (**4**) was generated [11]. The hydrogenation process was investigated by density functional theory (DFT) calculations and it was pointed out that the increase of binding energy between HBCs during reduction would be a possible cause of the experimentally observed limited hydrogenation [12]. Furthermore, a number of theoretical studies corroborate that the "edge" benzenes of HBC are less aromatic than those in the interior [13]. All these factors would account for the partial hydrogenation.

The HBC can be regarded as a building block for more giant graphitic polycyclic hydrocarbons of different shapes such as triangle (**5**), square (**6**), star shape (**7**), super-hexagon (**8**), and ribbon (**9**) (Figure 17.3) [14]. These molecules share common characteristics such as poor solubility (unless after attachment of solubilizing groups), strong tendency of aggregation, and superior chemical stability. The synthesis of these giant graphene fragments rely heavily on the oxidative cyclodehydrogenation of the branched oligophenylene precursors, thanks to the slippery slope phenomenon for this cyclization reaction to guarantee the effectiveness of ring cyclization [15].

It should be noted that all of these giant molecules are prepared from branched oligophenylenes. Few examples have demonstrated the direct π-extension on the armchair edge of HBCs with chemical methods such as a Diels–Alder reaction or oxidative ring cyclization, which are common for armchair edges on rylenes. This would be due to the lack of diene character of the armchair edge of HBC. While less explored, the reactivity of the armchair edges of PAHs larger than HBC, for example, the triangular PAHs based on HBC, could provide possible routes for synthetic chemists to construct giant graphene fragments (Figure 17.4).

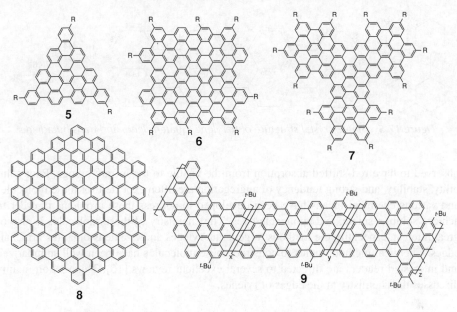

Figure 17.3 Giant graphene fragment with all armchair edges

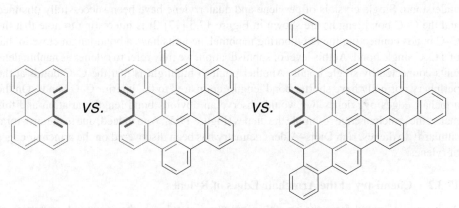

Figure 17.4 Comparison of the armchair edges in perylene, HBC and giant PAHs

17.3 Nanographenes with Both Armchair and Zigzag Edges

Presence of both armchair and zigzag edges in a nanographene fragment would lead to interesting physical and chemical properties. Rylenes are such molecules and intensively investigated for their intriguing optical and electronic properties. Based on the length of the armchair edges, rylene molecules are termed as perylene, terrylene, quaterrylene, pentarylene, hexarylene, and so on. By increasing the conjugation, this series of molecules is

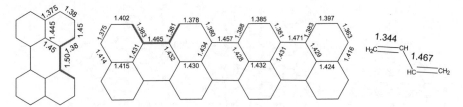

Figure 17.5 Single crystal structure of perylene, quaterrylene, and trans-butadiene

observed to have red-shifted absorption from the visible to the infrared range. Poor solubility, stability, and strong tendency of aggregation are often expected for these molecules and various chemical means have been developed to address these problems in order to achieve soluble, stable, and applicable dyes. The chemistry of these molecules is also of great interest, especially the methods to generate rylenes and π-extension at the armchair edges. Chemical methodologies to construct these molecules have been well summarized and interested readers are directed to several excellent reviews [16]. This section mainly discusses the chemistry at the edges of rylenes.

17.3.1 Structure of Rylenes

By examining the single crystal structure, the π-electron topology of rylenes can be better understood. Single crystals of perylene and quaterrylene have been successfully obtained and the C–C bond lengths are shown in Figure 17.5 [17]. It is interesting to note that the C–C bonds connecting the neighboring naphthalene units have a bond length close to that of a C–C single bond. At this level of sophistication, we may refer to rylenes as naphthalene units connected by single bonds. Another point to highlight is that the C–C bonds at the periphery normally have shorter bond lengths compared to the interior C–C bonds. On the armchair edges of rylenes, we would observe an obvious bond length variation and this makes the armchair edge resemble the dienes (Figure 17.5). And indeed, due to the structural similarity to dienes, rich Diels–Alder chemistry has been discovered on the armchair edges of rylenes.

17.3.2 Chemistry at the Armchair Edges of Rylenes

As we are informed from the crystal information of rylenes, the structural properties of the armchair edges of rylene allow various types of chemistry to be carried out on rylenes. Fusion of heteroatoms such as chalcogens, for example, has been practically achieved and perylene based heteroatom derivatives, **10**, **11**, and **12** have been synthesized [18]. The presence of chalcogen atoms at the bay region of perylene induces extra intermolecular interactions, including van deer Waals interactions and chalcogen–chalcogen interactions, which are essential for highly organized molecular packing and intermolecular charge transport. High charge carrier mobilities have been obtained for these novel materials (Figure 17.6).

The annulation of nitrogen atom on the armchair edges of rylenes (Figure 17.7) [19] is more intriguing due to two reasons. First, attachment of sp^2 hybridized electron-donating nitrogen atom modifies the electronic properties of perylenes which assists further reactions

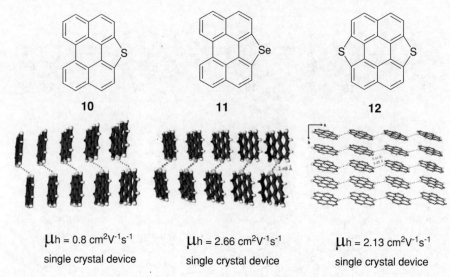

Figure 17.6 Chalcogen-annulated perylenes, their solid state packing and OFET mobility values. Adapted with permission from [18] © 2007 American Chemical Society

such as bromination and oxidative ring cyclization in many cases. Second, it provides a very useful site to attach solubilizing groups which are crucial to achieve soluble and characterizable higher order rylene ribbons. So far, soluble and stable quaterrylene (**14, 15**) and hexarylene ribbons (**16**) have been prepared with this strategy employed.

The Diels–Alder reaction is another widely explored reaction used to extend the π-conjugation at the armchair edge of rylenes. For instance, a two-fold Diels–Alder reaction was used to construct coronene (**21**) from perylene as shown in Scheme 17.3 [20]. Following

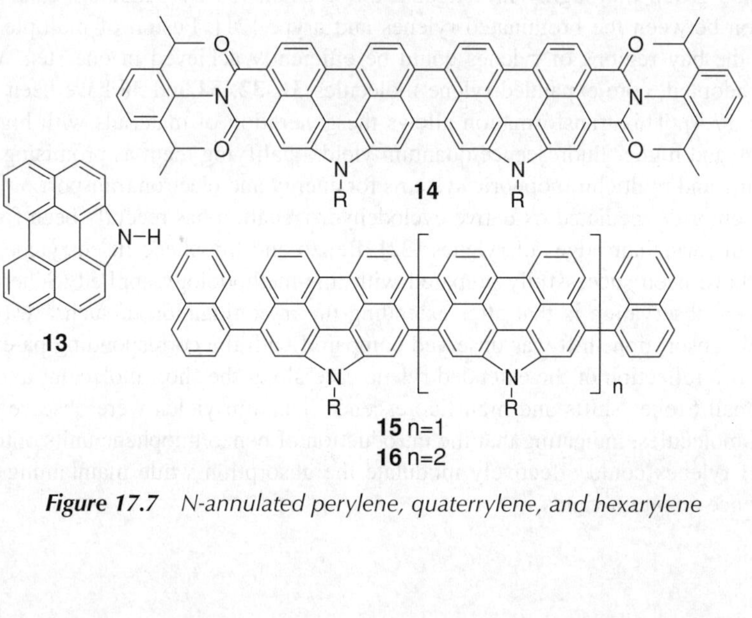

Figure 17.7 N-annulated perylene, quaterrylene, and hexarylene

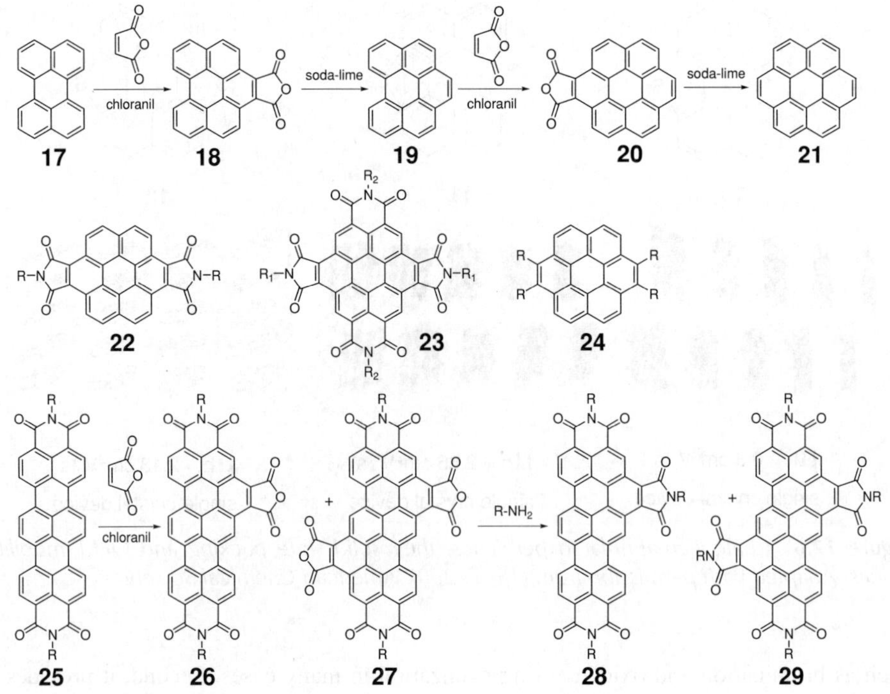

Scheme 17.3 Diels–Alder reaction on the armchair edge of rylenes

a similar strategy, various coronene based materials (**22, 23, 24**) could be synthesized with different dienes used. Functionalization of higher order rylenes to achieve π-extended rylenes (**28, 29**) is also possible by similar methods.

Müllen's group thoroughly investigated the π-extension by palladium-catalyzed ring annulation between the brominated rylenes and aryne [21]. Fusion of multiple benzene rings to the bay regions of rylenes could be efficiently achieved in one step. With this strategy adopted, core expanded rylene molecules **31, 32, 34** and **36** have been obtained (Scheme 17.4). This transformation allows the generation of materials with higher photostability and higher fluorescence quantum yield, qualifying them as promising dyes for biolabeling and multichromophoric systems for energy and electron transport.

Ferric chloride mediated oxidative cyclodehydrogenation has recently been found possible on the armchair edges of rylenes [22]. Benzo and thiophene fused rylenes (**38, 40** and **42**) have been successfully prepared with this methodology applied (Scheme 17.5). A common observation is that after extending the π-conjugation in such a way, a hypsochromic absorption shift was observed compared with the corresponding parent rylene diimide as a reflection of the extended rylene core along the short molecular axis. Meanwhile, small Stokes shifts and high fluorescence quantum yields were observed for this series of molecules, indicating that the introduction of benzo/thiophene units onto the bay region of rylenes could effectively modulate the absorption while maintaining the high fluorescence quantum yields.

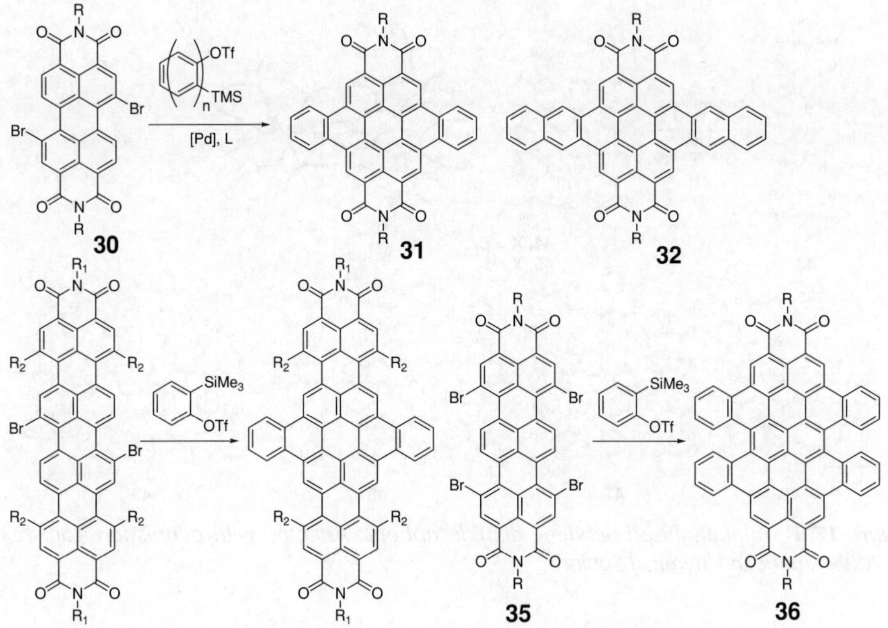

Scheme 17.4 Functionalization on the armchair edge of rylenes with arynes

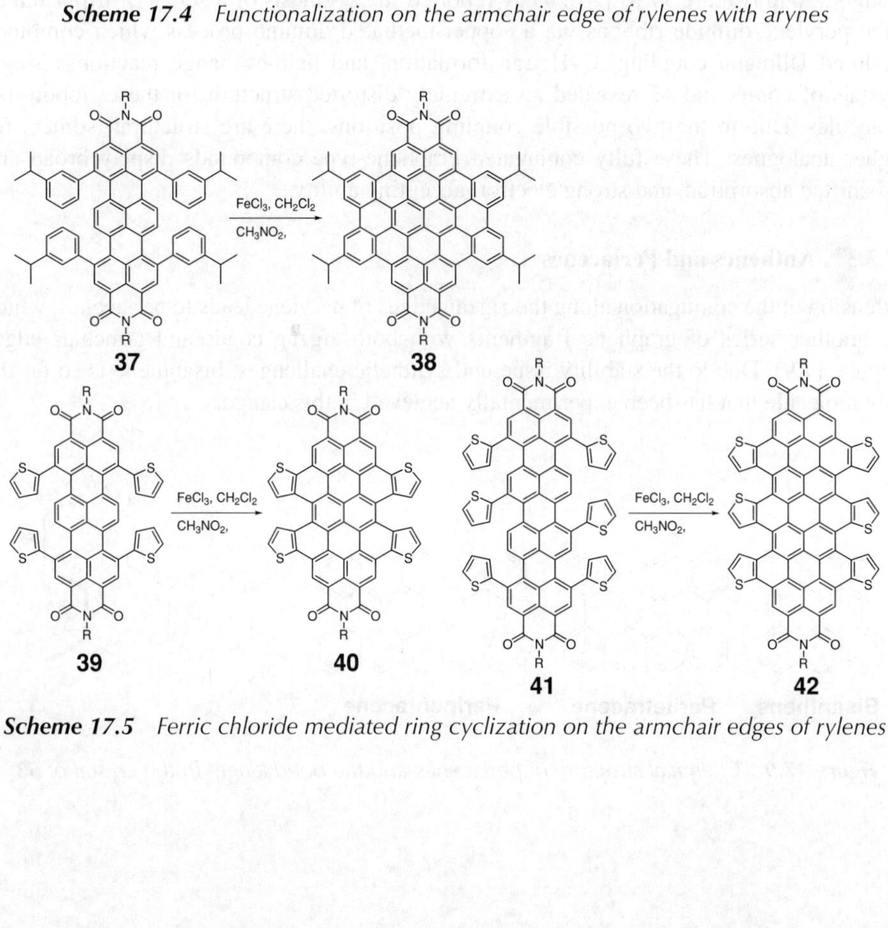

Scheme 17.5 Ferric chloride mediated ring cyclization on the armchair edges of rylenes

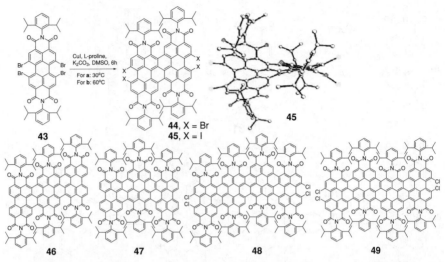

Figure 17.8 Laterally fused perylene diimide ribbons. Adapted with permission from [23a] © 2008 American Chemical Society

An interesting chemistry at the armchair edges of perylene diimide has been reported by Wang's group (Figure 17.8) [23]. They reported the synthesis of a series of triply linked oligo-perylene diimide ribbons via a copper-mediated domino process which combined modified Ullmann coupling, C–H transformation, and halo-exchange reactions. Single crystals of compound **45** revealed an extremely distorted structure for these ribbon-like molecules. Due to the two possible coupling positions, there are structural isomers for higher analogues. These fully conjugated graphene-type compounds display broad and red-shifted absorption, and strong electron-accepting ability.

17.3.3 Anthenes and Periacenes

Extension of the conjugation along the zigzag edges of perylene leads to periacenes, which are another series of graphene fragments with both zigzag edges and armchair edges (Figure 17.9). Due to the stability issue and synthetic challenges, bisanthene is so far the only molecule that has been experimentally achieved in this category.

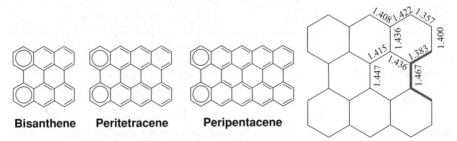

Figure 17.9 Chemical structure of periacenes and the bond length in the crystal of **53**

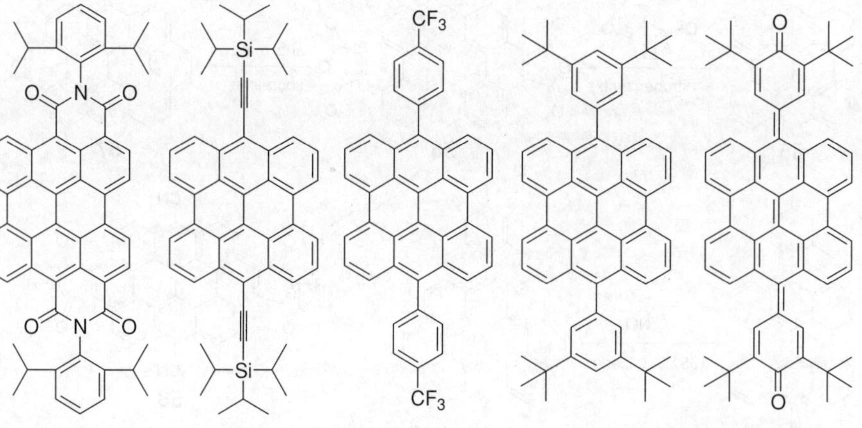

Figure 17.10 *Bisanthene derivatives with protecting groups located at the zigzag edges*

Only two aromatic sextet benzenoid rings can be drawn for periacenes and poor stability is expected for these molecules, especially for higher order periacenes. Similar to acenes, the most reactive sites are located at the zigzag edges. Therefore, kinetic protection by attachment of electron withdrawing or bulky groups on the zigzag edges is a commonly used functionalization method to achieve stable and soluble bisanthene derivatives. Several bisanthene based molecules have been prepared using this strategy (**50** to **54** in Figure 17.10) [24]. Molecules **50** to **53** all showed improved stability and solubility in comparison to the parent bisanthene. They also showed near-infrared absorption and emission with high to moderate fluorescence quantum yields. Amphoteric redox behavior was observed as well for this series of molecules. Another interesting bisanthene based sample is compound **54**, in which quinoidal character is introduced to the bisanthene core. This is a soluble and stable compound with absorption in the near IR region and it represents a rare example of quinoidal large polycyclic hydrocarbon derivatives.

As evidenced from the single crystal structure of bisanthene (Figure 17.9) [24b], the bay region, or the armchair edge of bisanthene possesses the diene character and shows similarity to that of rylenes. Chemistry that has been carried out on the armchair edges of rylenes is therefore possible on this series of molecules. The Diels–Alder reaction, for example, has been shown to be possible on the armchair edges of bisanthene (Scheme 17.6). Dienophiles, such as maleic anhydride, shielded acetylene (nitro ethane) and 1,4-naphthoquinone, undergo a Diels–Alder reaction with bisanthene to achieve core expanded PAHs (**57**, **58**, **60**, **62** and **63**) [25]. The exciting D–A story might go on for unrealized longer periacenes, as calculations at the B3LYP/6-31G* level of theory have pointed out that the Diels–Alder reaction would occur more easily at the bay region of longer periacenes [25e].

Extending the length of the armchair edges of bisanthene by fusing one more anthracene unit generates teranthene. Synthesis and isolation of teranthene is a big challenge in terms of solubility and stability of the target molecule. Only until recently the first teranthene derivative **64**, was achieved by Kubo's group (Figure 17.11) [26]. Attachment of two mesityl

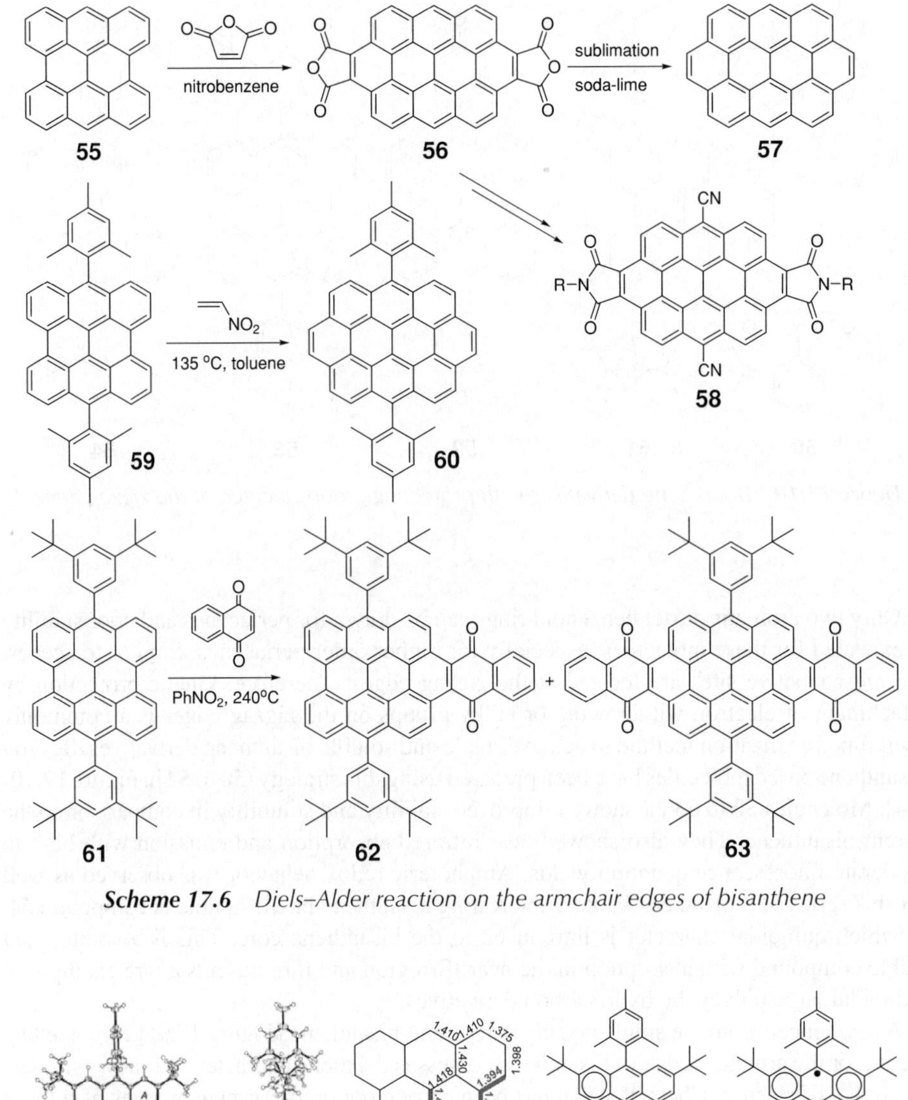

Scheme 17.6 Diels–Alder reaction on the armchair edges of bisanthene

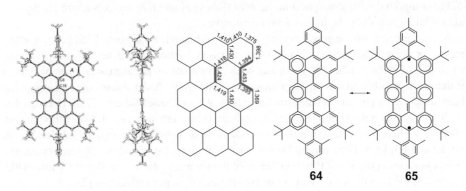

Figure 17.11 Crystal structure of teranthene. Adapted with permission from [26] © 2010 American Chemical Society

groups at the *meso*-positions and four *tert*-butyl groups rendered sufficient stability and solubility for the successful characterization and single crystal analysis. The green teranthene exhibited moderate stability in solution with a half-life of around 3 days upon exposure to air at room temperature. Like other open-shell PAHs with singlet biradical ground state, the ^{1}H NMR spectrum showed temperature dependence with a line broadening at room temperature.

The X-ray analysis revealed an effective D_{2h} symmetry for the teranthene core, with two mesityl groups nearly perpendicular to core plane. A prominent biradical character, as argued by Kubo *et al.*, was reflected by the geometry and bond length of the teranthene in comparison with the bisanthene. The bond length of C_8–C_{18} was determined to be 1.424 Å, which is significantly shorter than sp^2 C–C single bond (1.467 Å) and the corresponding one in bisanthene (1.447 Å), which was believed to result from the enforcement of biradical resonance contribution (**65**). This argument is less convincing if we consider a similar phenomenon found in rylenes molecules. As shown in Figure 17.5, from perylene to quaterrylene, a similar shortening effect of the C–C bonds connecting the naphthalene units can be observed (1.50 Å in perylene and 1.465 Å, 1.457 Å, 1.471 Å in quaterrylene). This, however, cannot be well explained by the presence of biradicals. Therefore, the shortening of the linking C–C bond would be due to topological change of the π electrons and/or the presence of biradicals. Nevertheless, the contribution from the biradical state can be confirmed if we take the bond length variation at the armchair edges into consideration. For teranthene, the bond length variation is about 0.06 Å, which is very similar to that of HBC core (0.057 Å) and quite different from that of perylene (0.12 Å). From this point of view, the properties of the armchair edges on teranthene would be more like that of HBC rather than perylene, which means that teranthene should have more benzenoid character but not diene character at the armchair edges. This can be realized by drawing the resonance structure as shown in **65**, in which biradical character is present.

17.4 Nanographene with Only Zigzag Edges

In contrast to nanographenes with armchair edges possessing closed-shell structures, the cutting of graphene along the zigzag edges will inevitably lead to a series of π-conjugated molecules with one or more unpaired electrons, or open-shell structures (Figure 17.12). Those molecules cannot be depicted in Kekulé structures, and are known as non-Kekulé polynuclear benzenoid molecules, or "open-shell nanographenes". The number of unpaired electrons increases with the increase of the molecular size, the smallest member, phenalenyl radical **66**, possesses one unpaired electron and with spin multiplicity $2S + 1 = 2$. Extension of π-conjugation results in triangulene **67** with two unpaired electrons and a triplet ground state, and further extension leads to larger high spin system **68**. All systems are featured by a large delocalization of spin densities over the entire molecules. The synthesis and characterization of those open-shell nanographenes have drawn much attention nowadays as the interest lies not only in the investigation of the electronic structures of these systems and the interactions of unpaired electrons, but also the possibility of them as molecule-based functional materials.

Despite the interesting properties and promising applications, the synthesis of those systems is very challenging for two reasons: firstly, construction of large fused π-electron

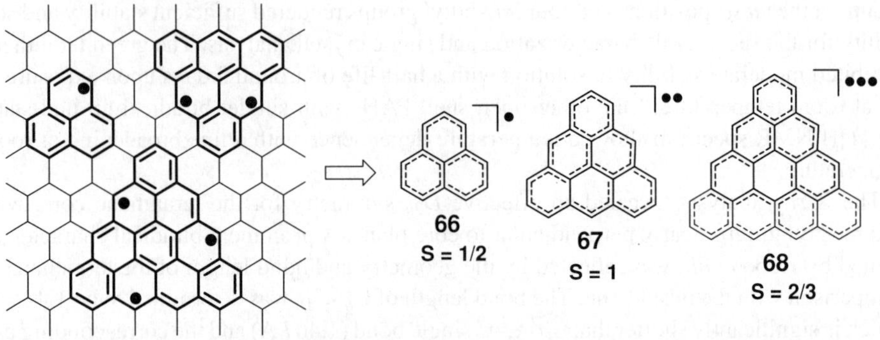

406 Graphene Chemistry

Figure 17.12 *Structures and spin-multiplicities of triangular polycyclic hydrocarbons*

systems usually in need of multiple-step sequences with adequate synthetic methods; secondly, the presence of free radicals in those open-shell systems make them extremely unstable, especially towards dimerization and oxidation, which impedes their synthesis and isolation. Nowadays, continuous efforts have been put by chemists to develop new synthetic methods to construct those systems, and the stability issues are partially overcome by ways such as steric protection. In the following section, the progress of synthesis and study of all zigzag polycyclic hydrocarbons will be introduced.

17.4.1 Phenalenyl-Based Open-Shell Systems

The phenalenyl radical **66** is the most fundamental and widely investigated member in this family, and it is characterized by a planar, rigid structure with the spin spread over the whole molecular skeleton, also known as "spin-delocalized nature" which is in contrast to typical stable neutral radicals such as TEMPO and α-nitronylnitroxide derivatives with a spin-localized nature. As interpreted in resonance structures shown in Figure 17.13, the spin density of **66** is predominately existed on its α positions, while the spin at the peripheral positions (β positions) is much smaller, which is also in agree with the non-bonding molecular orbital (NBMO) distribution obtained from Hückel-MO method. In addition, **66** is found to exhibit a high amphoteric redox ability with thermodynamically stable cation, neutral radical, and anion species, all of these features have aroused broad interest ranging from physical chemistry to synthetic chemistry.

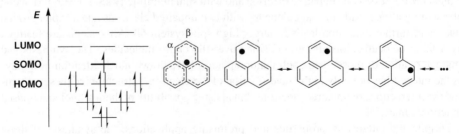

Figure 17.13 *Molecular orbital diagram and resonance structures of phenalenyl radical*

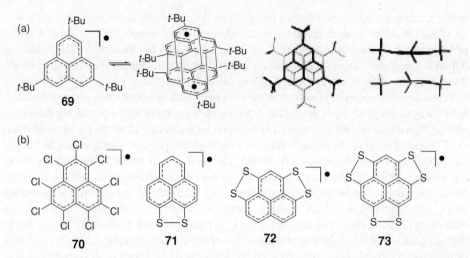

Figure 17.14 (a) tert-Butyl substituted phenalenyl radical and its dimer; (b) Heteroatom substituted phenalenyl radical. Reprinted with permission from [30] © 2006 American Chemical Society

The attempt to synthesize and isolate stable phenalenyl radical dated back to the 1950s, the most widely used approach was to oxidize phenalene or phenalenyl anion by different oxidants such as air or quinones [27]. However, the isolation of the parent phenalenyl radical turn out to be a failure due to rapid dimerization and oxidation, hence the phenalenyl chemistry in the last half-century was mainly performed in degassed solutions and under sealed conditions. Continuous efforts were carried out to isolate neutral phenalenyl radical in the solid state and progresses on stabilizing it were achieved by introducing primary alkyl groups, electron-donating and electron-withdrawing groups [28]. The first isolation of phenalenyl in the solid state was finally obtained in 1999 when Nakasuji et al. introduced three *tert*-butyl groups onto the β positions (Figure 17.14). The bulky substituents not only successfully shut down the σ-bond dimerization pathway, but also possessed a minimal perturbation to the electronic structure of the parent phenalenyl in that the connection positions were β positions with negligible spin densities [29].

The synthesis of **69** adopted a 10-step sequence, in which the key intermediate was a tri-*tert*-butyl substituted phenalene precursor. Treatment of the precursor with p-chloranil produced **69** as deep blue needles. This crystal showed high stability in the absence of air, while changed into phenalanone derivatives and other byproducts in one week in air. The high stability allowed a thorough characterization of **69** for the first time, such as spectroscopic study, X-ray single crystal analysis, and magnetic susceptibility measurement. The EPR spectrum showed a septet hyperfine structure with an observed g_e value of 2.0028, being consistent with genuine spin-doublet phenalenyl radical. The X-ray analysis revealed a nearly planar geometry with a slightly distorted D_{3h} symmetry, the molecule formed a π-dimeric pair in staggered alignment of the *tert*-butyl groups to avoid steric repulsion and maximize SOMO overlap. The dimeric pair adopted a herringbone packing motif with interplanar distance ranging from 3.201–3.323 Å, such a short distance indicated a strong antiferromagnetic interaction within the π-dimer. Interestingly, the dimeric behavior was also found in solution which led to the thermochromic phenomenon, the red-purple solution of **69** gradually turned blue upon cooling, in accordance to the increase of absorbance in 530–670 nm region, due to the formation of the diamagnetic dimer. The formation of the dimer was also detected by NMR and MS techniques [30].

Substitution of phenalenyl radical in the periphery with heteroatoms represents an alternative way to stabilize the radical. One representative example was a perchlorophenalenyl radical **70** with all α- and β-positions substituted by a chlorine atom prepared by Haddon et al. (Figure 17.14). The synthesis adopted a multi-step sequence with the reduction of the corresponding cation species as a final step. This radical was stable in air, and the X-ray analysis suggested that the single molecule was not planar anymore but ruffled into a propeller shape due to the demand of intramolecular Cl–Cl distance, and each of the molecules was widely separated by 3.78 Å, in sharp contrast to the *tert*-butyl phenalenyl **69** which was dimeric in the solid state. The reason for the long intermolecular distance of **70** was due to the non-planarity which inhibited the formation of 60° rotated stacking motif [31]. A series of phenalenyl radicals with a disulfide bridge across two neighboring active positions was also designed by Haddon, with the similar synthetic concept of reduction of the corresponding cation precursors. In spite of the absence of bulky substituents, no σ-dimerization was formed in the solid state and the radical **71** could survive in solid state in air for up to 24 h. The stabilization force came almost exclusively from the spin delocalization to the disulfide bridge. In the crystalline phase, **71** stacked in a sandwich herringbone motif of face-to-face π-dimers with 180° rotation between two radicals with intermolecular distances 3.13–3.22 Å. The S–S interactions between different π-dimers were also observed [32]. The synthesis of more spin-delocalized systems were also conducted, but the reduction of the corresponding cation precursor of **72** only resulted in formation of a closed-shell dimer with a S–S σ bond, and the synthesis of a three-fold symmetrical molecule **73** has not been reported so far [33].

Another intriguing molecular design to achieve stable phenalenyl-based open-shell systems is to link two phenalenyl moieties together to produce singlet biradical compounds having a Kekulé structure as a structural resonance (Figure 17.15). These systems usually exhibit a larger π-conjugation and a lower band gap compared to parent phenalenyl radical, and the stability comes from large delocalization of spin and aromatic stabilization in the biradical form (recovery of more sextet benzenoid rings). The radicals are weakly coupled both within and between molecules, leading to interesting electronic structures and intermolecular packing motif. Examples of such biradicals include a bis(phenalenyl) **74** linked by acetylene with a biradical character of 0.14 estimated by LUMO occupancy numbers obtained from a CASSCF(2,2) calculation [34]. Pentalenodiphenalene **75** was another low band gap bis(phenalenyl) compound exhibiting multistage redox properties, however, it was substantially destabilized by the electronic contribution of the 8π-electron antiaromatic pentalene subunit [35]. Other systems with aromatic bridges such as benzene, naphthalene, anthracene, and thiophene, lead to an interesting resonance between quinoidal structures and biradical structures, which will be discussed in detail in the subsequent sections.

In parallel to biradicals with a singlet ground state, a monoradical system can be stabilized by highly delocalization. One recent example was compound **77** in which two phenalenyl moieties were fused together by a five-membered ring [36]. The stability that arose from delocalization was strong enough to stabilize this system even in absence of steric protection.

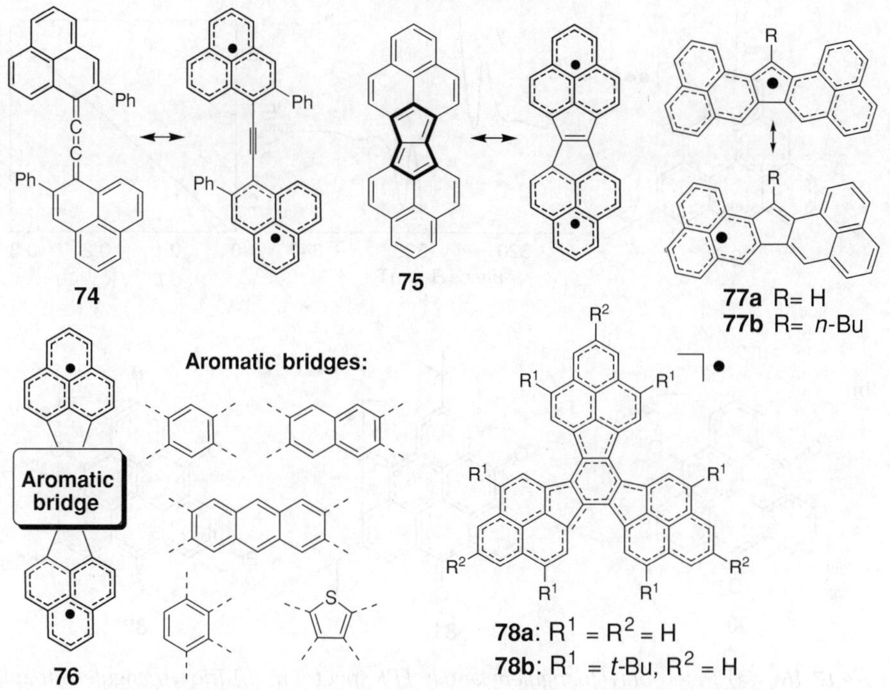

Figure 17.15 Structures of bis(phenalenyls) and tris(phenalenyls)

No formation of σ-dimer was observed which was evidenced by the unchanged EPR spectrum even at $-90°C$. The X-ray analysis of single crystal of **77b** indicated the formation of π dimers in the solid state, and the SOMO of the π dimer would be split into bonding and antibonding molecular orbitals due to the strong antiferromagnetic coupling of the unpaired electrons. Fusion of three phenalenyl moieties to one benzene ring led to a highly delocalized monoradical **78**, which is also a rare example of compounds with six-stage amphoteric redox behavior [37, 38].

Extension of the π-conjugation will lead to higher order analogue of phenalenyl radical with enhanced spin delocalized nature and spin multiplicities. Triangulene is the C_3-symmetric polycyclic hydrocarbon predicted to possess a triplet ground state due to the topological degeneracy of non-bonding molecular orbitals (NBMOs) with a non-disjointed nature [39]. Theoretical studies also predicted large spin densities at the edge sites, indicating its high reactivity in the neutral state. The first synthesis of triangulene diradical was conducted by Clar, but only a polymerized product was obtained due to the kinetic instability [40]. The pursuit of triangulene diradical continued and a closed-shell dianion of triangulene was achieved in 1977 and detected by NMR [41]. The first genuine triangulene diradical was detected in 2001 when Nakasuji et al. introduced three tert-butyl groups on three vertexes of the triangle, with the aim to increase the kinetic stability (Figure 17.16) [42]. The diradical species **79** was generated by treating dihydrotriangulene precursor with p-chloranil, unfortunately, the partial protection could not prevent the diradical from

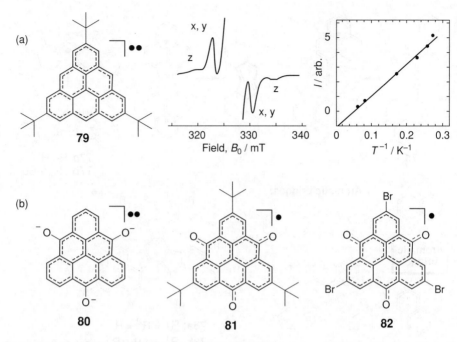

Figure 17.16 (a) Tri-tert-butyl triangulene and its EPR spectrum; (b) Trioxytriangulene trianion diradical and neutral radical. Reprinted with permission from [42] © 2001 American Chemical Society

polymerizing at room temperature, but the typical EPR spectrum attributed to the triplet species was obtained by rapid freezing of the sample to 123 K, and the linear dependence of the triplet signal intensity I on $1/T$ showed that the triplet state was the ground state.

Substitution of oxygen atoms at strategic positions of triangulene will lead to open-shell system termed "trioxytriangulene" **80**, which can be prepared by reduction of the corresponding diketone precursor. The stability of **80** is better than the partially protected **79**, as it is stable in degassed solution even at room temperature. The moderate stability is attributed to the extension of spin delocalization to the oxygen atom as well as the protection from dimerization. The ground state of **80** was determined to be a triplet by plotting the intensity of triplet component detected by EPR versus $1/T$ between 13 and 37 K [43, 44]. Although the triplet ground state species is too unstable for further study, the neutral trioxytriangulene monoradical systems, after certain substitution, have shown high stability and hold some promises as electrode-active materials in secondary batteries. Compound **81** gave a high capacity of 311 Ahkg^{-1}, exceeding those of Li-ion batteries in the first discharge process. Compared to **81**, the bromo-substituted **82** showed improved output voltage and cycle performance, indicating a tunable performance on a basis of chemical modification [45].

To sum up, the polycyclic hydrocarbons with all zigzag edges and non-Kekulé structures represent the most reactive members in the family of nanographenes. The synthesis and stabilization of the smallest unit, the phenalenyl radical, has been well established, and current research focusses more on the applications of them as functional materials. On the other hand, understanding of more extended systems, such as triangulene, is restricted to EPR detection at low temperatures due to the intrinsic low stability, genuine triangulene derivatives with enough stability are urgently required to further reveal its intriguing properties.

17.5 Quinoidal Nanographenes

Polycyclic hydrocarbons with quinoidal structure represent a unique class of π-conjugated molecules; they usually display a narrow HOMO-LUMO energy gap and intriguing electronic, optical, and magnetic properties. Generally there are two types of quinoidal structures, either on a basis of *o*-quinodimethane (*o*QDM) or *p*-quinodimethane (*p*QDM) units. In both cases a resonance structure exists between quinoidal structure and a singlet biradical structure, for the latter one aromatic benzenoid ring will be regained (Figure 17.17). Quinoidal compounds are fascinating both theoretically and experimentally, and various computational methods can be utilized to study the open-shell properties such as biradical character index, LUMO occupancy number, exchange interaction, spin density, and singlet-triplet gap. On the other hand, a number of experimental methods including nuclear magnetic resonance (NMR), electron paramagnetic resonance (EPR), superconducting quantum interference device (SQUID), X-ray single crystal analysis, Raman spectroscopy and so on are powerful tools to investigate the magnetic properties, the biradical characters and the intermolecular interactions. Like most of the open-shell systems, the utmost obstacle for the synthesis of quinoidal compounds is their relatively low stability.

The studies of compounds bearing *o*QDM subunits have lagged far behind their *p*QDM counterparts, mostly due to the lack of enough stability. Such derivatives as tetraphenyl-*o*QDM **83** [46] and pleiadene **84** [47,48] are highly reactive that could only be generated and detected in rigid glass matrices (Figure 17.18). Benefitting from modern synthetic methodologies, a number of stable *o*QDM derivatives **85** to **87** are prepared nowadays, allowing a glimpse of their physical properties [49–51]. It is noteworthy that all the compounds have relatively large HOMO-LUMO energy gaps and no definitive experimental evidence is available to support a biradical character. Nevertheless, some candidates such as **87** are quite promising as optoelectronic materials.

On the other hand, a plethora of extended and fused pQDM derivatives have been developed, such as Thiele/Tschitschibabin's hydrocarbons **88** [52], indenofluorene **89**, zethrenes **90** and bis(phenalenyls) **91** (Figure 17.19). Many of the candidates have shown an interesting biradical character in the ground state, and are stabilized by introducing bulky or electron-withdrawing substituents as well as by radical delocalization which allows a

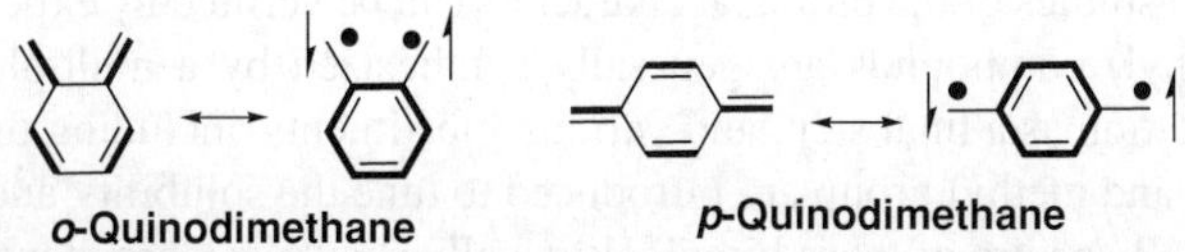

o-Quinodimethane   *p*-Quinodimethane

Figure 17.17 *Resonance structures of o- and p-quinodimethane*

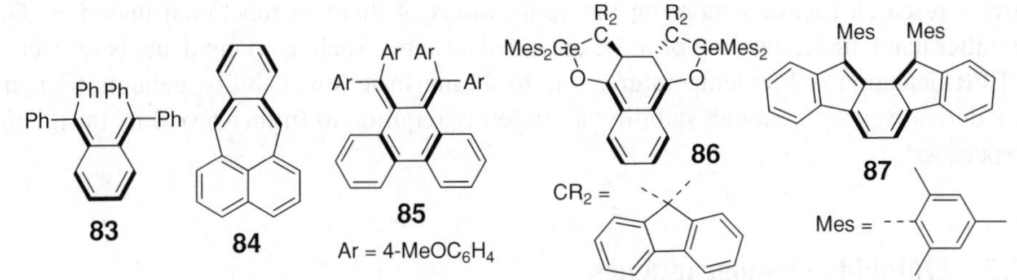

Figure 17.18 Structures of oQDM derivatives

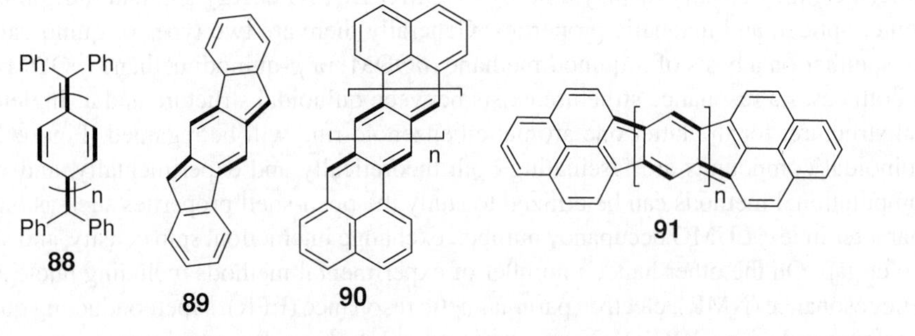

Figure 17.19 Structures of pQDM derivatives

thorough characterization and investigation. Moreover, some of them are found to be very promising optoelectronic materials.

17.5.1 Bis(Phenalenyls)

Connecting two phenalenyl radicals with benzene, naphthalene, and anthracene will produce a series of bis(phenalenyl) compounds **92–94**, which have been systematically investigated by Nakasuji and Kubo *et al.* The signature of this class of molecules is a resonance between closed-shell quinoidal structures and open-shell singlet biradical structures (Figure 17.20a). The biradical character y of those compounds could be estimated by the natural orbital occupancy number (NOON) of LUMO according to CASSCF(2,2)/6-31G calculations. With the increase in the conjugation length of the central linker, the biradical character will increase. The biradical character of anthracene linked bis(phenalenyl) **94** is estimated as $y = 0.68$, [53] compared to the benzene **92** ($y = 0.30$) [54] and naphthalene **93** ($y = 0.50$) [55] analogues. The larger biradical characters indicate weaker intramolecular coupling of two radicals, and the larger biradical contribution to the ground state due to the increase of aromatic stabilization energy from benzene to naphthalene.

The theoretical studies of the biradical characters can be verified by experimental results. The bis(phenalenyl) compounds are generally synthesized by a multiple step sequence with dehydrogenation as a final step, and various substituents, including the phenyl group, *tert*-butyl group, and methyl group are introduced to tune the solubility and intermolecular packing. The NMR spectra of these biradicals usually show peak broadening at room temperature and line sharpening upon cooling, due to the existence of the thermally accessible

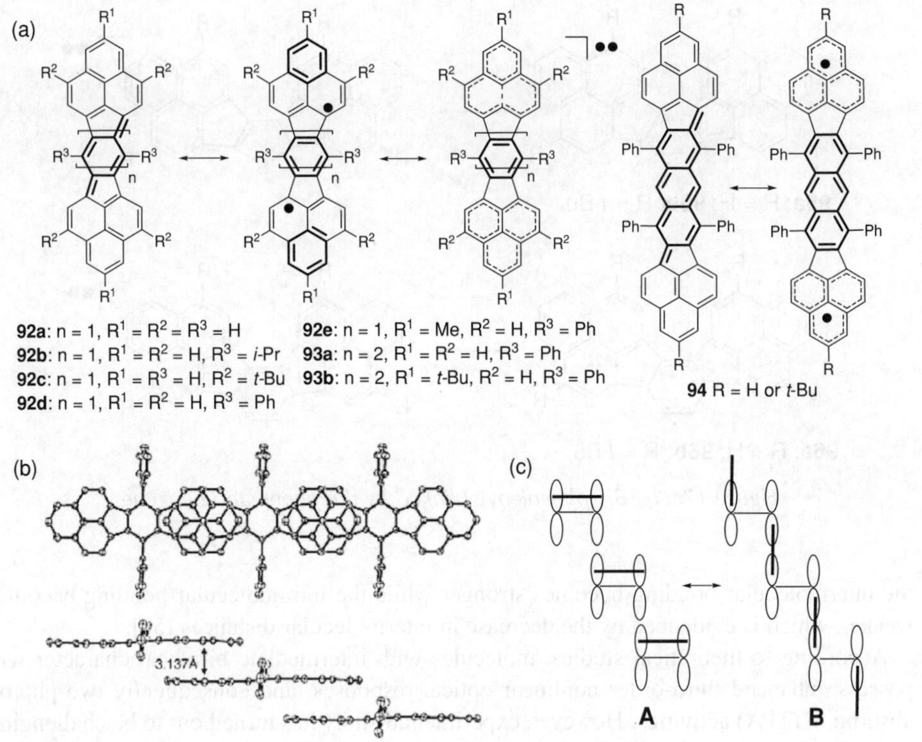

92a: n = 1, R¹ = R² = R³ = H
92b: n = 1, R¹ = R² = H, R³ = i-Pr
92c: n = 1, R¹ = R³ = H, R² = t-Bu
92d: n = 1, R¹ = R² = H, R³ = Ph
92e: n = 1, R¹ = Me, R² = H, R³ = Ph
93a: n = 2, R¹ = R² = H, R³ = Ph
93b: n = 2, R¹ = t-Bu, R² = H, R³ = Ph
94 R = H or t-Bu

Figure 17.20 (a) Resonance structures of bis(phenalenyls) (b) Crystal structure of **92d** (c) RVB model for **92e**. Reprinted with permission from [56] © 2005 Wiley-VCH Verlag GmbH & Co.KGaA, Weinheim

triplet species at higher temperatures and the reduction of population at lower temperatures; the small singlet-triplet gap could be determined by solid state EPR and SQUID measurements. One of the most salient features of singlet biradical compounds is the strong intermolecular interaction in the solid state, the bis(phenalenyls) always pack as one-dimensional chains in a staggered stacking mode with an average $\pi-\pi$ distance shorter than the van der Waals contacts of the carbon atoms (3.4 Å) (Figure 17.20). This packing will maximize the SOMO-SOMO overlapping between the radicals, leading to stabilized intermolecular orbitals corresponding to intermolecular covalency [56]. It was later found that the intermolecular interactions can be altered by varying the external conditions such as molecular structure, temperature, and pressure. For example, the intermolecular distance can be increased by changing R^1 from H to Me or by increasing temperature. A decreased $\pi-\pi$ distance would improve the intermolecular orbital overlap and strengthen the intermolecular bonding interaction. However, it would also weaken the intramolecular interaction by making unpaired electrons more localized. As a result, the electronic structure of the 1D chain can be depicted by the resonating valence bond (RVB) model as a superposition of a resonance balance between intramolecular bonding and intermolecular bonding (Figure 17.20c) [57]. Notably, with the linker change from benzene to anthracene,

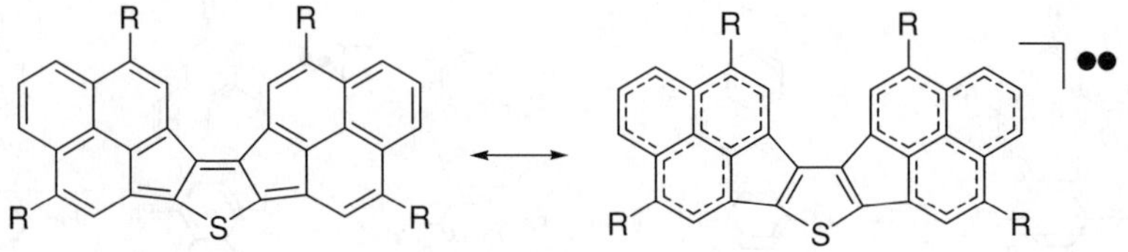

95a: R = H; **95b**: R = t-Bu

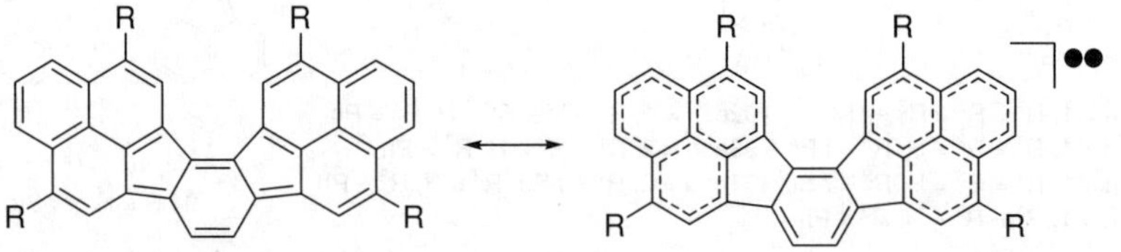

96a: R = H; **96b**: R = t-Bu

Figure 17.21 Bis(pheanlenyls) linked by thiophene and benzene

the intermolecular bonding becomes stronger while the intramolecular bonding becomes weaker, which is evidenced by the decrease in intermolecular distances [58].

According to theoretical studies, molecules with intermediate biradical character will possess enhanced third-order nonlinear optical responses, and consequently two-photon absorption (TPA) activities. However, experimental study has turned out to be challenging due to the high reactivity of those species. Fortunately, bis(phenalenyl) compounds provide stable candidates for testing this concept. Indeed, a maximum TPA cross section values up to 424 ± 64 GM at 1425 nm for **92d** and 890 ± 130 GM at 1500 nm for **93b** were obtained which are comparable to similar TPA chromophores with strong donor and/or accepter peripheral groups and are among the best for pure hydrocarbons without donor and acceptor substituents, providing new insights in designing TPA materials [59]. Moreover, the thin film of **92d** was also reported to display balanced ambipolar charge transport, due to the amphoteric redox behaviors and strong intermolecular communications. These results indicate a bright future for biradical compounds in materials science [60].

Fusion of phenalenyl moieties with alternative linkers will lead to other biradical systems (Figure 17.21). For example, fusion of phenalenyl to thiophene and alternative positions of benzene gives rise to compounds **95** [61] and **96** [62]. The compound **95** was found to form a dimeric pair in the crystalline phase with a bended structure for each monomer, while **96** appeared to be air-sensitive oil. An energy lowering of 7.08 kJ/mol and 35.38 kJ/mol from closed-shell form to open-shell form were calculated for **95a** and **96a**, respectively, suggesting that the ground state of these molecules is singlet biradical.

17.5.2 Zethrenes

Zethrene is the name given to an interesting class of polycylic hydrocarbons with Z-shape molecular skeleton, and it can also be viewed as two phenalenyl moieties fused in a "head-to-head" manner. Longitudinal homologues of zethrene are defined as heptazethrene and octazethrene with two phenalenyl rings separated by benzene or naphthalene unit,

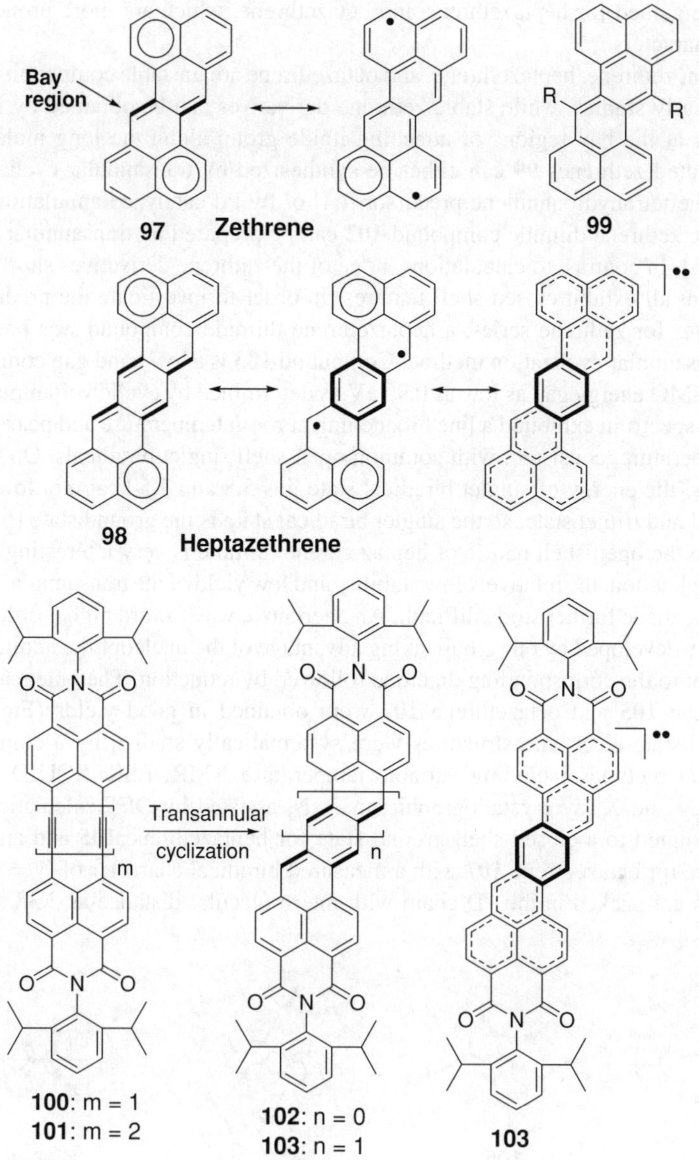

Figure 17.22 Structures of zethrene, heptazethrene, and their imide derivatives

respectively, in a similar Z-shape manner. It is predicted that zethrenes will possess a singlet biradical character as illustrated in Figure 17.22, The biradical character y for zethrene, heptazethrene, and octazethrene is calculated to be 0.407, 0.537 and 0.628, respectively, from the occupancy numbers of spin-unrestricted Hartree–Fock natural orbitals (UNOs) [63]. The increase in biradical character can be explained by the aromatic stabilization, since one sextet ring will be lost in the biradical form for zethrene, while one more sextet ring will be gained for heptazethrene and octazethrene, which are more prone to exhibit biradical characters.

The parent zethrene, heptazethrene, and octazethrene are unstable compounds as demonstrated by early studies, while stable zethrene derivatives can be obtained by introducing substituents at the bay regions or attaching imide group along the long molecular axis. The substituted zethrenes **99** can either be synthesized by transannular cyclization reaction from the tetrahydroannulene precursor [64] or by Pd-catalyzed annulation reactions, [65] and the zethrene diimide compound **102** can be prepared by transannular cyclization reaction [66]. In contrast to calculations, none of the zethrene derivatives show open-shell character but all exhibit closed-shell features. In order to investigate the predicted biradical character for zethrene series, a heptazethrene diimide compound was prepared by a similar transannular cyclization method. Compound **103** is a low band gap compound with HOMO-LUMO energy gap as low as 0.99 eV as determined by cyclic voltammetry, and its ^{1}H NMR spectrum exhibited a line broadening at room temperature and peak sharpening at low temperature, consistent with common open-shell singlet biradicals. On the basis of calculations, the energy of singlet biradical state lies 5.8 and 7.5 kcal/mol lower than the closed-shell and triplet state, so the singlet biradical state is the ground state [67].

Although the open-shell nature of heptazethrene diimide is very interesting in terms of material application, the relatively low stability and low yield of the transannular cyclization method have made further study difficult. An alternative way toward higher order zethrenes was recently developed by our group taking advantage of the nucleophilic addition of Grignard reagent to the corresponding diketone followed by reduction. The kinetically blocked heptazethrene **105** and octazethrene **107** were obtained in good yields (Figure 17.23). The ground-state electronic structures were systematically studied by a combination of experimental methods including variable temperature NMR, ESR, SQUID, FT-Raman spectroscopy and X-ray crystallographic analysis, assisted by DFT calculations. All the evidence pointed to a closed-shell ground state for heptazethrene **105** and an open-shell ground state for octazethrene **107** with a measured biradical character of ($y = 0.56$). Both compounds are packed in the 1D chain with intermolecular distance as 3.38 and 3.35 Å,

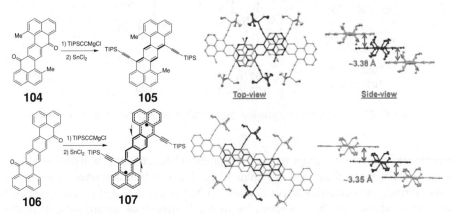

Figure 17.23 Synthesis and crystal structures of heptazethrene and octazethrene derivatives. See colour version in the colour plate section

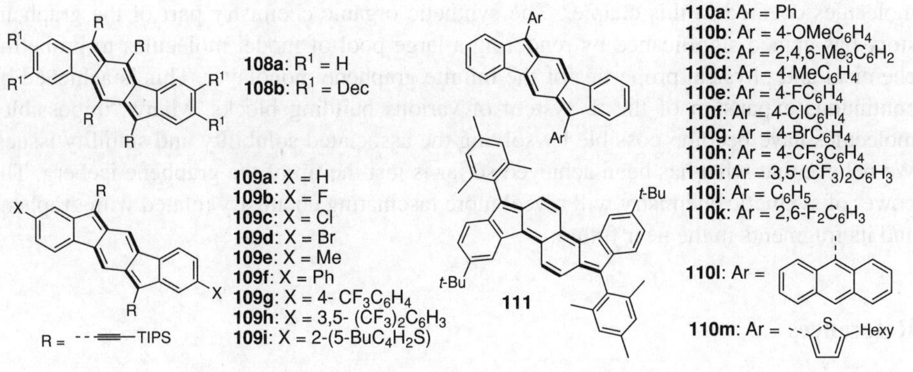

Figure 17.24 Structures of indenofluorene derivatives

much larger than the bis(phenalenyl) compounds (3.14 Å), and thus should be more adequately described as $\pi-\pi$ interaction instead of intramolecular covalent bonding. Moreover, both compounds showed good stability and large TPA cross sections (920 GM at 1250 nm for **105**, 1200 GM at 1250 nm for **107**), making them promising materials for nonlinear optics, ambipolar FETs, and spintronics [68].

17.5.3 Indenofluorenes

Fully conjugated indenofluorene derivatives are antiaromatic analogues of acenes, and are quite intriguing in terms of their bonding pictures (Figure 17.24). The indenofluorene derivatives can be prepared from the corresponding diketone precursors. The X-ray single crystal analysis indicated a bond length alternation in the central pQDM core but uniform in the periphery benzenes, and the molecules should be described as fully conjugated 20-π-electron hydrocarbon linked by a *s-trans* 1,3-diene [69, 70]. No open-shell character was observed for this system despite the existence of the pQDM moiety, while this system has shown promise as a semiconductor. An FET device was fabricated on vapor-deposited thin films of **110k** [71] and has shown electron mobility of 8.2×10^{-6} cm^2 V^{-1} s^{-1} and hole mobility of 1.9×10^{-5} cm^2 V^{-1} s^{-1}. A single crystal OFET with **110j** as the active component exhibited ambipolar behavior with hole and electron mobilities of 7×10^{-4} and 3×10^{-3} cm^2 V^{-1} s^{-1}, respectively [72]. A similar system involves a 24-π-electron antiaromatic system **111** with 2,6-naphthoquinodimethane unit. This compound is also a closed-shell molecule as indicated in NMR, EPR and X-ray analysis [73].

17.6 Conclusion

In this chapter, we mainly summarize the chemistry and properties of graphene nanofragments with various edge structures. The edges patterns and their properties of these polycyclic hydrocarbons have been found to be essential in interpreting the electronic and magnetic properties of the molecules as well as the reactivity at the edges. A lot of useful guiding principles and theoretical investigations have been established based on the molecules covered in this chapter. The synthetic organic chemistry part of the graphene story has gained significance by rendering a large pool of model molecules to probe the chemical and physical properties of the infinite graphene monolayer. This is achieved by continuous expansion of the π system of various building blocks. Many "impossible" molecules have become possible by solving the associated solubility and stability issues. We believe that what has been achieved so far is just the tip of the graphene iceberg. The power of synthetic chemistry will reveal more fascinating chemistry related with graphene and its fragments in the near future.

References

[1] K. S. Novoselov, A. K. Geim, S. V. Morozov, D. Jiang, Y. Zhang, S. V. Dubonos, et al., Electric field effect in atomically thin carbon films, *Science*, **306**, 666–669 (2004).

[2] (a) M. Bendikov, F. Wudl and D. F. Perepichka, Tetrathiafulvalenes, oligoacenes, and their buckminsterfullerene derivatives: the brick and mortar of organic electronics, *Chem. Rev.*, **104** (11), 4891–4946 (2004). (b) M. Bendikov, H. M. Duong, K. Starkey, K. N. Houk, E. A. Carter, and F. Wudl, Oligoacenes: theoretical prediction of open-shell singlet diradical ground states, *J. Am. Chem. Soc.*, **126**, 7416–7417 (2004). (c) X. Gao, J. L. Hodgson, D. Jiang, S. B. Zhang, S. Nagase, G. P. Miller and Z. Chen, Open-Shell singlet character of stable derivatives of nonacene, hexacene and teranthene, *Org. Lett.*, **13**, 3316–3319 (2011). (d) M. C. dos Santos, Electronic properties of acenes: Oligomer to polymer structure, *Phys. Rev. B*, **74**, 045426 (2006). (e) J. Hachmann, J. J. Dorando, M. Avilés and G. K.–L. Chan, The radical character of the acenes: A density matrix renormalization group study, *J. Chem. Phys.*, **127**, 134309 (2007). (f) D. Jiang and S. Dai, Electronic ground state of higher acenes, *J. Phys. Chem. A*, **112**, 332–335 (2008). (g) Z. Qu, D. Zhang, C. Liu and Y. Jiang, *J. Phys. Chem. A*, **113**, 7909–7914 (2009). (h) S. Kivelson and O. L. Chapman, Polyacene and a new class of quasi-one-dimensional conductors, *Phys. Rev. B*, **28**, 7236–7243 (1983). (i) H. F. Bettinger, Electronic structure of higher acenes and polyacene: The perspective developed by theoretical analyses, *Pure Appl. Chem.*, **82**, 905–915 (2010).

[3] (a) Y.–W. Son, M. L. Cohen and S. G. Louie, Half-metallic graphene nanoribbons, *Nature*, **444**, 347–349 (2006). (b) D. Jiang, X.–Q. Chen, W. Luo and W. A. Shelton, From trans-polyacetylene to zigzag-edged graphene nanoribbons, *Chem. Phys. Lett.*, **483**, 120–123 (2009). (c) D. Jiang, B. G. Sumpter and Sheng D, Unique chemical reactivity of a graphene nanoribbon's zigzag edge, *J. Chem. Phys.*, **126**, 134701 (2007). (d) K. Nakada, M. Fujita, G. Dresslhaus and M. S. Dresselhaus, Edge state in graphene ribbons: Nanometer size effect and edge shape dependence, *Phys. Rev. B*, **54**, 17954–17961 (1996). (e) Y.–W. Son, M. L. Cohen and S. G. Louie, Energy gaps in graphene nanoribbons, *Phys. Rev. Lett.*, **97**, 216803 (2006).

[4] (a) E. Clar, (1964), *Polycyclic Hydrocarbons, Vol. I–II*, Academic Press, London. (b) R. G. Harvey, (1997), *Polycyclic Aromatic Hydrocarbons*, Wiley-VCH Verlag, Weinheim.

[5] B. Purushothaman, M. Bruzek, S. R. Parkin, A.–F. Miller and J. E. Anthony, Synthesis and structural characterization of crystalline nonacenes, *Angew. Chem. Int. Ed.*, **50**, 7013–7017 (2011).

[6] L. R. Radovic and B. Bockrath, On the chemical nature of graphene edges: origin of stability and potential for magnetism in carbon materials, *J. Am. Chem. Soc.*, **127**, 5917–5927 (2005).

[7] (a) J. Wu, W. Pisula and K. Müllen, Graphenes as potential material for electronics, *Chem. Rev.*, **107**, 718–747 (2007). (b) M. M. Haley and R. R. Tykwinski (Eds) (2006), *Carbon-Rich Compounds: From Molecules to Materials, Chapter 3*, Wiley-VCH Verlag Gmah & KGaA, Weinheim. (c) J. M. Hughes, Y. Hernandez, D. Aherne, L. Doessel, K. Müllen, B. Moreton, et al., High quality dispersions of hexabenzocoronene in organic solvents, *J. Am. Chem. Soc.*, **134**, 12168–12179 (2012).

[8] R. Goddard, M. W. Naenel, W. C. Herndon, C. Kriiger and M. Zander, Crystallization of large planar polycyclic aromatic hydrocarbons: the molecular and crystal structures of hexabenzocoronene and benzodicoronene, *J. Am. Chem. Soc.*, **117**, 30–41 (1995).

[9] R. Rathore and C. L. Burns, A practical one-pot synthesis of soluble hexa-peri-hexabenzocoronene and isolation of its cation-radical salt, *J. Org. Chem.*, **68**, 4071–4074 (2003).

[10] L. Gherghel, J. D. Brand, M. Baumgarten and K. Müllen, Exceptional triplet and quartet states in highly charged hexabenzocoronenes, *J. Am. Chem. Soc.*, **121**, 8104–8105 (1999).

[11] M. D. Watson, M. G. Debije, J. M. Warman and K. Müllen, Peralkylated coronenes via regiospecific hydrogenation of hexa-*peri*-hexabenzocoronenes, *J. Am. Chem. Soc.*, **126**, 766–771 (2004).

[12] D. W. Boukhvalov, X. Feng and K. Müllen, First-principles modeling of the polycyclic aromatic hydrocarbons reduction, *J. Phys. Chem. C*, **115**, 16001–16005 (2011).

[13] D. Moran, F. Stahl, H. F. Bettinger, H. F. Schaefer III and P. v. R. Schleyer, Towards graphite: magnetic properties of large polybenzenoid hydrocarbons, *J. Am. Chem. Soc.*, **125**, 6746–6752 (2003).

[14] (a) X. Feng, M. Liu, W. Pisula, M. Takase, J. Li and K. Müllen, Supramolecular organization and photovoltaics of triangle-shaped discotic graphenes with swallow-tailed alkyl substituents, *Adv. Mater.*, **20**, 2684–2689 (2008). (b) C. D. Simpson, PhD Thesis, University of Mainz, 2003. (c) V. S. Iyer, M. Wehmeier, J. D. Brand, M. A. Keegstra and K. Müllen, From Hexa-*peri*-hexabenzocoronene to "Superacenes", *Angew. Chem. Int. Ed.*, **36**, 1604–1607 (1997). (d) J. Wu, L. Gherghel, M. D. Watson, J. Li, Z. Wang, C. D. Simpson, et al., From branched polyphenylenes to graphite ribbons, *Macromolecules*, **36**, 7082–7089 (2003).

[15] P. Rempala, J. Krolik and B. T. King, A slippery slope: mechanistic analysis of the intramolecular scholl reaction of hexaphenylbenzene, *J. Am. Chem. Soc.*, **126**, 15002–15003 (2004).

[16] (a) Z. Sun and J. Wu, Higher order acenes and fused acenes with near-infrared absorption and emission, *Aust. J. Chem.*, **64**, 519–528 (2011). (b) A. Herrmann and K. Müllen, From industrial colorants to single photon sources and biolabels: the fascination and function of rylene dyes, *Chem. Lett.*, **35**, 978–985 (2006).

[17] (a) D. Lewis and D. Peters, (1975) *Facts and Theories of Aromaticity*, The Macmillan Press Ltd, London. (b) E. Clar, (1964) *Polycyclic Hydrocarbons, Vol. I–II*, Academic Press, London. (c) K. A. Kerr, J. P. Ashmore and J. C. Speakman, The crystal and molecular structure of quaterrylene: a redetermination, *Proc. R. Soc. Lond. A*, **344**, 199–215 (1975).

[18] (a) Y. Sun, L. Tan, S. Jiang, H. Qian, Z. Wang, D. Yan, et al., High-performance transistor based on individual single-crystalline micrometer wire of perylo[1,12-b,c,d]thiophene, *J. Am. Chem. Soc.*, **129**, 1882–1883 (2007). (b) W. Jiang, Y. Zhou, H. Geng, S. Jiang, S. Yan, W. Hu, et al., Solution-processed, high-performance nanoribbon transistors based on dithioperylene, *J. Am. Chem. Soc.*, **133**, 1–3 (2011). (c) L. Tan, W. Jiang, L. Jiang, S. Jiang, Z. Wang, S. Yan and W. Hu, Single crystalline microribbons of perylo[1,12-b,c,d]selenophene for high performance transistors, *Appl. Phys. Lett.*, **94**, 153306 (2009).

[19] (a) W. Jiang, H. Qian, Y. Li and Z. Wang, Heteroatom-annulated perylenes: Practical synthesis, photophysical properties, and solid-state packing arrangement, *J. Org. Chem.*, **73**, 7369–7372 (2008). (b) C. Jiao, K.-W. Huang, J. Luo, K. Zhang, C. Chi and J. Wu, Bis-N-annulated quaterrylenebis(dicarboximide) as a new soluble and stable near-infrared dye, *Org. Lett.*, **11**, 4508–4511 (2009). (c) Y. Li, J. Gao, S. D. Motta, F. Negri and Z. Wang, Tri-N-annulated Hexarlene: An approach to well-defined graphene nanoribbons with large dipoles, *J. Am. Chem. Soc.*, **132**, 4208–4213 (2010).

[20] (a) E. Clar and M. Zander, Syntheses of coronene and 1:2-7:8-dibenzocoronene, *J. Chem. Soc.*, 4616–4619 (1957). (b) K. V. Rao and S. J. George, Synthesis and controllable self-assembly of a novel coronene bisimide amphiphile, *Org. Lett.*, **12**, 2656–2659 (2010). (c) S. Alibert-Fouet, I. Seguy, J. F. Bobo, P. Destruel and H. Bock, Liquid-crystalline and electron-deficient coronene oligocarboxylic esters and imides by twofold benzogenic Diels–Alder Reactions on perylenes, *Chem. Eur. J.*, **13**, 1746–1753 (2007). (d) H. Langhals and S. Poxleitner, Core-extended terrylenetetracarboxdiimides-novel strongly red fluorescent broadband absorbers, *Eur. J. Org. Chem.*, **5**, 797–800 (2008).

[21] (a) Y. Avlasevich, S. Müller, P. Erk and K. Müllen, Novel core-expanded rylenbis(dicarboximide) dyes bearing pentacene units: facile synthesis and photophysical properties, *Chem. Eur. J.*, **13**, 6555–6561 (2007). (b) C. L. Eversloh, Z. Liu, B. Müller, M. Stangl, C. Li and K. Müllen, Core-extended terrylene tetracarboxydiimide: synthesis and chiroptical characterization, *Org. Lett.*, **13**, 5528–5531 (2011).

[22] (a) Q. Bai, B. Gao, Q. Ai, Y. Wu and X. Ba, Core-extended terrylene diimide on the bay region: synthesis and optical and electrochemical properties, *Org. Lett.*, **13**, 6484–6487 (2011). (b) Y. Li, W. Xu, S. D. Motta, F. Negri, D. Zhu and Z. Wang, Core-extended rylene dyes via thiophene annulation, *Chem. Commun.*, **48**, 8204–8206 (2012).

[23] (a) Y. Shi, H. Qian, Y. Li, W. Yue and Z. Wang, Copper-mediated domino process for the synthesis of tetraiodinated di(perylene bisimide), *Org. Lett.*, **10**, 2337–2340 (2008). (b) H. Qian, F. Negri, C. Wang and Z. Wang, Fully conjugated tri(perylene bisimides): an approach to the construction of n-type graphene nanoribbons, *J. Am. Chem. Soc.*, **130**, 17970–17976 (2008).

[24] (a) J. H. Yao, C. Chi, J. Wu and K.-P. Loh, Bisanthracene bis(dicarboxylic imide)s as soluble and stable NIR dyes, *Chem. Eur. J.*, **15**, 9299–9302 (2009). (b) J. Li, K. Zhang, X. Zhang, K.-W. Huang, C. Chi and J. Wu, *meso*-Substituted bisanthenes as soluble and stable near-infrared dyes, *J. Org. Chem.*, **75**, 856–863 (2010). (c) K. Zhang, K.-W. Huang, J. Li, J. Luo, C. Chi and J. Wu, A soluble and stable quinoidal bisanthene with NIR absorption and amphoteric redox behavior, *Org. Lett.*, **11**, 4854–4857 (2009).

[25] (a) E. Clar, Synthesen von Benzologen des Perylens und Bisanthens, *Chem. Ber.*, **82**, 46–60 (1949). (b) J. Li, J.–J. Chang, H. S. Tan, H. Jiang, X. Chen, Z. Chen, et al., Disc-like 7,14-dicyano-ovalene-3,4:10,11-bis(dicarboximide) as a solution processible n-type semiconductor for air stable field-effect transistors, *Chem. Sci.*, **3**, 846–850 (2012). (c) E. H. Fort, M. S. Jeffreys and L. T. Scott, Diels–Alder cycloaddition of acetylene gas to a polycyclic aromatic hydrocarbon bay region, *Chem. Commun.*, **48**, 8102–8104 (2012). (d) J. Li, C. Jiao, K.–W. Huang, J. Wu, Lateral extension of π conjugation along the bay regions of bisanthene through a Diels–Alder cycloaddition reaction, *Chem. Eur. J.*, **17**, 14672–14680 (2011). (e) E. H. Fort and L. T. Scott, Carbon nanotubes from short hydrocarbon templates. Energy analysis of the Diels–Alder cycloaddition/rearomatization growth strategy, *J. Mater. Chem.*, **21**, 1373–1381 (2011).

[26] A. Konishi, Y. Hirao, M. Nakano, A. Shimizu, E. Botek, B. Champagne, et al., Synthesis and characterization of teranthene: a singlet biradical polycyclic aromatic hydrocarbon having Kekulé structures, *J. Am. Chem. Soc.*, **132**, 11021–11023 (2010).

[27] D. H. Reid, Stable π-electron systems and new aromatic structures. *Tetrahedron* **3**, 339–352. (1958)

[28] K. Nakasuji, M. Yamaguchi, I. Murata, K. Yamaguchi, T. Fueno, H. Ohya-Nishiguchi, et al., Synthesis and characterization of phenalenyl cations, radicals, and anions having donor and acceptor substituents: three redox states of modified odd alternant systems. *J. Am. Chem. Soc.* **111**, 9265–9267 (1989).

[29] K. Goto, T. Kubo, K. Yamamoto, K. Nakasuji, K. Sato, D. Shiomi, et al., A stable neutral hydrocarbon radical: synthesis, crystal structure, and physical properties of 2,5,8-tri-*tert*-butyl-phenalenyl. *J. Am. Chem. Soc.* **121**, 1619–1620 (1999).

[30] S. Suzuki, Y. Morita, K. Fukui, K. Sato, D. Shiomi, T. Takui and K. Nakasuji, Aromaticity on the pancake-bonded dimer of neutral phenalenyl radical as studied by MS and NMR spectroscopies and NICS analysis. *J. Am. Chem. Soc.* **128**, 2530–2531 (2006).

[31] P. A. Koutentis, Y. Chen, Y. Cao, T. P. Best, M. E. Itkis, L. Beer, et al., Perchlorophenalenyl radical. *J. Am. Chem. Soc.* **123**, 3864–3871 (2001).

[32] L. Beer, S. K. Mandal, R. W. Reed, R. T. Oakley, F. S. Tham, B. Donnadieu and R. C. Haddon, The first electronically stabilized phenalenyl radical: effect of substituents on solution chemistry and solid-state structure. *Cryst. Growth. Des.* **7**, 802–809 (2007).

[33] L. Beer, R. W. Reed, C. M. Robertson, R. T. Oakley, F. S. Tham and R. C. Haddon, Tetrathiophenalenyl radical and its disulfide-bridged dimer. *Org. Lett.* **10**, 3121–3123 (2008).

[34] T. Kubo, Y. Goto, M. Uruichi, K. Yakushi, M. Nakano, A. Fuyuhiro, et al., Synthesis and characterization of acetylene-linked bisphenalenyl and metallic-like behavior in its charge-transfer complex, *Chem. Asian. J.* **2**, 1370–1379 (2007).

[35] K. Nakasuji, K. Yoshida and I. Murata, Design and synthesis of a highly amphoteric condensed hydrocarbon with the highest reduction potential: pentaleno[1,2,3-*cd*:4,5,6-*c'd'*]diphenalene. *J. Am. Chem. Soc.* **105**, 5136–5137 (1983).

[36] T. Kubo, Y. Katada, A. Shimizu, Y. Hirao, K. Sato, T. Takui, et al., Synthesis, crystal structure, and physical properties of sterically unprotected hydrocarbon radicals. *J. Am. Chem. Soc.* **133**, 14240–14243 (2011).

[37] T. Kubo, K. Yamamoto, K. Nakasuji, T. Takui and I. Murata, Hexa-*tert*-butyltribenzodecacyclenyl: a six-stage amphoteric redox system. *Angew. Chem. Int. Ed. Engl.* **35**, 439–441 (1996).

[38] T. Kubo, K. Yamamoto, K. Nakasuji, T. Takui and I. Murata, 4,7,11,14,18,21-Hexa-*t*-butyltribenzodecacyclenyl radical: s six-stage amphoteric redox system. *Bull. Chem. Soc. Jpn.* **74**, 1999–2009 (2001).

[39] W. T. Borden and E. R. Davidson, Effects of electron repulsion in conjugated hydrocarbon diradicals. *J. Am. Chem. Soc.* **99**, 4587–4594 (1977).

[40] E. Clar and D. G. Stewart, Aromatic hydrocarbons. LXV. triangulene derivatives. *J. Am. Chem. Soc.* **75**, 2667–2672 (1953).

[41] O. Hara, K. Tanaka, K. Yamamoto, T. Nakazaw and Y. Murata, The chemistry of phenalenium systems: XXV the triangulenyl dianion. *Tetrahedron. Lett.* **18**, 2435–2436 (1977).

[42] J. Inoue, K. Fukui, T. Kubo, S. Nakazawa, K. Sato, D. Shiomi, et al., The first detection of a Clar's hydrocarbon, 2,6,10-tri-*tert*-butyltriangulene: a ground-state triplet of non-Kekulé polynuclear benzenoid hydrocarbon. *J. Am. Chem. Soc.* **123**, 12702–12703 (2001).

[43] G. Allinson, R. G. Bushby, J. L. Paillaud, ESR spectrum of a stable triplet π biradical: trioxy triangulene. *J. Am. Chem. Soc.* **115**, 2062–2064 (1993).

[44] G. Allinson, R. G. Bushby, J. L. Paillaud and M. Thornton-Pett, Synthesis of a derivative of triangulene; the first non-Kekulé polynuclear aromatic. *J. Chem. Soc. Perkin. Trans.* 385–390 (1995).

[45] Y. Morita, S. Nishida, T. Murata, M. Moriguchi, A. Ueda, M. Satoh, et al., Organic tailored batteries materials using stable open-shell molecules with degenerate frontier orbitals. *Nature Mater.* **10**, 947–951 (2011).

[46] G. Quinkert, W.-W. Wiersdorff, M. Finke and K. Opitz, Über das tetraphenyl-o-xylylen. *Tetrahedron Lett.*, 1966, **7**, 2193–2200 (1966).

[47] J. Kolc and J. Michl, Photochemical synthesis of matrix-isolated pleiadene. *J. Am. Chem. Soc.* **92**, 4147–4148 (1970).

[48] J. Kolc and J. Michl, pi,pi-Biradicaloid hydrocarbons. Pleiadene family. I. Photochemical preparation from cyclobutene precursors *J. Am. Chem. Soc.* **95**, 7391–7401 (1973).

[49] S. Iwashita, E. Ohta, H. Higuchi, H. Kawai, K. Fujiwara, K. Ono, et al., First stable 7,7,8,8-tetraaryl-*o*-quinodimethane: isolation, X-ray structure, electrochromic response of 9,10-bis(dianisylmethylene)-9,10-dihydrophenanthrene. *Chem. Commun.* 2076–2077 (2004).

[50] D. Ghereg, S. E.-C. El Kettani, M. Lazraq, H. Ranaivonjatovo, W. W. Schoeller, J. Escudie and H. Gornitzka, An isolable *o*-quinodimethane and its fixation of molecular oxygen to give an endoperoxide. *Chem. Commun.* 4821–4823 (2009).

[51] A. Shimizu and Y. Tobe, Indeno[2,1-*a*]fluorene: An air-stable *ortho*-quinodimethane derivative. *Angew. Chem. Int. Ed.* **50**, 6906–6910 (2011).

[52] L. K. Montgomery, J. C. Huffman, E. A. Jurczak and M. P. Grendze, The molecular structures of Thiele's and Chichibabin's hydrocarbons. *J. Am. Chem. Soc.* **108**, 6004–6011 (1986).

[53] A. Shimizu, Y. Hirao, K. Matsumoto, H. Kurata, T. Kubo, M. Uruichi and K. Yakushi, Aromaticity and π-bond covalency: prominent intermolecular covalent bonding

interaction of a Kekulé hydrocarbon with very significant singlet biradical character, *Chem. Commun.* **48**, 5629–5631 (2012).

[54] K. Ohashi, T. Kubo, T. Masui, K. Yamamoto, K. Nakasuji, T. Takui, et al., 4,8,12,16-Tetra-*tert*-butyl-*s*-indaceno[1,2,3-*cd*:5,6,7-*c'd'*]diphenalene: A four-stage amphoteric redox system. *J. Am. Chem. Soc.* **120**, 2018–2027 (1998).

[55] T. Kubo, A. Shimizu, M. Uruichi, K. Yakushi, M. Nakano, D. Shiomi, et al., Singlet biradical character of phenalenyl-based Kekulé hydrocarbon with naphthoquinoid structure. *Org. Lett.* **9**, 81–84 (2007).

[56] T. Kubo, A. Shimizu, M. Sakamoto, M. Uruichi, K. Yakushi, M. Nakano, et al., Synthesis, intermolecular interaction, and semiconductive behavior of a delocalized singlet biradical hydrocarbon. *Angew. Chem. Int. Ed.* **44**, 6564–6568 (2005).

[57] A. Shimizu, M. Uruichi, K. Yakushi, H. Matsuzaki, H. Okamoto, M. Nakano, et al., Resonance balance shift in stacks of delocalized singlet biradicals. *Angew. Chem. Int. Ed.* **48**, 5482–5486 (2009).

[58] A. Shimizu, T. Kubo, M. Uruichi, K. Yakushi, M. Nakano, D. Shiomi, et al., Alternating covalent bonding interactions in a one-dimensional chain of a phenalenyl-based singlet biradical molecule having Kekulé structures. *J. Am. Chem. Soc.* **132**, 14421–14428 (2010).

[59] K. Kamada, K. Ohta, T. Kubo, A. Shimizu, Y. Morita, K. Nakasuji, et al., Strong two-photon absorption of singlet diradical hydrocarbons. *Angew. Chem. Int. Ed.* **46**, 3544–3546 (2007).

[60] M. Chikamatsu, T. Mikami, J. Chisaka, Y. Yoshida, R. Azumi and K. Yase, Ambipolar organic field-effect transistors based on a low band gap semiconductor with balanced hole and electron mobilities. *Appl. Phys. Lett.* **91**, 043506 (2007).

[61] T. Kubo, M. Sakamoto, M. Akabane, Y. Fujiwara, K. Yamamoto, M. Akita, et al., Four-stage amphoteric redox properties and biradicaloid character of tetra-tertbutyldicyclopenta[*b;d*]thieno[1,2,3-*cd*;5,6,7-*c'd'*]diphenalene. *Angew. Chem. Int. Ed.* **43**, 6474–6479 (2004).

[62] T. Kubo, K. Yamamoto, K. Nakasuji and T. Takui, Tetra-*tert*-butyl-*as*-indaceno[1,2,3-*cd*:6,7,8-*c'd'*]diphenalene: a four-stage amphoteric redox system. *Tetrahedron Lett.* **42**, 7997–8001 (2001).

[63] M. Nakano, R. Kishi, A. Takebe, M. Nate, H. Takahashi, T. Kubo, et al., Second hyperpolarizability of zethrenes. *Comput. Lett.* **3**, 333–338 (2007).

[64] R. Umeda, D. Hibi, K. Miki and Y. Tobe, Tetradehydrodinaphtho[10]annulene: A hitherto unknown dehydroannulene and a viable precursor to stable zethrene derivatives. *Org. Lett.* **11**, 4104–4106 (2009).

[65] T. C. Wu, C. H. Chen, D. Hibi, A. Shimizu, Y. Tobe and Y. T. Wu, Synthesis, structure, and photophysical properties of dibenzo-[de,mn]naphthacenes. *Angew. Chem. Int. Ed.* **49**, 7059–7062 (2010).

[66] Z. Sun, K. W. Huang and J. Wu, Soluble and stable zethrenebis(dicarboximide) and its quinone. *Org. Lett.* **12**, 4690–4693 (2010).

[67] Z. Sun, K. W. Huang and J. Wu, Soluble and stable heptazethrenebis(dicarboximide) with a singlet open-shell ground state. *J. Am. Chem. Soc.* **133**, 11896–11899 (2011).

[68] Y. Li, W.-K. Heng, B. S. Lee, N. Aratani, J. L. Zafra, N. Bao, et al., Kinetically blocked stable heptazethrene and octazethrene: closed-shell or open-shell in the ground state? *J. Am. Chem. Soc.*, **134**, 14913–14922 (2012).

[69] D. T. Chase, B. D. Rose, S. P. McClintock, L. N. Zakharov and M. M. Haley, Indeno[1,2-*b*]fluorenes: Fully conjugated antiaromatic analogues of acenes. *Angew. Chem. Int. Ed.* **50**, 1127–1130 (2011).

[70] D. T. Chase, A. G. Fix, B. D. Rose, C. D. Weber, S. Nobusue, C. E. Stockwell, et al., Electron-accepting 6,12-diethynylindeno[1,2-*b*]fluorenes: synthesis, crystal structures, and photophysical properties. *Angew. Chem., Int. Ed.* **50**, 11103–11106 (2011).

[71] J. Nishida, S. Tsukaguchi and Y. Yamashita, Synthesis, crystal structures, and properties of 6,12-diaryl-substituted indeno[1,2-*b*]fluorenes. *Chem. Eur. J.* **18**, 8964–8970 (2012).

[72] D. T. Chase, A. G. Fix, S. J. Kang, B. D. Rose, C. D. Weber, Y. Zhong, et al., 6,12-Diarylindeno[1,2-*b*]fluorenes: syntheses, photophysics, and ambipolar OFETs. *J. Am. Chem. Soc.* **134**, 10349–10352 (2012).

[73] B. D. Rose, C. L. Vonnegut, L. N. Zakharov and M. M. Haley, Fluoreno[4,3-*c*]fluorene: A closed-shell, fully conjugated hydrocarbon. *Org. Lett.* **14**, 2426–2429 (2012).

18

Graphene Moiré Supported Metal Clusters for Model Catalytic Studies

Bradley F. Habenicht,[a] Ye Xu,[a] and Li Liu[b]
[a] Center for Nanophase Materials Sciences, Oak Ridge National Laboratory, USA
[b] Department of Chemistry, Texas A&M University, USA

18.1 Introduction

It has been known since the 1960s that single-layer graphene can be grown on transition metals [1–4]. Since then, graphene has been obtained and characterized on a number of single crystal transition metal surfaces, including Ru(0001) [5–7], Ir(111) [8,9], Rh(111) [10], Pt(111) [11], Pd(111) [12, 13], Co(0001) [14, 15], Ni(111) [16–18], and Cu(111) [19, 20]. While graphene itself continues to fascinate the research community with its unique electronic and mechanical properties [21, 22], transition metal-supported graphene is a class of interesting functional nano-material in its own right. The most salient feature of graphene on transition metals is that a long-range moiré superstructure can develop, on for example, Ru(0001), Rh(111), Ir(111), Pt(111), Pd(111), and Cu(111), a consequence of a sufficiently large mismatch between the lattice constants of graphene and the metal. The moiré is seen in scanning tunneling microscope (STM) as periodic arrays of regions with different contrasts (Figure 18.1), indicating different electronic coupling and possibly different geometric corrugation in the different regions of the graphene. Single-crystalline, defect-free domains of graphene extending over micron-size areas on metal surfaces have been obtained on Ru(0001) and Ir(111) [7, 23–26].

A consequence of the moiré superstructure is that the surface properties of the graphene become inhomogeneous. In effect, graphene moiré provides a high-quality, defect-free surface with a huge number of evenly spaced, periodic sites with identical reactivity, which

---

Graphene Chemistry: Theoretical Perspectives, First Edition. Edited by De-en Jiang and Zhongfang Chen.
© 2013 John Wiley & Sons, Ltd. Published 2013 by John Wiley & Sons, Ltd.

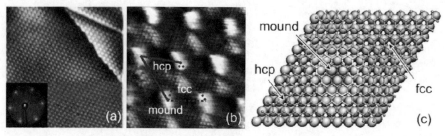

Figure 18.1 *(a) A typical STM image of graphene on Ru(0001) (100 × 100 nm^2, V_{sample} = +0.1 V, $I_{tunneling}$ = 0.2 nA); a typical LEED pattern is shown as an inset. Reprinted with permission from [27] © 2012 American Physical Society. (b) A close-up STM image (8 × 8 nm^2, V_{sample} = −0.3 V, $I_{tunneling}$ = 1.0 nA). These and the STM images below are taken at room temperature. Reprinted with permission from [28] © Royal Society of Chemistry. (c) The optimized structural model for (12 × 12)–C on (11 × 11)-Ru(0001). Gray and green spheres represent lower- and top-layer Ru atoms, respectively. Carbon atoms are shown in ball-and-stick model and colored by height (red = lowest; blue = highest) for clarity. The three different regions in the moiré unit cell are labeled in (b) and (c). See colour version in the colour plate section*

could serve as traps and reaction sites for atoms, clusters, and macromolecules. For instance, dense superlattices of evenly spaced metal clusters with narrow size distributions can be formed on graphene/Ir(111) surface under appropriate conditions [29, 30]. This suggests the intriguing possibility of using graphene moiré as a template to form a large number of nearly identical clusters with desired sizes, which would make model catalysts for studying catalytic reactions, including the size effect of nanocatalysts, given that catalytic activity can depend strongly on the size and shape of the catalyst particles, a concept proposed by Boudart [31] that has since been borne out by numerous studies of various catalytic reactions [32–35]. Furthermore, as graphene is a chemically inert carbon surface, only the clusters formed on it are expected to participate in catalytic reactions, thereby reducing the complexity of reaction mechanisms. This stands in contrast to oxide-supported catalysts, in which lattice cations, oxygen, and vacancies can all participate in catalytic reactions in addition to the metal [36]. Thus graphene-supported catalysts are ideal for studying the intrinsic catalytic activity of metals, particularly at low temperature.

In this chapter we use graphene moiré formed on Ru(0001) (henceforth denoted as g/Ru(0001) as an example to examine three aspects: (1) the geometric and electronic structures of g/Ru(0001); (2) the formation of metal clusters on g/Ru(0001); (3) the growth of 2D Au islands on g/Ru(0001) and their catalytic activity toward CO oxidation.

18.2 Graphene Moiré on Ru(0001)

Goodman and coworkers reported one of the earliest STM images of graphene on Ru(0001) in an investigation of carbonaceous deposits from methane decomposition, which crystallized into what was then termed "graphite monolayers" at above 1300 K [5]. More recent synthesis and studies of g/Ru(0001) have been reported by several groups [7, 23, 37, 38].

At present, two methods are primarily used to grow graphene on transition metal surfaces: chemical vapor deposition (CVD) and carbon surface segregation. In the CVD method typically C_2H_4 is used, which decomposes on the surface and releases H_2, with the carbon forming graphene when annealed to around 1300 K. Benzene can be used as an alternative to ethylene [39]. Our g/Ru(0001) samples are fabricated using the carbon surface segregation method. C_2H_4 or C_3H_6 is dissociated on a clean Ru(0001) single-crystal surface, which occurs at room temperature. The sample is then annealed to approximately 1400 K *in vacuo* to cause carbon atoms to dissolve into the bulk of the metals. When the temperature is lowered, carbon atoms from the bulk migrate back to the surface and nucleate graphene islands [40]. Further annealing heals defects in graphene [38]. Nucleation has been shown to preferentially occur at metal step edges [25, 39, 41]. With sufficient carbon coverage the substrate can become fully covered by single-crystalline graphene with very few defects.

The most salient feature of g/Ru(0001) in STM is the long-range, periodic arrays of high and low contrast regions (Figure 18.1) [6]. An energy-minimized structure of graphene on Ru(0001) based on density functional theory (DFT) calculations is illustrated in Figure 18.1(c) to show how the lattice mismatch between graphene and Ru(0001) leads to periodic shifting of the C atoms out of registry with the surface Ru atoms. High-resolution STM images (Figure 18.1b) reveal that there are three regions of different contrasts, and simulated STM images based on the proposed structure (Figure 18.1c) reproduce the same features in contrasts [42]. The bright, medium dark, and dark regions in STM (at negative sample bias) are identified as the mound, fcc, and hcp regions in the moiré unit cell. These regions are named according to how the two C sublattices of graphene are positioned relative to Ru (labeled in Figure 18.1c: In the fcc and hcp regions, the two C sublattices are located over the atop site and hcp (or fcc) site on the Ru(0001) surface, exposing the other surface site, fcc (or hcp), through C_6 rings. In the mound region the two C sublattices are located over the fcc and hcp surface threefold sites, exposing Ru atoms through C_6 rings. The inset of Figure 18.1(a) shows a typical LEED pattern of the g/Ru(0001) system. The (1×1) spots of Ru(0001) are surrounded by satellite spots, confirming the significantly longer-range hexagonal periodicity of the graphene moiré.

The exact periodicity of the graphene moiré on Ru(0001) has been a subject of debate. Goodman and coworkers analyzed the LEED pattern of g/Ru(0001) and found the graphene and Ru(0001) to come into coincidence at every 11 Ru and 12 C unit meshes (henceforth denoted as 12-on-11), which agrees with the observed periodicity of 3.0 nm in STM. More recent high-resolution STM generally agrees with the 12-on-11 interpretation [6, 43]. To minimize lattice strain in graphene purely based on the equilibrium constants of graphene and Ru, a 11-on-10 structure would be most favorable as was proposed by Vázquez de Parga et al. [23], but chemical interaction would likely lengthen C-C bonds and therefore favor larger structures. Jiang et al. performed DFT calculations to show that graphene has a stronger binding energy on Ru(0001) in the 12-on-11 structure (-0.021 eV/C) than in the 11-on-10 structure (-0.011 eV/C) [44]. A recent XRD study by Martoccia et al. [45] and LEED study by Moritz et al. [46] conclude that the coincidence structure in fact has a 25-on-23 periodicity consisting of four nearly identical 12.5-on-11.5 subunits. The [10$\bar{1}$0] directions of graphene and Ru(0001) are well aligned with very small amount of rotation detected [6].

Another aspect of graphene on Ru(0001) that remains in doubt is the geometric height of the corrugation of the moiré. It is observed that the apparent corrugation of the moiré

Table 18.1 Comparison of DFT-calculated geometric properties of (12 × 12)-graphene moiré on (11 × 11)-Ru(0001)

	theory level	corrugation[†] (Å)	adsorption energy[*] (meV/C)	height[#] (Å)
Jiang et al. [44]	GGA-PBE	1.7	−21	2.2
Wang et al. [49]	GGA-PBE	1.51	−27	2.22
Stradi et al. [53]	DFT + D/PBE	1.195	−206	2.195
this work[§]	GGA-PBE	1.54	−10	2.21
this work[§]	optB86b-vdW-DF	1.45	−173	2.18

[†]Maximum corrugation between highest and lowest C atoms.
[*]Adsorption energy versus flat freestanding graphene.
[#]Minimum height between lowest C atom and highest Ru atom. Experimentally determined height of graphene is 1.6~1.8 Å by STM [39] and TEM [54].
[§]Self-consistent DFT calculations were performed using VASP [55, 56]. Core electrons were described by the projector augmented wave method. Kinetic energy cutoff was 400 eV. Surface Brillouin zone was sampled at the Γ point only. Geometric optimization was converged to 0.05 eV/Å for each relaxed degree of freedom.

changes with the bias voltage in STM, being in excess of 1.0 Å at large negative bias but as low as 0.2 Å at large positive bias [23], which indicates electronic effects to be important. We measured the corrugation to be 1.5 Å at −0.3 V [28]. XRD analysis performed by Martoccia et al. [47] and LEED analysis performed by Moritz et al. [46] have determined that the geometric corrugation is 0.82 and 1.5 Å, respectively, while Borca et al. have employed He ion diffraction and concluded that it is only 0.15 Å [48]. The disagreement of the different experimental techniques remains to be resolved. Although the various experimental techniques have produced disparate measurements, different DFT calculations for the 12-on-11 structure have produced results that are in closer agreement (Table 18.1). The standard GGA predicts a geometric corrugation of 1.5–1.7 Å [44, 49]. Stradi et al. have included van der Waals interactions via the DFT + D formulation [50] and reported a smaller corrugation of 1.195 Å. We have used the optB86b van der Waals density functional (optB86b-vdW-DF) [51], which predicts a corrugation of 1.45 Å. All the predicted values are therefore at the high end of the experimentally reported range of values. For our computational work, we take the 12-on-11 structure as depicted in Figure 18.1(c) as the basis, which is closely in line with the predominance of the available experimental evidence, and is expected to introduce negligible errors [46]. It has been suggested that the height of corrugation in graphene moiré is directly related to the strength of chemical bonding at the interface [52]. In terms of the distance from the metal surface; binding energy; and electronic structure (discussed next), graphene is strongly coupled to Ru(0001), unlike on substrates such as Pt(111) and Ir(111) [26].

As has been pointed out [42, 44, 49], a spontaneous sp^2 to sp^3 rehybridization occurs forming C–Ru bonds in the low (fcc and hcp) regions, where the C atoms in one of the two C sublattices align with the surface Ru atoms. This rehybridization can be visualized in an electron density plot such as shown in Figure 18.2. The C–Ru bonding is counterbalanced by the lattice strain in graphene due to the lattice mismatch, which causes a portion of the graphene to be decoupled from the surface and form the protruding mound region.

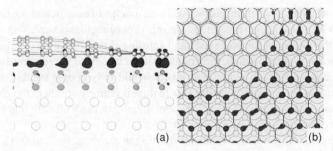

Figure 18.2 DFT-calculated charge density difference ($\Delta\rho = \rho_{total} - \rho_{Ru} - \rho_{graphene}$) for graphene moiré on Ru(0001): (a) side view and (b) top view. The isosurfaces shown are ± 0.008 e/Å^3 (dark = density accumulation; light = density depletion). Ru atoms are shown as large spheres, and the graphene moiré is shown as a hexagonal wire network. Reproduced with permission from [81] © 2013 American Institute of Physics

The mound and the fcc/hcp regions may therefore be characterized as chemically different species. This is reflected in the STM, where all carbon atoms in the mound region are visible but only half the C atoms in the fcc and hcp regions are resolved [6]. The rehybrization of the C orbital causes band dispersion near the Fermi level, which reduces the density of states there [57]. Angle-resolved photoemission experiments further demonstrate the strong coupling between graphene and Ru(0001), where a large π band gap opens at $\bar{K}$ and graphene becomes n-doped [58]. In comparison, on less reactive transition metals such as Ir(111) the band structure of graphene is barely perturbed [59].

We have investigated the band structure of monolayer graphene on Ru(0001) along the $\bar{\Gamma}$ to $\bar{K}$ in detail using angle-resolved ultraviolet photoemission spectroscopy (ARUPS; Figures 18.3b–c) [27]. As is consistent with graphene, the Brillouin zone edge along $\bar{\Gamma}$ to $\bar{K}$ occurs at 1.7 Å^{-1} [58]. The fact that this feature is decoupled from the Ru(0001) band structure suggests that graphene is out of registry with the substrate. Our spectra (Figure 18.3b–d) taken at 45 eV photon energy are able to resolve a splitting along $\bar{\Gamma}$ to $\bar{K}$ in the bands that are associated with graphene.

Spin-orbit coupling between graphene and the substrate could cause the observed band splitting. However, due to the low Z value of Ru, spin-orbit coupling should be negligibly

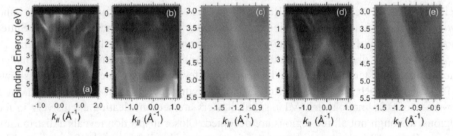

Figure 18.3 Angle-resolved ultraviolet photoemission spectroscopy spectra of: (a) clean Ru(0001); (b) g/Ru(0001); (c) area indicated by the white dashed rectangle in (b) enlarged; (d) g/O/Ru(0001); and (e) area indicated by the white dashed rectangle in (d) enlarged. (a) is taken at 22 eV photon energy and (b–e) at 45 eV photon energy along the $\bar{\Gamma}$-$\bar{K}$ direction. Reprinted with permission from [27] © 2012 by American Physical Society

small [60]. Moreover, the splitting is observed over a range of energy, and not just around the Fermi level, which also suggests against spin-orbit coupling that is wave vector-dependent. Another possible origin of the observed band splitting is the formation of bilayer or multilayer graphene. In such a situation, the graphene layer closest to the substrate would couple strongly to the Ru(0001) and additional layers would only couple weakly, which would result in the formation of distinct bands. This feature is entirely absent in the photoemission data. Furthermore, the measured work function, 4.2 eV, agrees well with other experiments on single layer graphene grown on Ru.

To further explore the nature of the band splitting in graphene on Ru(0001), graphene on oxidized Ru(0001) was studied. At room temperature g/Ru(0001) is stable in air because graphene passivates Ru toward oxidation [61, 62]. We heated a g/Ru(0001) sample at 700 K in 1×10^{-7} Torr of O_2 for 5 min. to introduce chemisorbed O onto the Ru(0001) surface. This results in a graphene layer on oxidized Ru(0001) (denoted O/Ru(0001)). In the g/Ru(0001) system, the C atoms should gain charge from the substrate [7, 58]. On the other hand, graphene on chemisorbed oxygen on Ru(0001) would be slightly cationic. The resulting change in the coupling of graphene to the support material has both energetic and geometric manifestations. The splitting of the band running from $\bar{\Gamma}$ to $\bar{K}$ is reduced from 800 meV to 400 meV (Figure 18.3d), and the corrugation of the graphene also decreases from around 1 Å in g/Ru(0001) to 0.3 Å in g/O/Ru(0001). The decrease of corrugation upon O intercalation between graphene and Ru(0001) has also been noted by Sutter et al. [54]. The variable splitting of the graphene bands further supports the idea that the corrugation is a direct result of the coupling between graphene and the substrate and chemically differentiates the high and low regions of the graphene moiré.

18.3 Metal Cluster Formation on g/Ru(0001)

The periodic change in the chemical reactivity of the graphene moiré surface lends itself well to template-driven synthesis of metal nanoparticles. This is illustrated by the dense superlattices of Ir, Pt, Re, W, and Au that have been grown on g/Ir(111) [29, 30]. To test the suitability of g/Ru(0001) as such a template, we deposited five catalytically relevant metals on g/Ru(0001), including Rh, Pt, Pd, Co, and Au, to study their nucleation and growth behavior [38]. Pt deposition on g/Ru(0001) has also been studied by Pan et al. [63] and Donner et al. [39], and Ru deposition on g/Ru(0001) has been recently reported by Sutter et al. [54]. Filling of 100% of the moiré cells is shown to be possible by Donner et al.

The five metals displayed disparate cluster morphologies, with Rh and Pt forming smaller, more isolated clusters and Pd, Co and Au forming a few large aggregates. Figure 18.4 shows STM images of Rh clusters on g/Ru(0001) at varying coverages of Rh deposition. At a coverage of 0.05 monolayers (ML), small clusters are seen nucleating exclusively in fcc regions, although not all fcc regions are occupied. Closer inspection reveals a distribution in size. Figure 18.5 shows a histogram of the diameter distribution of the Rh clusters. The clusters have heights varying from 3–12 Å, which correspond to 1–4 atomic layers of Rh. These clusters coexist even at 0.05 ML Rh coverage and most of them grow in a 3D mode. Further inspection of the STM images and the histogram (Figures 18.4–18.5) indicates that the growth of the Rh clusters on g/Ru(0001) occurs in a self-limiting fashion in terms

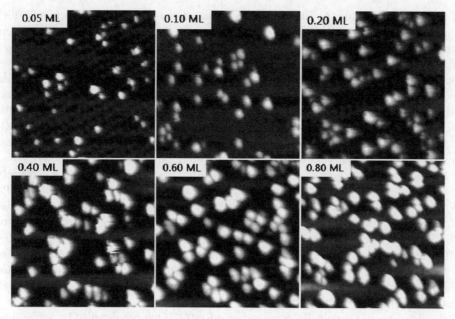

Figure 18.4 STM images (100 × 100 nm^2, $V_{sample} = +1.0$ V, $I_{tunneling} = 0.1$ nA) of 0.05, 0.10, 0.20, 0.40, 0.60 and 0.80 ML of Rh deposited on g/Ru(0001) at room temperature. Reprinted with permission from [38] © 2010 Elsevier

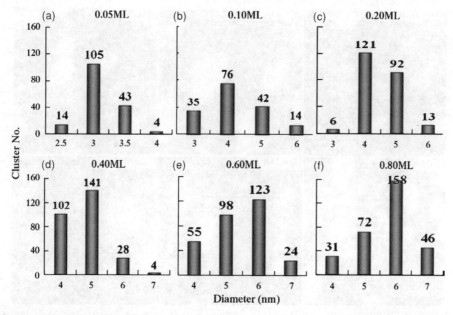

Figure 18.5 Histograms of the number of clusters as a function of diameter at different Rh coverages shown in Figure 18.4. Reprinted with permission from [38] © 2010 Elsevier

of the lateral dimension. As Rh coverage is increased, the cluster size and height profiles continue to increase, while the cluster number density rises only marginally. Rh clusters formed at 0.8 ML of Rh coverage on g/Ru(0001) maintained their original morphology after annealing up to 900 K *in vacuo*, or up to 700 K in 1 Torr of CO for 10 min. Further increase of the annealing temperature to 1100 K causes the size of the clusters to increase and the number density of the clusters to decrease. Pt behaved comparably to Rh. At less than 0.1 ML coverage Pt clusters were highly dispersed and preferred to nucleate in the fcc region, with clusters of 1–4 atomic layers coexisting. At higher coverage below 1 ML Pt clusters with a fairly narrow size distribution and diameters of less than 5 nm were observed, which is consistent with the findings of other experimental work [63, 64].

Like Rh and Pt, the nucleation of Co, Pd, and Au starts in the fcc regions at extremely low coverages. However, they display much lower nucleation densities than Rh and Pt. When Pd was deposited at coverages of 0.1 and 0.4 ML, clusters with an average diameter of 8 and 14 nm formed (Figure 18.6a–b), with a far lower number density than Rh or Pt. Likewise Co forms a few large aggregates on g/Ru(0001) at coverages of 0.2 and 0.4 ML (Figure 18.6c–d). The clusters are similar to those of Pd, with slightly different size distributions of 10 and 12 nm at 0.2 and 0.4 ML. Au behaves differently from Pd and Co in that it forms large 2D islands instead of 3D particles (Figure 18.6e–f), which we will discuss in more detail in the next section.

Besides these five metals, Ru deposited on g/Ru(0001) has been studied by Sutter et al. [54], and its nucleation and growth behavior is similar to those of Rh and Pt. For 0.02 and

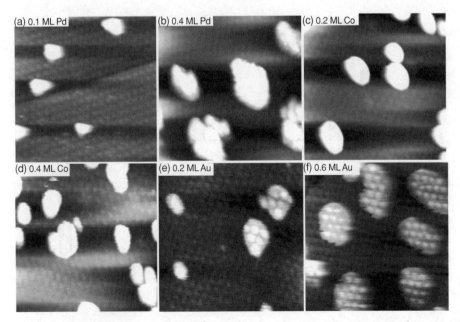

Figure 18.6 STM images (50 × 50 nm^2, $V_{sample} = +1.0$ V, $I_{tunneling} = 0.1$ nA) of (a) 0.1 ML Pd, (b) 0.4 ML Pd, (c) 0.2 ML Co, (d) 0.4 ML Co, (e) 0.2 ML Au, and (f) 0.6 ML Au deposited on g/Ru(0001) at room temperature. Reprinted with permission from [38] © 2010 Elsevier

0.3 ML of Ru deposited at 200 K, well dispersed Ru clusters are obtained that mostly have a diameter of 20 Å and heights of 1~2 atomic layers. The clusters are likewise observed to be predominantly adsorbed in the fcc regions.

Overall, therefore, the smallest clusters of Rh, Pt, Ru, Pd, Co, and Au are all seen to nucleate in the fcc regions of g/Ru(0001), but while Rh, Pt, and Ru form a mixture of single and multiple atomic layer high clusters at low coverages and remain dispersed [38, 54], Pd and Co on g/Ru(0001) rapidly form large 3D aggregates [38], and Au form large 2D islands. Various factors have been suggested to explain the disparate cluster nucleation and growth behavior of the different metals on graphene moiré. A high cohesive energy, extended valence orbitals of the metal atom, and a small mismatch in lattice constant with Ir(111) have been proposed by N'Diaye et al. to be necessary characteristics of the deposited metals to enable the formation of dense superlattices of clusters on g/Ir(111) [30].

The preference of metal clusters containing more than a few atoms for 2D planar geometry on graphene moiré suggests strong interactions with the graphene. Intuitively, a competition between the metal-graphene and metal-metal interactions should determine the shape and size of the metal clusters as they grow on graphene moiré, controlling the transition from 2D to 3D geometry. As long as the metal–C interaction is stronger than the metal-metal interaction, a 2D growth mode will persist (i.e., metal wetting the graphene). If the metal-metal interactions are stronger than the metal–C interaction, growth will shift to a 3D mode.

The strength of the metal-C bond, which can form upon the sp^3 rehybridization of the C atoms, increases in the order of Co < Pd < Rh < Pt < Ir < Ru if we take the bond enthalpies of gas-phase metal monocarbide molecules as an indication [65, 66]. The metal-C bond strength on the graphene moiré is, however, not invariant, but is a function of position on the moiré surface. Feibelman explained the critical size at which Ir clusters on g/Ir(111) transition from one atomic layer to two atomic layers, by calculating that at a certain size (26 atoms) additional Ir atoms preferentially settle on top of the existing cluster instead of continuing to attach to the periphery of the cluster because Ir atoms are less strongly bound to C the further outside the fcc region they are located [67].

The other key parameter is the cohesive energy of the metal, or the metal–metal bond energy. The cohesive energy of bulk metals decreases in the order of Ir > Ru > Pt > Rh > Co > Pd [68]. Therefore, metals such as Co and Pd, which have weak metal–C bonds tend to aggregate into larger clusters, while metals with stronger metal–C bonds tend to form smaller, more dispersed clusters on g/Ru(0001). The cohesive energy of a metal cluster, however, also depends on the size of the cluster. It has been demonstrated for metal clusters that the cohesive energy decreases with decreasing size [69–71], so one cannot assume the cohesive energies of the small clusters that initially form on the graphene moiré to be the same as those of the bulk metals.

The bonding between clusters and graphene is further affected by the substrate metal on which the graphene is grown. For instance, Pt clusters have been grown on both g/Ru(0001) and g/Ir(111), but the resulting clusters are qualitatively different. N'Diaye et al. [30] have shown that depositing submonolayer amounts of Pt results in dense cluster formation exclusively in the hcp regions on g/Ir(111), and around 0.1 ML and below most of the Pt clusters are 2D, mono-atomic in height. In contrast, Pan et al. have reported that Pt clusters predominantly nucleate in the fcc regions on g/Ru(0001) and form a sparse cluster lattice [63], and 1–3 atomic layer high Pt clusters are found to coexist at the lowest Pt coverage studied (0.02 ML). While there is clear evidence that graphene interacts much more strongly with Ru(0001) than with Ir(111), it is unclear through what mechanisms this stronger interaction affects the clustering behavior of metals deposited on the graphene moiré.

To provide a quantitatively correct explanation of cluster nucleation and growth behavior for a given metal on a graphene moiré surface and to verify the validity of the argument based on metal–C and metal–metal interactions would therefore require a significant amount of additional information about the size-dependent surface bond energies which it is difficult to obtain experimentally. Furthermore, since the kinetics of surface mass transport has not been taken into account, the effect of the deposition temperature cannot be fully understood, but we know it influences the nucleation site and the number density and size distribution of the clusters [30, 39]. Although the exact nucleation and growth mechanisms are still unclear, what is known about cluster formation on g/Ru(0001), g/Ir(111), and g/Rh(111) shows that metal-supported graphene moiré has the potential of being a viable template for designing cluster materials with tunable properties.

18.4 Two-dimensional Au Islands on g/Ru(0001) and its Catalytic Activity

As mentioned earlier, the smallest Au clusters are observed to nucleate in the fcc regions, the same as the other metals (not shown). However, beginning at slightly higher coverage Au exclusively forms a few large islands that are several nm in diameter up to approximately 1 ML equivalent of Au dosage (Figure 18.7), at room temperature [28, 72]. This is different from Au deposited on g/Ir(111), which forms large 3D clusters [30]. All the Au clusters on g/Ru(0001) display a moiré pattern that matches the periodicity of the underlying graphene moiré (Figure 18.7).

Line scan has been performed on different surface motifs for the 0.25 ML of Au deposited (Figure 18.8). The step height of graphene on Ru(0001) is measured to be 2.1 Å, which corresponds to the height of a mono-atomic step on Ru(0001) [6]. The height of a typical Au

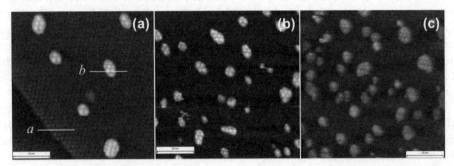

Figure 18.7 STM images (100 × 100 nm^2, $V_{sample} = +1.0$ V, $I_{tunneling} = 0.1$ nA) of (a) 0.25 ML, (b) 0.5 ML, (c) 0.75 ML of Au deposited on g/Ru(0001) at room temperature. The lines and labels in (a) indicate the positions of the line scan shown in Figure 18.8. (a), (c) Reprinted with permission from [72] © 2011 Elsevier

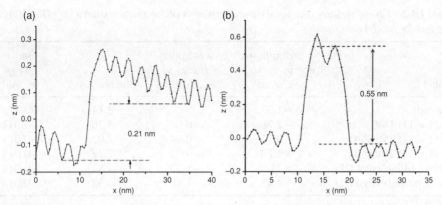

Figure 18.8 Two line scan profiles as indicated in Figure 18.7(a): (a) across a Ru step; (b) across an Au island. Reprinted with permission from [28] © The Royal Society of Chemistry

island is measured to be 5.5 Å, which corresponds to twice of the height of a mono-atomic step on Au(111) (2.5 Å) and suggests a bilayer structure. No bias-dependent change in the image or the apparent height of the Au islands is observed when the tip voltage is varied from -1.5 V to $+1.5$ V, indicating that the measured height is the geometric height of the Au islands and that the Au is geometrically corrugated just like the graphene moiré.

Au on g/Ru(0001) thus exhibits a growth mode that is very different from the other metals. The flatness of the Au clusters and the conformation to the graphene moiré suggest that these are flexible 2D islands. Two-dimensional Au structures reported in the existing literature are stabilized by strong interaction with the substrates [73–76]. Au forms epitaxial overlayers because it has a relatively low cohesive energy and therefore wets reactive metal and oxide surfaces. Our STM and Auger analyses indicate no defects or contaminants such as oxygen on the graphene that could serve as anchor sites for Au. This means that the Au–g/Ru(0001) interaction is intrinsically strong and at least on the same order of magnitude as Au–Au interaction, in order to stabilize the 2D islands at room temperature. This is, however, inconsistent with the fact that Au–C bonds are generally quite weak, so Au deposited on graphite and other graphitic carbon surfaces forms large 3D particles [77]. Indeed the calculated adsorption energy of an Au adatom on g/Ru(0001) at the center of the fcc region is -1.57 eV, significantly weaker than the bulk cohesive energy of Au, 2.99 eV (calculated in GGA-PBE; see Table 18.2). The peculiarities in Au clustering behavior on g/Ru(0001) are therefore the subject of ongoing research.

Although the STM-measured height of 5.5 Å suggests a bilayer structure, the Au–graphene distance is not necessarily the same as the Au–Au interlayer distance, in which case the 5.5 Å height might need to be interpreted differently. To determine if a bilayer structure is consistent with the STM measurements, and to shed light on the bonding mechanism between the Au islands and g/Ru(0001), we have performed DFT calculations for several model flat Au structures on g/Ru(0001).

It is worth noting that the Au–Au distance in Au nanostructures can be considerably different from that in bulk Au. The GGA-PBE optimized in-plane Au–Au distance is 2.76 and 2.78 Å for a freestanding monolayer and bilayer of Au, respectively, which are

Table 18.2 Characteristics of several different models of Au nanostructures on g/Ru(0001). See text for model descriptions

model	adsorption energy vs $Au_{(g)}^{\dagger}$ (eV/Au)	adsorption energy vs $Au_L{}^*$(eV/Au)	height[#] (Å)	Bader charge (e^-/Au)
adatom	−1.57	–	2.14	−0.172
A (11 × 11) 1ML	−2.76	−0.08	3.96	−0.011
B (11 × 11) 2ML-AA	−2.77	−0.04	7.02	−0.011
C (11 × 11) 2ML-AB	−2.78	−0.03	7.27	−0.011
A (in optB86b-vdW-DF)	−5.00	−1.47	3.52	−0.016
C (in optB86b-vdW-DF)	−4.45	−1.40	6.58	−0.014

Self-consistent DFT calculations were performed using VASP [55, 56] in GGA-PBE except where noted. Core electrons were described by the projector augmented wave method. Kinetic energy cutoff was 400 eV. Surface Brillouin zone was sampled at the Γ point only. Geometric optimization was converged to 0.05 eV/Å for each relaxed degree of freedom.
[†] Adsorption energy versus free gas-phase Au atom averaged by total number of Au atoms. For comparison, the bulk lattice constant and cohesive energy are 4.175 Å and 2.99 eV in GGA-PBE, and 4.121 Å and 3.60 eV in optB86b-vdW-DF.
*Adsorption energy versus freestanding Au monolayer or bilayer averaged over 121 contacting Au atoms.
[#] Height between highest Au atom in each model and highest C atom on g/Ru(0001) without Au, except for the adatom for which the optimized Au–C bond length is indicated.

contracted from the calculated bulk value of 2.95 Å due to the under-coordination of the Au atoms, a quantum size effect with respect to the thickness of the Au. This distance is close to the calculated in-plane lattice constant of Ru of 2.73 Å. Thus, nano-sized Au in the freestanding limit may have a similar surface packing density to Ru(0001).

We first investigated a hexagonal close-packed *monolayer* of Au on g/Ru(0001): Model A consists of (11 × 11)-Au on g/Ru(0001), for a total of 121 Au atoms per (11 × 11)-g/Ru(0001) unit cell. The energy-minimized structure of Model A is shown in Figure 18.9.

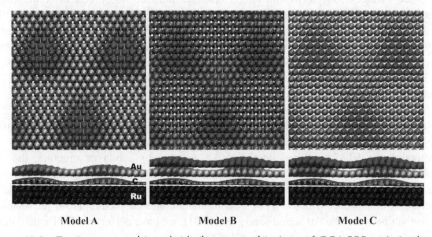

Figure 18.9 Top (upper panels) and side (lower panels) views of GGA-PBE-optimized structures for: Model A, (11 × 11)-Au monolayer; Model B, (11 × 11)-Au bilayer (AA-stacking); Model C, (11 × 11)-Au bilayer (AB-stacking), adsorbed on g/Ru(0001). Au atoms are gray-coded according to height as an aid to the eye

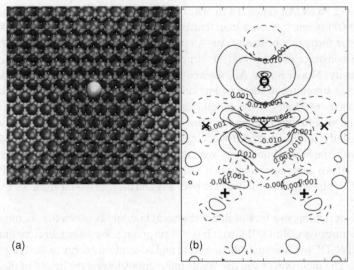

Figure 18.10 (a) Snapshot and (b) charge density difference contours ($\Delta\rho = \rho_{total} - \rho_{g/Ru} - \rho_{Au}$) for an isolated Au adatom on g/Ru(0001). In (b), the plane is cut through the Au adatom along the [01$\bar{1}$0] direction of Ru(0001). The contours included are ± 0.1, ± 0.01, and ± 0.001 e/Å^3. Solid and dashed lines represent positive (density gain) and negative (density loss) values, respectively. "O" indicates Au atoms, "×" indicate C atoms, and "+" indicate Ru atoms. Reprinted with permission from [28] © The Royal Society of Chemistry

We also optimized (10 × 10) and (12 × 12)-Au monolayer on the (11 × 11)-g/Ru(0001) surface. The (10 × 10) structure relaxes to open up a large hole in the mound region, whereas many Au atoms are squeezed out of the monolayer in the mound region in the (12 × 12) structure. The average adsorption energy is −2.53 and −2.65 eV/Au for the (10 × 10) and (12 × 12)-Au respectively, both of which are less stable than the (11 × 11)-Au (−2.76 eV/Au). These findings suggest that (11 × 11) is the optimal packing density for Au monolayer on the (11 × 11) g/Ru(0001) surface. The optimized Au monolayer conforms to the underlying corrugation of the graphene with a maximum height of 3.96 Å, which is less than the measured height of 5.5 Å for the Au islands, suggesting that the observed 2D Au is unlikely to be monoatomic in height.

It is instructive to compare the energy of an isolated Au adatom adsorbed on g/Ru(0001) (Figure 18.10a) to that of an Au atom in the Au monolayer. A single Au adatom prefers to adsorb atop a C atom that is located at the center of the fcc region above a surface hcp site, with an Au-C distance of 2.14 Å and an adsorption energy of −1.57 eV, which is in close agreement with a previously reported atomic Au adsorption energy on g/Ru(0001) [49]. A charge density difference plot (Figure 18.10b) shows evidence of bonding between the Au adatom and Ru through the graphene. The adsorption of the Au monolayer is qualitatively different. First, the Au monolayer (Model A) is located at nearly 4 Å above the graphene, considerably higher than the bonding distance of 2.14 Å for an individual Au adatom. Second, although the adsorption energy for the monolayer, with respect to a freestanding Au monolayer, is merely −0.08 eV/Au, the average energy per Au atom with respect to a free

Au atom is -2.76 eV/Au (Table 18.2). Thus the chemical bond between the Au monolayer and g/Ru(0001) is much weaker than that between an Au adatom and g/Ru(0001), in part because of the formation of strong Au–Au bonds, even in just a monolayer of Au.

Next we construct two Au *bilayer* structures on g/Ru(0001) based on the (11×11) packing density: Model B is an AA-stacked bilayer, and Model C is an AB-stacked bilayer. The optimized structures of these two models are shown in Figure 18.9. In both models as in the monolayer of Model A, the Au bilayers conform to the corrugation of the underlying graphene and should thus display the same moiré as the graphene does. The two bilayers are energetically similar, with AB-stacking being slightly more stable (Table 18.2). Like the monolayer, the adsorption energy with respect to a freestanding Au bilayer is miniscule per interfacial Au atom compared to an isolated Au adatom adsorbed on g/Ru(0001). The calculated heights of the two bilayers are over 7 Å and diverge too far from the experimental value of 5.5 Å.

The evidence points to a lack of direct chemical bonding between the Au monolayer and bilayer structures on g/Ru(0001), which is not surprising. We have therefore employed the optB86b-vdW–DF to recalculate the structure and adsorption energy of the bilayer structure of Model C. The inclusion of van der Waals interactions lowers the height of the Au bilayer to 6.58 Å, in closer agreement with experiment than the GGA-PBE results. Furthermore, the average energy of an Au atom in the bilayer is calculated to be more stable than that in bulk Au (Table 18.2). While the accuracy of the vdW–DF functionals [51, 78, 79] remains to be established for different classes of material, our results suggest that the 2D Au islands on g/Ru(0001) are likely stabilized by van der Waals interactions to a significant extent, owing to the extended size and the flexible nature of the Au layers, and the smoothness of the graphene moiré surface. This stands in complete contrast to Au stabilized on reactive surfaces primarily due to chemical bonding and charge transfer [74, 75, 80]. The vdW–DF results suggest that Au bilayers are probable models for the observed Au islands on g/Ru(0001) but do not establish how the bilayers form to the exclusion of monolayers and thicker layers. It can be seen in Table 18.2 that the Au bilayer has no particular energetic advantage over the Au monolayer or vice versa. We will begin to address the kinetic reasons for the nucleation of the Au bilayers in forthcoming publications.

It is well known that nano-sized Au, including bilayer structures, is active for CO oxidation at low to moderate temperatures [74, 82, 83]. We have therefore investigated the adsorption of CO on the 2D Au islands on g/Ru(0001) using polarization modulation infrared reflection absorption spectroscopy (PM-IRAS) (Figure 18.11). When the Ru(0001) surface is completely covered with graphene, the peak associated with CO absorbed on Ru(0001) at 2063 cm^{-1} [84] disappears, indicating that the Ru surface has been completely passivated. However, after 0.5 ml of Au is deposited on the g/Ru(0001) surface, a CO stretch reappears at 2095 cm^{-1}. A survey of literature suggests that the CO stretch on neutral Au is around 2120 cm^{-1}, whereas it blue-shifts on Au that is electron poor and red-shifts on Au that is electron rich [85]. 2095 cm^{-1} for CO absorbed on the 2D Au islands therefore suggests that the Au is negatively charged. This observation is in agreement with Bader charge analyses [86–88], according to which the Au bilayers are negatively charged (Table 18.2).

We investigated the catalytic activity of CO absorbed on the 2D Au islands by titrating the system with O_2 at 85 K and monitoring the CO stretching peak via PM-IRAS (Figure 18.12). Upon increasing O_2 exposure, the intensity of the CO stretch is observed to decrease,

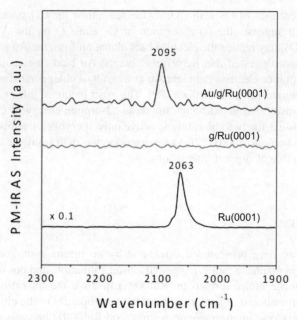

Figure 18.11 PM-IRAS spectra after a CO exposure of 15 L on Ru(0001), g/Ru(0001), and 0.5 ml Au/g/Ru(0001) at 85 K. Reprinted with permission from [72] © 2011 Elsevier

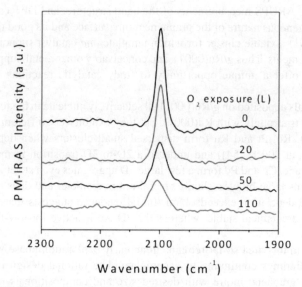

Figure 18.12 PM-IRAS spectra after each indicated O_2 exposure on 1.0 ml Au/g/Ru(0001) surface pre-covered by CO at 85 K. Reprinted with permission from [72] © 2011 Elsevier

consistent with reactions of O_2 with CO. At the same time the CO stretch broadens and blue-shifts, which suggests the co-absorption of O_2 and CO on the Au since the co-adsorption with O_2 may reduce the electron back-donation from the Au to the $2\pi^*$ orbital of CO. It has been suggested that negatively charged Au binds molecular oxygen more strongly than neutral or electron-deficient Au as a result of charge transfer from the Au to O_2 to form a superoxo-like intermediate [89]. The exact nature of the surface O_2 and CO species is under investigation. Although the weak adsorption energy of CO suggests that the 2D Au islands are likely to be catalytic active only at cryogenic temperatures, the 2D Au islands formed on g/Ru(0001) offer a unique opportunity to study the catalytic activity of nano-sized Au free of support interactions.

18.5 Summary

In this chapter we have presented a brief, and by no means complete, survey of the current research in graphene moiré-templated cluster formation, and our own contribution in utilizing graphene moiré formed on Ru(0001) to drive the formation of clusters of several catalytic metals and in exploring the catalytic properties of the clusters. The moiré superstructure develops when graphene is grown on Ru(0001) because of the mismatch between the Ru and graphene lattice constants. The moiré has periodically alternating regions of strong and weak interactions with the Ru(0001) substrate. The C atoms undergo *sp*^{3} rehybridization in the strongly interacting regions (fcc and hcp regions), which results in chemically different regions on the same graphene surface, a phenomenon that is evidenced by STM images, ARUPS measurements of the band structure, and DFT calculations. The periodic, heterogeneous nature of the graphene moiré surface and its good thermal stability makes g/Ru(0001) a viable cluster formation template *and* catalyst support that is free of other reactive elements. Thus g/Ru(0001), and conceivably other metal-supported graphene moiré surfaces, offer a unique opportunity to study catalytic reactions free of support interactions.

Different metals deposited on g/Ru(0001) all selectively nucleate in the fcc regions where graphene interacts strongly with Ru(0001), but they exhibit very different nucleation and growth behavior. Rh, Pt and Ru form dispersed small clusters when deposited at room temperature (Ru at 200 K [54]) and undergo a 2D-to-3D transition in morphology with increasing coverage. Co and Pd form a few large 3D aggregates even at very low coverages, whereas Au forms large bilayer islands several nm in diameter and does not transition to 3D growth until deposition exceeds 1 ml. Initial evidence suggests that the Rh clusters are thermally and chemical stable, whereas the 2D Au is active for low-temperature CO oxidation.

The research in this area so far remains in an early exploration phase, with interesting and surprising findings continuing to accumulate. The rational design of metal cluster superlattices on graphene moiré with desired size and composition is not yet possible because of a lack of quantitative understanding for the nucleation and growth mechanisms. With continued research, however, we are optimistic that the necessary predictive abilities will be gained and that the full potential of graphene moiré-templated clusters for model catalytic studies will be realized.

Acknowledgments

This work was supported by the Center for Atomic Level Catalyst Design, an Energy Frontier Research Center funded by the US Department of Energy (US-DOE), Office of Science, Office of Basic Energy Sciences under Award Number DE-SC0001058. The computational work used resources of the National Energy Research Scientific Computing Center, which is supported by US-DOE Office of Science under Contract DE-AC02-05CH11231; and of the Oak Ridge Leadership Computing Facility, which is supported by US-DOE Office of Science under Contract DE-AC05-00OR22725, and was performed at Center for Nanophase Materials Sciences, which is sponsored at Oak Ridge National Laboratory by the Scientific User Facilities Division, Office of Basic Energy Sciences, US Department of Energy.

References

[1] S. Hagstrom, H. B. Lyon and G. A. Somorjai, Surface structures on clean platinum(100) surface, *Phys. Rev. Lett.*, **15**, 491–493 (1965).
[2] H. B. Lyon and G. A. Somorjai, Low energy electron-diffraction study of clean (100) (111) and (110) faces of platinum, *J. Chem. Phys.*, **46**, 2539–2550 (1967).
[3] A. E. Morgan and G. A. Somorjai, Low energy electron diffraction studies of gas adsorption on platinum (100) single crystal surface, *Surf. Sci.*, **12**, 405–425 (1968).
[4] J. W. May, Platinum surface LEED rings, *Surf. Sci.*, **17**, 267–270 (1969).
[5] M.-C. Wu, Q. Xu and D. W. Goodman, Investigations of graphitic overlayers formed from methane decomposition on Ru(0001) and Ru(1120) catalysts with scanning tunneling microscopy and high-resolution electron energy loss spectroscopy, *J. Phys. Chem.*, **98**, 5104–5110 (1994).
[6] S. Marchini, S. Gunther and J. Wintterlin, Scanning tunneling microscopy of graphene on Ru(0001), *Phys. Rev. B*, **76**, 075429 (2007).
[7] P. W. Sutter, J. I. Flege and E. A. Sutter, Epitaxial graphene on ruthenium, *Nat. Mater.*, **7**, 406–411 (2008).
[8] A. T. N'Diaye, J. Coraux, T. N. Plasa, C. Busse and T. Michely, Structure of epitaxial graphene on Ir(111), *New J. Phys.*, **10**, 043033 (2008).
[9] E. Loginova, N. C. Bartelt, P. J. Feibelman and K. F. McCarty, Factors influencing graphene growth on metal surfaces, *New J. Phys.*, **11**, 063046 (2009).
[10] M. Sicot, S. Bouvron, O. Zander, U. Rudiger, Y. S. Dedkov and M. Fonin, Nucleation and growth of nickel nanoclusters on graphene moiré on Rh(111), *Appl. Phys. Lett.*, **96**, 093115 (2010).
[11] T. A. Land, T. Michely, R. J. Behm, J. C. Hemminger and G. Comsa, STM investigation of single layer graphite structures produced on Pt(111) by hydrocarbon decomposition, *Surf. Sci.*, **264**, 261–270 (1992).
[12] S.-Y. Kwon, C. V. Ciobanu, V. Petrova, V. B. Shenoy, J. Bareño, V. Gambin, I. Petrov and S. Kodambaka, Growth of semiconducting graphene on palladium, *Nano Lett.*, **9**, 3985–3990 (2009).
[13] J.-H. Gao, N. Ishida, I. Scott and D. Fujita, Controllable growth of single-layer graphene on a Pd(111) substrate, *Carbon*, **50**, 1674–1680 (2012).

[14] D. Eom, D. Prezzi, K. T. Rim, H. Zhou, M. Lefenfeld, S. Xiao, et al., Structure and electronic properties of graphene nanoislands on Co(0001), *Nano Lett.*, **9**, 2844–2848 (2009).

[15] A. Varykhalov and O. Rader, Graphene grown on Co(0001) films and islands: Electronic structure and its precise magnetization dependence, *Phys. Rev. B*, **80**, 035437 (2009).

[16] Y. Gamo, A. Nagashima, M. Wakabayashi, M. Terai and C. Oshima, Atomic structure of monolayer graphite formed on Ni(111), *Surf. Sci.*, **374**, 61–64 (1997).

[17] H. Kawanowa, H. Ozawa, T. Yazaki, Y. Gotoh and R. Souda, Structure analysis of monolayer graphite on Ni(111) surface by Li^+-impact collision ion scattering spectroscopy, *Jpn. J. Appl. Phys. 1*, **41**, 6149–6152 (2002).

[18] R. Addou, A. Dahal and M. Batzill, Graphene on ordered Ni-alloy surfaces formed by metal (Sn, Al) intercalation between graphene/Ni(111), *Surf. Sci.*, **606**, 1108–1112 (2012).

[19] X. Chen, S. Liu, L. Liu, X. Liu, X. Liu and L. Wang, Growth of triangle-shape graphene on Cu(111) surface, *Appl. Phys. Lett.*, **100**, 163106–163103 (2012).

[20] W. Kim, K. Yoo, E. K. Seo, S. J. Kim and C. Hwang, Scanning tunneling microscopy study on a graphene layer grown on a single-crystal Cu(111) surface by using chemical vapor deposition, *J. Korean Phys. Soc.*, **59**, 71–74 (2011).

[21] K. S. Novoselov, A. K. Geim, S. V. Morozov, D. Jiang, Y. Zhang, S. V. Dubonos, et al., Electric field effect in atomically thin carbon films, *Science*, **306**, 666–669 (2004).

[22] A. K. Geim, Graphene: Status and prospects, *Science* **324**, 1530–1534 (2009).

[23] A. L. Vázquez de Parga, F. Calleja, B. Borca, M. C. G. Passeggi, J. J. Hinarejos, F. Guinea and R. Miranda, Periodically rippled graphene: Growth and spatially resolved electronic structure, *Phys. Rev. Lett.*, **100**, 056807 (2008).

[24] J. Coraux, A. T. N'Diaye, C. Busse and T. Michely, Structural coherency of graphene on Ir(111), *Nano Lett.*, **8**, 565–570 (2008).

[25] J. Coraux, A. T. N'Diaye, M. Engler, C. Busse, D. Wall, N. Buckanie, et al., Growth of graphene on Ir(111), *New J. Phys.*, **11**, 023006 (2009).

[26] J. Wintterlin and M. L. Bocquet, Graphene on metal surfaces, *Surf. Sci.*, **603**, 1841–1852 (2009).

[27] K. Katsiev, Y. Losovyj, Z. Zhou, E. Vescovo, L. Liu, P. A. Dowben and D. W. Goodman, Graphene on Ru(0001): Evidence for two graphene band structures, *Phys. Rev. B*, **85**, 195405 (2012).

[28] Y. Xu, L. Semidey-Flecha, L. Liu, Z. Zhou and D. W. Goodman, Exploring the structure and chemical activity of 2-D gold islands on graphene moiré/Ru(0001), *Faraday Discuss.*, **152**, 267–276 (2011).

[29] A. T. N'Diaye, S. Bleikamp, P. J. Feibelman and T. Michely, Two-dimensional Ir cluster lattice on a graphene moire on Ir(111), *Phys. Rev. Lett.*, **97**, 215501 (2006).

[30] A. T. N'Diaye, T. Gerber, C. Busse, J. Myslivecek, J. Coraux and T. Michely, A versatile fabrication method for cluster superlattices, *New J. Phys.*, **11**, 103045 (2009).

[31] M. Boudart, Catalysis by supported metals, *Adv. Catal.*, **20**, 153–166 (1969).

[32] C. C. Chusuei, X. Lai, K. Luo and D. W. Goodman, Modeling heterogeneous catalysts: metal clusters on planar oxide supports, *Top. Catal.*, **14**, 71–83 (2001).

[33] Y. F. Han, D. Kumar and D. W. Goodman, Particle size effects in vinyl acetate synthesis over Pd/SiO_2, *J. Catal.*, **230**, 353–358 (2005).

[34] K. M. Bratlie, H. Lee, K. Komvopoulos, P. D. Yang and G. A. Somorjai, Platinum nanoparticle shape effects on benzene hydrogenation selectivity, *Nano Lett.*, **7**, 3097–3101 (2007).

[35] S. H. Joo, J. Y. Park, J. R. Renzas, D. R. Butcher, W. Y. Huang and G. A. Somorjai, Size effect of ruthenium nanoparticles in catalytic carbon monoxide oxidation, *Nano Lett.*, **10**, 2709–2713 (2010).

[36] P. Mars and D. W. Van Krevelen, Oxidations carried out by means of vanadium oxide catalysts, *Chem. Eng. Sci. Spec. Suppl.*, **3**, 41–59 (1954).

[37] Y. Pan, H. G. Zhang, D. X. Shi, J. T. Sun, S. X. Du, F. Liu and H. J. Gao, Highly ordered, millimeter-scale, continuous, single-crystalline graphene monolayer formed on Ru(0001), *Adv. Mater.*, **21**, 2777–2780 (2009).

[38] Z. Zhou, F. Gao and D. W. Goodman, Deposition of metal clusters on single-layer graphene/Ru(0001): Factors that govern cluster growth *Surf. Sci.*, **604**, L31–L38 (2010).

[39] K. Donner and P. Jakob, Structural properties and site specific interactions of Pt with the graphene/Ru(0001) moiré overlayer, *J. Chem. Phys.*, **131**, 164701 (2009).

[40] K. F. McCarty, P. J. Feibelman, E. Loginova and N. C. Bartelt, Kinetics and thermodynamics of carbon segregation and graphene growth on Ru(0001), *Carbon*, **47**, 1806–1813 (2009).

[41] S. Saadi, F. Abild-Pedersen, S. Helveg, J. Sehested, B. Hinnemann, C. C. Appel and J. K. Nørskov, On the role of metal step-edges in graphene growth, *J. Phys. Chem. C*, **114**, 11221–11227 (2010).

[42] B. Wang, M. L. Bocquet, S. Marchini, S. Gunther and J. Wintterlin, Chemical origin of a graphene moiré overlayer on Ru(0001), *Phys. Chem. Chem. Phys.*, **10**, 3530–3534 (2008).

[43] Y. Pan, D. X. Shi and H. J. Gao, Formation of graphene on Ru(0001) surface, *Chinese Phys.*, **16**, 3151–3153 (2007).

[44] D. E. Jiang, M. H. Du and S. Dai, First principles study of the graphene/Ru(0001) interface, *J. Chem. Phys.*, **130**, 074705 (2009).

[45] D. Martoccia, P. R. Willmott, T. Brugger, M. Bjorck, S. Gunther, C. M. Schleputz, et al., Graphene on Ru(0001): A 25 × 25 supercell, *Phys. Rev. Lett.*, **101**, 126102 (2008).

[46] W. Moritz, B. Wang, M. L. Bocquet, T. Brugger, T. Greber, J. Wintterlin and S. Gunther, Structure determination of the coincidence phase of graphene on Ru(0001), *Phys. Rev. Lett.*, **104**, 136102 (2010).

[47] D. Martoccia, M. Bjorck, C. M. Schleputz, T. Brugger, S. A. Pauli, B. D. Patterson, et al., Graphene on Ru(0001): a corrugated and chiral structure, *New J. Phys.*, **12**, 043028 (2010).

[48] B. Borca, S. Barja, M. Garnica, M. Minniti, A. Politano, J. M. Rodriguez-Garcia, et al., Electronic and geometric corrugation of periodically rippled, self-nanostructured graphene epitaxially grown on Ru(0001), *New J. Phys.*, **12**, 093018 (2010).

[49] B. Wang, S. Gunther, J. Wintterlin and M. L. Bocquet, Periodicity, work function and reactivity of graphene on Ru(0001) from first principles, *New J. Phys.*, **12**, 043041 (2010).

[50] S. Grimme, Semiempirical GGA-type density functional constructed with a long-range dispersion correction, *J. Comput. Chem.*, **27**, 1787 (2006).

[51] J. Klimeš, D. R. Bowler and A. Michaelides, Van der Waals density functionals applied to solids, *Phys. Rev. B*, **83**, 195131 (2011).

[52] A. B. Preobrajenski, M. L. Ng, A. S. Vinogradov and N. Martensson, Controlling graphene corrugation on lattice-mismatched substrates, *Phys. Rev. B*, **78**, 073401 (2008).

[53] D. Stradi, S. Barja, C. Díaz, M. Garnica, B. Borca, J. J. Hinarejos, et al., Role of Dispersion forces in the structure of graphene monolayers on Ru surfaces, *Phys. Rev. Lett.*, **106**, 186102 (2011).

[54] E. Sutter, P. Albrecht, B. Wang, M. L. Bocquet, L. J. Wu, Y. M. Zhu and P. Sutter, Arrays of Ru nanoclusters with narrow size distribution templated by monolayer graphene on Ru, *Surf. Sci.*, **605**, 1676–1684 (2011).

[55] G. Kresse and J. Furthmuller, Efficiency of ab-initio total energy calculations for metals and semiconductors using a plane-wave basis set, *Comp. Mater. Sci.*, **6**, 15–50 (1996).

[56] G. Kresse and J. Furthmuller, Efficient iterative schemes for ab initio total-energy calculations using a plane-wave basis set, *Phys. Rev. B*, **54**, 11169–11186 (1996).

[57] D. Tománek, S. G. Louie, H. J. Mamin, D. W. Abraham, R. E. Thomson, E. Ganz and J. Clarke, Theory and observation of highly asymmetric atomic structure in scanning-tunneling-microscopy images of graphite, *Phys. Rev. B*, **35**, 7790–7793 (1987).

[58] T. Brugger, S. Gunther, B. Wang, J. H. Dil, M. L. Bocquet, J. Osterwalder, et al., Comparison of electronic structure and template function of single-layer graphene and a hexagonal boron nitride nanomesh on Ru(0001), *Phys. Rev. B*, **79**, 045407 (2009).

[59] I. Pletikosic, M. Kralj, P. Pervan, R. Brako, J. Coraux, A. T. N'Diaye, et al., Dirac cones and minigaps for graphene on Ir(111), *Phys. Rev. Lett.*, **102**, 056808 (2009).

[60] E. E. Krasovskii and E. V. Chulkov, Rashba polarization of bulk continuum states, *Phys. Rev. B*, **83**, 155401 (2011).

[61] H. Zhang, Q. Fu, Y. Cui, D. L. Tan and X. H. Bao, Growth mechanism of graphene on Ru(0001) and O2 adsorption on the graphene/Ru(0001) surface, *J. Phys. Chem. C*, **113**, 8296–8301 (2009).

[62] B. Borca, F. Calleja, J. J. Hinarejos, A. L. V. de Parga and R. Miranda, Reactivity of periodically rippled graphene grown on Ru(0001), *J. Phys.-Condens. Mat.*, **21**, 134002 (2009).

[63] Y. Pan, M. Gao, L. Huang, F. Liu and H. J. Gao, Directed self-assembly of monodispersed platinum nanoclusters on graphene moiré template, *Appl. Phys. Lett.*, **95**, 093106 (2009).

[64] H. Zhang, Q. Fu, Y. Cui, D. L. Tan and X. H. Bao, Fabrication of metal nanoclusters on graphene grown on Ru(0001), *Chinese Sci. Bull.*, **54**, 2446–2450 (2009).

[65] J. A. Martinho Simoes and J. L. Beauchamp, Transition-metal hydrogen and metal-carbon bond strengths - The keys to catalysis, *Chem. Rev.*, **90**, 629–688 (1990).

[66] D. Tzeli and A. Mavridis, Electronic structure of cobalt carbide, CoC, *J. Phys. Chem. A*, **110**, 8952–8962 (2006).

[67] P. J. Feibelman, Onset of three-dimensional Ir islands on a graphene/Ir(111) template, *Phys. Rev. B*, **80**, 085412 (2009).

[68] C. Kittel, *Introduction to Solid State Physics*, John Wiley & Sons, Inc., New York (1996).

[69] H. K. Kim, S. H. Huh, J. W. Park, J. W. Jeong and G. H. Lee, The cluster size dependence of thermal stabilities of both molybdenum and tungsten nanoclusters, *Chem. Phys. Lett.*, **354**, 165–172 (2002).

[70] C. T. Campbell, S. C. Parker and D. E. Starr, The effect of size-dependent nanoparticle energetics on catalyst sintering, *Science*, **298**, 811–814 (2002).

[71] Y. Xu, W. A. Shelton and W. F. Schneider, Effect of particle size on the oxidizability of platinum clusters, *J. Phys. Chem. A*, **110**, 5839–5846 (2006).

[72] L. Liu, Z. Zhou, Q. Guo, Z. Yan, Y. Yao and D. W. Goodman, The 2-D growth of gold on single-layer graphene/Ru(0001): Enhancement of CO adsorption, *Surf. Sci.*, **605**, L47–L50 (2011).

[73] R. Q. Hwang, J. Schroder, C. Gunther and R. J. Behm, Fractal growth of 2-dimensional islands – Au on Ru(0001), *Phys. Rev. Lett.*, **67**, 3279–3282 (1991).

[74] M. S. Chen and D. W. Goodman, The structure of catalytically active gold on titania, *Science*, **306**, 252–255 (2004).

[75] Y. Zhang, L. Giordano and G. Pacchioni, Gold nanostructures on TiO_x/Mo(112) thin films, *J. Phys. Chem. C*, **112**, 191–200 (2008).

[76] X. Lin, N. Nilius, H. J. Freund, M. Walter, P. Frondelius, K. Honkala and H. Hakkinen, Quantum well states in two-dimensional gold clusters on MgO thin films, *Phys. Rev. Lett.*, **102**, 206801 (2009).

[77] Z. Ma, C. D. Liang, S. H. Overbury and S. Dai, Gold nanoparticles on electroless-deposition-derived MnOx/C: Synthesis, characterization, and catalytic CO oxidation, *J. Catal.*, **252**, 119–126 (2007).

[78] M. Dion, H. Rydberg, E. Schroder, D. C. Langreth and B. I. Lundqvist, Van der Waals density functional for general geometries, *Phys. Rev. Lett.*, **92**, 246401 (2004).

[79] J. Klimeš, R. B. David and M. Angelos, Chemical accuracy for the van der Waals density functional, *J. Phys.-Condens. Mat.*, **22**, 022201 (2010).

[80] D. Ricci, A. Bongiorno, G. Pacchioni and U. Landman, Bonding trends and dimensionality crossover of gold nanoclusters on metal-supported MgO thin films, *Phys. Rev. Lett.*, **97**, 036106 (2006).

[81] L. Semidey-Flecha, D. Teng, B. F. Habenicht, D. S. Sholl and Y. Xu, Adsorption and diffusion of the Rh and Au adatom on graphene moiré/Ru(0001), *J. Chem. Phys.*, **138**, 184710 (2013).

[82] M. Haruta, T. Kobayashi, H. Sano and N. Yamada, Novel gold catalysts for the oxidation of carbon-monoxide at a temperature far below 0°C, *Chem. Lett.*, 405–408 (1987).

[83] M. Valden, X. Lai and D. W. Goodman, Onset of catalytic activity of gold clusters on titania with the appearance of nonmetallic properties, *Science*, **281**, 1647–1650 (1998).

[84] H. Pfnür, D. Menzel, F. M. Hoffmann, A. Ortega and A. M. Bradshaw, High resolution vibrational spectroscopy of CO on Ru(001): The importance of lateral interactions, *Surf. Sci.*, **93**, 431–452 (1980).

[85] M. S. Chen and D. W. Goodman, Catalytically active gold: From nanoparticles to ultrathin films, *Accounts Chem. Res.*, **39**, 739–746 (2006).

[86] R. F. W. Bader, *Atoms in Molecules – A Quantum Theory*, Oxford University Press, Oxford (1990).

[87] G. Henkelman, A. Arnaldsson and H. Jónsson, A fast and robust algorithm for Bader decomposition of charge density, *Comput. Mat. Sci.*, **36**, 254 (2006).
[88] W. Tang, E. Sanville and G. Henkelman, A grid-based Bader analysis algorithm without lattice bias, *J. Phys.: Condens. Matter*, **21**, 084204 (2009).
[89] U. Landman, B. Yoon, C. Zhang, U. Heiz and M. Arenz, Factors in gold nanocatalysis: Oxidation of CO in the non-scalable size regime, *Top. Catal.*, **44**, 145 (2007).

Index

Note: Page numbers in *italics* refer to Figures; those in **bold** to Tables.

acenes
  experimental synthesis, 2–3, 394
  reactivity related to length, 32, *32*
  see also oligocenes
adsorption
  ambient molecules, and device performance, 209–10
  of atomic oxygen on carbon nanotubes, 82, *86*, 86–8
  defects and edges as binding sites, 224
  on graphene
    inert (closed-shell) adsorbates, 217–20
    metal adatoms and clusters, 192, 222–3, 357, 435–8, *437*
    open-shell adsorbates, 215–17
    single-stranded DNA, 224
    sites, position labeling, 239, *239*
    water and ice, **213**, 213–15, *214*
  at graphene nanoribbon surfaces, 60–63, *62*
  of hydrogen
    with applied electric field, 224–5
    atomic hydrogen, 383–6, *385*
    effects of metal dopants, 374–7
    related to interlayer distance, 373–4, *374*, 379–80
  migration barriers of adsorbates, 210
  permanganate oxidants (MnO_4^-), 88–91, *90*
  porous graphene sites (for lithium), 139–40
  quantitative modeling for gases in graphdiyne, 145, **145**
adsorption energy, E_a, 212, 213, **213**, 360, 372
alternant structures, 102, 114, 115–16
aluminum-doped graphene, CO adsorption, 222, *223*, 225
ammonia (NH_3)
  adsorption on graphene, 217–18, *218*, 222, 225
  interaction with graphene oxide, 247
  in synthesis of N-doped graphene, 196, **197**
anthracene, reactivity, 32
antidot lattices, 45, 335
anti-ferromagnetic (AFM) state, nanoribbons, 9–10, 18, 54, 326–9, *327*
armchair carbon nanotubes
  atomic oxygen adsorption, *86*, 86–7, 88
  orientation selectivity, oxygen pair unzipping, 90, 91, *91*
armchair graphene nanoribbons (AGNRs)
  aromaticity and mean bond lengths, *38*, 38–9
  band structure related to width, 37, *37*, 39, 56–8, *57*
    using tight binding calculations, 57, 326, *326*

Graphene Chemistry: Theoretical Perspectives, First Edition. Edited by De-en Jiang and Zhongfang Chen.
© 2013 John Wiley & Sons, Ltd. Published 2013 by John Wiley & Sons, Ltd.

armchair graphene nanoribbons (AGNRs)
(*Continued*)
  Clar structures and formulas, 34–7, *35, 36*
  gas adsorption at edges, 224
  structural deformation, 64–5
  surface adsorption of metal atoms, 60, 61–2
  thermoelectric performance, 337
aromaticity
  analysis methods, 36–7
  armchair graphene nanoribbons, *37,* 37–9, *38*
  aromatic bridges, phenalenyl-based compounds, 408, *409*
  Clar sextet representations, PAHs, 31–3, *32*
  concept definition, 30–31
  finite carbon nanotubes, 34
  graphene, 42–4, *44,* 102
  stabilization in zethrenes, 415–16
  zigzag graphene nanoribbons, 40–41, *41*
atomic force microscopy (AFM), 63, 80, 236
attachment limited growth (ALG), 281–2

band gaps
  absence, in graphene nanosheets, 29, 185, 233
  effect of edge reconstruction, ZGNRs, 17–18, **18,** 20
  engineering, for electronic devices, 61–3, 236, 330–31, 335
  fullerenes, their polymers and hybrids, 155, 167
  periodicity, as function of GNR width, 37, *37,* 56–8, *57*
  spin state modulation with applied electric field, 54–5, *55*
benzene, electronic structure, 2, *2,* 102, **102**
benzoperylenes, structures, 33, *33*
biotechnology materials, 224, 236
biradical compounds, stability, 408, 412–14

bisanthenes
  chemical reactions and derivatives, 403, *403, 404*
  magnetism, 119–20, *121*
  structure, 402, *402*
bond dissociation energy, ZGNR edges, 41–2, *42*
bond lengths
  change on hydrogenation to graphane, 239
  distortion with fluorination, 95
  at graphene edges, 292
  HBC (coronene-based nanographene), 394, *395*
  in nitrogen and phosphorus doped graphene, 191–2, **192**
  oscillation patterns in nanographenes
    hexangulenes, 105–7, *106,* 109–10, *110*
    triangulenes, 111, *111*
  in physisorption and chemisorption, 210
  rylenes, compared, *398,* 405
  variation in graphene nanoribbons, *35, 36, 36,* 107, *107*
boron, as graphene dopant, 185, 188–91, 222, *223*
boron nitride nanotubes
  nanojunctions and superarchitectures, *171,* 171–3, *172*
  properties compared with carbon nanotubes, 171
buckminsterfullerene (buckyballs), 153, 154, 168

calcium-doped graphene/nanoribbons, 376, *377*
carbon monoxide (CO)
  catalytic oxidation
    graphene-based catalytic materials, 356–9, *361,* 361–2, 438–40, *439*
    performance improvement methods, 359–60, *360*
    reaction mechanisms, 357, *358, 359,* 361

desorption during oxidation of
graphene, 293, 294, *295*
gas sensor materials, 222, 247
carbon nanobuds (CNBs), 165, *167*,
167–8
carbon nanoscrolls (CNS), 379, *379*
carbon nanotubes (CNTs)
Clar theory explanation of properties,
33–4, *34*
cutting to create nanoribbons, 3–4, *4*,
85–91, *86*, 92–5
effects of compression, 88, *89*
hybrid nanostructures, 163–8, *165*, *167*,
169–71, *170*
periodicity of HOMO-LUMO gaps, 39,
39
polymerization, compared with
bundles, 156–7, *157*
properties and applications, 156, 372
N-doped, for biosensors, 201
superarchitectures, 157–60, *158*, *159*,
379–80, *380*
carbon surface segregation, 427
carbonyl pairs (CP)
formation from epoxy chains, 83, *84*,
301, *301*
at tears in graphene sheet edge, 80, *81*,
301, *302*
CASSCF (complete active space
self-consistent field) calculations,
13–14, 408, 412
catalysts
for carbon monoxide oxidation, 356–62
effect of applied electric field, 360,
360
effect of mechanical strain, 359, *360*
poisoning and tolerance, 361
for combustion of nitromethane, 362–3,
363
for CVD graphene synthesis
insertion of C atoms at growing
graphene edge, 275–8, *277*
orientation determination, *278*,
278–80, *279*, *280*
selection of transition metal, 258–9,
259, 261

terraces/step edges, and nucleation
clusters, 266–71, *267*, *269*,
270
upright nanoribbon formation,
271–3, *272*
electrochemical, 200, 347–53
gold-embedded graphene for propene
epoxidation, 362, *363*
photocatalysts, 353–6
potential uses of graphene-based
nanomaterials, 347, 362, 426
characterization techniques, 184, *198*,
198–200, 407
charge carrier mobility, 209, 215, 243,
255, 398
chemically modified graphene (CMG)
catalysis of nitromethane combustion,
362–3, *363*
fabrication by exfoliation techniques,
234–5, *235*, 291–2, *292*
properties and device applications,
236–7, 245–7
research activity, 233–4, *234*
chemical vapor deposition (CVD)
carbon feedstocks, 259, 260–61
catalyst selection, 258–9, **259**, 261
challenges for optimum quality
production, 260–61, 281–2,
329
compared with other graphene
synthesis methods, 257, **257**,
259–60
graphene orientation, 261, 278–80
single step *(in situ)* N-doping, 196,
197
stages of process, 257–8, *258*
continuous growth, 271–8
nucleation, 261–71
precipitated and diffusive
mechanisms, 260, 262, *262*
chemisorption, 210, 213, 215–16
on metal–nitrogen edge sites, 351–2,
352
of oxygen on graphene edges, 292–4,
293, *294*
chimeric magnetism, 115–16, *117*

chiral nanotubes
  oxygen adsorption and unzipping, 87, 87–8
  roll-up vectors and Clar formulas, 33–4, 34
Clar theory
  Clar rings in armchair hexangulenes, 108–10, 109, 110
  formulas, applied to explain properties
    armchair graphene nanoribbons, 34–9, 35, 36, 38
    carbon nanotubes, 33–4, 34
    graphene, 43, 43
    HBC (coronene derivative), 394
    zigzag graphene nanoribbons, 40, 40–42, 42
  migration of π-sextets, and stability, 31–3, 32, 40
  sextet rule
    applied to linear oxidative unzipping, 296, 296–7, 297
    original introduction for PAH representation, 31
    research applications, 44–5
    structure representation conventions, 31, 31
cluster (molecular) models
  computational codes, 238, 382
  etching of armchair and zigzag edges, 292–4
  fluorination of graphene, 95
  for graphene oxide reduction mechanisms, 302–11, 303
  oxidative unzipping, 81–2, 90, 294–7
coronene
  derivatives, 399–400, 400
  linear oxidation mechanism, 294–6, 295
covalently bonded graphenes (CBGs), see superarchitectures, carbon-based
critical size, of clusters for nucleation, 262–3, 263
cutting
  catalyzed by metal nanoparticles, 92, 92–5, 96

  control, using applied external strain, 83–5, 85
  by fluorination, 95, 95
  methods, range of approaches, 4, 79–80, 91–2
  oxidative mechanisms
    coronene and graphene flakes, 294–7
    in graphene sheets, 80–83, 300–301
    unzipping carbon nanotubes, 86, 86–91, 90, 296
decarboxylation, cluster models and kinetics, 308–9, 309, 310
defects
  edge and bulk disorder, 332
  effects on electronic transmission, 322, 331, 332
  extended line (ELD) nanowire structures, 194, **195**, 195, 201
  point and linear types, 255–6, 256
  as sites for adsorption, 224, 244, 351–2, 352
  stability of catalytic metal atoms/clusters, 357–60
  stabilization by doping (Stone–Wales defects), 186, 186
  suppression of phonon thermal conductance, 337
density functional theory (DFT)
  calculation packages, 12, 18, 103–4
  effects of functional approximation choice, 54–5, 211–12, 382
  long-range corrected (LC-DFT) methods, 10, 21–2
  van der Waals correction, 132–3, 211, 212, 438
  methods used for doped systems, 187–8, 237–8
  modeling for graphene moiré structures, 428, **428**, 435–8, **436**
  value and limitations for new materials study, 320, 381–2
density matrix renormalization group (DMRG) method, 11–12, 13, 13

density of states (DOS), electronic
  graphene moiré, mound region, 428–9,
      429
  graphene sheets
    by density functional theory, 2, *3*
    nearest neighbour tight binding
        model, 185, *185*
  molecules adsorbed on graphene, *216*,
      216–17, *217, 218*
  nanosieve and nanofunnel hybrid
      structures, 169, *170*
  related to size, finite ribbons/nanodots,
      60, *61*
  zigzag graphene nanoribbons
    doped and undoped systems, *195*
    undisturbed and edge-reconstructed,
        19–20, *20*
deuterium gas (D_2)
  kinetic molecular separation from H_2,
      132, 140, *140*, 141–3
  partition function calculation, 137–8
diamond, doping mechanism, 220, *221*
Diels–Alder reactions, 396, 399–400, *400*,
    403, *404*
diffusion limited growth (DLG), 271,
    281–2
diimide compounds
  armchair-edge ribbon rylenes, 402,
      402
  zethrene derivatives, *415*, 416
dipole layers, effects on doping, 218–20,
    219
dissociative adsorption, 225, 244, 294,
    298–9, *299*
doping
  charge transfer mechanisms in
      adsorption, 215–21
  effects of dopant concentration, 188,
      189, 194
  effects on electronic properties, 332–4,
      333, 355, 355–6
  electrochemical surface transfer
      (diamond, graphene), 220–21,
      221
  enhancement of hydrogen adsorption,
      384
  inducing edge asymmetry in ZGNRs,
      56
  lithium adsorption
    in bandgap engineering, 61–2, *62*
    on graphene, CNTs, and porous
        graphene compared, 139–40
  n- and p-types, *187*, 187–8
  position of dopants, 188, *190*, 191
  techniques for graphene sheets, 185–7,
      186
  see also nitrogen-doped graphene

edge disorder, nanoribbons, 332, 337–8,
    338
edge epitaxy, 279
edge reconstruction
  in graphene growth on metal surfaces,
      273–5, *274*
  nanoribbons (Rc-ZGNRs), 9–10,
      17–22, 331
effective valence bond model (EVB)
  computational methods, 10–11
  outcomes compared with DFT
      methods, 13, 18, **18**, 20–22, *21*
electric fields, applied
  and bending of zigzag graphene
      nanoribbons, 64
  effects on band structure, 54–5, *55*
  to enhance adsorption on graphene,
      224–5, 360, *360*
electrochemical catalysts
  catalytic materials, 350–52, *352, 353*
  electroreduction of oxygen,
      mechanisms, **349**, 349–50, *351*
  importance for fuel cells, 347–8
electronic transport calculation, nanoscale,
    322–4, *323*
electron injection, photoinduced, 354, *354*
epitaxial growth, 278, 279, 435
epoxy groups
  alignment and unzipping mechanism,
      83, *84, 295*, 295–6
  carbon nanotube formation, preferred
      sites, *86*, 86–8, *87*
  coadsorbed with hydroxyl groups, *243*,
      243–4

epoxy groups (*Continued*)
  de-epoxidation, reaction models
    with heat, 313, *314*
    with hydrazine, 302–7, *304, 306,*
      312–13, 313
  formation and diffusion kinetics, *81,*
    81–3, *82*
etching
  at armchair and zigzag edges, 292–4
  catalyzed by metal nanoparticles, *92,*
    93
  of vacancy defects, 298–9
ether trimers (oxidative cutting
  nucleation), *82,* 83
exfoliation, for functionalized graphene
  production, 234–5, *235,* 291, *292,*
  374
exohedral atomic doping, 186, *186.* See
  also adsorption.*ee also* adsorption
extended line of defects, 5-8-5 (ELD),
  194, **195,** *195,* 331, *335*

ferromagnetism
  fullerene polymers, 156
  fused-azulenes, 15, 17
  induced by adsorption, functionalized
    graphene, 241, 246
  zigzag graphene nanoribbons, 9–10,
    19–22, 30, 326–9, *327*
field effect transistors (FET)
  band-gap tuning by applied strain, 64
  N-doped graphene nanoribbon devices,
    200
  using graphene sheets, 183, *187,* 192
  using nanopeapod (metallofullerene)
    structures, 165
fluorination
  as graphene cutting method, 95, *95*
  intercalation compounds, for hydrogen
    storage, *380,* 380–81
foams, carbon
  armchair structures, 161–2, *162, 164*
  hexagonal and triangular structures,
    162, *163, 164*
  properties and uses, 160, 379
  zigzag structures, 161, *161*

fuel cell catalysts, 200, 347–53
fullerenes
  C_{60} structure and derivative
    applications, 153, 154–5, 372
  encapsulation in carbon nanotubes,
    164–5, *165, 166*
  polymer synthesis, 156
  surface-bonded hybrids (nanobuds),
    165, *167,* 167–9, *168, 169*
  theoretical polymers, *155,* 155–6, *156*
functionalization, chemical
  of carbon superarchitectures, 173
  graphene nanoribbon edges, 56, 334
  as molecular doping, 186, *186,* 222–4
  nitrogen and metals for porous
    graphene, 140–44
  production methods for graphene
    sheets, 234–6, 333, 374, *375*
  theoretical study approaches, 237–8,
    248–9
functionalized graphene, *see* chemically
  modified graphene
gasification of graphene fragments, 293–4,
  295
gas sensing
  advantages of graphene use, 209–10,
    236
  polymer functionalization of graphene,
    247
  sensitivity improvement by doping,
    62–3, 222
gas separation, *see* molecular sieving
Gibbs free energy
  and CVD nucleation cluster size, 266,
    268–9, *269*
  in hydrazine de-epoxidation of
    graphene oxide, 303, **305,** 306–7,
    310
gold, nucleation and growth on graphene
  moiré, *434,* 434–40
grain boundaries, graphene, *256,* 258, 261,
  265–6, 331
graphane
  capacity for hydrogen chemical
    storage, 382–3

dehydrogenation, 60
structure/properties compared with
    graphene, 239–40, *240*, 333, *383*
graphdiyne
  gas diffusion and selectivity (for
      hydrogen purification), 144–6,
      146
  structure, 131, *132*, 144
graphene
  aromaticity, 42–4, *44*, 102
  characterization and structure, 184, 314
  de-epoxidation, PBC models, 312–13,
      313
  foam superarchitectures, 160, *161*,
      161–2, *162*, 164
  growth kinetics, 266–71, 276–8, *277*
  moiré superstructures, uses, 425–6
  nanocomposites, 162–3, 237
  oxidation models, 292–301
  potential device applications, 29–30,
      52, 233
  properties, remarkable features, 51–2,
      63, 160, 183, 255
  sheets
    corrugated (folded) surfaces, 225–6,
        236–7
    doping, types and effects on
        properties, 185–94, *186*,
        215–21
    electronic properties, 184–5, *185*,
        324–5, *325*
    hybrids with buckyballs (nanobuds),
        168, 168–9, *169*
    hydrogen storage capacity, 373–4,
        374
    metal-embedded, catalysis, 356–7,
        358, 359–60, 361–2
    nanosieve/nanofunnel structures,
        169–71, *170*
    oxidative cutting, 80–85
    permeability to gases, 130
    surface adsorption of atoms, 61,
        62–3, 192, 212–15, 234
    theoretical junction types, *160*,
        160–61, *163*
  solubility, 209, 233, 394

  synthesis methods compared, 257, **257**,
      259–60
  see also chemically modified graphene;
      nitrogen-doped graphene; porous
      graphene
graphene hydroxide (GOH), 243, *244*
graphene nanodots (GNDs)
  bonding and pi-electrons, 101–2, **102**,
      405–6
  elasticity, 65
  electronic properties, 58–60, *59*, *61*
  magnetism, 112–16, 122
  structural geometry, 104–11, *397*
  synthesis, 103, 396
graphene nanoribbons (GNRs)
  comparison of armchair and zigzag
      forms
    edge structures, *30*, 52, *52*
    effects of applied strains, 63–4
    electronic and magnetic properties,
        29–30, 53–8, 59, 325–9
    metal atom adsorption, 61, 376,
        377
    thermal conductance, 337
  device applications, 200
  fabrication
    control of properties, 52, 85, 330–34
    top-down and bottom-up
        approaches, 79–80, 329–30,
        330
  finite length (nanodots), 58–60, *59*, *61*,
      65–6
  formation energies, at metal catalyst
      surfaces, 271–3, *272*
  mechanical/electromechanical
      properties, 63–6
  nanowiggles (GNWs), *334*, 334–5,
      338
  stacking, 336
  substitutional doping, 194, **195**, *195*,
      332–3
  surface chemical adsorption, 60–63, *62*,
      247–8
  see also armchair graphene
      nanoribbons; zigzag graphene
      nanoribbons

graphene oxide (GO)
   interaction with ammonia, 247
   metal-embedded, catalysis of CO
      oxidation, 358–9, *359*
   properties and applications, 223,
      240–41
      catalytic activity, 347, 348
      surface structure models, functional
         groups, 240, *242*, 291–2
         residual groups after reduction, 311,
            311, *312*
      synthesis of reduced graphene oxide
         (rGO)
         with exfoliation, for functionalized
            graphene, 234–5, 291–2, *292*
         N-doped sheets, 192–3
         reduction by reaction with
            hydrazine, 302–7, *304*, **305**
         thermal reduction, 307–9
graphite
   liquid-phase direct exfoliation, 235,
      235
   mechanical peeling, 257, **257**
   oxidation, in functionalized graphene
      synthesis, 291–2, *292*
graphite intercalation compounds (GIC),
   380, 380–81
graphitic carbon nitrides (g-C_3N_4), 355,
   355–6, *356*
graphyne, 352, *353*
Green's functions, 323, 324
growth front, graphene
   edge reconstruction, zigzag and
      armchair edges, 273–5, *274*
   shape determination and growth rate,
      275, 275–8, *277*
   upright nanoribbons, formation, 271–3,
      272

haeckelite lattices, 331, *335*
HBC (hexa-*peri*-hexabenzocoronene)
   radical cation and anions, 394–6, *395*
   regioselective hydrogenation, 396, *396*
   structure related to properties, 394, *395*
helium, isotopic mixture separation, 132,
   140, 141

heterodoping
   effects on electronic properties,
      188–92, *189*, *190*, *191*
   principle and synthesis, 186, *186*
hexangulenes
   geometry
      armchair-edged, 107–10, *108*, *109*
      zigzag-edged, *104*, 104–7, *105*
   magnetism, 112–14, 115–16, *117*
highest occupied molecular orbital
   (HOMO), *see* molecular orbitals,
   frontier
high-resolution transmission electron
   microscopy (HRTEM), 165, *198*,
   199, 331
hopping
   adsorbed oxygen, energy barriers, *81*,
      82
   electronic, 21, 64, 185
      nearest neighbour hopping integrals,
         321, *321*
Hückel Molecular Orbital (HMO)
   analysis, 12, 17, 18, 104, 406
hybrid carbon nanostructures, 163–71,
   200–201, 348
hydrazine (N_2H_4), de-epoxidation agent,
   302–7, *304*, *306*, 312–13, *313*
hydrogen
   adsorption on graphene
      effect of electric field, 224–5
      mechanisms and effects on
         properties, 238–40, 333–4
      spillover from metal catalyst
         clusters, 383–6, *384*, *385*
   in CVD, influence on growth of
      graphene, 281
   as energy source, 371
   isotopic mixture (H_2/D_2) separation,
      131–2, 140
      energy pathways, asymmetrically
         doped membranes, *141*, 141–3
      partition function calculation, 137–8
   nanomeshes for purification from
      syngas
      graphdiyne, 131, 144–6
      porous graphene, 139, 140–41

in nanoparticle-catalyzed CNT
   unzipping, 92–5, *93*
partial hydrogenation of HBC, 396,
   396
photocatalytic production from water,
   201, 355–6
storage materials
   alternatives and requirements,
      371–3
   binding energy computation
      methods, 381–2
   chemical storage, 382–6
   covalent carbon and graphene
      nanostructures, 168–9, 377–81
   metal-doped graphene, 139–40,
      374–7
   spaced layers of graphene sheets,
      373–4, *374, 375*
hydroxyl groups
   adsorption and mobility on graphene,
      241–3, *242*, 244–5, *385*, 385–6
   coadsorbed with epoxy groups, *243*,
      243–4
   thermal dehydroxylation of graphene
      oxide, *307*, 307–8, *308*
   effect of temperature, 310, *310*

indenofluorenes, 417, *417*
isotope separation
   helium isotopes (^{3}He/^{4}He), 132, 141
   hydrogen and deuterium (H$_2$/D$_2$), 132,
      137–8, 141–3, *144*
   principles of quantum sieving, 129,
      140, *140*
   traditional methods, 131–2

Kekulé structures, related to aromaticity,
   31, 34, 38
kinetic Wulff construction, crystal growth,
   275, *275*
Kohn–Sham (KS) energy levels
   acenes and bisanthenes, 118, *118*,
      119–20, *120*
   in DFT computation, 103
   graphene nanodots, 113, *113*, 114, *115*,
      117

Landauer theory (nanoscale electronic
   transport), 322–4
Lennard–Jones pairwise interactions,
   133–4
Lieb's theorem, 60, 321
linear defects
   extended 5–8–5 line (ELD), 194, **195,**
      195
   types and effects on properties, 256,
      256, 300
lithium
   adsorbed on zigzag graphene
      nanoribbons, 61–2, *62*
   doped porous graphene for hydrogen
      storage, 139–40, 375–6, *376*
   kinetics of adsorption, *141,* 142–3,
      143
long-range corrected density functional
   theory (LC-DFT), 10, 21–2
low energy electron diffraction (LEED)
   analysis, *426,* 427, 428
lowest unoccupied molecular orbital
   (LUMO), *see* molecular orbitals,
   frontier

magic size, carbon clusters (C$_{21}$), 264–6,
   266, 273
magnetization, related to spin
   bisanthenes, 119–20, *121*
   hexangulenes, 112–14, *114,* 115–16,
      117
   oligocenes, 117–19, *118, 119*
   square/rectangular nanographenes, 122
   triangulenes, 114–15, *116*
   see also ferromagnetism
mean bond length (MBL), 36, *37, 38,*
   38–9
membranes
   artificial semipermeable nanosieves,
      171
   gas permeation rate by transition state
      theory, *134,* 134–9
   mechanical support for nanomeshes,
      147
   porous, gas and isotope separation
      technologies, 129–32, *131*

metal nanoparticles
 catalysts for hydrogen CNT unzipping, 92, 92–5
 catalytic performance for CO oxidation, 356–62
 dispersed on N-doped graphene, for fuel cells, 348, 350–52
 template-driven synthesis
  2-D islands (gold), *434,* 434–40
  3-D nanoclusters, 430–34, *431, 432*
metal–organic frameworks (MOFs), 371–2
minimum energy pathways
 membrane permeability, *141,* 141–3, 145, *146*
 thermal de-epoxidation of graphene, 313, *314*
Möbius nanorings, *335,* 336
moiré patterns, metal-supported graphene
 on iridium(111) surface, *277,* 277–8, 426
 nanomaterial properties and uses, 425–6
 on ruthenium surface, g/Ru(0001)
  band structure, *429,* 429–30
  2-D gold islands *434,* 434–40
  geometric periodicity, 427–9, **428**
  synthesis methods, 426–7
  template-driven metal cluster synthesis, 430–34, *431, 432*
molecular dynamics, 171, 248
 ab initio simulations, 95, 238, 312
 with tight-binding calculations, 157, 164
molecular magnets, 60
molecular orbitals, frontier (FMOs: HOMO/LUMO)
 azulenes, 15–17, *16*
 benzene, 2
 charge transfer mechanism in adsorption, 216–18
 gap periodicity in finite carbon nanotubes, 39, *39*
 nanodots
  exchange interactions and magnetism, 112
  size dependence of energy gaps, 58–9, *59*

planar PAHs, 33
zigzag graphene nanoribbons, 18–19, *19*
molecular sieving
 selectivity, 130–31
 use of single atomic layer carbon membranes, 129–30, *131*
Monte Carlo simulations
 graphene zigzag edge growth, 278
 nanochannel catalytic etching, *92, 93*
 physisorption of hydrogen on graphene, 239
nanobuds
 carbon (CNBs), 165, *167,* 167–8
 graphene (GNBs), *168,* 168–9, *169*
nanocomposites, 162–3, 237, 350–52, 353–4
nanoelectromechanical systems (NEMS), 63, *65,* 236–7
nanofunnels, 169, *170*
nanographenes, *see* graphene nanodots
nanomeshes
 fabrication from graphene, 130, *131*
 gas transmission dynamics, transition state theory, 134–6
 calculation of partition functions for H_2/D_2 separation, 133–4, 137–8
 graphene-like polymers, 130–31, *132*
 performance improvement and potential applications, 147
 see also porous graphene
nanopeapods, 163–5, *165, 166*
nanosieves, 169, *170,* 171
naphthalene
 migrating Clar π-sextet, 31–2, *32*
 as subunit of rylenes, 398
 as thermal isomer of azulene, 14
nine-membered ring ether–trimer (NET), 299–300, *300*
nitric oxide (NO), 216–17, *217*
nitrogen dioxide (NO_2), adsorption, 216, *216,* 222, *223,* 247

nitrogen-doped graphene (NG)
  applications, *184,* 200–201, 348
  band structure related to dopant
      distribution, 188–91, *189,*
      190
  catalytic active sites, 349–50, *350*
  characterization, *198,* 198–200
  structures, cluster models, *348,* 348–9,
      349, 351
  surface adsorption of atoms and
      molecules, 222, 350–52,
      352
  synthesis methods, 196, *196,* **197**
nitrogen functionalized porous graphene
    (NPG)
  asymmetric doping with metals
      lithium, 142–3, **143**
      titanium, *131,* 143, **144,** *144*
  performance analysis for gas/isotope
      separation, 140–42, *141,*
      142
  preparation of membranes, 130
nitromethane combustion, catalysis,
    362–3, *363*
Nobel Prize awards, 1, 183
non-bonding molecular orbitals
    (NBMOs), triangulene edges, 111,
    114–15, *115,* 409
nucleation
  in chemical vapor deposition
      carbon feedstock dissociation at
          metal surface, 262
      catalyst terraces and step edges,
          266–71, *267, 269,* 273
      graphene islands, possible isomers,
          263–4, *264*
      importance for graphene quality
          control, 261
      initial formation of C chains and
          nanoarches, 262–3, *263*
      magic cluster size, core shell
          geometry, 264–6, *265, 266*
      rate calculation, 269–70, *270*
  metal clusters on graphene moiré,
      430–34, *431, 432*
  nucleation barrier, G^*, 266, 268–9, *269,*
      270–71

nucleus-independent chemical shifts
    (NICS) analysis, 32, *32,* 33–4
Nudged Elastic Band (NEB) method, 142,
    237–8

oligocenes (oligoacenes)
  electron spin coupling, 12–14, *13, 14*
  spin polarization and magnetism, DFT
      models, *118,* 118–19, *119*
  structural geometry, *104,* 117, 122
open carbon frameworks (OCFs), 381, *381*
oxidation
  of carbon monoxide
      catalytic materials, 356–9, 361,
          438–40, *439*
      effects of strain and electric field on
          catalysts, 359–60, *360*
  of graphene
      effects on shape and size, 80
      introduction of defects, 241, 297–8,
          298
  of graphite, 291–2, *292*
  resistance, of boron nitride nanotubes,
      171
  of zigzag graphene nanoribbon edges,
      55–6
  see also oxidative cutting
oxidative cutting
  carbon nanotubes, 85–91, 296
  coronene and graphene flakes, 294–7
  graphene sheets, 80–85, 300–301
oxygen reduction reactions (ORR),
    catalysis
  alternatives to platinum
      graphyne, 352, *353*
      other metals on porphyrin-like
          subunits, 351–2, *352*
  N-doped graphene, 200
  platinum nanoparticle/N-graphene
      composites, 200, 348, 350–51
  reduction pathways, **349,** 349–50, *351*

paramagnetic instability, ZGNRs, 326–7,
    327
partition functions
  calculation approaches, for nanopore
      kinetics, 133–4

partition functions (*Continued*)
  fully coupled 4D calculations, 137–8, *138*
  harmonic approximation, 138–9
  one dimensional approximation, 135–6
  well-separated 'reactant' system, 136
pentacene, 32
periodic (PBC, slab) models
  computational codes, 238
  creation and etching of vacancy defects, 297–9
  de-epoxidation of graphene, 312–13
  oxidative unzipping, 82–3, 299–301
perylene, *397*, 398, *398*, 405
phenalenyl-based compounds
  biradical and monoradical stability, 408–9
  bis(phenalenyls), *409*, 412–14, *413*, *414*
  resonance structures of spin-delocalized radical, 406, *406*
  synthesis and steric radical stabilization, *407*, 407–8
  see also triangulenes
photocatalysts
  graphene–semiconductor nanocomposites, 201, 353–4
  graphitic carbon nitrides, optical tuning, *355*, 355–6, *356*
physisorption
  adsorption sites and orientations, **213**, *213*, 213–15
  compared with chemisorption, 210, 213
  effect of surface corrugations, 225–6
  of hydrogen on graphene, 238–9, 373–4
pillared graphene, 374, 379–80, *380*
point defects, types, 255–6, *256*
polyacenes, 12–14, *14*, 17
polyazulene, *14*, 14–17, *16*
polycyclic aromatic hydrocarbons (PAHs)
  Clar structure representations, 31, *32*, 32–3, *33*
  geometry, armchair-edged, 396, *397*
  as models for graphene oxide reduction, 302, *303*, 308

structural relationship to graphene, 42, 101, 396
synthesis, 103, 393, 396
see also graphene nanodots; HBC; phenalenyl-based compounds; rylenes
polyphenylene, 376, *376*
porous graphene
  antidot lattices, electronic tailoring, 335
  isotope mixture separation, 140, *140*
  lithium doped, for hydrogen storage, 139–40, 375–6, *376*
  nitrogen functionalized, 140–43, **142**, *144*
  pore sizes and structure, 4, *4*, 130, *132*
  selectivity in gas separation, 130–31, *131*, 139
projector augmented wave (PAW) method, 103
propene epoxidation, catalysis, 362, *363*
pyridine-like doping, 186, *186*, 191, *191*, 201

quantum chemical models, methodology, 10–12, 138, 237–8
quantum confinement effects, 153–4
  in graphene nanoribbons, 53, 56, 325
  in nanodots, 58, 59
quantum dots, graphene, *see* graphene nanodots
quantum sieving, 129, 132
quantum tunneling, 132, 141
quinoidal structures
  in bisanthene derivatives, 403
  hexangulene GNDs, 104, 105, *105*, 107
  in indenofluorenes, 417
  quinodimethyl (QDM) resonance structures, types, 411, *411*
  stability and synthesis, 411–12, *412*

radicals
  of HBC, 394–6, *395*
  partial, as edge states of zigzag GNRs, 41–2, *42*
  phenalenyl, spin-delocalized, *406*, 406–8

Raman spectroscopy, for doping
    characterization, 193, *198*, 199
reactive force field simulations, 94–5,
    238
rhodium, nanocluster formation, 430–32,
    431
ruthenium, support for graphene moiré
    characterization of surface interactions,
        426–30
    metal cluster nucleation and growth,
        430–34
rylenes
    armchair edge chemistry
        Diels–Alder reactions, 399–400,
            400
        diimide ribbon molecule synthesis,
            402, *402*
        heteroatom annulation, 398–9, *399*
        reactions modulating optical
            properties, 400, *401*
    characteristics, 397–8
    single crystal structures, 398, *398*

scanning tunneling microscopy (STM)
    images
    appearance of graphene moiré
        structures, 425, *426*, 427
    of doped graphene, *193*, 194, *198*, 199
    high-resolution, nanographene
        fragments, 43, *44*, 272, 273
    simulated, *35*, 36, *36*, 43
Seebeck coefficient, 336
semipermeable membranes, artificial, 171
singly occupied molecular orbitals
    (SOMOs)
    overlap in phenalenyls 407, 413, *413*
    in zigzag graphene nanoribbons,
        18–19, *19*
solar cells, use of N-doped carbon thin
    films, 200–201
solubility
    graphene, 209, 233
    tuning by synthetic chemistry, PAHs,
        394, 396, 403, 412
spillover mechanism, chemical storage,
    383–6, *384*, 385

spin density distribution
    catalytic active sites, 348–9
    fused-azulenes, *16*, 17
    hydroxyl groups adsorbed on graphene,
        241, *242*
    phenalenyl radical and derivatives,
        405–6, *406*, 409
    polyacenes, 14, *14*
    zigzag graphene nanoribbons
        (ZGNRs), 20–22, *21*, *22*
spintronic devices, 246, 327–8
stacking, graphene nanoribbons, 336
Stone–Wales defects
    adsorption energy (hydroxyl groups),
        244, *245*
    sheet distribution for Haeckelite
        lattices, 331, *335*
    stabilized by doping, 186, *186*
strain, applied mechanical
    band structure effects in nanoribbons,
        63–4
    for cutting control, 83–5, *85*
    effect on catalytic CO oxidation, 359,
        360
    prevention of transition metal
        clustering, 377
substitutional doping (method), 185–6,
    186, 196, *196*, 221–2
superarchitectures, carbon-based
    boron nitride nanotubes and
        monolayers, 171–3, *172*
    carbon nanotubes, 156–60, *158*, *159*
    design motivations, 154, 173
    graphene, *160*, 160–63, *163*, 378–81,
        381
    hybrid designs, 163–71, 379–80, *380*
    nanoribbon superlattices and rings,
        335, 335–6
surface transfer doping, 192–4, *193*,
    215
    electrochemical, 220–21, *221*
syngas purification, 131

'tear-from-the-edge' oxidative cutting,
    80–81, *81*, 301, *302*
teranthene, 403, *404*, 405

thermoelectric effect and devices, 336–8, *337, 338*
tight-binding models
  applied to graphene nanoribbons, 53, 326–7, *327*
  in description of electronic properties, 320–22, 323
titanium
  complexes for hydrogen storage, 372, 377, *378*
  dopant for N-functionalized porous graphene, *141,* 143, **144,** *144*
  TiO_2–graphene nanocomposite as photocatalyst, 353–4, *354*
transition metals
  atomic clustering on carbon nanostructures, 372
  as cataylsts and templates, CVD, 258, 261–3, *265,* 273–5, *274*
  cluster formation on g/Ru(0001), 430–34, *431, 432*
  with graphene, for enhanced hydrogen storage, 377, 379
  graphene superstructures, moiré patterns, 425–6, *426*
  see also metal nanoparticles
transition state (TS) theory, *134,* 134–9, *138,* 141
triangulenes
  delocalized spin systems, 405–6, *406,* 409–10, *410*
  geometry, zigzag-edged, *108,* 110–11
  magnetism, 114–15, *116*
trioxytriangulene, 410, *410*
two-photon absorption (TPA), 414

ultracapacitors, 200, 236
unzipping, *see* cutting

vacancy defects
  etching and Carlsson mechanism combustion, 298–9, *299*
  interaction with etching nanoparticles, 93
  introduced by oxidation, 241, 297–8, *298*
  types, 256, *256*
valence bond (VB) theory
  applied to alternant systems, 102
  and benzenoid patterns, nanographenes, 108
  carbon nanotube Clar structure models, 33–4, *34*
  effective valence bond model, 10–11
  resonating models, bis(phenalenyls), *413,* 413–14
van der Waals interactions
  correction for, in density functional theory, 132–3, 211, *212,* 438
  effects on carbon nanostructure assemblages, 154
  and graphene orientation in CVD growth, 261, 272–3
variational transition state theory (VTST), 141–2
Vienna *ab initio* simulation package (VASP), 12, 18, 103–4

water
  adsorption sites on graphene, **213,** 213–15, *214*
  hydroxyl groups and wettability of graphene, 244–5
  between silanol and graphene, effect on band structure, 218–19, *219*
wiggle-type nanoribbons (GNWs), *334,* 334–5, 338
  synthesis, 330, *330*
work functions
  graphene/semiconductor difference, 353
  shift, induced by adsorption, 60, 220–21
  solar cell operating mechanism, 200–201

X-ray photoelectron spectroscopy (XPS), for doping characterization, *198,* 198–9

Young's modulus
   carbon nanotubes, 156, 160
   graphene, 51, 63, 255
   nanodots, 65

zethrenes, 414–17, *415, 416*
zigzag carbon nanotubes
   atomic oxygen adsorption, *86,* 87, 88
   oxygen pair unzipping, orientation selectivity, 90–91, *91*
zigzag graphene nanoribbons (ZGNRs)
   bending, in applied electric field, 64
   Clar formulas, 40, *40*
   edge states, reconstructed and undisturbed
      chemical functionalization, 56
      energy gaps, 17–18, **18**
      frontier molecular orbitals, 18–19, *19*
      oxidized edges, 55–6
      projected density of states, 19–20, *20,* 53
      room-temperature ferromagnetism, 9–10
      spin density distribution, 20–22, *21,* 54, *54*
   electric field tuning of magnetic properties, 54–5, *55*
   magnetic states and electronic conductance, 326–9, *327, 328*
   reactivity related to aromaticity, 40–42, *41, 42*
   surface adsorption of metal atoms, 61
ZT values (thermoelectric conversion efficiency), 336–8, *337, 338*